Frontiers in Nutritional Science

This series of books addresses a wide range of topics in nutritional science. The books are aimed at advanced undergraduate and graduate students, researchers, university teachers, policy makers and nutrition and health professionals. They offer original syntheses of knowledge, providing a fresh perspective on key topics in nutritional science. Each title is written by a single author or by groups of authors who are acknowledged experts in their field. Titles include aspects of molecular, cellular and whole-body nutrition and cover humans and wild, captive and domesticated animals. Basic nutritional science, clinical nutrition and public health nutrition are each addressed by titles in the series.

Titles available

1. Nutrition and Immune Function
   Edited by P.C. Calder, C.J. Field and H.S. Gill
2. Fetal Nutrition and Adult Disease: Programming of Chronic Disease through Fetal Exposure to Undernutrition
   Edited by S.C. Langley-Evans
3. The Psychology of Food Choice
   Edited by R. Shepherd and M. Raats
4. Peptides in Energy Balance and Obesity
   Edited by G. Frühbeck

To my family for teaching me perseverance in projects and endeavours of every kind.

Peptides in Energy Balance and Obesity

Edited by

Gema Frühbeck

Department of Endocrinology
Metabolic Research Laboratory
Clínica Universitaria de Navarra
University of Navarra

and

CIBER Fisiopatología de la Obesidad y Nutrición
Instituto de Salud Carlos III
Pamplona
Spain

www.cabi.org

in association with
The Nutrition Society

CABI is a trading name of CAB International

CABI Head Office
Nosworthy Way
Wallingford
Oxfordshire OX10 8DE
UK

Tel: +44 (0)1491 832111
Fax: +44 (0)1491 833508
E-mail: cabi@cabi.org
Website: www.cabi.org

CABI North American Office
875 Massachusetts Avenue
7th Floor
Cambridge, MA 02139
USA

Tel: +1 617 395 4056
Fax: +1 617 354 6875
E-mail: cabi-nao@cabi.org

A catalogue record for this book is available from the British Library, London, UK.

Library of Congress Cataloging-in-Publication Data

Peptides in energy balance and obesity / edited by Gema Frühbeck.
p. ; cm. -- (Frontiers in nutritional science ; no. 4)
Includes bibliographical references and index.
ISBN 978-1-84593-407-1 (alk. paper)
1. Peptide hormones. 2. Obesity--Pathophysiology. 3. Energy metabolism.
I. Frühbeck, Gema. II. Nutrition Society (Great Britain) III. Title. IV. Series.
[DNLM: 1. Peptides--metabolism. 2. Appetite Regulation. 3. Energy Metabolism--physiology. 4. Homeostasis. 5. Obesity--physiopathology. QU 68 P42437 2009]

QP572.P4.P47 2009
612.4'05--dc22

2008037565

ISBN-13: 978 1 84593 407 1

Typeset by AMA Dataset Ltd, UK
Printed and bound in the UK by MPG Books Group

The paper used for the text pages in this book is FSC certified. The FSC (Forest Stewardship Council) is an international network to promote responsible management of the world's forests.

Contents

Contributors

Abizaid, Alfonso, *Institute for Neuroscience, Carleton University, Ottawa, Ontario, Canada.*

Aloe, Luigi, *Institute of Neurobiology and Molecular Medicine, National Research Council-European Brain Research Institute, NGF Section, Rome, Italy.*

Berlan, Michel, *Unité de Recherches sur les obésités, Unité Inserm-UPS 586, Université Paul Sabatier, Institut Louis Bugnard, Hôpital Rangueil, TSA 50032, 31059 Toulouse cedex 9, France.*

Bloom, Stephen R., *Department of Metabolic Medicine, Division of Investigative Science, Imperial College London, Hammersmith Campus, Du Cane Road, London W12 ONN, UK.*

Catalán, Victoria, *Metabolic Research Laboratory, Clínica Universitaria de Navarra, University of Navarra, 31008, Pamplona, Spain.*

Chaldakov, George, *Division of Cell Biology, Medical University, BG-9002 Varna, Bulgaria.*

Clément, Karine, *INSERM, Nutriomique, U872, Paris, France; University Paris 6, F-75006 Paris, France; CHRU Pitié Salpétrière, Hôtel-Dieu Nutrition Department, Centre de Recherche en Nutrition Humaine, F-75013 Paris, France.*

Del Parigi, Angelo, *National Institute of Diabetes and Digestive and Kidney Diseases, National Institute of Health, Department of Health and Human Services, Phoenix, Arizona, USA.*

Dugail, Isabelle, *INSERM, U872, F-75006 Paris, France; Centre de Recherche des Cordeliers, Université Pierre et Marie Curie – Paris 6, UMR S 872, F-75006 Paris, France; Université Paris Descartes, UMR S 872, F-75006 Paris, France.*

Figlewicz Lattemann, Dianne, *VA Puget Sound Health Care System (151), 1660 So. Columbian Way, Seattle, WA 98108, USA and Department of Psychiatry and Behavioral Sciences, University of Washington, Box 356560, Seattle, WA 98195-6560, USA.*

Fiore, Marco, *Institute of Neurobiology and Molecular Medicine, National Research Council-European Brain Research Institute, NGF Section, Rome, Italy.*

Frühbeck, Gema, *Metabolic Research Laboratory and Department of Endocrinology, Clínica Universitaria de Navarra, University of Navarra, 31008, Pamplona, Spain.*

Galitzky, Jean, *Unité de Recherches sur les obésités, Unité Inserm-UPS 586, Université Paul Sabatier, Institut Louis Bugnard, Hôpital Rangueil, TSA 50032, 31059 Toulouse cedex 9, France.*

Gómez-Ambrosi, Javier, *Metabolic Research Laboratory, Clínica Universitaria de Navarra, University of Navarra, 31008, Pamplona, Spain.*

Guerre-Millo, Michèle, *INSERM, U872, F-75006 Paris, France; Centre de recherche des Cordeliers, Université Pierre et Marie Curie – Paris 6, UMR S 872, F-75006 Paris, France; Université Paris Descartes, UMR S 872, F-75006 Paris, France.*

Hansen, Michelle J., *Department of Pharmacology, University of Melbourne, Victoria 3010, Australia.*

Harrold, Joanne A., *Neuroendocrine and Obesity Biology Unit, Department of Medicine, University of Liverpool, Duncan Building, Liverpool L69 3GA, UK.*

Hauner, Hans, *Else Kröner-Fresenius Centre for Nutritional Medicine, Technical University of Munich, Klinikum Rechts der Isar, 81675 Munich, Germany.*

Henegar, Corneliu, *INSERM, Nutriomique, U872, Paris, France; University Paris 6, F-75006 Paris, France; CHRU Pitié Salpétrière, Hôtel-Dieu Nutrition Department, Centre de Recherche en Nutrition Humaine, F-75013 Paris, France.*

Horvath, Tamas L., *Department of Obstetrics, Gynecology and Reproductive Sciences and Department of Neurobiology, Yale University School of Medicine, New Haven, Connecticut, USA.*

Hristova, Mariyana G., *Division of Cell Biology, Medical University, BG-9002 Varna, Bulgaria.*

Inui, Akio, *Department of Behavioural Medicine, Kagoshima University Graduate School of Medical and Dental Sciences, Kagoshima, 890-8520, Japan.*

Lafontan, Max, *Unité de Recherches sur les obésités, Unité Inserm-UPS 586, Université Paul Sabatier, Institut Louis Bugnard, Hôpital Rangueil, TSA 50032, 31059 Toulouse cedex 9, France.*

Langin, Dominique, *INSERM, U858, Laboratoire de Recherches sur les Obésités, Institut de Médecine Moléculaire de Rangueil, F-31062 Toulouse, France.*

Le, Duc Son NT, *National Institute of Diabetes and Digestive and Kidney Diseases, National Institute of Health, Department of Health and Human Services, Phoenix, Arizona, USA.*

Madani, Rana, *Adipokines and Metabolism Research Group, Centre for Clinical Pharmacology, University College London, 5 University Street, London WC1E 6JJ, UK.*

Mantovani, Giovanni, *Department of Medical Oncology, University of Cagliari, Cagliari, Italy.*

Mohamed-Ali, Vidya, *Adipokines and Metabolism Research Group, Centre for Clinical Pharmacology, University College London, 5 University Street, London WC1E 6JJ, UK.*

Moro, Cédric, *Unité de Recherches sur les obésités, Unité Inserm-UPS 586, Université Paul Sabatier, Institut Louis Bugnard, Hôpital Rangueil, TSA 50032, 31059 Toulouse cedex 9, France.*

Morris, Margaret J., *School of Medical Sciences, University of New South Wales, NSW 2052, Australia.*

Ogston, Nicola C., *Adipokines and Metabolism Research Group, Centre for Clinical Pharmacology, University College London, 5 University Street, London WC1E 6JJ, UK.*

Pancheva, Rouzha, *Nutrigenomics Centre, Medical University, BG-9002 Varna, Bulgaria.*

Pannacciulli, Nicola, *National Institute of Diabetes and Digestive and Kidney Diseases, National Institute of Health, Department of Health and Human Services, Phoenix, Arizona, USA.*

Perboni, Simona, *Department of Medical Oncology, University of Cagliari, Cagliari, Italy.*

Rancic, Gorana, *Department of Histology and Embryology, Medical Faculty, Niŝ, Serbia.*

Rodríguez, Amaia, *Metabolic Research Laboratory, Clínica Universitaria de Navarra, University of Navarra, 31008, Pamplona, Spain.*

Sanders, Nicole M., *VA Puget Sound Health Care System (151), 1660 So. Columbian Way, Seattle, WA 98108, USA and Department of Psychiatry and Behavioral Sciences, University of Washington, Box 356560, Seattle, WA 98195-6560, USA.*

Sengenes, Coralie, *Unité de Recherches sur les obésités, Unité Inserm-UPS 586, Université Paul Sabatier, Institut Louis Bugnard, Hôpital Rangueil, TSA 50032, 31059 Toulouse cedex 9, France.*

Sipols, Alfred J., *Institute for Experimental and Clinical Medicine, Department of Medicine, University of Latvia, Riga LV-1004, Latvia.*

Small, Caroline J., *Department of Metabolic Medicine, Division of Investigative Science, Imperial College London, Hammersmith Campus, Du Cane Road, London W12 ONN, UK.*

Taleb, Soraya, *INSERM, Nutriomique, U872, Paris, France; University Paris 6, F-75006 Paris, France; CHRU Pitié Salpétrière, Hôtel-Dieu Nutrition Department, Centre de Recherche en Nutrition Humaine, F-75013 Paris, France.*

Tataranni, P. Antonio, *National Institute of Diabetes and Digestive and Kidney Diseases, National Institute of Health, Department of Health and Human Services, Phoenix, Arizona, USA.*

Tonchev, Anton B., *Division of Cell Biology and Nutrigenomics Centre, Medical University, BG-9002 Varna, Bulgaria.*

Ueno, Naohiko, *Division of Diabetes, Digestive and Kidney Diseases, Department of Clinical Molecular Medicine, Kobe University Graduate School of Medicine, Kobe, 650-0017, Japan.*

Williams, Gareth, *Neuroendocrine and Obesity Biology Unit, Department of Medicine, University of Liverpool, Duncan Building, Liverpool L69 3GA, UK and School of Medicine and Dentistry, University of Bristol, Bristol, UK.*

Wynne, Katie, *Department of Metabolic Medicine, Division of Investigative Science, Imperial College London, Hammersmith Campus, Du Cane Road, London W12 ONN, UK.*

Zucker, Jean-Daniel, *INSERM, Nutriomique, U872, Paris, France; University Paris 6, F-75006 Paris, France; CHRU Pitié Salpêtrière, Hôtel-Dieu Nutrition Department, Centre de Recherche en Nutrition Humaine, F-75013 Paris, France.*

Preface

> Thou seest I have more flesh than another man, and therefore more frailty.
> (Falstaff, in William Shakespeare's King Henry IV, Part I, Act III, Scene II)

The prevalence of overweight and obesity has reached alarming proportions worldwide, placing this problem as one of the most relevant public health concerns. Given the current obesity epidemic, it is of paramount relevance to understand better the mechanisms underlying energy balance regulation. The survival of higher organisms is dependent on the ability to procure, use and conserve energy efficiently. From an evolutionary point of view, animals feed to satisfy their immediate caloric and nutritional requirements from meal to meal, but also to allow energy and nutrients to be stored in anticipation of high energy demands or seasonal food shortages. Thus, food intake control involves the integration of external environmental cues with multiple internal physiological signals, as well as external social elements and hedonic influences.

It is now evident that energy balance is achieved through highly integrated interactions involving the brain and the periphery, which are influenced by both genetic and environmental factors. The past decades have witnessed an explosion in the identification and characterization of the many bioactive peptides involved in energy homeostasis. In addition, most of the peptides related to body-weight control have been shown to participate in other pathophysiological manifestations, providing a molecular basis for obesity-associated diseases such as type 2 diabetes mellitus, coronary heart disease, hypertension, dyslipidaemia, stroke, osteoarthritis, sleep apnea and cancer, among others.

The aim of this book is to provide an updated, detailed and comprehensive account of the field through a cutting-edge analysis by leading experts in the area. To achieve this, the book is divided into three parts, focusing on the peptides operating both centrally and peripherally at the same time as providing an integral and integrated perspective of the multifaceted and complex regulation of energy balance homeostasis.

Part I contains three chapters covering the central pathways involved in the control of food intake. The first of these is devoted to the orexigenic neuropeptides, i.e. those that increase or stimulate appetite, while the second is a description of the peptides with anorexigenic effects, i.e. those that decrease or stop food intake. Since this is a rapidly evolving field, the third chapter concentrates on emerging and newly identified factors and their interaction with the already well-known peptides.

Part II encompasses six chapters that deal with the peripheral signals participating in energy homeostasis and their control in health and disease. Regulation of body weight was once considered a simple feedback control system in which the hypothalamus modulated food intake and energy expenditure to compensate for fluctuations in body weight. The existing body of evidence has fostered the transition from the classic adipostat, a sensor of body adiposity that informs the hypothalamus about the abundance of energy stores, to a more dynamic and multifactorial model including signals emerging from several different organs such as the gut, the liver, the pancreas and the vascular system. The underlying molecular mechanisms by which adipose tissue enlargement and the subsequent increase in adipokines contribute to the pathophysiological events in the gastrointestinal, hepatic, pancreatic, musculoskeletal, cardiovascular and immune systems are now beginning to be better understood and are covered in detail in this section of the book.

Part III contains six chapters providing an integrative approach to current knowledge in energy balance regulation. Adipose tissue biology and the hierarchy of the neural circuitry controlling energy homeostasis deserve special attention, as does the relevance of food reward signals and the links between the homeostatic and hedonic systems. Specific chapters address the available advances in technology to analyse these complex issues, including functional neuroimaging and the whole range of the 'omics' strategies. The final chapter takes a fresh and innovative look at future potential approaches to obesity management.

I would like to express my gratitude to The Nutrition Society for the confidence and vision in this project, as well as to Professor Philip C. Calder, the Series Editor, for his guidance and helpful suggestions. My thanks are also due to Chris McEnnerney for her diligent and meticulous editing skills. Christopher Holt is also gratefully acknowledged. Special thanks go to Sarah Hulbert and the staff at CABI for their immense patience and understanding of all my limitations and interfering circumstances. Without Sarah's helpfulness and support it would have been impossible to complete this enterprise. Finally, I would like to thank each of the authors for contributing such insightful chapters, despite their many commitments and busy agendas. This book could not have become a reality were it not for everybody's dedicated efforts.

Gema Frühbeck

1 Orexigenic Peptides

Margaret J. Morris[1] and Michelle J. Hansen[2]

[1]School of Medical Sciences, University of New South Wales, Australia;
[2]Department of Pharmacology, University of Melbourne, Australia

The Obesity Problem

The problem of obesity has been recognized by the World Health Organization (WHO) as an epidemic, with most countries experiencing a dramatic increase in incidence in the last decades (WHO, 1997, 2003). Moreover, obesity constitutes a major risk factor for other diseases and is estimated to account for 5–11% of health care costs (Finkelstein *et al.*, 2005; Fry and Finley, 2005; Allender and Rayner, 2007; Runge, 2007).

Obesity results from a chronic imbalance between energy intake and energy expenditure, characterized by increased adipose tissue stores. The recent rise in the prevalence of obesity most likely reflects lifestyle changes, with a reduction in physical activity and increased availability of cheap, highly palatable and energy dense foods (Chopra *et al.*, 2002). Building on previous reviews of hypothalamic appetite regulation (Flier and Maratos-Flier, 1998; Woods *et al.*, 1998; Inui, 1999; Schwartz *et al.*, 2000; Seeley and Woods, 2003; Woods *et al.*, 2004; Horvath, 2005; Woods, 2005; Abizaid *et al.*, 2006; Goldstone, 2006; Morton *et al.*, 2006; Leibowitz, 2007), this chapter reviews the critical role of hypothalamic orexigenic neuropeptides in the control of food intake and examines the evidence for changes in these mediators in diet-induced obesity.

Control of Food Intake and Energy Homeostasis

Hypothalamic circuits involved in energy homeostasis

While the regulation of energy homeostasis involves several brain regions including the cortex, brainstem, parts of the limbic system and the amygdala, it is the hypothalamus that integrates the humoral and neural signals involved in the control of food intake (Schwartz *et al.*, 2000; Williams *et al.*, 2001; Wilding,

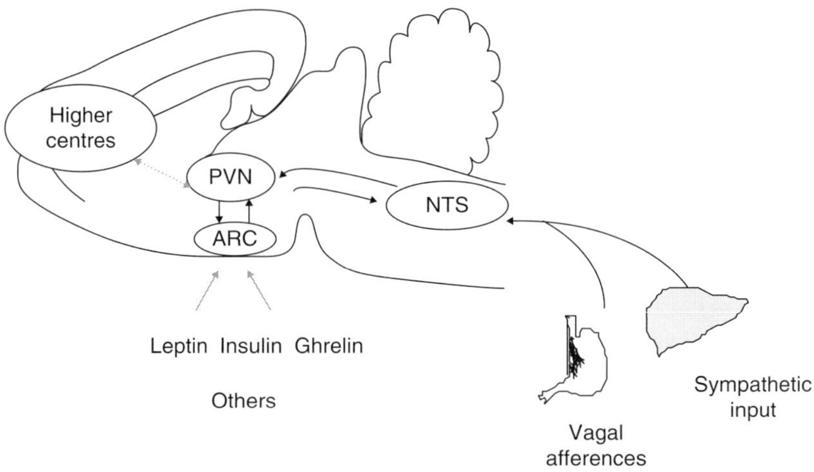

Fig. 1.1. Central control of appetite involves hormonal input from the periphery, as well as neural signals from the gut that are integrated in the medulla. Ascending input to the hypothlamus regulates the activity of hypothalamic circuits involved in feeding. ARC, arcuate nucleus; NTS, nucleus tractus solitarius; PVN, paraventricular nucleus.

2002; Woods, 2005; Wynne *et al.*, 2005; Abizaid *et al.*, 2006; Goldstone, 2006; Morton *et al.*, 2006; Gao and Horvath, 2007, 2008; Valassi *et al.*, 2008). The arcuate nucleus (ARC) in the ventral hypothalamus has extensive reciprocal projections with other hypothalamic nuclei, including the paraventricular nucleus (PVN), ventromedial nucleus (VMH), dorsomedial nucleus (DMH) and lateral hypothalamus. The PVN is densely innervated by the ARC and lateral hypothalamus. The lateral hypothalamus has reciprocal projections with the ARC, PVN and the caudal brainstem. These hypothalamic circuits consist of distinct neuronal cell populations, which may be regulated differentially by energy status and a number of circulating hormones (Fig. 1.1).

Peripheral signals involved in energy homeostasis

Food intake can be regulated by peripherally derived signals, which can be classified as short-term or long-term signals (reviewed in Woods *et al.*, 1998; Spiegelman and Flier, 2001; Wilding, 2002; Ahima, 2006; Moran, 2006; Coll *et al.*, 2007; Cummings and Overduin, 2007; Klok *et al.*, 2007). Short-term signals, both neural and humoral, influence the size of a single meal and either initiate or terminate a meal. These signals are generated by the pancreas, liver or gastrointestinal tract as either afferent sensory relays (vagal, splanchnic or spinal) or hormones that access the central nervous system (CNS).

Long-term (adiposity) signals provide information to the brain about the status of energy stores and are proposed to underpin adaptive CNS responses to restore energy (Ahima *et al.*, 2006; Pliquett *et al.*, 2006). These hormonal signals include leptin and insulin, which circulate at levels proportional to body adiposity. Leptin, the highly conserved 167-amino acid protein product of the *ob* gene, is predominantly synthesized and secreted into the circulation from white adipose tissue (Zhang *et al*., 1994). Leptin administration reduces food intake and increases energy expenditure, resulting in body weight loss (Pelleymounter *et al*., 1995). Insulin is secreted from pancreatic β-cells in response to increases in blood glucose. In the presence of insulin, ingested glucose is taken up by insulin-sensitive tissues, where it is metabolized and excess energy stored (reviewed in Baskin *et al*., 1999b; Niswender and Schwartz, 2003; Pliquett *et al*., 2006). Insulin also exerts effects in the CNS, and insulin signalling within the brain participates in the regulation of food intake and body weight. Anorectic effects of both leptin and insulin are partly mediated by regulating the CNS actions of the orexigenic neuropeptides considered below.

These short- and long-term signals modulate the activity of brain neuropeptide circuits, thereby adjusting food intake and energy metabolism. The discussion below focuses on those signals that increase food intake.

Ghrelin

Ghrelin, a circulating 28-amino acid peptide, endogenous ligand for the growth hormone secretagogue receptor (GHS-R), stimulates growth hormone secretion potently both *in vivo* and *in vitro* (Kojima *et al*., 1999). While first decribed in the periphery, it is now clear that ghrelin is also produced in the hypothalamus. Ghrelin is produced predominantly by the stomach and circulating levels respond to changes in energy intake. Lower amounts of ghrelin are derived from the bowel, pancreas, kidney, testis, placenta, pituitary and hypothalamus (reviewed in Horvath *et al*., 2001; Muccioli *et al*., 2002; Gualillo *et al*., 2006; García *et al*., 2007; López *et al*., 2007). GHS-R were shown to be expressed in hypothalamic nuclei implicated in body weight regulation including the PVN, ARC and VMH (Guan *et al*., 1997).

Within the CNS, ghrelin immunoreactivity was shown in the ARC (Kojima *et al*., 1999) and, subsequently, ghrelin mRNA was found to be expressed in a previously uncharacterized group of neurones in the hypothalamus adjacent to the third ventricle (Cowley *et al*., 2003). Central and peripheral administration of ghrelin increases food intake and body weight dose-dependently (Tschöp *et al*., 2000). Ghrelin increases the body weight of growth hormone-deficient dwarf rats, suggesting the adipogenic and orexigenic actions of ghrelin are independent of its ability to stimulate growth hormone secretion (Tschöp *et al*., 2000).

Peripheral ghrelin administration was shown to induce c-fos immunoreactivity in the ARC, which contains a large proportion of neurones that coexpress the potent orexigenic peptides, neuropeptide Y (NPY) and agouti-related peptide (AgRP) (Dickson and Luckman, 1997). Approximately 90% of NPY/AgRP neurones in the ARC express GHS-R mRNA, suggesting ghrelin may influence energy homeostasis via NPY/AgRP signalling (Willesen *et al*., 1999). In line with this, both NPY and AgRP mRNA expression in the ARC were increased following acute and chronic intracerebroventricular (ICV) administration of ghrelin (Kamegai *et al*.,

2001; Nakazato *et al.*, 2001). Furthermore, administration of antisera to NPY or AgRP, Y1 or Y5 NPY receptor antagonists or α-melanocyte-stimulating hormone (α-MSH), a melanocortin agonist, blocked ghrelin-induced hyperphagia (Nakazato *et al.*, 2001; Shintani *et al.*, 2001; Lawrence *et al.*, 2002b). However, NPY-deficient mice respond to the orexigenic effect of ghrelin (Tschöp *et al.*, 2000). The observation that ghrelin reduced the inhibitory actions of leptin on food intake and ARC NPY mRNA expression led to the suggestion that ghrelin might oppose the actions of leptin (Nakazato *et al.*, 2001; Shintani *et al.*, 2001).

Neuropeptide Y (NPY)

NPY is a 36-amino acid peptide member of the pancreatic polypeptide family that includes peptide YY (PYY) and pancreatic polypeptide (Tatemoto *et al.*, 1982). NPY is distributed abundantly and widely in both the central and peripheral nervous systems. Within the CNS, NPY is synthesized predominantly in the locus coeruleus, the nucleus of the solitary tract and the ARC (Chronwall *et al.*, 1985). NPY is co-localized with noradrenaline and adrenaline in medullary and pontine areas. Within the hypothalamus, there is a dense NPY-containing projection from the ARC to the PVN, an area important in the control of feeding behaviour (Bai *et al.*, 1985; Chronwall *et al.*, 1985). NPY-containing neurones in the ARC also send projections to the lateral hypothalamus, DMH, medial preoptic area, perifornical area and within the ARC (Broberger *et al.*, 1998).

NPY receptors

To date, six NPY receptor subtypes (Y1–Y5 and y6), sharing modest (30–50%) sequence homology have been reported; all are members of the G protein-coupled receptor superfamily (Blomqvist and Herzog, 1997; Brain and Cox, 2006). Recently, Y7 has been discovered in fish, amphibians and chicken (Bromée *et al.*, 2006). Y1, Y2, Y4 and Y5 receptors are expressed in both the human and rat CNS (Parker *et al.*, 2002; Thorsell and Heilig, 2002) and have been located in a number of hypothalamic nuclei, including the ARC, PVN, lateral hypothalamus, supraoptic, VMH and DMH (Parker and Herzog, 1999). Over 80% of NPY-containing neurones in the ARC coexpress NPY Y2 receptor mRNA and this pre-synaptic receptor has been proposed to regulate the release of NPY (Broberger *et al.*, 1998; King *et al.*, 1999).

Role of NPY in control of food intake and energy balance

NPY is one of the most potent orexigenic agents known, with central administration shown to stimulate food intake dose-dependently in satiated rats (Clark *et al.*, 1984). Injection of NPY into specific hypothalamic nuclei highlighted the particularly potent orexigenic action of NPY within the PVN. Chronic NPY administration produces many of the metabolic/endocrine hallmarks of obesity, such as hyperphagia, weight gain, increased adiposity, hyperinsulinaemia and hypertriglyceridaemia (Zarjevski *et al.*, 1993). Both Y1 and Y5 NPY receptors are proposed to be involved primarily in mediating the effects of NPY on food intake (Gerald *et al.*, 1996; Kanatani *et al.*, 1996). Central administration of selective Y1 or Y5 receptor agonists increases food intake in rodents (Cabrele *et al.*, 2000; Parker *et al.*, 2000; Mullins *et al.*, 2001), whereas administration of selective Y1 or Y5 receptor antagonists partially decreases NPY-induced hyperphagia and/or fasting-induced food intake (Kanatani *et al.*, 1996; Polidori *et al.*, 2000; Chaffer and Morris, 2002). Y5 receptor antisense oligonucleotides reduced spontaneous and NPY-induced food intake significantly (Tang-Christensen *et al.*, 1998). Much interest has been directed at developing receptor selective antagonists for NPY as therapeutic approaches to treating obesity.

NPY-deficient mice had normal spontaneous food intake, body weight and adiposity, suggesting a high degree of developmental redundancy in the control of food intake (Erickson *et al.*, 1996b). Subsequently, NPY-deficient mice were shown to have an attenuated re-feeding response to 24- and 48-h fasting (Bannon *et al.*, 2000). Moreover, NPY-deficient mice were shown to be hyperresponsive to the anorectic effects of leptin (Hollopeter *et al.*, 1998). Surprisingly, the Y1 receptor-deficient mouse has increased body weight and adiposity, without hyperphagia (Kushi *et al.*, 1998; Pedrazzini *et al.*, 1998). These mice have reduced dark-phase locomotor activity, which may contribute to the mild obesity observed. Additional evidence highlighting the importance of the Y1 receptor in NPY-induced food intake was the observation that administration of a Y1 receptor antagonist had no effect on feeding in Y1 receptor-deficient mice, but decreased food intake in Y5 receptor-deficient mice and wild-type control mice (Kanatani *et al.*, 2000).

While Y5 receptor-deficient mice grow normally, they develop mild late-onset obesity, which may be due to hyperphagia (Marsh *et al.*, 1998; Frühbeck and Gómez-Ambrosi, 2003). NPY-induced food intake was attenuated at high NPY doses and blocked completely in Y5 receptor-deficient mice treated with a Y1 receptor antagonist, confirming the importance of Y1 and Y5 receptor subtypes in NPY-induced hyperphagia (Marsh *et al.*, 1998). Y2 receptor-deficient mice have increased body weight, food intake and adiposity and a reduced response to leptin, but normal feeding responses to NPY and fasting (Naveilhan *et al.*, 1999). In contrast, Sainsbury *et al.* (2002) reported reduced body weight and white adipose tissue (WAT) in hypothalamus-specific Y2-deleted mice, despite increased food intake. Moreover, there was a synergistic action on this lean phenotype in Y2 and Y4 double-knockout mice (Sainsbury *et al.*, 2003). Thus, Y1 and Y5 receptors may have a major role in mediating the actions of NPY on food intake, while the Y2 and Y4 receptors may play a lesser, perhaps indirect role in NPY-induced hyperphagia. Interestingly, an association between a defect in the Y2 receptor gene and severe obesity has been described (Ma *et al.*, 2005).

Modulation of hypothalamic levels of NPY

The level of NPY signalling is strongly influenced by nutritional status (Williams *et al.*, 2001; Parker *et al.*, 2002). Fasting or food restriction increased hypothalamic expression of NPY mRNA in the ARC and of NPY peptide in the ARC and PVN (Sahu *et al*., 1988). During lactation, a state of high energy demand accompanied by hyperphagia, NPY mRNA expression was increased in the ARC and DMH (Smith, 1993).

Hypothalamic NPY-containing neurones express leptin receptors, providing a mechanism by which leptin modulates NPY levels (Mercer *et al*., 1996; Baskin *et al*., 1999a). Central leptin administration attenuated the increase in ARC NPY mRNA expression associated with fasting (Schwartz *et al*., 1996). Elevated NPY mRNA expression in the ARC has been described in genetic models of obesity linked to defective leptin signalling, including *ob/ob* and *db/db* mice and *fa/fa* Zucker rats (Sanacora *et al*., 1990; Chua *et al*., 1991; Wilding *et al*., 1993). Normalization of leptin levels restores NPY mRNA expression in these genetic models of obesity, except in the *db/db* mouse that has a mutated leptin receptor (Stephens *et al*., 1995; Schwartz *et al*., 1996). Furthermore, leptin was shown to reduce *in vitro* NPY overflow from rat hypothalamic explants (Lee and Morris, 1998). NPY is one of the major downstream mediators of leptin, and genetic deletion of NPY partially ameliorates the obese phenotype of the leptin-deficient *ob/ob* mouse (Erickson *et al*., 1996a).

Insulin has also been implicated in the modulation of hypothalamic NPY. Uncontrolled type 1 diabetes in the rat is associated with profound hyperphagia, increased NPY mRNA expression and peptide levels in ARC and NPY peptide release in PVN, all of which are prevented by insulin treatment (Sahu *et al*., 1990; White *et al*., 1990; Gozali *et al*., 2002). Moreover, insulin has been shown to inhibit NPY mRNA expression (Schwartz *et al*., 1992). Thus, insulin may induce hypophagia, at least in part, by reducing NPY signalling.

Hypothalamic levels of NPY in obesity

Variable results have been obtained from studies investigating the effect of diet-induced obesity on hypothalamic NPY (Table 1.1), possibly due to differences in methodology, such as age and strain of the animals, and diet duration and composition. Although in rodents the effect of diet-induced obesity on brain NPY levels remains controversial, in general, evidence for a reduction in NPY peptide appears to predominate (Table 1.1). Our laboratory has studied extensively the effect of cafeteria diet-induced obesity on hypothalamic NPY peptide levels. In response to a high-fat, palatable cafeteria diet, rats double their caloric intake, increasing body weight by around 25%, and show a progressive reduction in hypothalamic NPY peptide with increasing duration of dietary intervention. The reduction in hypothalamic NPY content was accompanied by reduced basal NPY overflow from the hypothalamus and increased orexigenic responsiveness to exogenous NPY administration, suggesting reduced activity of the hypothalamic NPY system in animals rendered obese by a high-fat diet (Hansen *et al*., 2001, 2004). Nutritional manipulation early in life also influenced PVN NPY overflow in adult offspring, suggesting programming effects on hypothalamic NPY responsiveness (Kozak *et al*., 2005).

Despite high leptin levels, most obese humans and rodents lack responsiveness to its appetite-suppressing effects. Leptin has been shown to modulate

Table 1.1. Effect of dietary obesity on hypothalamic NPY mRNA, peptide and receptor expression.

Reference	Duration	% Fat of diet	Measure	Effect on NPY
Rat studies				
Wilding *et al.*, 1992	7 weeks	13	Peptide mRNA	↔ Total hypothalamus ↑ PVN, ARC, MPO, AHA
Widdowson *et al.*, 1997	8 weeks	13	Binding	↑ Y5 and/or Y2 dorsal and LH
Stricker-Krongrad *et al.*, 1998	5 months	72.5	Peptide	↓ PVN, ARC
Widdowson *et al.*, 1999	8 weeks	13	mRNA	↓ total hypothalamus
Hansen *et al.*, 2001	5 months	30	Peptide	↓ PVN
Torri *et al.*, 2002	2 months	28	IHC	↔ PVN
Hansen *et al.*, 2004	9–17 weeks	30	Peptide	↓ AHA, PO, PVN, ARC
Wang *et al.*, 2007	13 weeks	40	mRNA	↑ hypothalamic NPY, Y1, Y2 and Y5
Mouse studies				
Guan *et al.*, 1998	6 months	60	mRNA	↑ DMH, ↓ ARC
Lin *et al.*, 2000	8, 19 weeks	58.7	mRNA	↓ ARC
Huang *et al.*, 2003	22 weeks	40	mRNA	↑ ARC; ↔ ARC Y1, Y2 and Y5
Rahardjo *et al.*, 2007	22 weeks	40	Binding	↔ ARC, DMH, LH, VMH Y2
Chen *et al.*, 2007	7 weeks	32	Peptide	↓ AHA, PVN
Morris *et al.*, 2008	10 weeks	32	Peptide	↓ AHA, ARC, PH

Note: Representative studies with dietary intervention over 2 weeks are included. ↑ Indicates increased levels, ↓ indicates reduced levels, ↔ indicates no significant change. AHA, anterior hypothalamic area; ARC, arcuate nucleus; DMH, dorsomedial hypothalamus; LH, lateral hypothalamus; MPO, medial preoptic area; PH, posterior hypothalamus; PO, preoptic area; VMH, ventromedial hypothalamus. Y1, Y2 and Y5 refer to NPY receptor subtypes.

NPY/AgRP and α-MSH secretion from the ARC of lean mice. Although high-fat diet-induced obese mice have normal functional leptin receptor levels and increased suppressor of cytokine signalling 3 (SOCS3) levels, leptin fails to modulate peptide secretion and any element of the leptin signalling cascade system (Enriori *et al.*, 2007). Despite this leptin resistance, the melanocortin system downstream of the ARC in diet-induced obesity (DIO) mice is overresponsive to melanocortin agonists, probably due to upregulation of melanocortin 4 receptor (MC4R). In addition, by decreasing the fat content of the mouse's diet, leptin responsiveness of NPY/AgRP and pro-opiomelanocortin (POMC) neurones recovered simultaneously, with mice regaining normal leptin sensitivity and glycaemic control. These findings highlight the physiological importance of leptin sensing in the melanocortin circuits, showing that their loss of leptin sensing likely contributes to the pathology of leptin resistance (Enriori *et al.*, 2007). Interestingly, early in the course of high-fat diet-induced weight gain, a period of central leptin hypersensitivity has been described, which is followed by the development of central leptin insensitivity as the high-fat feeding is maintained over time (Fam *et al.*, 2007). This leptin insensitivity does not appear to be explained by a downregulation of functional leptin receptor protein levels, reduced leptin signalling, an increase in either SOCS3 or NPY expression or reduced function of the melanocortin system (Fam *et al.*, 2007). Thus, the influence of high-fat feeding on other actions of leptin, such as its effect on the endocannabinoid system, should be investigated further.

Melanin-concentrating Hormone (MCH)

Melanin-concentrating hormone (MCH) was isolated originally from salmon pituitary as a cyclic 17-amino acid polypeptide with a cysteine–cysteine disulfide bond (Kawauchi *et al.*, 1983). Subsequently, MCH was isolated from rat hypothalamus (Vaughan *et al.*, 1989) and identified as a cyclic 19-amino acid neuropeptide in both human and rat (Presse *et al.*, 1990). The N-terminus of mammalian MCH is extended by two amino acids and the loop structure is highly conserved (Presse *et al.*, 1990). MCH is expressed predominantly in the magnocellular neurones of the lateral hypothalamus and the subzona incerta (Skofitsch *et al.*, 1985; Bittencourt *et al.*, 1992), which project widely to areas implicated in the control of feeding (Zamir *et al.*, 1986).

The first MCH receptor subtype identified was the previously cloned orphan G protein-coupled receptor, somatostatin-like receptor-1 (SLC-1), now known as MCH-1 receptor (Bachner *et al.*, 1999; Chambers *et al.*, 1999). The MCH-1 receptor has a broad expression pattern throughout the brain and is found in areas known to control motivation, emotion, olfaction and possibly memory (reviewed in Pissios and Maratos-Flier, 2003). Within the hypothalamus, MCH-1 receptor is found in the ARC, PVN, VMH and DMH, areas implicated in energy homeostasis (Hervieu *et al.*, 2000; Kokkotou *et al.*, 2001). In addition, MCH-1 receptor-deficient mice have decreased adiposity, increased activity and decreased susceptibility to diet-induced obesity (Marsh *et al.*, 2002). Chronic administration of the MCH-1 receptor antagonist, SNAP-7941, to diet-induced obese rats resulted in a marked and sustained decrease in body weight (Borowsky *et al.*, 2002). On the other hand, the observation of increased responsiveness to a melanocortin agonist, in the face of a high-fat diet, suggests melanocortin analogues may have potential for the pharmacological treatment of obesity (Hansen *et al.*, 2005).

Due to the fact that MCH failed to induce body weight gain and hyperphagia in MCH-1 receptor-deficient mice, this receptor was believed to be primarily responsible for the stimulatory effect of MCH on food intake (Marsh *et al.*, 2002). Increased MCH-1 receptor staining was observed in the hypothalamic infundibular nucleus post-mortem in cachectic patients (Unmehopa *et al.*, 2005), supporting a role for this receptor in energy balance. However, in a large study, no significant association between MCH-1 receptor gene single nucleotide polymorphisms (SNPs) and obesity was evident (Wermter *et al.*, 2005). A second MCH receptor subtype was identified in human and monkey brains, but this

MCH-2 receptor has been reported not to exist in rodents (Hill *et al.*, 2001; Rodriguez *et al.*, 2001; Wang *et al.*, 2001).

Central administration of MCH has been shown to stimulate food intake in both rats and mice (Qu *et al.*, 1996). Although repeated injections over 1 week did not alter body weight (Rossi *et al.*, 1997), continuous infusion of MCH increased body weight and adiposity (Gomori *et al.*, 2003; Ito *et al.*, 2003). Targeted deletion of the MCH gene led to hypophagia, low body weight (~25% lighter) and an inappropriately increased metabolic rate (Shimada *et al.*, 1998). More recently, MCH deletion has been shown to lead to resistance to diet-induced obesity in a strain-specific manner (Kokkotou *et al.*, 2005). Conversely, transgenic mice overexpressing MCH were hyperphagic and developed obesity when placed on a high-fat diet (Ludwig *et al.*, 2001). Little is known about the effect of a high-fat diet on hypothalamic MCH expression and content. It has been reported previously that after 2 months of high-fat diet in rats (28% kcal as fat), lateral hypothalamic MCH mRNA expression was not altered (Torri *et al.*, 2002). Further, our laboratory has shown no alteration in the feeding responsiveness to exogenous MCH administration in high-fat fed rats (Hansen and Morris, unpublished data). We have evidence that MCH may stimulate feeding, in part, by eliciting NPY release (Chaffer and Morris, 2002) and this observation may be consistent with our reported hyperresponsiveness to NPY administration in diet-induced obesity.

MCH mRNA expression was elevated in obese *ob/ob* and *db/db* mice, the obese Zucker rat and in response to fasting (Qu *et al.*, 1996; Tritos *et al.*, 2001). Treatment with leptin reduced the fasting-induced elevation in MCH mRNA expression (Tritos *et al.*, 2001). Moreover, MCH-1 receptor mRNA expression in the CNS was shown to be elevated by fasting and, in the *ob/ob* mouse, this was normalized by leptin treatment (Kokkotou *et al.*, 2001). Thus, a reduction in MCH signalling likely contributes to the inhibition of food intake induced by leptin. MCH has shown little or no interaction with NPY or hypocretin in inducing food intake when injected together into the third ventricle (Sahu, 2002). Interestingly, a mouse model of MCH neurone ablation exhibits a phenotype that highly resembles that of mice lacking only the MCH gene with reduced food intake and increased energy expenditure (Alon and Friedman, 2006). Moreover, the ablation of MCH neurones in mice with an *ob/ob* mutant background improved obesity and glucose tolerance (Gao and Horvath, 2008). These findings suggest that the function of MCH cells in energy regulation is limited to the MCH system itself, rather than to other aspects of the cells as the classic neurotransmitter function or synaptic plasticity, which are distinct from NPY cells.

Orexin A and B/Hypocretin 1 and 2

Orexins A and B, also known as hypocretins 1 and 2, are proteolytically cleaved from the same precursor, prepro-orexin, that is expressed in a small group of neurones in the perifornical region of the lateral hypothalamus (Sakurai *et al.*, 1998). Orexin-containing neurones project to a number of sites, including the PVN, ARC, nucleus of the solitary tract and dorsal motor nucleus of the vagus, as well as the olfactory bulb (Peyron *et al.*, 1998; Shibata *et al.*, 2008). Orexins act at two G protein-coupled receptor subtypes, orexin-1 and orexin-2, that are expressed differentially within the brain (Sakurai *et al.*, 1998). Both orexin A and B stimulate food intake; however, orexin A acting at the orexin-1 receptor appears to be more potent (Sakurai *et al.*, 1998). Chronic central administration of orexin to rats does not result in obesity (Yamanaka *et al.*, 1999). A Y1 NPY antagonist completely blocked the stimulation of food intake by orexin A and B, suggesting the hyperphagia induced by orexin may be mediated partly by NPY (Jain *et al.*, 2000). Furthermore, following central administration of orexin, *fos* expression was increased in NPY-containing neurones in the ARC (Yamanaka *et al.*, 2000). Orexin mRNA is upregulated by fasting and hypoglycaemia (Cai *et al.*, 1999). Furthermore, most orexin neurones coexpress the leptin receptor Ob-Rb (Hakansson *et al.*, 1999), and leptin administration has been shown to reduce orexin mRNA expression and inhibit the fasting-induced increase in orexin mRNA and orexin-1 receptor (López *et al.*, 2000). In addition, orexin expression is reduced by α-MSH (Coll *et al.*, 2007).

Targeted deletion of the orexin gene resulted in narcolepsy, suggesting orexin's involvement in general arousal (Chemelli *et al.*, 1999). More recently, the prominent role of orexins as critical components in maintaining and regulating the stability of arousal has been established (Boutrel and de Lecea, 2008). Furthermore, hypocretin-producing neurones have been suggested to be part of the circuitries that mediate the hypothalamic response to acute stress. Intracerebral administration of hypocretin leads to a dose-related reinstatement of drug- and food-seeking behaviours (Boutrel and de Lecea, 2008). Moreover, stress-induced reinstatement can be blocked with hypocretin receptor 1 antagonism. These findings, together with recent data showing that hypocretin is critically involved in cocaine sensitization through the recruitment of NMDA receptors in the ventral tegmental area, strongly suggest that activation of hypocretin neurones plays a critical role in the development of the addiction process. The activity of hypocretin neurones may affect addictive behaviour by contributing to brain sensitization or by modulating the brain reward system (Boutrel and de Lecea, 2008).

Galanin and Galanin-like Peptide

Galanin

Galanin is a 29 (30 in human)-amino acid residue neuropeptide, isolated originally from small intestine, that is conserved across species and distributed widely in various brain regions (Melander *et al.*, 1986; Merchenthaler *et al.*, 1993). Galanin is expressed in a number of hypothalamic areas, including the PVN, lateral hypothalamus and ARC. Central administration of galanin increases food intake in satiated rats rapidly, with little effect on macronutrient preference (Kyrkouli *et al.*, 1986; Smith *et al.*, 1996; Crawley, 1999; Leibowitz and Wortley, 2004). Microinjection studies have revealed galanin can exert orexigenic effects in the PVN, lateral hypothalamus and VMH (Kyrkouli *et al.*, 1990; Schick *et al.*, 1993). The underlying macronutrient preference appears to be important in determining the feeding response to galanin (Smith *et al.*, 1996). The fact that levels of galanin peptide and gene expression in the PVN correlate with natural feeding preferences has led to the suggestion that endogenous galanin may respond to signals related to the metabolism of fat (Leibowitz and Wortley, 2004). Galanin, like NPY and MCH, was shown to be a target of leptin signalling in the hypothalamus (Sahu, 1998).

The behavioural, metabolic and neuroendocrine actions of galanin are mediated by distinct G protein-coupled galanin receptor subtypes; GalR1, GalR2 and GalR3 have been cloned and characterized in rat, mouse and human (Branchek *et al.*, 2000; Gundlach, 2002). GalR1 and GalR2 mRNA are expressed widely in the CNS, including the hypothalamus, whereas GalR3 mRNA is restricted to several hypothalamic areas (Mennicken *et al.*, 2002). Defining the relative roles each galanin receptor subtype plays in the feeding action of galanin has been somewhat hampered by the lack of potent and specific antagonist molecules.

Compared to NPY, changes in feeding status, such as fasting or food restriction, appear to have more limited impact on galanin mRNA levels in the hypothalamus (Brady *et al.*, 1990). While endogenous galanin has been reported to be upregulated in obese rats fed a high-fat diet (Leibowitz and Wortley, 2004), transgenic galanin overexpression in the brain had no effect on body weight or feeding behaviour (Hohmann *et al.*, 2003). Galanin knockout mice were more sensitive to the effects of leptin on body weight and adiposity (Hohmann *et al.*, 2003).

Galanin-like peptide (GALP)

A second member of the galanin peptide family was identified based on its ability to activate GalR *in vitro*. Galanin-like peptide (GALP) is a 60-amino acid residue neuropeptide that shares partial sequence identity with galanin (Ohtaki *et al.*, 1999). GALP mRNA expression shows a different pattern to galanin, being restricted to neurones in the arcuate nucleus/median eminence and posterior pituitary (Takatsu *et al.*, 2001; Gundlach, 2002; Kageyama *et al.*, 2005). GALP-containing projections have been described from ARC to PVN and, more recently, neuronal interactions between GALP and orexin and MCH neurones in the lateral hypothalamus have been reported (Takenoya *et al.*, 2005). In the rat, GALP mRNA expression in the ARC was downregulated by fasting for 48 h (Jureus *et al.*, 2000).

ICV administration of GALP stimulates feeding in the rat rapidly and potently (Matsumoto *et al.*, 2002; Kageyama *et al.*, 2005), and some groups have reported a reduction in food intake at 24 h (Lawrence *et al.*, 2002a), although this was not observed in our studies (Tan *et al.*, 2005). There is evidence that GALP may elicit hyperphagia by stimulating NPY release (Seth *et al.*, 2003) and, in line with this, we have demonstrated inhibition of GALP-induced feeding by prior ICV administration of the NPY Y_1 antagonist, BIBO3304. GALP feeding effects are species-specific. In rats, ICV injection of GALP has dichotomous actions on energy balance, stimulating feeding over the first hour, but reducing food intake and body weight at 24 h, as well as causing an increase in core body temperature (Man and Lawrence, 2008). In mice, GALP induces only an anorexic action (Krasnow *et al.*, 2003), including *ob/ob* (Kageyama *et al.*, 2005), while its effects on core body temperature have not been established.

Studies examining GALP expression in obese animals demonstrated reduced GALP expression in *ob/ob*, *db/db* mice and *fa/fa* rats (Kumano *et al.*, 2003; Saito *et al.*, 2004; Shen and Gundlach, 2004). We observed exaggerated feeding effects of GALP in rats rendered obese by a palatable high-fat diet, even when corrected for their basal hyperphagia, suggesting that adaptive changes in GALP or its receptors might occur in response to prolonged weight gain and hyperleptinaemia (Tan *et al.*, 2005). The question of whether dietary obesity upregulates GALP production warrants investigation, particularly given the observation that GALP mRNA appears to be upregulated by leptin treatment (Jureus *et al.*, 2000).

While GALP exhibits agonistic activity at GalR1 and GalR2 *in vitro*, the receptor subtype involved in its feeding actions is unclear. There appear to be some differences in the receptors responsible for the feeding effects of galanin and GALP, and the two peptides induce differential effects on *fos* induction in the hypothalamus, suggesting they may activate different receptors (Fraley *et al.*, 2003). Central administration of a GalR2/3 agonist in rats did not induce *c-fos* in any of the brain regions that expressed this protein after GALP injection and had no effect on food intake, body weight and body temperature in rats or mice. These data suggest that GALP induces differential effects on energy balance and brain activity in mice compared to rats, which are unlikely to be due to activation of the GalR2 or GalR3 receptor (Man and Lawrence, 2008). Thus, it is unlikely that GALP signals solely through galanin receptors, and the existence of a yet-to-be-identified GALP-specific receptor is suggested.

The Melanocortin Antagonists – Agouti-related Peptide (AgRP) and Agouti

Melanocortin receptors

Five G-protein-coupled melanocortin receptor (MC1–5R) subtypes have been identified with sequence homology ranging from 35 to 60% (Fisher *et al.*, 1999). MC3R and MC4R are distributed differentially in the CNS. The MC3R has a relatively restricted distribution, with greatest density in the ARC, VMH, lateral hypothalamus and preoptic nucleus (Roselli-Rehfuss *et al.*, 1993). Interestingly, MC3R in the ARC are expressed selectively on POMC- and AgRP-containing neurones, suggesting an autoinhibitory role for MC3R in melanocortin signalling (Bagnol *et al.*, 1999; Jegou *et al.*, 2000). The precise physiological role of MC3R in the control of food intake is unclear. The MC4R is distributed widely in the CNS and MC4R mRNA is expressed in areas implicated in the control of food intake and body weight, such as the PVN, ARC, median preoptic area, VMH and DMH (Mountjoy *et al.*, 1994; Kishi *et al.*, 2003).

POMC and MC4R knockout mice have profound hyperphagia and mature-onset obesity (Huszar *et al.*, 1997; Frühbeck and Gómez-Ambrosi, 2003). On the other hand, MC3R-deficient mice have normal body weight and food intake but have increased adiposity and feed efficiency (Butler *et al.*, 2000; Chen *et al.*, 2000). Genetic deletion of MC4R in mice and humans reportedly results in severe hyperphagic obesity (Coll *et al.*, 2007). MC4R mutations are responsible for up to 5% of cases of severe childhood obesity and between 0.5 and 2.5% of adult obesity. Estimates of its frequency in the general population of the UK suggest a mutational frequency of 1/1000, making MC4R deficiency one of the most common single gene disorders, with mutations of the MC4R estimated to account for approximately 4% of morbid obesity in humans (Wisse and Schwartz, 2001; Coll *et al.*, 2007). The phenotypic features of MC4R deficiency include hyperphagia, an increase in fat and lean mass and an increase in bone mineral density (Farooqi *et al.*, 2003).

Until recently, it was uncertain as to whether each feature of the complex phenotype of MC4R deficiency could be ascribed to different regions of the brain, with one site controlling food intake and another regulating energy expenditure. There are now clear data demonstrating there is indeed functional divergence of the melanocortin pathway (Coll *et al.*, 2007). Using cre-lox technology, Balthasar *et al.* (2005) partially 'rescued' MC4R-deficient mice by re-expressing MC4R only in the PVN and the amygdala. Although still heavier than control littermates, these mice were significantly lighter than mice globally lacking MC4Rs. This was due entirely to a normalization in food intake, since the reduced energy intake typical of mice globally lacking MC4R was unaffected by this targeted re-expression. Thus, MC4R in the PVN and/or amygdala appear to control food intake, but MC4R expressed elsewhere have a role in the control of energy expenditure.

Interestingly, the participation of the CNS–MCR system in the control of adiposity through effects on nutrient partitioning and cellular lipid metabolism independently of nutrient intake has been reported (Nogueiras *et al.*, 2007). Both the pharmacological inhibition of MCR in rats and the genetic disruption of MC4R in mice have been shown to promote lipid uptake directly and potently, triglyceride synthesis and fat accumulation in white adipose tissue, with increased CNS–MCR signalling triggering lipid mobilization. These effects preceded changes in adiposity and were independent of food intake (Nogueiras *et al.*, 2007). Furthermore, decreased CNS–MCR signalling promoted increased insulin sensitivity and glucose uptake in white adipose tissue, while decreasing glucose utilization in muscle and brown adipose tissue. This central melanocortinergic control of peripheral nutrient partitioning depended on a functional sympathetic nervous system (SNS) and was enhanced by synergistic effects on liver triglyceride synthesis (Nogueiras *et al.*, 2007). Thus, even in the absence of hyperphagia, an enhanced adiposity resulting from decreased melanocortin signalling may be observed, which is consistent with feeding-independent changes in substrate utilization observed in chronic MCR blockade in rodents and in humans with loss-of-function mutations in MC4R.

The data that define the physiological function of MC3R in energy metabolism are not as clear-cut as those for MC4R, although it is likely these two receptors serve non-redundant roles (Coll *et al.*, 2007). Mice lacking MC3R have a unique phenotype; despite an increased fat mass, their total body weight is similar to wild-type mice due to a reduction in lean mass. MC3R may influence feed efficiency (weight gain per kcal consumed) and the partitioning of fuel stores into fat.

Melanocortin antagonists

The melanocortin system has two endogenous antagonists, agouti and agouti-related peptide (AgRP). Dominant mutations of the agouti gene give rise to mice (A^{vy}/) with the oldest known model of mature-onset obesity, characterized by hyperphagia, increased adiposity, hyperglycaemia, hyperinsulinaemia and increased somatic growth. The obesity of the A^{vy}/-mouse is a result of ectopic expression of agouti in the brain, causing antagonism of MC3/4R. Following the cloning of the agouti gene in 1992, a homologous gene encoding a 132-amino acid protein, AgRP, was identified by database searching (Ollmann *et al.*, 1997; Shutter *et al.*, 1997).

AgRP is a competitive antagonist at MC3R and MC4R and possibly acts as an inverse agonist at the MC4R (Ollmann *et al.*, 1997; Yang *et al.*, 1999; Nijenhuis *et al.*, 2001). Within the CNS, the synthesis of AgRP is restricted to ARC neurones, that coexpress NPY and project to a number of hypothalamic areas implicated in the control of feeding, such as the PVN, DMH and lateral hypothalamus (Broberger *et al.*, 1998; Hahn *et al.*, 1998; Bagnol *et al.*, 1999). Thus, these AgRP/NPY positive neurones form a parallel but distinct population alongside neurones containing α-MSH (Broberger *et al.*, 1998).

Transgenic overexpression of AgRP in mice resulted in hyperphagia and severe obesity (Ollmann *et al.*, 1997), while ablation of AgRP neurones in mice led to a lean phenotype (Bewick *et al.*, 2005). Central administration of AgRP stimulated food intake dose-dependently and increased body weight, and this effect was shown to persist for up to a week (Rossi *et al.*, 1998; Hagan *et al.*, 2000). The increased food intake following central administration of AgRP highlights the importance of the tonic inhibitory balance at the MC4R in the control of feeding (Fan *et al.*, 1997). The long-lasting effect of AgRP cannot be explained readily by competitive antagonism and an alternative mechanism has been suggested to sustain the hyperphagia induced by AgRP (Hagan *et al.*, 2000). Ancillary proteins such as mahogany and syndecan-3 modulate the interaction of the antagonists with melanocortin receptors. Although their biological effects remain unclear, they provide further links between pigmentation, obesity and the immune system by affecting the balance between agonist and antagonist at receptors on melanocytes, altering behaviour and basal metabolic rate, and mediating an interaction between activated T-cells and macrophages (Dinulescu and Cone, 2000; Reizes *et al.*, 2001, 2003; Strader *et al.*, 2004).

Effects of melanocortin antagonists on food intake and energy balance

Central administration of the MC3/4R antagonist, SHU9119, caused a potent dose-dependent increase in food intake (Fan *et al.*, 1997; Giraudo *et al.*, 1998), and chronic SHU9119 administration increased body weight and adiposity (Adage *et al.*, 2001). This suggests POMC neurones, via release of α-MSH to activate MC4R, exert an inhibitory tone on feeding (Fan *et al.*, 1997). Additional evidence supporting a role for the MC4R in feeding comes from work using the MC4R antagonist, HS014, which has 85-fold selectivity for MC4R over MC3R (Schioth *et al.*, 2002). HS014 increased food intake dose-dependently and blocked the anorectic effect of α-MSH (Kask *et al.*, 1998b; Vergoni *et al.*, 1998). After 2 weeks of central HS014 administration via osmotic minipumps, body weight was increased by 20%, as a result of increased fat deposits (Kask *et al.*, 1999).

Central administration of SHU9119 and HS014 reduced the anorectic action of leptin (Kask *et al.*, 1998a; Satoh *et al.*, 1998) and attenuated the leptin-induced expression of *fos* in the PVN (Seeley *et al.*, 1997). Furthermore, both MC4R-deficient and A^{vy}/-mice were resistant to peripheral and central administration of leptin, suggesting the anorectic effect of leptin was mediated downstream by MC4R signalling (Marsh *et al.*, 1999).

Modulation of hypothalamic AgRP levels

AgRP neurones in the ARC are regulated differentially by leptin and energy state. The Ob-Rb is coexpressed with AgRP mRNA in ARC neurones (Baskin *et al.*, 1999a; Wilson *et al.*, 1999). AgRP mRNA expression is decreased by leptin and increased by fasting (Hahn *et al.*, 1998; Mizuno and Mobbs, 1999). Animals with defective leptin signalling, such as *ob/ob* mice, *db/db* mice and obese Zucker rats, have elevated AgRP mRNA in the ARC (Shutter *et al.*, 1997; Mizuno *et al.*, 1998; Kim *et al.*, 2000), which may contribute to the hyperphagia observed in these genetic models of obesity.

Few investigations have characterized changes in AgRP expression following exposure to a high-fat diet. Total hypothalamic AgRP content has been reported previously to be elevated after 2 and 8 weeks of a palatable diet, whereas POMC and α-MSH peptide content were not altered in rats (Harrold *et al.*, 1999). In contrast, AgRP mRNA expression was decreased after 2 days of high-fat diet in mice, but after 1 week there was no difference (Ziotopoulou *et al.*, 2000). The variability of the results may be accounted for by differences in methodology, as outlined previously.

Interestingly, germline deletion of NPY or AgRP have no major effect on body weight (Coll *et al.*, 2007), although mice lacking AgRP do become modestly lean late on in life due to an increase in energy expenditure. Genetic ablation of NPY/AgRP neurones in adult life leads to profound, life-threatening hypophagia (Flier, 2006). Interestingly, when these neurones are ablated in the newborn period, the effects on body weight and food intake are much more modest, suggesting that network-based compensatory mechanisms can develop in neonates but do not occur readily in adults. It is as yet unclear whether the effects of ablation of these neurones on food intake are due solely to the loss of NPY and AgRP or whether the loss of other neurotransmitters is involved (Coll *et al.*, 2007).

Overview and Future Perspectives

The regulation of food intake involves a highly complex system of peripheral and central signals that are processed in a number of brain regions. The control of food intake involves integrated, redundant pathways, which allows extensive interactions between agents that stimulate or inhibit food intake. These integrative aspects are dealt with in Part III of this book. The ARC contains two prominent neuronal populations, a medial population that coexpresses the orexigenic mediators NPY and AgRP, while more lateral neurones express the anorexigenic neuropeptides, POMC and cocaine- and amphetamine-regulated transcript (CART), and both of these neuronal populations coexpress the leptin receptor. The orexigenic peptides, GALP and ghrelin, are also expressed in the ARC and participate in appetite control.

The lateral hypothalamus expresses MCH and orexins A and B, neuropeptides that are regulated negatively by leptin signalling. Cell bodies in the ARC and lateral hypothalamus project to the PVN. In turn, the ARC also receives input from the lateral hypothalamus; thus MCH and/or orexin may interact with NPY/AgRP neurones and there is pharmacological evidence for interactions between these mediators (Fig. 1.2). Hypothalamic neurones communicate with the brainstem via central autonomic pathways, where peripheral meal-related signals (cholecystokinin, gastric distension, afferent sensory relays) and descending hypothalamic signals are integrated to terminate a meal. Ascending projections from the nucleus of the solitary tract appear to be involved in long-term hypothalamic adaptive changes in food intake.

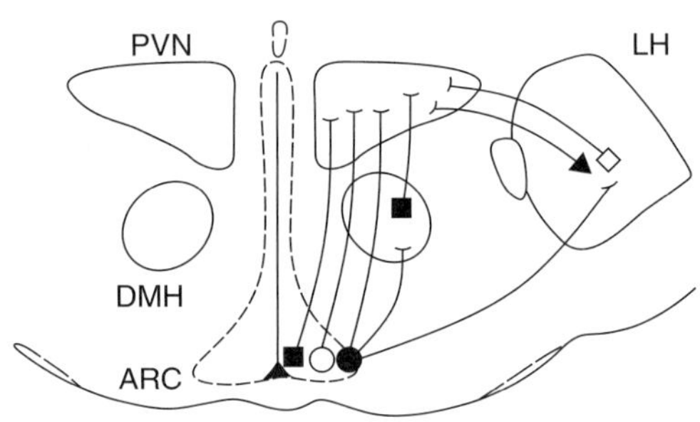

Fig. 1.2. Simplified schematic view of some of the hypothalamic circuits involved in increasing food intake. These orexigenic circuits are regulated differentially by leptin and can also influence each other. Galanin (solid square), GALP (open circle) and NPY/AgRP (solid circle) neurones project to the PVN. NPY/AgRP neurones also project to the LH and may make contact with orexin (triangle) and MCH (open square) neurones. Note that in addition to the projections from DMH to PVN shown, reciprocal connections occur between the hypothalamic nuclei involved in appetite regulation, including the VMH (not shown). For example, orexin-containing cells in the LH project to NPY/AgRP cells in the ARC. Further, these orexigenic pathways make contact with anorexigenic circuits discussed in Chapter 2. ARC, arcuate nucleus; DMH, dorsomedial hypothalamus; LH, lateral hypothalamus; PVN, paraventricular nucleus; VMH, ventromedial hypothalamus.

Conversely, obesity can also be induced by neurobiological disorders and drugs. The aetiology of obesity is complex and includes biology, behaviour and environment. Physicians are faced with the need to manage obesity while strategies for prevention and sustained weight reduction are limited. Present treatment options comprise lifestyle modification, diet, pharmacotherapy and bariatric surgery. Considerable headway has been made in elucidating the neurobiological underpinnings of obesogenic behaviour. There is now a growing understanding of the metabolic, hormonal and behavioural circuitries that contribute to the complex and redundant system for energy balance (Knecht *et al.*, 2008). Changing the net balance of this system to prevent or reduce obesity requires multimodal and long-term interventions.

Investigation of the role of hypothalamic orexigenic neuropeptides in the regulation of food intake in humans and alterations in obesity is currently limited. The neuronal circuits controlling food intake are highly conserved; thus, animal models are a useful tool in understanding the pharmacology and physiology of appetite regulation. An integrated approach exploiting multiple experimental techniques (pharmacological, biochemical, physiological and genetic) appears the most effective option for identifying aberrant hypothalamic signalling and therefore potential therapeutic targets for the treatment of obesity.

References

Abizaid, A., Gao, Q. and Horvath, T.L. (2006) Thoughts for food: brain mechanisms and peripheral energy balance. *Neuron* 51, 691–702.

Adage, T., Scheurink, A.J., de Boer, S.F., de Vries, K., Konsman, J.P., Kuipers, F., Adan, R.A., Baskin, D.G., Schwartz, M.W. and van Dijk, G. (2001) Hypothalamic, metabolic, and behavioral responses to pharmacological inhibition of CNS melanocortin signaling in rats. *Journal of Neuroscience* 21, 3639–3645.

Ahima, R.S. (2006) Adipose tissue as an endocrine organ. *Obesity* 14 (Suppl. 5), 242S–249S.

Ahima, R.S., Qi, Y. and Singhal, N.S. (2006) Adipokines that link obesity and diabetes to the hypothalamus. *Progress in Brain Research* 153, 155–174.

Allender, S. and Rayner, M. (2007) The burden of overweight and obesity-related ill health in the UK. *Obesity Reviews* 8, 467–473.

Alon, T. and Friedman, J.M. (2006) Late-onset leanness in mice with targeted ablation of melanin concentrating hormone neurons. *Journal of Neuroscience* 26, 378–389.

Bachner, D., Kreienkamp, H., Weise, C., Buck, F. and Richter, D. (1999) Identification of melanin concentrating hormone (MCH) as the natural ligand for the orphan somatostatin-like receptor 1 (SLC-1). *FEBS Letters* 457, 522–524.

Bagnol, D., Lu, X.Y., Kaelin, C.B., Day, H.E., Ollmann, M., Gantz, I., Akil, H., Barsh, G.S. and Watson, S.J. (1999) Anatomy of an endogenous antagonist: relationship between agouti-related protein and proopiomelanocortin in brain. *Journal of Neuroscience* 19, RC26 (1–7).

Bai, F.L., Yamano, M., Shiotani, Y., Emson, P.C., Smith, A.D., Powell, J.F. and Tohyama, M. (1985) An arcuato-paraventricular and -dorsomedial hypothalamic neuropeptide Y-containing system which lacks noradrenaline in the rat. *Brain Research* 331, 172–175.

Balthasar, N., Dalgaard, L.T., Lee, C.E., Yu, J., Funahashi, H., Williams, T., Ferreira, M., Tang, V., McGovern, R.A., Kenny, C.D., Christiansen, L.M., Edelstein, E., Choi, B.,

Boss, O., Aschkenasi, C., Zhang, C.Y., Mountjoy, K., Kishi, T., Elmquist, J.K. and Lowell, B.B. (2005) Divergence of melanocortin pathways in the control of food intake and energy expenditure. *Cell* 123, 493–505.

Bannon, A.W., Seda, J., Carmouche, M., Francis, J.M., Norman, M.H., Karbon, B. and McCaleb, M.L. (2000) Behavioral characterization of neuropeptide Y knockout mice. *Brain Research* 868, 79–87.

Baskin, D.G., Breininger, J.F. and Schwartz, M.W. (1999a) Leptin receptor mRNA identifies a subpopulation of neuropeptide Y neurons activated by fasting in rat hypothalamus. *Diabetes* 48, 828–833.

Baskin, D.G., Figlewicz Lattemann, D., Seeley, R.J., Woods, S.C., Porte, D. Jr and Schwartz, M.W. (1999b) Insulin and leptin: dual adiposity signals to the brain for the regulation of food intake and body weight. *Brain Research* 848, 114–123.

Bewick, G.A., Gardiner, J.V., Dhillo, W.S., Kent, A.S., White, N.E., Webster, Z., Ghatei, M.A. and Bloom, S.R. (2005) Postembryonic ablation of AgRP neurons in mice leads to a lean, hypophagic phenotype. *FASEB Journal* 19, 1680–1682.

Bittencourt, J.C., Presse, F., Arias, C., Peto, C., Vaughan, J., Nahon, J.L., Vale, W. and Sawchenko, P.E. (1992) The melanin-concentrating hormone system of the rat brain: an immuno- and hybridization histochemical characterization. *Journal of Comparative Neurology* 319, 218–245.

Blomqvist, A.G. and Herzog, H. (1997) Y-receptor subtypes – how many more? *Trends in Neurosciences* 20, 294–298.

Borowsky, B., Durkin, M.M., Ogozalek, K., Marzabadi, M.R., DeLeon, J., Lagu, B., Heurich, R., Lichtblau, H., Shaposhnik, Z., Daniewska, I., Blackburn, T.P., Branchek, T.A., Gerald, C., Vaysse, P.J. and Forray, C. (2002) Antidepressant, anxiolytic and anorectic effects of a melanin-concentrating hormone-1 receptor antagonist. *Nature Medicine* 8, 825–830.

Boutrel, B. and de Lecea, L. (2008) Addiction and arousal: the hypocretin connection. *Physiology and Behavior* 93, 947–951.

Brady, L.S., Smith, M.A., Gold, P.W. and Herkenham, M. (1990) Altered expression of hypothalamic neuropeptide mRNAs in food-restricted and food-deprived rats. *Neuroendocrinology* 52, 441–447.

Brain, S.D. and Cox, H.M. (2006) Neuropeptides and their receptors: innovative science providing novel therapeutic targets. *British Journal of Pharmacology* 147 (Suppl. 1), S202–S211.

Branchek, T.A., Smith, K.E., Gerald, C. and Walker, M.W. (2000) Galanin receptor subtypes. *Trends in Pharmacological Sciences* 21, 109–117.

Broberger, C., Johansen, J., Johansson, C., Schalling, M. and Hokfelt, T. (1998) The neuropeptide Y/agouti gene-related protein (AGRP) brain circuitry in normal, anorectic, and monosodium glutamate-treated mice. *Proceedings of the National Academy of Sciences of the United States of America* 95, 15043–15048.

Bromée, T., Sjödin, P., Fredriksson, R., Boswell, T., Larsson, T.A., Salaneck, E., Zoorob, R., Mohell, N. and Larhammar, D. (2006) Neuropeptide Y-family receptors Y6 and Y7 in chicken. Cloning, pharmacological characterization, tissue distribution and conserved synteny with human chromosome region. *The FEBS Journal* 273, 2048–2063.

Butler, A.A., Kesterson, R.A., Khong, K., Cullen, M.J., Pelleymounter, M.A., Dekoning, J., Baetscher, M. and Cone, R.D. (2000) A unique metabolic syndrome causes obesity in the melanocortin-3 receptor-deficient mouse. *Endocrinology* 141, 3518–3521.

Cabrele, C., Langer, M., Bader, R., Wieland, H.A., Doods, H.N., Zerbe, O. and Beck-Sickinger, A.G. (2000) The first selective agonist for the neuropeptide YY5 receptor increases food intake in rats. *Journal of Biological Chemistry* 275, 36043–36048.

Cai, X.J., Widdowson, P.S., Harrold, J., Wilson, S., Buckingham, R.E., Arch, J.R., Tadayyon, M., Clapham, J.C., Wilding, J. and Williams, G. (1999) Hypothalamic orexin expression: modulation by blood glucose and feeding. *Diabetes* 48, 2132–2137.

Chaffer, C.L. and Morris, M.J. (2002) The feeding response to melanin-concentrating hormone is attenuated by antagonism of the NPY Y(1)-receptor in the rat. *Endocrinology* 143, 191–197.

Chambers, J., Ames, R.S., Bergsma, D., Muir, A., Fitzgerald, L.R., Hervieu, G., Dytko, G.M., Foley, J.J., Martin, J., Liu, W.S., Park, J., Ellis, C., Ganguly, S., Konchar, S., Cluderay, J., Leslie, R., Wilson, S. and Sarau, H.M. (1999) Melanin-concentrating hormone is the cognate ligand for the orphan G-protein-coupled receptor SLC-1. *Nature* 400, 261–265.

Chemelli, R.M., Willie, J.T., Sinton, C.M., Elmquist, J.K., Scammell, T., Lee, C., Richardson, J.A., Williams, S.C., Xiong, Y., Kisanuki, Y., Fitch, T.E., Nakazato, M., Hammer, R.E., Saper, C.B. and Yanagisawa, M. (1999) Narcolepsy in orexin knockout mice: molecular genetics of sleep regulation. *Cell* 98, 437–451.

Chen, A.S., Marsh, D.J., Trumbauer, M.E., Frazier, E.G., Guan, X.M., Yu, H., Rosenblum, C.I., Vongs, A., Feng, Y., Cao, L., Metzger, J.M., Strack, A.M., Camacho, R.E., Mellin, T.N., Nunes, C.N., Min, W., Fisher, J., Gopal-Truter, S., MacIntyre, D.E., Chen, H.Y. and Van der Ploeg, L.H. (2000) Inactivation of the mouse melanocortin-3 receptor results in increased fat mass and reduced lean body mass. *Nature Genetics* 26, 97–102.

Chen, H., Hansen, M.J., Jones, J.E., Vlahos, R., Anderson, G.P. and Morris, M.J. (2007) Detrimental metabolic effects of combining long-term cigarette smoke exposure and high-fat diet in mice. *American Journal of Physiology Endocrinology and Metabolism* 293, E1564–E1571.

Chopra, M., Galbraith, S. and Darnton-Hill, I. (2002) A global response to a global problem: the epidemic of overnutrition. *Bulletin World Health Organization* 80, 952–958.

Chronwall, B.M., DiMaggio, D.A., Massari, V.J., Pickel, V.M., Ruggiero, D.A. and O'Donohue, T.L. (1985) The anatomy of neuropeptide-Y-containing neurons in rat brain. *Neuroscience* 15, 1159–1181.

Chua, S.C. Jr, Brown, A.W., Kim, J., Hennessey, K.L., Leibel, R.L. and Hirsch, J. (1991) Food deprivation and hypothalamic neuropeptide gene expression: effects of strain background and the diabetes mutation. *Brain Research – Molecular Brain Research* 11, 291–299.

Clark, J.T., Kalra, P.S., Crowley, W.R. and Kalra, S.P. (1984) Neuropeptide Y and human pancreatic polypeptide stimulate feeding behavior in rats. *Endocrinology* 115, 427–429.

Coll, A.P., Farooqi, I.S. and O'Rahilly, S. (2007) The hormonal control of food intake. *Cell* 129, 251–262.

Cowley, M.A., Smith, R.G., Diano, S., Tschop, M., Pronchuk, N., Grove, K.L., Strasburger, C.J., Bidlingmaier, M., Esterman, M., Heiman, M.L., Garcia-Segura, L.M., Nillni, E.A., Mendez, P., Low, M.J., Sotonyi, P., Friedman, J.M., Liu, H., Pinto, S., Colmers, W.F., Cone, R.D. and Horvath, T.L. (2003) The distribution and mechanism of action of ghrelin in the CNS demonstrates a novel hypothalamic circuit regulating energy homeostasis. *Neuron* 37, 649–661.

Crawley, J.N. (1999) The role of galanin in feeding behavior. *Neuropeptides* 33, 369–375.

Cummings, D.E. and Overduin, J. (2007) Gastrointestinal regulation of food intake. *Journal of Clinical Investigation* 117, 13–23.

Dickson, S.L. and Luckman, S.M. (1997) Induction of c-fos messenger ribonucleic acid in neuropeptide Y and growth hormone (GH)-releasing factor neurons in the rat arcuate nucleus following systemic injection of the GH secretagogue, GH-releasing peptide-6. *Endocrinology* 138, 771–777.

Dinulescu, D.M. and Cone, R.D. (2000) Agouti and agouti-related protein: analogies and contrasts. *Journal of Biological Chemistry* 275, 6695–6698.

Enriori, P.J., Evans, A.E., Sinnayah, P., Jobst, E.E., Tonelli-Lemos, L., Billes, S.K., Glavas, M.M., Grayson, B.E., Perello, M., Nillni, E.A., Grove, K.L. and Cowley, M.A. (2007) Diet-induced obesity causes severe but reversible leptin resistance in arcuate melanocortin neurons. *Cell Metabolism* 5, 181–194.

Erickson, J.C., Hollopeter, G. and Palmiter, R.D. (1996a) Attenuation of the obesity syndrome of *ob/ob* mice by the loss of neuropeptide Y. *Science* 274, 1704–1707.

Erickson, J.C., Clegg, K.E. and Palmiter, R.D. (1996b) Sensitivity to leptin and susceptibility to seizures of mice lacking neuropeptide Y. *Nature* 381, 415–421.

Fam, B.C., Morris, M.J., Hansen, M.J., Kebede, M., Andrikopoulos, S., Proietto, J. and Thorburn, A.W. (2007) Modulation of central leptin sensitivity and energy balance in a rat model of diet-induced obesity. *Diabetes, Obesity and Metabolism* 9, 840–852.

Fan, W., Boston, B.A., Kesterson, R.A., Hruby, V.J. and Cone, R.D. (1997) Role of melanocortinergic neurons in feeding and the agouti obesity syndrome. *Nature* 385, 165–168.

Farooqi, I.S., Keogh, J.M., Yeo, G.S., Lank, E.J., Cheetham, T. and O'Rahilly, S. (2003) Clinical spectrum of obesity and mutations in the melanocortin 4 receptor gene. *New England Journal of Medicine* 348, 1085–1095.

Finkelstein, E.A., Ruhm, C.J. and Kosa, K.M. (2005) Economic causes and consequences of obesity. *Annual Review of Public Health* 26, 239–257.

Fisher, S.L., Yagaloff, K.A. and Burn, P. (1999) Melanocortin-4 receptor: a novel signalling pathway involved in body weight regulation. *International Journal of Obesity and Related Metabolic Disorders* 23 (Suppl. 1), 54–58.

Flier, J.S. (2006) AgRP in energy balance: will the real AgRP please stand up? *Cell Metabolism* 3, 83–85.

Flier, J.S. and Maratos-Flier, E. (1998) Obesity and the hypothalamus: novel peptides for new pathways. *Cell* 92, 437–440.

Fraley, G.S., Shimada, I., Baumgartner, J.W., Clifton, D.K. and Steiner, R.A. (2003) Differential patterns of Fos induction in the hypothalamus of the rat following central injections of galanin-like peptide and galanin. *Endocrinology* 144, 1143–1146.

Frühbeck, G. and Gómez-Ambrosi, J. (2003) Control of body weight: a physiologic and transgenic perspective. *Diabetologia* 46, 143–172.

Fry, J. and Finley, W. (2005) The prevalence and costs of obesity in the EU. *Proceedings of the Nutrition Society* 64, 359–362.

Gao, Q. and Horvath, T.L. (2007) Neurobiology of feeding and energy expenditure. *Annual Reviews of Neuroscience* 30, 367–398.

Gao, Q. and Horvath, T.L. (2008) Neuronal control of energy homeostasis. *FEBS Letters* 582, 132–141.

García, M.C., López, M., Alvarez, C.V., Casanueva, F., Tena-Sempere, M. and Diéguez, C. (2007) Role of ghrelin in reproduction. *Reproduction* 133, 531–540.

Gerald, C., Walker, M.W., Criscione, L., Gustafson, E.L., Batzl-Hartmann, C., Smith, K.E., Vaysse, P., Durkin, M.M., Laz, T.M., Linemeyer, D.L., Schaffhauser, A.O., Whitebread, S., Hofbauer, K.G., Taber, R.I., Branchek, T.A. and Weinshank, R.L. (1996) A receptor subtype involved in neuropeptide-Y-induced food intake. *Nature* 382, 168–171.

Giraudo, S.Q., Billington, C.J. and Levine, A.S. (1998) Feeding effects of hypothalamic injection of melanocortin 4 receptor ligands. *Brain Research* 809, 302–306.

Goldstone, A.P. (2006) The hypothalamus, hormones, and hunger: alterations in human obesity and illness. *Progress in Brain Research* 153, 57–73.

Gomori, A., Ishihara, A., Ito, M., Mashiko, S., Matsushita, H., Yumoto, M., Tanaka, T., Tokita, S., Moriya, M., Iwaasa, H. and Kanatani, A. (2003) Chronic intracerebroventricular infusion of MCH causes obesity in mice. Melanin-concentrating hormone. *American Journal of Physiology – Endocrinology and Metabolism* 284, E583–588.

Gozali, M., Pavia, J.M. and Morris, M.J. (2002) Involvement of neuropeptide Y in glucose sensing in the dorsal hypothalamus of streptozotocin diabetic rats – *in vitro* and *in vivo* studies of transmitter release. *Diabetologia* 45, 1332–1339.

Gualillo, O., Lago, F., Casanueva, F.F. and Diéguez, C. (2006) One ancestor, several peptides post-translational modifications of preproghrelin generate several peptides with antithetical effects. *Molecular and Cellular Endocrinology* 256, 1–8.

Guan, X.M., Yu, H., Palyha, O.C., McKee, K.K., Feighner, S.D., Sirinathsinghji, D.J., Smith, R.G., Van der Ploeg, L.H. and Howard, A.D. (1997) Distribution of mRNA encoding the growth hormone secretagogue receptor in brain and peripheral tissues. *Brain Research Molecular Brain Research* 48, 23–29.

Guan, X.M., Yu, H., Trumbauer, M., Frazier, E., Van der Ploeg, L.H. and Chen, H. (1998) Induction of neuropeptide Y expression in dorsomedial hypothalamus of diet-induced obese mice. *Neuroreport* 9, 3415–3419.

Gundlach, A.L. (2002) Galanin/GALP and galanin receptors: role in central control of feeding, body weight/obesity and reproduction? *European Journal of Pharmacology* 440, 255–268.

Hagan, M.M., Rushing, P.A., Pritchard, L.M., Schwartz, M.W., Strack, A.M., Van Der Ploeg, L.H., Woods, S.C. and Seeley, R.J. (2000) Long-term orexigenic effects of AgRP-(83–132) involve mechanisms other than melanocortin receptor blockade. *American Journal of Physiology – Regulatory, Integrative and Comparative Physiology* 279, R47–52.

Hahn, T.M., Breininger, J.F., Baskin, D.G. and Schwartz, M.W. (1998) Coexpression of Agrp and NPY in fasting-activated hypothalamic neurons. *Nature Neuroscience* 1, 271–272.

Hakansson, M., de Lecea, L., Sutcliffe, J.G., Yanagisawa, M. and Meister, B. (1999) Leptin receptor- and STAT3-immunoreactivities in hypocretin/orexin neurones of the lateral hypothalamus. *Journal of Neuroendocrinology* 11, 653–663.

Hansen, M.J., Ball, M.J. and Morris, M.J. (2001) Enhanced inhibitory feeding response to alpha-melanocyte stimulating hormone in the diet-induced obese rat. *Brain Research* 892, 130–137.

Hansen, M.J., Jovanovska, V. and Morris, M.J. (2004) Adaptive responses in hypothalamic neuropeptide Y in the face of prolonged high-fat feeding in the rat. *Journal of Neurochemistry* 88, 909–916.

Hansen, M.J., Schiöth, H.B. and Morris, M.J. (2005) Feeding responses to a melanocortin agonist and antagonist in obesity induced by a palatable high-fat diet. *Brain Research* 1039, 137–145.

Harrold, J.A., Williams, G. and Widdowson, P.S. (1999) Changes in hypothalamic agouti-related protein (AGRP), but not alpha-MSH or pro-opiomelanocortin concentrations in dietary-obese and food-restricted rats. *Biochemical and Biophysical Research Communications* 258, 574–577.

Hervieu, G.J., Cluderay, J.E., Harrison, D., Meakin, J., Maycox, P., Nasir, S. and Leslie, R.A. (2000) The distribution of the mRNA and protein products of the melanin-concentrating hormone (MCH) receptor gene, *slc*-1, in the central nervous system of the rat. *European Journal of Neuroscience* 12, 1194–1216.

Hill, J., Duckworth, M., Murdock, P., Rennie, G., Sabido-David, C., Ames, R.S., Szekeres, P., Wilson, S., Bergsma, D.J., Gloger, I.S., Levy, D.S., Chambers, J.K. and Muir, A.I. (2001) Molecular cloning and functional characterization of MCH2, a novel human MCH receptor. *Journal of Biological Chemistry* 276, 20125–20129.

Hohmann, J.G., Krasnow, S.M., Teklemichael, D.N., Clifton, D.K., Wynick, D. and Steiner, R.A. (2003) Neuroendocrine profiles in galanin-overexpressing and knockout mice. *Neuroendocrinology* 77, 354–366.

Hollopeter, G., Erickson, J.C., Seeley, R.J., Marsh, D.J. and Palmiter, R.D. (1998) Response of neuropeptide Y-deficient mice to feeding effectors. *Regulatory Peptides* 75–76, 383–389.

Horvath, T.L. (2005) The hardship of obesity: a soft-wired hypothalamus. *Nature Neuroscience* 8, 561–565.

Horvath, T.L., Diano, S., Sotonyi, P., Heiman, M. and Tschöp, M. (2001) Minireview: ghrelin and the regulation of energy balance – a hypothalamic perspective. *Endocrinology* 142, 4163–4169.

Huang, X.F., Han, M. and Storlien, L.H. (2003) The level of NPY receptor mRNA expression in diet-induced obese and resistant mice. *Brain Research – Molecular Brain Research* 115, 21–28.

Huszar, D., Lynch, C.A., Fairchild-Huntress, V., Dunmore, J.H., Fang, Q., Berkemeier, L.R., Gu, W., Kesterson, R.A., Boston, B.A., Cone, R.D., Smith, F.J., Campfield, L.A., Burn, P. and Lee, F. (1997) Targeted disruption of the melanocortin-4 receptor results in obesity in mice. *Cell* 88, 131–141.

Inui, A. (1999) Feeding and body-weight regulation by hypothalamic neuropeptides – mediation of the actions of leptin. *Trends in Neurosciences* 22, 62–67.

Ito, M., Gomori, A., Ishihara, A., Oda, Z., Mashiko, S., Matsushita, H., Yumoto, M., Sano, H., Tokita, S., Moriya, M., Iwaasa, H. and Kanatani, A. (2003) Characterization of MCH-mediated obesity in mice. *American Journal of Physiology – Endocrinology and Metabolism* 284, E940–945.

Jain, M.R., Horvath, T.L., Kalra, P.S. and Kalra, S.P. (2000) Evidence that NPY Y1 receptors are involved in stimulation of feeding by orexins (hypocretins) in sated rats. *Regulatory Peptides* 87, 19–24.

Jegou, S., Boutelet, I. and Vaudry, H. (2000) Melanocortin-3 receptor mRNA expression in pro-opiomelanocortin neurones of the rat arcuate nucleus. *Journal of Neuroendocrinology* 12, 501–505.

Jureus, A., Cunningham, M.J., McClain, M.E., Clifton, D.K. and Steiner, R.A. (2000) Galanin-like peptide (GALP) is a target for regulation by leptin in the hypothalamus of the rat. *Endocrinology* 141, 2703–2706.

Kageyama, H., Takenoya, F., Kita, T., Hori, T., Guan, J.L. and Shioda, S. (2005) Galanin-like peptide in the brain: effects on feeding, energy metabolism and reproduction. *Regulatory Peptides* 126, 21–26.

Kamegai, J., Tamura, H., Shimizu, T., Ishii, S., Sugihara, H. and Wakabayashi, I. (2001) Chronic central infusion of ghrelin increases hypothalamic neuropeptide Y and Agouti-related protein mRNA levels and body weight in rats. *Diabetes* 50, 2438–2443.

Kanatani, A., Ishihara, A., Asahi, S., Tanaka, T., Ozaki, S. and Ihara, M. (1996) Potent neuropeptide Y Y1 receptor antagonist, 1229U91: blockade of neuropeptide Y-induced and physiological food intake. *Endocrinology* 137, 3177–3182.

Kanatani, A., Mashiko, S., Murai, N., Sugimoto, N., Ito, J., Fukuroda, T., Fukami, T., Morin, N., MacNeil, D.J., Van der Ploeg, L.H., Saga, Y., Nishimura, S. and Ihara, M. (2000) Role of the Y1 receptor in the regulation of neuropeptide Y-mediated feeding: comparison of wild-type, Y1 receptor-deficient, and Y5 receptor-deficient mice. *Endocrinology* 141, 1011–1016.

Kask, A., Rago, L., Wikberg, J.E. and Schioth, H.B. (1998a) Evidence for involvement of the melanocortin MC4 receptor in the effects of leptin on food intake and body weight. *European Journal of Pharmacology* 360, 15–19.

Kask, A., Rago, L., Mutulis, F., Pahkla, R., Wikberg, J.E. and Schioth, H.B. (1998b) Selective antagonist for the melanocortin 4 receptor (HS014) increases food intake in free-feeding rats. *Biochemical and Biophysical Research Communication* 245, 90–93.

Kask, A., Pahkla, R., Irs, A., Rago, L., Wikberg, J.E. and Schioth, H.B. (1999) Long-term administration of MC4 receptor antagonist HS014 causes hyperphagia and obesity in rats. *Neuroreport* 10, 707–711.

Kawauchi, H., Kawazoe, I., Tsubokawa, M., Kishida, M. and Baker, B.I. (1983) Characterization of melanin-concentrating hormone in chum salmon pituitaries. *Nature* 305, 321–323.

Kim, E.M., O'Hare, E., Grace, M.K., Welch, C.C., Billington, C.J. and Levine, A.S. (2000) ARC POMC mRNA and PVN alpha-MSH are lower in obese relative to lean zucker rats. *Brain Research* 862, 11–16.

King, P.J., Widdowson, P.S., Doods, H.N. and Williams, G. (1999) Regulation of neuropeptide Y release by neuropeptide Y receptor ligands and calcium channel antagonists in hypothalamic slices. *Journal of Neurochemistry* 73, 641–646.

Kishi, T., Aschkenasi, C.J., Lee, C.E., Mountjoy, K.G., Saper, C.B. and Elmquist, J.K. (2003) Expression of melanocortin 4 receptor mRNA in the central nervous system of the rat. *Journal of Comparative Neurology* 457, 213–235.

Klok, M.D., Jakobsdottir, S. and Drent, M.L. (2007) The role of leptin and ghrelin in the regulation of food intake and body weight in humans: a review. *Obesity Reviews* 8, 21–34.

Knecht, S., Ellger, T. and Levine, J.A. (2008) Obesity in neurobiology. *Progress in Neurobiology* 84, 85–103.

Kojima, M., Hosoda, H., Date, Y., Nakazato, M., Matsuo, H. and Kangawa, K. (1999) Ghrelin is a growth-hormone-releasing acylated peptide from stomach. *Nature* 402, 656–660.

Kokkotou, E.G., Tritos, N.A., Mastaitis, J.W., Slieker, L. and Maratos-Flier, E. (2001) Melanin-concentrating hormone receptor is a target of leptin action in the mouse brain. *Endocrinology* 142, 680–686.

Kokkotou, E., Jeon, J.Y., Wang, X., Marino, F.E., Carlson, M., Trombly, D.J. and Maratos-Flier, E. (2005) Mice with MCH ablation resist diet-induced obesity through strain-specific mechanisms. *American Journal of Physiology – Regulatory, Integrative and Comparative Physiology* 289, R117–124.

Kozak, R., Richy, S. and Beck, B. (2005) Persistent alterations in neuropeptide Y release in the paraventricular nucleus of rats subjected to dietary manipulation during early life. *European Journal of Neuroscience* 21, 2887–2892.

Krasnow, S.M., Fraley, G.S., Schuh, S.M., Baumgartner, J.W., Clifton, D.K. and Steiner, R.A. (2003) A role for galanin-like peptide in the integration of feeding, body weight regulation, and reproduction in the mouse. *Endocrinology* 144, 813–822.

Kumano, S., Matsumoto, H., Takatsu, Y., Noguchi, J., Kitada, C. and Ohtaki, T. (2003) Changes in hypothalamic expression levels of galanin-like peptide in rat and mouse models support that it is a leptin-target peptide. *Endocrinology* 144, 2634–2643.

Kushi, A., Sasai, H., Koizumi, H., Takeda, N., Yokoyama, M. and Nakamura, M. (1998) Obesity and mild hyperinsulinemia found in neuropeptide Y-Y1 receptor-deficient mice. *Proceedings of the National Academy of Sciences of the United States of America* 95, 15659–15664.

Kyrkouli, S.E., Stanley, B.G. and Leibowitz, S.F. (1986) Galanin: stimulation of feeding induced by medial hypothalamic injection of this novel peptide. *European Journal of Pharmacology* 122, 159–160.

Kyrkouli, S.E., Stanley, B.G., Seirafi, R.D. and Leibowitz, S.F. (1990) Stimulation of feeding by galanin: anatomical localization and behavioral specificity of this peptide's effects in the brain. *Peptides* 11, 995–1001.

Lawrence, C.B., Baudoin, F.M. and Luckman, S.M. (2002a) Centrally administered galanin-like peptide modifies food intake in the rat: a comparison with galanin. *Journal of Neuroendocrinology* 14, 853–860.

Lawrence, C.B., Snape, A.C., Baudoin, F.M. and Luckman, S.M. (2002b) Acute central ghrelin and GH secretagogues induce feeding and activate brain appetite centers. *Endocrinology* 143, 155–162.

Lee, J. and Morris, M.J. (1998) Modulation of neuropeptide Y overflow by leptin in the rat hypothalamus, cerebral cortex and medulla. *Neuroreport* 9, 1575–1580.

Leibowitz, S.F. (2007) Overconsumption of dietary fat and alcohol: mechanisms involving lipids and hypothalamic peptides. *Physiology and Behavior* 91, 513–521.

Leibowitz, S.F. and Wortley, K.E. (2004) Hypothalamic control of energy balance: different peptides, different functions. *Peptides* 25, 473–504.

Lin, S., Storlien, L.H. and Huang, X.F. (2000) Leptin receptor, NPY, POMC mRNA expression in the diet-induced obese mouse brain. *Brain Research* 875, 89–95.

López, M., Seoane, L., Garcia, M.C., Lago, F., Casanueva, F.F., Señaris, R. and Diéguez, C. (2000) Leptin regulation of prepro-orexin and orexin receptor mRNA levels in the hypothalamus. *Biochemical and Biophysical Research Communications* 269, 41–45.

López, M., Tovar, S., Vázquez, M.J., Williams, L.M. and Diéguez, C. (2007) Peripheral tissue–brain interactions in the regulation of food intake. *Proceedings of the Nutrition Society* 66, 131–155.

Ludwig, D.S., Tritos, N.A., Mastaitis, J.W., Kulkarni, R., Kokkotou, E., Elmquist, J., Lowell, B., Flier, J.S. and Maratos-Flier, E. (2001) Melanin-concentrating hormone overexpression in transgenic mice leads to obesity and insulin resistance. *Journal of Clinical Investigation* 107, 379–386.

Ma, L., Tataranni, P.A., Hanson, R.L., Infante, A.M., Kobes, S., Bogardus, C. and Baier, L.J. (2005) Variations in peptide YY and Y2 receptor genes are associated with severe obesity in Pima Indian men. *Diabetes* 54, 1598–1602.

Man, P.S. and Lawrence, C.B. (2008) The effects of galanin-like peptide on energy balance, body temperature and brain activity in the mouse and rat are independent of the GALR2/3 receptor. *Journal of Neuroendocrinology* 20, 128–137.

Marsh, D.J., Hollopeter, G., Kafer, K.E. and Palmiter, R.D. (1998) Role of the Y5 neuropeptide Y receptor in feeding and obesity. *Nature Medicine* 4, 718–721.

Marsh, D.J., Hollopeter, G., Huszar, D., Laufer, R., Yagaloff, K.A., Fisher, S.L., Burn, P. and Palmiter, R.D. (1999) Response of melanocortin-4 receptor-deficient mice to anorectic and orexigenic peptides. *Nature Genetics* 21, 119–122.

Marsh, D.J., Weingarth, D.T., Novi, D.E., Chen, H.Y., Trumbauer, M.E., Chen, A.S., Guan, X.M., Jiang, M.M., Feng, Y., Camacho, R.E., Shen, Z., Frazier, E.G., Yu, H., Metzger, J.M., Kuca, S.J., Shearman, L.P., Gopal-Truter, S., MacNeil, D.J., Strack, A.M., MacIntyre, D.E., Van der Ploeg, L.H. and Qian, S. (2002) Melanin-concentrating hormone 1 receptor-deficient mice are lean, hyperactive, and hyperphagic and have altered metabolism. *Proceedings of the National Academy of Sciences of the United States of America* 99, 3240–3245.

Matsumoto, Y., Watanabe, T., Adachi, Y., Itoh, T., Ohtaki, T., Onda, H., Kurokawa, T., Nishimura, O. and Fujino, M. (2002) Galanin-like peptide stimulates food intake in the rat. *Neuroscience Letters* 322, 67–69.

Melander, T., Hokfelt, T. and Rokaeus, A. (1986) Distribution of galanin-like immunoreactivity in the rat central nervous system. *Journal of Comparative Neurology* 248, 475–517.

Mennicken, F., Hoffert, C., Pelletier, M., Ahmad, S. and O'Donnell, D. (2002) Restricted distribution of galanin receptor 3 (GalR3) mRNA in the adult rat central nervous system. *Journal of Chemical Neuroanatomy* 24, 257–268.

Mercer, J.G., Hoggard, N., Williams, L.M., Lawrence, C.B., Hannah, L.T., Morgan, P.J. and Trayhurn, P. (1996) Coexpression of leptin receptor and preproneuropeptide Y mRNA in arcuate nucleus of mouse hypothalamus. *Journal of Neuroendocrinology* 8, 733–735.

Merchenthaler, I., Lopez, F.J. and Negro-Vilar, A. (1993) Anatomy and physiology of central galanin-containing pathways. *Progress in Neurobiology* 40, 711–769.

Mizuno, T.M. and Mobbs, C.V. (1999) Hypothalamic agouti-related protein messenger ribonucleic acid is inhibited by leptin and stimulated by fasting. *Endocrinology* 140, 814–817.

Mizuno, T.M., Kleopoulos, S.P., Bergen, H.T., Roberts, J.L., Priest, C.A. and Mobbs, C.V. (1998) Hypothalamic pro-opiomelanocortin mRNA is reduced by fasting and [corrected] in *ob/ob* and *db/db* mice, but is stimulated by leptin. *Diabetes* 47, 294–297.

Moran, T.H. (2006) Gut peptide signaling in the controls of food intake. *Obesity* 14 (Suppl. 5), 250S–253S.

Morris, M.J., Chen, H., Watts, R., Shulkes, A. and Cameron-Smith, D. (2008) Brain neuropeptide Y and CCK and peripheral adipokine receptors: temporal response in obesity induced by palatable diet. *International Journal of Obesity* 32, 249–258.

Morton, G.J., Cummings, D.E., Baskin, D.G., Barsh, G.S. and Schwartz, M.W. (2006) Central nervous system control of food intake and body weight. *Nature* 443, 289–295.

Mountjoy, K.G., Mortrud, M.T., Low, M.J., Simerly, R.B. and Cone, R.D. (1994) Localization of the melanocortin-4 receptor (MC4-R) in neuroendocrine and autonomic control circuits in the brain. *Molecular Endocrinology* 8, 1298–1308.

Muccioli, G., Tschop, M., Papotti, M., Deghenghi, R., Heiman, M. and Ghigo, E. (2002) Neuroendocrine and peripheral activities of ghrelin: implications in metabolism and obesity. *European Journal of Pharmacology* 440, 235–254.

Mullins, D., Kirby, D., Hwa, J., Guzzi, M., Rivier, J. and Parker, E. (2001) Identification of potent and selective neuropeptide Y Y(1) receptor agonists with orexigenic activity *in vivo*. *Molecular Pharmacology* 60, 534–540.

Nakazato, M., Murakami, N., Date, Y., Kojima, M., Matsuo, H., Kangawa, K. and Matsukura, S. (2001) A role for ghrelin in the central regulation of feeding. *Nature* 409, 194–198.

Naveilhan, P., Hassani, H., Canals, J.M., Ekstrand, A.J., Larefalk, A., Chhajlani, V., Arenas, E., Gedda, K., Svensson, L., Thoren, P. and Ernfors, P. (1999) Normal feeding behavior, body weight and leptin response require the neuropeptide Y Y2 receptor. *Nature Medicine* 5, 1188–1193.

Nijenhuis, W.A., Oosterom, J. and Adan, R.A. (2001) AgRP(83-132) acts as an inverse agonist on the human-melanocortin-4 receptor. *Molecular Endocrinology* 15, 164–171.

Niswender, K.D. and Schwartz, M.W. (2003) Insulin and leptin revisited: adiposity signals with overlapping physiological and intracellular signaling capabilities. *Frontiers in Neuroendocrinology* 24, 1–10.

Nogueiras, R., Wiedmer, P., Perez-Tilve, D., Veyrat-Durebex, C., Keogh, J.M., Sutton, G.M., Pfluger, P.T., Castaneda, T.R., Neschen, S., Hofmann, S.M., Howles, P.N., Morgan, D.A., Benoit, S.C., Szanto, I., Schrott, B., Schürmann, A., Joost, H.G., Hammond, C., Hui, D.Y., Woods, S.C., Rahmouni, K., Butler, A.A., Farooqi, I.S., O'Rahilly, S., Rohner-Jeanrenaud, F. and Tschöp, M.H. (2007) The central melanocortin system directly controls peripheral lipid metabolism. *Journal of Clinical Investigation* 117, 3475–3488.

Ohtaki, T., Kumano, S., Ishibashi, Y., Ogi, K., Matsui, H., Harada, M., Kitada, C., Kurokawa, T., Onda, H. and Fujino, M. (1999) Isolation and cDNA cloning of a novel galanin-like peptide (GALP) from porcine hypothalamus. *Journal of Biological Chemistry* 274, 37041–37045.

Ollmann, M.M., Wilson, B.D., Yang, Y.K., Kerns, J.A., Chen, Y., Gantz, I. and Barsh, G.S. (1997) Antagonism of central melanocortin receptors *in vitro* and *in vivo* by agouti-related protein. *Science* 278, 135–138.

Parker, E.M., Balasubramaniam, A., Guzzi, M., Mullins, D.E., Salisbury, B.G., Sheriff, S., Witten, M.B. and Hwa, J.J. (2000) [D-Trp(34)] neuropeptide Y is a potent and selective neuropeptide Y Y(5) receptor agonist with dramatic effects on food intake. *Peptides* 21, 393–399.

Parker, E., Van Heek, M. and Stamford, A. (2002) Neuropeptide Y receptors as targets for anti-obesity drug development: perspective and current status. *European Journal of Pharmacology* 440, 173–187.

Parker, R.M. and Herzog, H. (1999) Regional distribution of Y-receptor subtype mRNAs in rat brain. *European Journal of Neuroscience* 11, 1431–1448.

Pedrazzini, T., Seydoux, J., Kunstner, P., Aubert, J.F., Grouzmann, E., Beermann, F. and Brunner, H.R. (1998) Cardiovascular response, feeding behavior and locomotor activity in mice lacking the NPY Y1 receptor. *Nature Medicine* 4, 722–726.

Pelleymounter, M.A., Cullen, M.J., Baker, M.B., Hecht, R., Winters, D., Boone, T. and Collins, F. (1995) Effects of the obese gene product on body weight regulation in *ob/ob* mice. *Science* 269, 540–543.

Peyron, C., Tighe, D.K., van den Pol, A.N., de Lecea, L., Heller, H.C., Sutcliffe, J.G. and Kilduff, T.S. (1998) Neurons containing hypocretin (orexin) project to multiple neuronal systems. *Journal of Neuroscience* 18, 9996–10015.

Pissios, P. and Maratos-Flier, E. (2003) Melanin-concentrating hormone: from fish skin to skinny mammals. *Trends in Endocrinology and Metabolism* 14, 243–248.

Pliquett, R.U., Führer, D., Falk, S., Zysset, S., von Cramon, D.Y. and Stumvoll, M. (2006) The effects of insulin on the central nervous system – focus on appetite regulation. *Hormone and Metabolic Research* 38, 442–446.

Polidori, C., Ciccocioppo, R., Regoli, D. and Massi, M. (2000) Neuropeptide Y receptor(s) mediating feeding in the rat: characterization with antagonists. *Peptides* 21, 29–35.

Presse, F., Nahon, J.L., Fischer, W.H. and Vale, W. (1990) Structure of the human melanin concentrating hormone mRNA. *Molecular Endocrinology* 4, 632–637.

Qu, D., Ludwig, D.S., Gammeltoft, S., Piper, M., Pelleymounter, M.A., Cullen, M.J., Mathes, W.F., Przypek, R., Kanarek, R. and Maratos-Flier, E. (1996) A role for melanin-concentrating hormone in the central regulation of feeding behaviour. *Nature* 380, 243–247.

Rahardjo, G.L., Huang, X.F., Tan, Y.Y. and Deng, C. (2007) Decreased plasma peptide YY accompanied by elevated peptide YY and Y2 receptor binding densities in the medulla oblongata of diet-induced obese mice. *Endocrinology* 148, 4704–4710.

Reizes, O., Lincecum, J., Wang, Z., Goldberger, O., Huang, L., Kaksonen, M., Ahima, R., Hinkes, M.T., Barsh, G.S., Rauvala, H. and Bernfield, M. (2001) Transgenic expression of syndecan-1 uncovers a physiological control of feeding behavior by syndecan-3. *Cell* 106, 105–116.

Reizes, O., Benoit, S.C., Strader, A.D., Clegg, D.J., Akunuru, S. and Seeley, R.J. (2003) Syndecan-3 modulates food intake by interacting with the melanocortin/AgRP pathway. *Annals of the New York Academy of Sciences* 994, 66–73.

Rodriguez, M., Beauverger, P., Naime, I., Rique, H., Ouvry, C., Souchaud, S., Dromaint, S., Nagel, N., Suply, T., Audinot, V., Boutin, J.A. and Galizzi, J.P. (2001) Cloning and molecular characterization of the novel human melanin-concentrating hormone receptor MCH2. *Molecular Pharmacology* 60, 632–639.

Roselli-Rehfuss, L., Mountjoy, K.G., Robbins, L.S., Mortrud, M.T., Low, M.J., Tatro, J.B., Entwistle, M.L., Simerly, R.B. and Cone, R.D. (1993) Identification of a receptor for gamma melanotropin and other proopiomelanocortin peptides in the hypothalamus and limbic system. *Proceedings of the National Academy of Sciences of the United States of America* 90, 8856–8860.

Rossi, M., Choi, S.J., O'Shea, D., Miyoshi, T., Ghatei, M.A. and Bloom, S.R. (1997) Melanin-concentrating hormone acutely stimulates feeding, but chronic administration has no effect on body weight. *Endocrinology* 138, 351–355.

Rossi, M., Kim, M.S., Morgan, D.G., Small, C.J., Edwards, C.M., Sunter, D., Abusnana, S., Goldstone, A.P., Russell, S.H., Stanley, S.A., Smith, D.M., Yagaloff, K., Ghatei, M.A. and Bloom, S.R. (1998) A C-terminal fragment of agouti-related protein increases feeding and antagonizes the effect of alpha-melanocyte stimulating hormone *in vivo*. *Endocrinology* 139, 4428–4431.

Runge, C.F. (2007) Economic consequences of the obese. *Diabetes* 56, 2668–2672.

Sahu, A. (1998) Evidence suggesting that galanin (GAL), melanin-concentrating hormone (MCH), neurotensin (NT), proopiomelanocortin (POMC) and neuropeptide Y (NPY) are targets of leptin signalling in the hypothalamus. *Endocrinology* 139, 795–798.

Sahu, A. (2002) Interactions of neuropeptide Y, hypocretin-I (orexin A) and melanin-concentrating hormone on feeding in rats. *Brain Research* 944, 232–238.

Sahu, A., Kalra, P.S. and Kalra, S.P. (1988) Food deprivation and ingestion induce reciprocal changes in neuropeptide Y concentrations in the paraventricular nucleus. *Peptides* 9, 83–86.

Sahu, A., Sninsky, C.A., Kalra, P.S. and Kalra, S.P. (1990) Neuropeptide-Y concentration in microdissected hypothalamic regions and *in vitro* release from the medial basal hypothalamus-preoptic area of streptozotocin-diabetic rats with and without insulin substitution therapy. *Endocrinology* 126, 192–198.

Sainsbury, A., Schwarzer, C., Couzens, M., Fetissov, S., Furtinger, S., Jenkins, A., Cox, H.M., Sperk, G., Hokfelt, T. and Herzog, H. (2002) Important role of hypothalamic Y2 receptors in body weight regulation revealed in conditional knockout mice. *Proceedings of the National Academy of Sciences of the United States of America* 99, 8938–8943.

Sainsbury, A., Baldock, P.A., Schwarzer, C., Ueno, N., Enriquez, R.F., Couzens, M., Inui, A., Herzog, H. and Gardiner, E.M. (2003) Synergistic effects of Y2 and Y4 receptors on adiposity and bone mass revealed in double knockout mice. *Molecular and Cellular Biology* 23, 5225–5233.

Saito, J., Ozaki, Y., Kawasaki, M., Ohnishi, H., Okimoto, N., Nakamura, T. and Ueta, Y. (2004) Galanin-like peptide gene expression in the hypothalamus and posterior pituitary of the obese *fa/fa* rat. *Peptides* 25, 967–974.

Sakurai, T., Amemiya, A., Ishii, M., Matsuzaki, I., Chemelli, R.M., Tanaka, H., Williams, S.C., Richardson, J.A., Kozlowski, G.P., Wilson, S., Arch, J.R., Buckingham, R.E., Haynes, A.C., Carr, S.A., Annan, R.S., McNulty, D.E., Liu, W.S., Terrett, J.A., Elshourbagy, N.A., Bergsma, D.J. and Yanagisawa, M. (1998) Orexins and orexin receptors: a family of hypothalamic neuropeptides and G protein-coupled receptors that regulate feeding behaviour. *Cell* 92, 573–585.

Sanacora, G., Kershaw, M., Finkelstein, J.A. and White, J.D. (1990) Increased hypothalamic content of preproneuropeptide Y messenger ribonucleic acid in genetically obese Zucker rats and its regulation by food deprivation. *Endocrinology* 127, 730–737.

Satoh, N., Ogawa, Y., Katsuura, G., Numata, Y., Masuzaki, H., Yoshimasa, Y. and Nakao, K. (1998) Satiety effect and sympathetic activation of leptin are mediated by hypothalamic melanocortin system. *Neuroscience Letters* 249, 107–110.

Schick, R.R., Samsami, S., Zimmermann, J.P., Eberl, T., Endres, C., Schusdziarra, V. and Classen, M. (1993) Effect of galanin on food intake in rats: involvement of lateral and ventromedial hypothalamic sites. *American Journal of Physiology – Regulatory, Integrative and Comparative Physiology* 264, R355–361.

Schioth, H.B., Bouifrouri, A.A., Rudzish, R., Muceniece, R., Watanobe, H., Wikberg, J.E. and Larhammar, D. (2002) Pharmacological comparison of rat and human melanocortin 3 and 4 receptors *in vitro*. *Regulatory Peptides* 106, 7–12.

Schwartz, M.W., Sipols, A.J., Marks, J.L., Sanacora, G., White, J.D., Scheurink, A., Kahn, S.E., Baskin, D.G., Woods, S.C., Figlewicz, D.P. and Porte, D. Jr (1992) Inhibition of hypothalamic neuropeptide Y gene expression by insulin. *Endocrinology* 130, 3608–3616.

Schwartz, M.W., Seeley, R.J., Campfield, L.A., Burn, P. and Baskin, D.G. (1996) Identification of targets of leptin action in rat hypothalamus. *Journal of Clinical Investigation* 98, 1101–1106.

Schwartz, M.W., Woods, S.C., Porte, D. Jr, Seeley, R.J. and Baskin, D.G. (2000) Central nervous system control of food intake. *Nature* 404, 661–671.

Seeley, R.J. and Woods, S.C. (2003) Monitoring of stored and available fuel by the CNS: implications for obesity. *Nature Reviews Neuroscience* 4, 901–909.

Seeley, R.J., Yagaloff, K.A., Fisher, S.L., Burn, P., Thiele, T.E., van Dijk, G., Baskin, D.G. and Schwartz, M.W. (1997) Melanocortin receptors in leptin effects. *Nature* 390, 349.

Seth, A., Stanley, S., Dhillo, W., Murphy, K., Ghatei, M. and Bloom, S. (2003) Effects of galanin-like peptide on food intake and the hypothalamo–pituitary–thyroid axis. *Neuroendocrinology* 77, 125–131.

Shen, J. and Gundlach, A.L. (2004) Galanin-like peptide mRNA alterations in arcuate nucleus and neural lobe of streptozotocin-diabetic and obese Zucker rats. Further evidence for leptin-dependent and independent regulation. *Neuroendocrinology* 79, 327–337.

Shibata, M., Mondal, M.S., Date, Y., Nakazato, M., Suzuki, H. and Ueta, Y. (2008) Distribution of orexins-containing fibres and contents of orexins in the rat olfactory bulb. *Neuroscience Research* 61, 99–105.

Shimada, M., Tritos, N.A., Lowell, B.B., Flier, J.S. and Maratos-Flier, E. (1998) Mice lacking melanin-concentrating hormone are hypophagic and lean. *Nature* 396, 670–674.

Shintani, M., Ogawa, Y., Ebihara, K., Aizawa-Abe, M., Miyanaga, F., Takaya, K., Hayashi, T., Inoue, G., Hosoda, K., Kojima, M., Kangawa, K. and Nakao, K. (2001) Ghrelin, an endogenous growth hormone secretagogue, is a novel orexigenic peptide that antagonizes leptin action through the activation of hypothalamic neuropeptide Y/Y1 receptor pathway. *Diabetes* 50, 227–232.

Shutter, J.R., Graham, M., Kinsey, A.C., Scully, S., Luthy, R. and Stark, K.L. (1997) Hypothalamic expression of ART, a novel gene related to agouti, is upregulated in obese and diabetic mutant mice. *Genes and Development* 11, 593–602.

Skofitsch, G., Jacobowitz, D.M. and Zamir, N. (1985) Immunohistochemical localization of a melanin concentrating hormone-like peptide in the rat brain. *Brain Research Bulletin* 15, 635–649.

Smith, B.K., York, D.A. and Bray, G.A. (1996) Effects of dietary preference and galanin administration in the paraventricular or amygdaloid nucleus on diet self-selection. *Brain Research Bulletin* 39, 149–154.

Smith, M.S. (1993) Lactation alters neuropeptide-Y and proopiomelanocortin gene expression in the arcuate nucleus of the rat. *Endocrinology* 133, 1258–1265.

Spiegelman, B.M. and Flier, J.S. (2001) Obesity and the regulation of energy balance. *Cell* 104, 531–543.

Stephens, T.W., Basinski, M., Bristow, P.K., Bue-Valleskey, J.M., Burgett, S.G., Craft, L., Hale, J., Hoffmann, J., Hsiung, H.M., Kriauciunas, A., Mackellar, W., Rosteck, P.R. Jr, Schoner, B., Smith, D., Tinsley, F.C., Zhang, X.-Y. and Heiman, M. (1995) The role of neuropeptide Y in the antiobesity action of the obese gene product. *Nature* 377, 530–532.

Strader, A.D., Reizes, O., Woods, S.C., Benoit, S.C. and Seeley, R.J. (2004) Mice lacking the syndecan-3 gene are resistant to diet-induced obesity. *Journal of Clinical Investigation* 114, 1354–1360.

Stricker-Krongrad, A., Cumin, F., Burlet, C. and Beck, B. (1998) Hypothalamic neuropeptide Y and plasma leptin after long-term high-fat feeding in the rat. *Neuroscience Letters* 254, 157–160.

Takatsu, Y., Matsumoto, H., Ohtaki, T., Kumano, S., Kitada, C., Onda, H., Nishimura, O. and Fujino, M. (2001) Distribution of galanin-like peptide in the rat brain. *Endocrinology* 142, 1626–1634.

Takenoya, F., Hirayama, M., Kageyama, H., Funahashi, H., Kita, T., Matsumoto, H., Ohtaki, T., Katoh, S., Takeuchi, M. and Shioda, S. (2005) Neuronal interactions between galanin-like peptide- and orexin- or melanin-concentrating hormone-containing neurons. *Regulatory Peptides* 126, 79–83.

Tan, H.M., Gundlach, A.L. and Morris, M.J. (2005) Exaggerated feeding response to central galanin-like peptide administration in diet-induced obese rats. *Neuropeptides* 39, 333–336.

Tang-Christensen, M., Kristensen, P., Stidsen, C.E., Brand, C.L. and Larsen, P.J. (1998) Central administration of Y5 receptor antisense decreases spontaneous food intake and attenuates feeding in response to exogenous neuropeptide Y. *Journal of Endocrinology* 159, 307–312.

Tatemoto, K., Carlquist, M. and Mutt, V. (1982) Neuropeptide Y – a novel brain peptide with structural similarities to peptide YY and pancreatic polypeptide. *Nature* 296, 659–660.

Thorsell, A. and Heilig, M. (2002) Diverse functions of neuropeptide Y revealed using genetically modified animals. *Neuropeptides* 36, 182–193.

Torri, C., Pedrazzi, P., Leo, G., Muller, E.E., Cocchi, D., Agnati, L.F. and Zoli, M. (2002) Diet-induced changes in hypothalamic pro-opio-melanocortin mRNA in the rat hypothalamus. *Peptides* 23, 1063–1068.

Tritos, N.A., Mastaitis, J.W., Kokkotou, E. and Maratos-Flier, E. (2001) Characterization of melanin concentrating hormone and preproorexin expression in the murine hypothalamus. *Brain Research* 895, 160–166.

Tschöp, M., Smiley, D.L. and Heiman, M.L. (2000) Ghrelin induces adiposity in rodents. *Nature* 407, 908–913.

Unmehopa, U.A., van Heerikhuize, J.J., Spijkstra, W., Woods, J.W., Howard, A.D., Zycband, E., Feighner, S.D., Hreniuk, D.L., Palyha, O.C., Guan, X.M., Macneil, D.J., Van der Ploeg, L.H. and Swaab, D.F. (2005) Increased melanin-concentrating hormone receptor type I in the human hypothalamic infundibular nucleus in cachexia. *Journal of Clinical Endocrinology and Metabolism* 90, 2412–2419.

Valassi, E., Scacchi, M. and Cavagnini, F. (2008) Neuroendocrine control of food intake. *Nutrition, Metabolism and Cardiovascular Diseases* 18, 158–168.

Vaughan, J.M., Fischer, W.H., Hoeger, C., Rivier, J. and Vale, W. (1989) Characterization of melanin-concentrating hormone from rat hypothalamus. *Endocrinology* 125, 1660–1665.

Vergoni, A.V., Bertolini, A., Mutulis, F., Wikberg, J.E. and Schioth, H.B. (1998) Differential influence of a selective melanocortin MC4 receptor antagonist (HS014) on melanocortin-induced behavioral effects in rats. *European Journal of Pharmacology* 362, 95–101.

Wang, C., Yang, N., Wu, S., Liu, L., Sun, X. and Nie, S. (2007) Difference of NPY and its receptor gene expressions between obesity and obesity-resistant rats in response to high-fat diet. *Hormone and Metabolic Research* 39, 262–267.

Wang, S., Behan, J., O'Neill, K., Weig, B., Fried, S., Laz, T., Bayne, M., Gustafson, E. and Hawes, B.E. (2001) Identification and pharmacological characterization of a novel human melanin-concentrating hormone receptor, mch-r2. *Journal of Biological Chemistry* 276, 34664–34670.

Wermter, A.K., Reichwald, K., Buch, T., Geller, F., Platzer, C., Huse, K., Hess, C., Remschmidt, H., Gudermann, T., Preibisch, G., Siegfried, W., Goldschmidt, H.P., Li, W.D., Price, R.A., Biebermann, H., Krude, H., Vollmert, C., Wichmann, H.E., Illig, T., Sorensen, T.I., Astrup, A., Larsen, L.H., Pedersen, O., Eberle, D., Clement, K., Blundell, J., Wabitsch, M., Schafer, H., Platzer, M., Hinney, A. and Hebebrand, J. (2005) Mutation analysis of the MCHR1 gene in human obesity. *European Journal of Endocrinology* 152, 851–862.

White, J.D., Olchovsky, D., Kershaw, M. and Berelowitz, M. (1990) Increased hypothalamic content of preproneuropeptide-Y messenger ribonucleic acid in streptozotocin-diabetic rats. *Endocrinology* 126, 765–772.

WHO (1997) Obesity: preventing and managing the global epidemic. Report of a WHO Consultation on Obesity, 3–5 June. World Health Organization, Geneva, Switzerland.

WHO (2003) Diet, nutrition and the prevention of chronic diseases. Report of a Joint WHO/FAO Expert Consultation. WHO Technical Report Series No 916. World Health Organization, Geneva, Switzerland.

Widdowson, P.S., Upton, R., Henderson, L., Buckingham, R., Wilson, S. and Williams, G. (1997) Reciprocal regional changes in brain NPY receptor density during dietary restriction and dietary-induced obesity in the rat. *Brain Research* 774, 1–10.

Widdowson, P.S., Henderson, L., Pickavance, L., Buckingham, R., Tadayyon, M., Arch J.R. and Williams, G. (1999) Hypothalamic NPY status during positive energy balance and the effects of the NPY antagonist, BW 1229U91 on the consumption of highly palatable energy-rich diet. *Peptides* 20, 367–372.

Wilding, J.P. (2002) Neuropeptides and appetite control. *Diabetic Medicine* 19, 619–627.

Wilding, J.P., Gilbey, S.G., Mannan, M., Aslam, N., Ghatei, M.A. and Bloom, S.R. (1992) Increased neuropeptide Y content in individual hypothalamic nuclei, but not neuropeptide Y mRNA, in diet-induced obesity in rats. *Journal of Endocrinology* 132, 299–304.

Wilding, J.P., Gilbey, S.G., Bailey, C.J., Batt, R.A., Williams, G., Ghatei, M.A. and Bloom, S.R. (1993) Increased neuropeptide-Y messenger ribonucleic acid (mRNA) and decreased neurotensin mRNA in the hypothalamus of the obese (*ob/ob*) mouse. *Endocrinology* 132, 1939–1944.

Willesen, M.G., Kristensen, P. and Romer, J. (1999) Co-localization of growth hormone secretagogue receptor and NPY mRNA in the arcuate nucleus of the rat. *Neuroendocrinology* 70, 306–316.

Williams, G., Bing, C., Cai, X.J., Harrold, J.A., King, P.J. and Liu, X.H. (2001) The hypothalamus and the control of energy homeostasis: different circuits, different purposes. *Physiology and Behavior* 74, 683–701.

Wilson, B.D., Bagnol, D., Kaelin, C.B., Ollmann, M.M., Gantz, I., Watson, S.J. and Barsh, G.S. (1999) Physiological and anatomical circuitry between agouti-related protein and leptin signalling. *Endocrinology* 140, 2387–2397.

Wisse, B.E. and Schwartz, M.W. (2001) Role of melanocortins in control of obesity. *Lancet* 358, 857–859.

Woods, S.C. (2005) Signals that influence food intake and body weight. *Physiology and Behavior* 86, 709–716.

Woods, S.C., Seeley, R.J., Porte, D. Jr and Schwartz, M.W. (1998) Signals that regulate food intake and energy homeostasis. *Science* 280, 1378–1383.

Woods, S.C., D'Alessio, D.A., Tso, P., Rushing, P.A., Clegg, D.J., Benoit, S.C., Gotoh, K., Liu, M. and Seeley, R.J. (2004) Consumption of a high-fat diet alters the homeostatic regulation of energy balance. *Physiology and Behavior* 83, 573–578.

Wynne, K., Stanley, S., McGowan, B. and Bloom, S. (2005) Appetite control. *Journal of Endocrinology* 184, 291–318.

Yamanaka, A., Sakurai, T., Katsumoto, T., Yanagisawa, M. and Goto, K. (1999) Chronic intracerebroventricular administration of orexin-A to rats increases food intake in daytime, but has no effect on body weight. *Brain Research* 849, 248–252.

Yamanaka, A., Kunii, K., Nambu, T., Tsujino, N., Sakai, A., Matsuzaki, I., Miwa, Y., Goto, K. and Sakurai, T. (2000) Orexin-induced food intake involves neuropeptide Y pathway. *Brain Research* 859, 404–409.

Yang, Y.K., Thompson, D.A., Dickinson, C.J., Wilken, J., Barsh, G.S., Kent, S.B. and Gantz, I. (1999) Characterization of agouti-related protein binding to melanocortin receptors. *Molecular Endocrinology* 13, 148–155.

Zamir, N., Skofitsch, G. and Jacobowitz, D.M. (1986) Distribution of immunoreactive melanin-concentrating hormone in the central nervous system of the rat. *Brain Research* 373, 240–245.

Zarjevski, N., Cusin, I., Vettor, R., Rohner-Jeanrenaud, F. and Jeanrenaud, B. (1993) Chronic intracerebroventricular neuropeptide-Y administration to normal rats mimics hormonal and metabolic changes of obesity. *Endocrinology* 133, 1753–1758.

Zhang, Y., Proenca, R., Maffei, M., Barone, M., Leopold, L. and Friedman, J.M. (1994) Positional cloning of the mouse obese gene and its human homologue. *Nature* 372, 425–432.

Ziotopoulou, M., Mantzoros, C.S., Hileman, S.M. and Flier, J.S. (2000) Differential expression of hypothalamic neuropeptides in the early phase of diet-induced obesity in mice. *American Journal of Physiology – Endocrinology and Metabolism* 279, E838–845.

2 Anorexigenic Peptides

Simona Perboni,[1] Naohiko Ueno,[2] Giovanni Mantovani[1] and Akio Inui[3]

[1]*Department of Medical Oncology, University of Cagliari, Italy;* [2]*Division of Diabetes, Digestive and Kidney Diseases, Department of Clinical Molecular Medicine, Kobe University Graduate School of Medicine, Japan;* [3]*Department of Behavioural Medicine, Kagoshima University Graduate School of Medical and Dental Sciences, Japan*

Introduction

Normal-weight individuals keep a constant balance between energy intake and energy expenditure. When the intake is greater than the expenditure, the excess calories are stored in adipocytes, leading to obesity. According to the *lipostatic theory* for energy homeostasis, healthy animals tend to modulate energy intake to keep their fat mass stable (Ahima, 2006; Coll *et al.*, 2007; Knecht *et al.*, 2008). For most people, the amount and composition of food eaten varies considerably from meal to meal and from day to day. Yet, over time, energy intake tends to be matched to expenditure and body weight is tightly conserved. Thus, food intake and both meal frequency and meal size must also be highly regulated (Druce *et al.*, 2004). In the proposed model, at baseline, catabolic effectors are activated in response to physiological concentrations of adiposity signals, such as insulin and leptin: this activation is essential to prevent excessive weight gain. In contrast, anabolic pathways are those that stimulate food intake and decrease energy expenditure and are strongly inhibited by the same basal concentrations of adiposity signals. Therefore, under physiological conditions, catabolic pathways are activated while anabolic pathways are largely inhibited. The response to weight loss includes both activation of anabolic and inhibition of catabolic pathways and thus is inherently more vigorous than the response to weight gain. This model intends to understand better the apparent bias of the energy homeostasis system in favour of weight gain present in Western society, in which plenty of palatable foods are available. From an evolutionary perspective, it seems likely that the ability to survive and reproduce in environments with limited access to food provides a strong selection bias in favour of regulatory systems that vigorously defend against deficits of body fat (Schwartz *et al.*, 2003; Abizaid *et al.*, 2006; Gao and Horvath, 2008). In this study, attention is focused on the central mechanisms that regulate energy homeostasis and on anorexigenic peptides in particular.

Signals of Long-term and Short-term Energy Stores

Long-term and short-term hormonal signals from the periphery act on the central nervous system (CNS) to influence feeding behaviour (see Fig. 2.1).

Signals from the gastrointestinal tract and the liver are involved primarily in short-term regulation and satiety (Moran, 2006; Cummings and Overduin, 2007; Klok *et al.*, 2007; López *et al.*, 2007). The presence of nutrients in the stomach and in the proximal intestine activates baro- and chemo-receptors that convey signals to the enteric nervous system, from which the afferent vagal fibres project to the nucleus of the solitary tract (NST) in the brainstem. Nutrients arriving via the portal vein may also trigger vagal afferent signals from the liver. Glucose can modulate food intake by acting on glucose-responsive neurones in the CNS. In response to nutrient stimulation, the proximal intestine releases cholecystokinin (CCK), a potent anorexigenic peptide, which reaches the liver via the portal vein and reaches the CNS via the systemic circulation and via vagal fibres. In the terminal small intestine, glucagon-like peptide-1 (GLP-1) is released by endocrine cells and inhibits feeding, most likely at a hepatic site or by inhibiting gastric emptying (Konturek *et al.*, 2004). The short-term signals by themselves do not produce sustained alterations in energy intake and body adiposity.

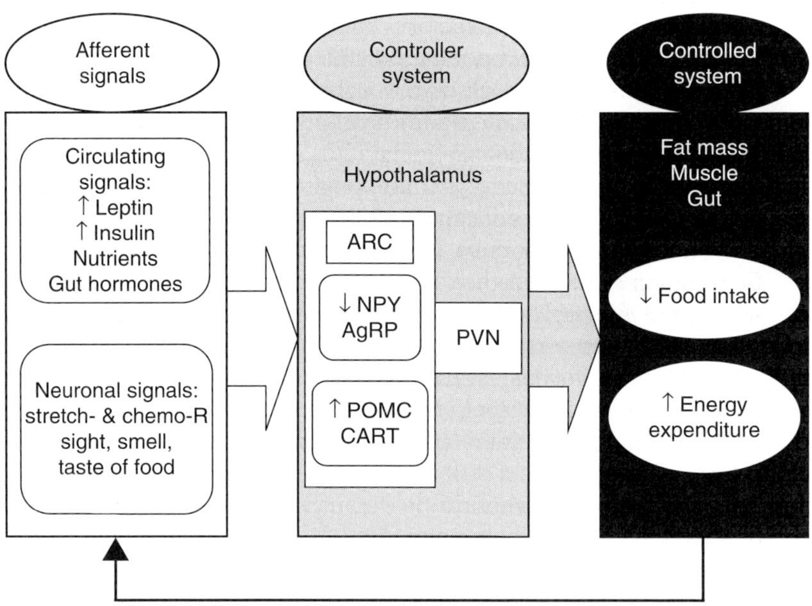

Fig. 2.1. A simplified model of interaction between peripheral signals and hypothalamic nuclei and body weight regulation. ARC, arcuate nucleus; NPY, neuropeptide Y; AgRP, agouti-related peptide; PVN, paraventricular nucleus; POMC, proopiomelanocortin; CART, cocaine- and amphetamine-regulated transcript.

Insulin and leptin are considered the most important long-term regulators of energy balance (Niswender and Schwartz, 2003; Seeley and Woods, 2003; Benoit *et al.*, 2004; Woods, 2005; Pliquett *et al.*, 2006). Both act in the CNS to inhibit food intake and to increase energy expenditure, most likely by activating the sympathetic nervous system (SNS). Insulin is secreted from pancreatic β-cells in response to circulating nutrients (glucose and amino acids) and to the incretin hormones, glucose-dependent insulinotropic polypeptide (GIP) and GLP-1, which are released during meal ingestion and absorption. Insulin can also act indirectly by stimulating leptin production from adipose tissue via increased glucose metabolism. There is also evidence that leptin can inhibit insulin secretion from the pancreas. The gastric hormone, ghrelin, increases food intake and decreases fat oxidation in rodents, may have an anabolic role in the regulation of energy homeostasis and appears to set sensitivity to the satiety-producing effects of the short-term signals such as CCK. The long-term signals involved in the control of adiposity appear to be few in number and play a highly specialized role (Havel *et al.*, 2001; Ahima, 2006; Valassi *et al.*, 2008).

The Hypothalamus as an Integrator of Energy Balance

It has been accepted for decades that the hypothalamus plays a critical role in the regulation of feeding (Horvath, 2005; Abizaid *et al.*, 2006; Morton *et al.*, 2006; Gao and Horvath, 2007, 2008). An array of orexigenic and anorexigenic peptides that constitute a major part of the neural circuitry regulating feeding behaviour and body weight are produced primarily by the neurones localized in hypothalamus areas, such as the arcuate nucleus (ARC), ventromedial (VMH), dorsomedial (DMN) and paraventricular (PVN) nuclei of the hypothalamus and the lateral hypothalamic (LH) area (Elmquist *et al.*, 1998). In particular, the ARC integrates signals from the periphery (it is accessible to circulating factors because the blood–brain barrier is incomplete here) and probably from the brainstem.

The ARC contains two distinct subsets of neurones controlling food intake. One contains neuropeptide Y (NPY) and agouti-related peptide (AgRP) and acts as a stimulus to feeding. The second subset of neurones contains α-melanocyte-stimulating hormone (α-MSH) and cocaine- and amphetamine-regulated transcript (CART) and acts as an inhibitor on food intake. In physiological conditions, when one of these subsets is activated, the other is inhibited (Druce *et al.*, 2004; Abizaid *et al.*, 2006; Flier, 2006; Morton *et al.*, 2006; Gao and Horvath, 2007, 2008). The neurones are adjacent to the highly permeable median eminence and are regulated by circulating hunger and satiety signals, such as glucose, ghrelin and CCK. They are also modulated by signals of long-term body energy stores, such as leptin and insulin, by monitoring blood levels of these substances. ARC neurones are kept informed of whether or not the body has sufficient calories and nutrients, and can modify feeding accordingly (Sahu, 2004). Some ARC neurones contain NPY and project to the PVN to regulate its influence on feeding. When activated, NPY positive neurones can produce remarkable increases in food intake that result in obesity. Although NPY deficiency has no significant effect in mice fed a normal rodent diet, energy expenditure is elevated during fasting, and hyperphagia and weight gain are blunted during re-feeding in these rodents (Patel *et al.*, 2006). Hypothalamic expression of AgRP is increased in NPY knockouts (NPYko) compared to wild-type mice, but unlike wild types, there is no further increase in AgRP when NPYko mice are fasted. Moreover, NPYko mice have higher oxygen consumption and uncoupling protein-1 (UCP1) expression in brown adipose tissue during fasting. The failure of an increase in orexigenic peptides and a higher thermogenesis may contribute to attenuation of weight gain when NPYko mice are re-fed. NPY deficiency also makes the mice less susceptible to diet-induced obesity as a result of reduced feeding and increased energy expenditure (Patel *et al.*, 2006). The resistance to diet-induced obesity in NPYko mice is associated with a reduction in nocturnal feeding and increased expression of anorexigenic hypothalamic peptides.

The PVN coordinates outputs for energy homeostasis via the endocrine axes and the SNS. These neurones project to preganglionic neurones in the intermediolateral horn (T1-L2) of the spinal cord, which control the SNS, to preganglionic neurones in the dorsal motor nucleus of the vagus, which regulate the parasympathetic control of gastrointestinal motility and insulin secretion, and to the NST and hypoglossal nucleus, which regulate swallowing and tongue movement. PVN also receives inputs from the melanin-concentrating hormone (MCH) projections from the LH (which is a stimulator of feeding), the brainstem and the amygdala (Druce *et al.*, 2004).

The brain detects alterations in environmental temperature and diet and, through hypothalamic signalling, controls energy expenditure, increasing lipolysis and thermogenesis, or both (Sahu, 2004). In rodents, leptin activates pro-opiomelanocortin (POMC) neurones in ARC, which project directly to sympathetic preganglionic neurones in the spinal cord (Schwartz *et al.*, 2003) and to neurones in the PVN that control sympathetic outflow to the peripheral tissues, such as brown adipose tissue, white adipose tissue and muscle. Abundant evidence indicates that many rodent models of obesity have reduced energy expenditure and that this contributes importantly to the development of obesity. The most compelling evidence comes from mice lacking leptin (*ob/ob* mice) or mice lacking the receptor for leptin (*db/db* mice), which exhibit both increased food intake and decreased energy expenditure. The role of reduced energy expenditure in promoting human obesity is less clear, given the heterogeneity and multifactorial influences of human height and body composition. If rates of energy expenditure are normalized to lean body mass, lean and obese subjects, in general, have similar rates of energy expenditure.

Catabolic Peptides

Leptin

Leptin is a hormone produced by adipose tissue that circulates as a 16-kDa protein in rodent and human plasma. It exerts a lipostatic negative feedback signal to regulate energy balance (Ahima *et al.*, 2006). The plasma levels of leptin are highly correlated with adipose tissue mass and fall in both humans and mice after weight loss (Maffei *et al.*, 1995). In the presence of increasing fat stores, circulating leptin levels increase to curb appetite and facilitate energy utilization. Blood-borne leptin exerts its anorexigenic actions in the brain, which it can access via a saturable transport mechanism at the blood–brain barrier. The direct appetite-suppressant effects of leptin are thought to be mediated predominantly by activation of the long isoform of the leptin receptor (LRb) in periventricular sites which regulate feeding, including the ARC, DMN and VMH nuclei of the hypothalamus and the dorsal vagal complex of the brainstem (Prolo *et al.*, 1998).

The leptin receptor (LR) contains a single transmembrane domain and is a member of the class I cytokine receptor family (Tartaglia *et al.*, 1995). Among the six known LR splice variants, only LRb, the signalling form, has been shown to be essential for normal energy homeostasis (Friedman and Halaas, 1998). Unlike the other LR isoforms, LRb contains a 302-amino acid cytoplasmatic domain that includes motifs for the binding of intracellular signalling molecules such as Janus kinase (JAK). Leptin binding to its receptor induces receptor dimerization, activation of JAK2 and JAK2-mediated phosphorylation and activation of the signal transducer and activator of transcription 3 (STAT3). Activated STAT3 dimerized, translocates to the nucleus and *trans*-activates target genes, including suppressor of cytokine signalling 3 protein (SOCS3), NPY and POMC (Sahu, 2004; Frühbeck, 2006).

Among several STAT proteins, leptin increases only STAT3 phosphorylation and STAT3 DNA-binding activity in the hypothalamus, particularly in the ARC, LH, VMN and DMN. Animal models have shown that LRb–STAT signalling is regulated for both the stimulation of POMC and full repression of AgRP transcription. Leptin-induced inhibition of NPY gene expression is mediated by a distinct signal, such as the phosphatidylinositol 3-kinase (PI3K) cascade (Bates *et al.*, 2003).

Leptin signalling through the JAK2–STAT3 pathway is thought to be under the negative feedback control of SOCS3 proteins. Another negative regulator of leptin receptor signalling is protein tyrosine phosphatase 1B (PTP1B), which co-localizes in the hypothalamic areas with LRb. PTP1B-deficient mice show an enhanced response towards leptin-mediated weight loss and suppression of feeding. The hypothalamus of these mice also displays markedly increased leptin-induced STAT3 phosphorylation. Leptin-activated JAK2, but not STAT3 or LR, is a substrate of PTP1B. These results suggest that PTP1B regulates leptin signalling negatively and provides a mechanism by which it may regulate obesity. The interaction between SOCS3 and PTP1B may be critical during normal leptin signalling, as well as during the development of leptin resistance. Thus, PTP1B emerges as a potential novel target to treat leptin resistance in obesity (Cheng *et al.*, 2002; Zabolotny *et al.*, 2002). In addition, leptin might activate the PI3K-dependent pathway and reduce cAMP levels in the hypothalamus. The cAMP levels seem to be related with increased food intake, with leptin modifying cAMP response element-mediated gene expression such as NPY neurones in the hypothalamus.

The PI3K–cAMP pathway interacting with the JAK2–STAT3 pathway constitutes a critical component of leptin signalling in the hypothalamus. This may be a common pathway for leptin and insulin signalling in the hypothalamus. Furthermore, leptin and ghrelin reportedly also regulate AMP-activated protein kinase K (AMPK) (Anderson *et al.*, 2004). It is possible that defects in either one or both of the signalling pathways may be responsible for the development of leptin resistance seen in obesity (Niswender and Schwartz, 2003; Benoit *et al.*, 2004; Sahu, 2004; Woods, 2005; Pliquett *et al.*, 2006).

Leptin may facilitate satiety not only as a tonic adiposity signal but also as a short-term energy balance signal in the CNS and in the periphery. In rodents, in contrast to the previous literature, a study has demonstrated that acute increases in central leptin levels may increase postprandial satiety potently and influence body fluid homeostasis (Zorrilla *et al.*, 2005). This finding is consistent with ample clinical evidence showing that leptin reduces meal frequency but not average meal size, by prolonging the inter-meal interval. Humans with congenital leptin deficiency are constantly hungry and demand food continuously. The administration of leptin increases postprandial satiety or the absence of motivation to reinitiate feeding after meal completion (Montague *et al.*, 1997). In the periphery, leptin is secreted by endocrine cells of the gastric mucosa in response to ingestion of a meal. Leptin acts on vagal afferent fibres that sense the satiety signal to the hindbrain (Peters *et al.*, 2005, 2007). Although leptin is not considered a primary satiety factor in humans because changes in food intake do not induce short-term increases in blood leptin concentrations (Jequier, 2002), it may have a permissive effect on satiety by inhibiting orexigenic inputs to the CNS sufficiently to allow satiety signals from gut hormones and baro-receptors to affect eating behaviour (McDuffie *et al.*, 2004).

Some studies provide evidence that a specific brain system involving connections between the basolateral amygdala (BLA) and the LH is crucial for allowing learned cues to override satiety and promote eating in satiated rats (Petrovich *et al.*, 2002). Leptin and LRb mRNA increased in BLA following conditioned taste aversion, indicating that leptin and its receptors may take part in conditioned taste aversion learning and operate as a mediating factor between feeding and taste in rats. The neuronal projections from the amygdala to the ARC represent possible neuroanatomic substrates for this effect (Han *et al.*, 2003).

In rodents, it is known that leptin increases energy expenditure through the induction of the mitochondrial uncoupling proteins, UCP-1, UCP-2 and UCP-3, through the SNS (Inui, 2000). In rodents, leptin increases gene expression of UCP-2 in white adipose tissue and of UCP-1 and UCP-3 in brown fat (Scarpace *et al.*, 1998). Increasing evidence from human studies suggests that leptin influences the human energy balance predominantly through appetite, but appears not to be involved in regulating energy expenditure. None of the expected factors such as resting metabolic rate, total diurnal energy expenditure or dietary induced thermogenesis has been related to blood leptin concentrations (Hukshorn and Saris, 2004).

Leptin and the neuronal circuit regulating weight

The hypothalamus is the major site of leptin action in energy homeostasis (Halaas and Friedman, 1997). Leptin receptors have been found in several hypothalamic nuclei, including the ARC, VMH, LH, DMN and PVN involved in energy homeostasis. Leptin acts through or in concert with several neuropeptides, monoamines and other transmitter substances that affect food intake in the brain–gut axis (Banks *et al.*, 1996). Among the leptin sensitive neurones, NPY/AgRP is orexigenic (see Chapter 1) and POMC/CART is anorexigenic; leptin reduces NPY/AgRP activity, while stimulating POMC/CART neuronal activity. NPY is regarded as one of the central molecules in appetite regulation (Inui, 2003; Kamiji and Inui, 2007). The expression of the LRb in NPY neurones modulates the activity of these neurones that are activated by fasting. Interestingly, the fasting-induced upregulation of NPY expression can be blunted by the administration of leptin (Schulz and Lehnert, 2004).

Pinto *et al.* (2004) have shown that in ARC, leptin may modulate the hypothalamic neuronal populations not only by direct action on LR but also as a consequence of the differential effects of the hormone on synaptic input to NPY and POMC neurones. Leptin can modulate both synapse number and activity of NPY and POMC neurones in the hypothalamus of *ob/ob* mice, resulting in alteration of the NPY and POMC hypothalamic tone. Remarkable similarity to the phenomenon of long-term potentiation, linking learning and memory in the hippocampus, has suggested that a long-term programming of hypothalamic neurones by leptin may underlie the theoretical body weight 'set-point' aimed at maintaining body weight (Harrold, 2004). Leptin action on POMC neurones may also involve reduction of γ-amino-butyric acid (GABA) and AgRP release from NPY/AgRP neurones.

Although the selective ablation of LR in POMC neurones of the arcuate nucleus results in obesity (Balthasar *et al.*, 2005), it is less pronounced than in mice that globally lack LR. In fact, LR on other neurones, such as those that express steroidogenic factor 1 in the ventromedial hypothalamus, is required for normal body weight (Dhillon *et al.*, 2006). Mice lacking LR on these neurones in the VMH become obese, despite having no discernible increase in food intake. Interestingly, when placed on a high-fat diet, these mice appear unable to suppress food intake and stimulate energy expenditure adequately. Thus, LR in steroidogenic factor 1 neurones of the VMH may play an important role in the adaptive changes critical for resisting diet-induced obesity. Although mice with a combined deficiency of LR on both steroidogenic factor 1 and POMC weighed more than loss of either alone, they were still less obese than mice that globally lack LR, indicating that there must be other sites important for leptin action. Two such sites may be the caudal brainstem and the ventral tegmental area in the midbrain. The caudal brainstem contains both leptin-responsive, LRb-expressing neurones (Coll *et al.*, 2007) and a population of POMC neurones, just like the ARC. However, some relevant differences between the brainstem and the ARC are evident. Although fasting induces a fall in POMC mRNA in both regions, in contrast to ARC, the reduction seen in the brainstem is not reversed by leptin administration. Furthermore, leptin does not cause STAT3 phosphorylation or *c-fos* activation within brainstem POMC neurones, suggesting that leptin signalling via POMC-derived peptides in the CNS occurs entirely via hypothalamic POMC neurones (Huo *et al.*, 2006).

A significant role of POMC/CART neurones in mediating leptin actions is further evident from the fact that POMC neurones are glucose responsive and express K^+-ATP channels that are activated by leptin. Although it has been

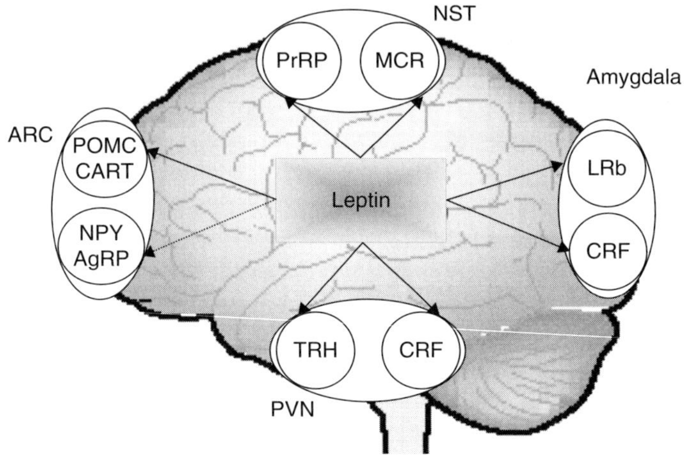

Fig. 2.2. Interactions between leptin and neuropeptides in the central nervous system. Continuous arrows show stimulatory inputs while the dotted line shows inhibitory inputs. ARC, arcuate nucleus; POMC, proopiomelanocortin; CART, cocaine- and amphetamine-regulated transcript; NPY, neuropeptide Y; AgRP, agouti-related peptide; NST, nucleus solitary tract; PrRP, prolactin-releasing peptide; MCR, melanocortin receptor; PVN, paraventricular nucleus; TRH, thyroid-releasing hormone; CRF, corticotropin-releasing factor; LRb, leptin receptor isoform b.

considered that melanocortin signalling is localized downstream to leptin, data have accumulated to support the concept of a leptin-independent melanocortin signalling system (Shimizu *et al.*, 2007).

Among other leptin-target neurones, MCH, galanin (GAL), galanin-like peptide (GALP), orexin, neurotensin (NT) and corticotropin-releasing factor (CRF)-producing neurones are notable (see Fig. 2.2). Leptin increases levels of CRF mRNA in the PVN and stimulates the release of CRF from perfusion slices of both amygdala and the PVN (Sahu, 2004). Leptin decreases MCH, GAL and orexin gene expression and increases GALP, NT and CRF gene expression in the hypothalamus (Valassi *et al.*, 2008). CCK potentiates the anorectic effects of leptin. Anorectic prolactin-releasing peptide (PrRP) neurones express LR and interact with leptin to reduce food intake. Ghrelin and leptin interact functionally in that ghrelin blocks the effects of leptin on feeding and prior leptin administration attenuates the action of ghrelin on NPY neurones. The regulation of the effects of ghrelin on hypothalamic neurones may be one of the important mechanisms of leptin signalling in the hypothalamus (Inui, 2001).

Leptin and neuroendocrine functions

Congenital leptin deficiency due to mutations in the leptin gene or receptor is a rare cause of severe early-onset obesity exhibiting a wide array of endocrine disturbances in both rodents and humans (Lahlou *et al.*, 2002; Jequier and Tappy, 1999; Farooqi and O'Rahilly, 2007; Farooqi *et al.*, 2007; Farooqi, 2008). Leptin therapy has shown to have dramatically beneficial effects on appetite, weight, fat mass, hyperinsulinaemia and dyslipidaemia, as well as on neuroendocrine phenotypes and the immune function (Farooqi *et al.*, 2002; Gibson *et al.*, 2004; Licinio *et al.*, 2004; Tilg and Moschen, 2006). According to the lipostatic theory, a state of 'perceived starvation' occurs in these subjects and results in a chronic stimulation of excessive food intake (Montague *et al.*, 1997). Leptin treatment blunts the changes in circulating thyroid hormone and corticosterone levels that are normally associated with food deprivation. It has been suggested that the inhibition of thyroid hormone secretion may have evolved to limit energy expenditure and prevent protein catabolism during starvation. Leptin deficiency has been associated with impairment of the thyrotrope response to thyroid-releasing hormone (TRH) stimulation, while leptin replacement in leptin-deficient humans and during food restriction reverses the suppression of triiodotyronine, thyroid-stimulating hormone and TRH mRNA levels in PVN (Ahima and Osei, 2004). The effect of leptin on circulating thyroid hormone can be explained at least in part by the high expression of LR in the ARC and by the known projection of the ARC to the PVN, where the TRH neurones are localized. Leptin induces TRH gene expression selectively in the PVN, probably mediated by α-MSH. Central administration of α-MSH mimics the effects of leptin, increasing the TRH and CRF gene expression in the PVN (Sarkar *et al.*, 2002).

Starvation is also associated with an impaired immune response, which leptin is able to improve. Leptin stimulates proliferation of CD4+ T-cells and increases production of cytokines by T-helper-1 cells. These results indicate that leptin is also a key link between nutritional state and immunity (La Cava and Matarese, 2004; Tilg and Moschen, 2006).

Total leptin deficiency or insensitivity is associated with hypothalamic hypogonadism in humans and rodents. Leptin treatment restores LH secretion and pubertal development in leptin-deficient patients, confirming its critical role in reproduction. A link between leptin levels and the onset of puberty in humans has been suggested by demonstrating a transient increase in leptin levels before the onset of puberty in boys. It was proposed that the high levels of leptin observed in children might reflect leptin resistance, as seen in obesity, aimed at maintaining a sufficient amount of food intake and growth, and prevent the onset of premature puberty. Centrally, leptin administration decreases the expression of NPY in the ARC and consequently removes the inhibitory action of NPY on growth-hormone-releasing hormone (GnRH) release. Leptin stimulates the synthesis and release of luteinizing hormone and follicle-stimulating hormone in animals. Ovarian follicular cells are regulated directly by leptin, indicating that it is able to control the hypothalamic–pituitary–gonadal axis at multiple levels (Moran and Phillip, 2003; Ahima and Osei, 2004). These results show that leptin is not only an adipostat signal but that it also acts as a metabolic switch, informing the brain when fat reserves are adequate to direct energy expenditure towards other biological activities (Banks, 2004; Blüher and Mantzoros, 2007).

Leptin and the development of the hypothalamic circuitry

Leptin plays an important role in regulating energy homeostasis in adults, but it is now clear that it also acts as a trophic signal for the development of the hypothalamic pathways for feeding. Different neonatal nutrition and maternal factors have long-term effects on obesity, but little is known about how the neonatal environment influences central mechanisms regulating food intake and energy balance. During neonatal development, food intake must be maximized to support growth, yet plasma leptin levels are relatively high. The general thinking has been that the neonatal brain was relatively insensitive to leptin and might exhibit leptin resistance. Leptin receptors are present and functional in the ARC during the postnatal period and suggest that the leptin insensitivity observed during this period may be due to a failure of these cells to relay leptin signals to other parts of the hypothalamus. In *ob/ob* mice, the neuronal projection pathways from the ARC are disrupted permanently. Treatment with exogenous leptin rescues the development of ARC projections in neonates, but not in adult mice. These findings suggest that leptin plays a neurotrophic role during the development of the hypothalamus and that this activity is restricted to a neonatal critical period that precedes the acute regulation of food intake in adults. The postnatal surge in circulating leptin coincides with the development of ARC projections and supports this hypothesis (Bouret and Simerly, 2004; Bouret *et al.*, 2004).

Leptin resistance and obesity

After the discovery of leptin, the initial hypothesis that human obesity might result from a deficiency in leptin was discarded. Only a few individuals with severe obesity have been identified as having congenital leptin deficiency. Obese humans have high plasma leptin concentrations related to the size of adipose tissue, but this elevated leptin signal does not induce the expected response. This fact suggests that obese humans are resistant to the effects of endogenous leptin. The resistance is also shown by the lack of effect of exogenous administration to induce weight loss in obese patients (Jequier, 2002). Leptin resistance may be defined as reduced sensitivity, or complete insensitivity, to leptin action, as it occurs for insulin hormone in type 2 diabetes (Sahu, 2004).

Human and rodent studies indicate that the major cause of this resistance arises from an inability of leptin to cross the blood–brain barrier, with additional roles played by receptor and postreceptor defects (Banks, 2003). The leptin transporter is a saturable system; beyond a certain plasma leptin level, an increased production by the growing fat mass would be futile. Furthermore, severe hyperleptinaemia downregulates the leptin transporters, worsening the situation (Caro *et al.*, 1996). This mechanism may explain why the exogenous administration of leptin to treat obesity might be ineffective if endogenous leptin has already saturated its transporters. However, the brain–blood barrier resistance is acquired, and to some extent is reversible, with weight loss (Banks, 2004).

In rodents, the downregulation of LR is a well-established mechanism of leptin resistance (Martin *et al.*, 2000). Caro *et al.* (1996) showed that, in humans, the abundance and sequence of the hypothalamic LR was not downregulated in obese patients. Probably, the existence of a saturable system at the blood–brain barrier translates in only moderately elevated hypothalamic interstitial leptin concentrations, compared with serum levels. Further, mechanisms underlying leptin resistance are related to a chronic elevation of the hypothalamic leptin tone, as studied in rats (Sahu, 2004), as well as by decreased postreceptor signalling of the JAK2–STAT3, PI3K–cAMP or SOCS3 cascades (Frühbeck, 2006). Interestingly, in spite of the known leptin resistance associated with obesity, the melanocortin system downstream of the ARC in diet-induced obese mice is over-responsive to melanocortin agonists, probably due to upregulation of MC4R (Enriori *et al.*, 2007). Moreover, by decreasing the fat content of the rodent's diet, leptin responsiveness of NPY/AgRP and POMC neurones recovered simultaneously, with mice regaining normal leptin sensitivity and glycaemic control. These findings highlight the physiological importance of leptin sensing in the melanocortin circuits also.

CRF

The CRF system includes CRF, at least two different CRF receptor subtypes, a CRF-binding protein and endogenous CRF receptor ligands such as urocortins. CRF is a 41-amino acid and is expressed abundantly in the PVN neurones that project to the median eminence to stimulate the secretion of adrenocorticotropic hormone. CRF plays an important role in a variety of endocrine systems, such as stress and endocrine, autonomic and behavioural responses.

CRF is expressed widely throughout the brain and peripheral tissues. In the CNS, the major sites of expression are the PVN, cortex, cerebellum and amygdala–hippocampus complex. The broad distribution of CRF neurones conforms to the many expected functions of this peptide. CRF decreases feeding stimulated by GABA agonists, norepinephrine, dynorphin and NPY. CRF can bind to at least two different CRF receptor subtypes, CRF1 and CRF2 receptor, with its three splice variants that have been cloned and characterized. CRF receptors are members of a seven-transmembrane receptor family that signal by coupling to stimulatory G proteins. In the brain, CRF1 receptors are distributed in the cortical, hypothalamic, limbic, cerebellar regions and the pituitary gland. In the PVN, CRF1 mRNA is not detected under basal conditions but can be induced acutely by stressful stimuli (Richard *et al.*, 2002). The involvement of the CRF1 receptor in anxiogenic behaviours, depression and anorexia has been reported. In humans, the CRF2α receptor is mainly a brain receptor, while the CRF2γ receptor has been found solely in the human brain. The CRF2 receptor is involved primarily in the feeding-suppressive and thermogenic responses to CRF and CRF-related peptides. Mice lacking CRF2 receptors show an increased nocturnal food intake in relation to meal size, rather than meal frequency (Tabarin *et al.*, 2007). Following acute restraint stress, CRF2 knockout mice showed an intact immediate anorectic response with increased latency to eat and decreased meal size. However, CRF2 deletion abolished the prolonged phase of restraint-induced anorexia. CRF2 knockout mice did not differ from wild-type controls in feeding responses to food deprivation or injection of ghrelin receptor agonists. A potent ghrelin analogue stimulated food intake dose-dependently by increasing meal size and meal number. These findings suggest that the CRF2 receptor is involved in the control of meal size during the active phase of eating and following acute exposure to stress (Chen *et al.*, 2005; Tabarin *et al.*, 2007).

Urocortin II and urocortin III constitute two specific endogenous ligands for the CRF2 receptors. These receptors are particularly distributed in the PVN, lateral septum, amygdala, hippocampus and retina (Dautzenberg and Hauger, 2002). CRF receptors also bind urocortin, urocortin II and urocortin III. Urocortins are homologous to CRF and urotensin I and show a particularly high specificity for CRF2 receptors. Urocortin is a 40-amino acid peptide with 43–45% of amino acid homologous with CRF. In mammalian brain, expression of urocortin mRNA and protein is restricted to the Edinger–Westphal locus, the hypothalamic area and a small population of neurones in the forebrain. Urocortin II in mouse and the homologous stress-related peptide in humans bind selectively the CRF2 receptor. They are highly expressed in the PVN, supraoptic and ARC, locus coeruleus and motor nuclei of the brainstem and spinal cord. Urocortin III in mouse and its homologue stresscopin in humans bind selectively to the CRF2 receptor. They are highly expressed in the rostral perifornical area of the hypothalamus, the posterior part of the bed nucleus of the stria terminalis, the lateral septum and the medial amygdaloid nucleus. CRF-binding protein has high affinity for rat/human CRF, urotensin I and urocortin. It is distributed widely throughout the brain, where it is expressed in the cortex, amygdala and hypothalamus. This protein could increase the availability of CRF or urocortin for CRF receptors.

Peptides of the CRF family are anorexigenic agents. They decrease food intake in rodents and stimulate energy expenditure, likely by stimulating the SNS. The CRF system plays a physiological role in energy balance regulation. The anorexigenic effects of urocortin are seen at doses lower than those producing measurable anxiety or taste aversion. The sites of the anorexigenic and thermogenic actions of the peptides from the CRF family have yet to be delineated fully. There is evidence that the urocortins can act either centrally or peripherally to elicit anorexigenic effects. In the brain, PVN also has been reported as one of the major sites for the anorexigenic effects of CRF. There is also evidence that urocortin evokes anorexia when injected in the lateral septum, while in the parabranchial nucleus, CRF induces dehydration. The medial preoptic area has been reported as a site for the thermogenic action of CRF. Some experiments indicate that the perifornical area of the lateral hypothalamus and Edinger–Westphal nucleus could be the brain source of CRF and urocortin neurones potentially involved in the regulation of energy balance (Richard *et al.*, 2002).

It has been demonstrated that leptin exerts its effects on food intake and energy expenditure at least in part via CRF receptor-mediated pathways. Conversely, the effects of changes in nutritional status on CRF neurones require leptin. The anatomical basis for the effects of leptin on CRF is the localization of LR on CRF neurones in various nuclei of the hypothalamus, including ARC and PVN (Schulz and Lehnert, 2004).

The progress made in recent years corroborates the potential of the CRF system as a target for antiobesity drugs with unspecific activation of CRF circuitries producing anxiogenic, hypophysiotropic and other potentially undesirable actions. Synthetic ligands for the CRF2 and CRF1 receptors offer alternative possibilities for developing molecules that elicit effects related specifically to the regulation of energy balance (Richard *et al.*, 2000).

Melanocortins

The melanocortinergic signalling system in the brain is an important member of the family of catabolic central pathways, as supported by solid genetic and pharmacological evidence (Huszar *et al.*, 1997; Inui, 2004; Coll *et al.*, 2007). It has been shown that leptin exerts its action partly by activation of the melanocortin system in the brain (Seeley *et al.*, 1997; Choi *et al.*, 2003; Gao and Horvath, 2007, 2008).

Melanocortins are a family of peptides, including α-MSH and corticotropin, which are cleaved from POMC precursors. In the mammalian brain, POMC is expressed by neurones of the ARC, adjacent to NPY-producing cells, and by the neurones of the NST (Inui, 2000). These neurones release α-MSH from axon terminals, where it can bind and activate melanocortin receptors (MCR) on post-synaptic membrane surfaces.

Among the five MCR subtypes identified, MC4R is strongly implicated in food intake and possibly in energy expenditure, with knocking out of this receptor subtype causing hyperphagia and obesity in mice (Huszar *et al.*, 1997), central administration of an MC4R agonist producing anorexia, whereas administration of an antagonist of MC4R stimulates feeding (Schwartz *et al.*, 1999). MC4R is a seven-transmembrane G protein-coupled receptor (GPCR) encoded by a single exon gene localized on chromosome 18q22. MC4R is highly expressed in PVN, which contains POMC and AgRP fibres. MC4R is stimulated by α-MSH and antagonized by AgRP, which is the endogenous antagonist of the melanocortin system (Barb *et al.*, 2004).

Mice with targeted disruption of the *POMC* gene are obese (Yaswen *et al.*, 1999). In humans, MC4R mutations have been reported as the most common single genetic cause of obesity in some populations, being responsible for about 4% of early-onset obesity (Miraglia Del Giudice *et al.*, 2002; Farooqi *et al.*, 2003). Mutations in MC4R result in a distinct obesity syndrome that is inherited in a co-dominant manner (Fig. 2.3). Mutations leading to complete loss of function are associated with a more severe phenotype. The correlation between the signalling properties of these mutant receptors and energy intake emphasizes the key role of this receptor in the control of eating behaviour in humans (Mergen *et al.*, 2001; Coll *et al.*, 2007).

MC4R activity affects meal size and meal choice but not meal frequency, with the type of diet affecting the efficacy of MC4R agonists to reduce food intake. The central sites involved in the different aspects of feeding behaviour that are affected by MC4R signalling are being unravelled (Adan *et al.*, 2006). The PVN plays an important role in food intake per se, whereas melanocortin signalling in the lateral hypothalamus is associated with the response to a high-fat diet. MC4R signalling in the brainstem has been shown to affect meal size. Further genetic, behavioural and brain region-specific studies need to clarify how the MC4R agonists affect feeding behaviour in order to determine which obese individuals would benefit most from treatment with these drugs. Application of MCR agonists in humans has already revealed side effects, such as penile erection, which may complicate introduction of these drugs in the treatment of obesity.

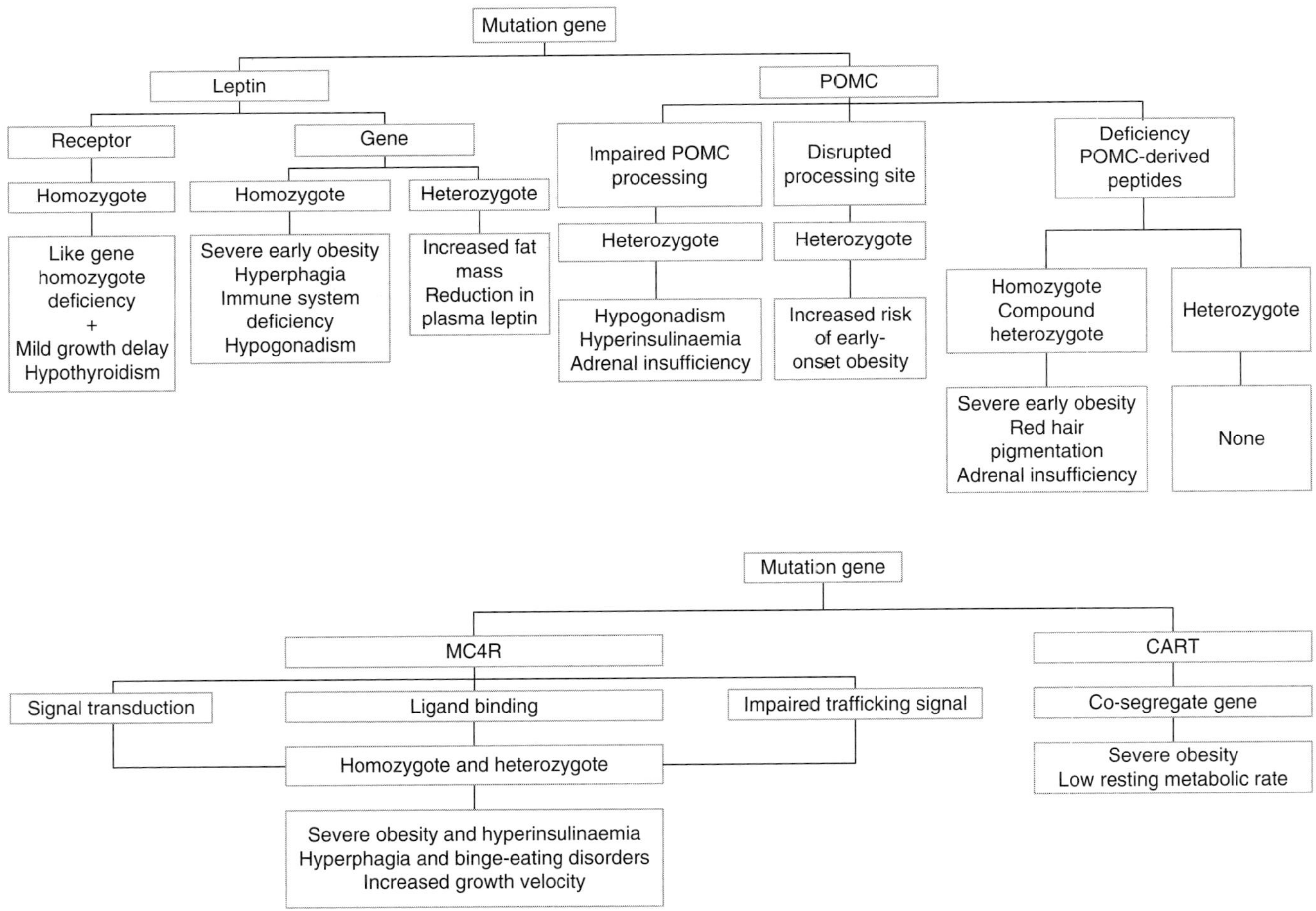

Fig. 2.3. Monogenic causes of obesity in humans. Note that the schematic representation is limited primarily to peptides described in the text (Inui, 2003, 2004; Korner and Aronne, 2003).

The role of the CNS-MCR system in the control of adiposity through effects on nutrient partitioning and cellular lipid metabolism independent of nutrient intake has been established (Nogueiras *et al.*, 2007). Pharmacological inhibition of MCR in rats and genetic disruption of MCR4 in mice promote lipid uptake, triglyceride synthesis and fat accumulation in white adipose tissue directly and potently, while increased CNS-MCR signalling triggers lipid mobilization. These effects have been shown to be independent of food intake and precede changes in adiposity. In addition, decreased CNS-MCR signalling promotes increased insulin sensitivity and glucose uptake in white adipose tissue, while decreasing glucose utilization in muscle and brown adipose tissue. Interestingly, this CNS control of peripheral nutrient partitioning depends on functionality of the SNS and is enhanced by synergistic effects on liver triglyceride synthesis (Nogueiras *et al.*, 2007). The reported findings offer an explanation for enhanced adiposity resulting from decreased melanocortin signalling, even in the absence of hyperphagia, and are consistent with feeding-independent changes in substrate utilization, as reflected by the respiratory quotient, which is increased with chronic MCR blockade in rodents and in humans with loss-of-function mutations in MC4R.

Transgenic mice overexpressing MSH show reduced weight gain and adiposity, improved glucose tolerance and insulin sensitivity. These results are observed in diverse backgrounds such as lean and genetically obese mice. Additional studies are necessary to determine if MSH overexpression might protect against more common forms of obesity such as diet-induced obesity and the associated metabolic changes. Long-term melanocortinergic activation has been targeted as a potential strategy for antiobesity and/or antidiabetic therapy (Savontaus *et al.*, 2004).

Brain-derived neurotrophic factor

A number of recent studies have provided insight into the mediators and signalling mechanisms that lie beyond the melanocortin receptors (Coll *et al.*, 2007). Brain-derived neurotrophic factor (BDNF), a member of the neurotropin family, is expressed highly in the VMH and moderately in the PVN and LH. Several lines of evidence suggest an important role for BDNF in energy homeostasis. Chronic central BDNF infusion suppresses appetite dose-dependently and induces body weight loss in rats. Central and peripheral administration of BDNF decreases food intake and increases energy expenditure in *db/db* mice. Rodents with conditional deletion of BDNF or BDNF heterozygous mice develop hyperphagia and adult-onset obesity (Rios *et al.*, 2001). Moreover, selective deletion of BDNF in the VMH and DMH of adult mice results in hyperphagic behaviour and obesity (Unger *et al.*, 2007). Food deprivation reduces BDNF expression in the VMH (Xu *et al.*, 2003). Thus, BDNF expression is regulated by nutritional status and also by MC4R signalling.

The BDNF receptor TrkB is localized in the hypothalamus and *TrkB*-mutant mice develop hyperphagia and severe obesity. Mice with a hypomorphic mutation in TrkB (resulting in 25% normal levels of expression) closely resemble mice lacking MC4R in that they develop hyperphagia and obesity, increased body length and excessive weight gain on a high-fat diet. BDNF expression in the VMH is reduced in the AgRP-overexpressing mutant mice, while the MC4 agonist, MTII, increases the level of BDNF mRNA significantly in this brain area of food-deprived mice. Further, central infusion of BDNF into mice with deficient MC4R signalling suppresses the hyperphagia and excessive weight gain observed on high-fat diets. In humans, genetic disruption of the neurotrophin receptor TrkB (Yeo *et al.*, 2004) and its ligand BDNF (Gray *et al.*, 2006, 2007) causes severe hyperphagia and obesity, developmental delay, impaired short-term memory and unusually hyperactive behaviour. These observations suggest that BDNF is an important effector through which MC4R signalling controls energy balance. Altogether, it is conceivable that BDNF could be a missing piece in the puzzle behind the mechanism of the development of obesity seen after VMH lesion (Sahu, 2004).

CART

CART originally was identified as a hypothalamic neuropeptide upregulated by cocaine and amphetamine treatment. Central administration of CART induces *c-fos* expression in several hypothalamic nuclei related with feeding control (Tsuruta *et al.*, 2002). CART is expressed widely in the brain, including hypothalamic areas such as the ARC, PVN and DMN. It is a neuropeptide with potent but short-lived anorectic effects (Schwartz *et al.*, 1999). Recombinant CART fragments decrease food intake in rodents, whereas anti-CART antibodies increase it.

POMC and CART were found to co-localize in the ARC. Most POMC/CART and NPY/AgRP neurones express the LRb, through which leptin conveys different messages to each type of neurone. CART and AgRP expression is modulated directly by leptin, which upregulates CART mRNA expression and downregulates AgRP mRNA (Kristensen *et al.*, 1998).

GLP-1 and OXM

Preproglucagon is cleaved in a tissue-specific manner by prohormone convertase-1 and -2, giving rise to a number of products with a variety of functions in both the CNS and peripheral tissues. In the intestine and CNS, the major posttranslational products of preproglucagon cleavage are GLP-1, glucagon-like peptide-2 (GLP-2), glicentin (also known as enteroglucagon) and oxyntomodulin (OXM; also known as glucagon-37).

GLP-1 is a peptide product of the proglucagon gene, released from the L-cells of the small intestine in response to food ingestion (Drucker, 2006). GLP-1 is a potent inducer of glucose-dependent insulin release. This has led to the development of GLP-1 agonists, which have clinical utility in the treatment of type 2 diabetes mellitus (Drucker, 2006). GLP-1 can also influence food intake with the GLP-1 analogue, exenatide, capable of lowering both blood glucose and body weight in obese type 2 diabetic subjects. The effects on body weight may be as a result of the induction of satiety via inhibition of gastric emptying, but there is also evidence that GLP-1 can influence feeding behaviour by acting at the NST in the brainstem and the PVN of the hypothalamus (Coll *et al.*, 2007). GLP-2 has also been found to be an inhibitor of food intake in the rat. Glicentin acts within peripheral tissues, with its roles including the inhibition of gastric acid secretion in rats. No effect of glicentin in the CNS has been reported to date (Darkin, 2001).

OXM is a product of proglucagon processing in the intestine and CNS, particularly in the neurones of the NST of the brainstem. OXM is a 37-amino acid peptide that contains the 29-amino acid sequence of glucagons, followed by an 8-amino acid carboxyterminal extension. Bataille *et al.* (1981) elucidated the structure of OXM and showed that this peptide stimulated cAMP accumulation in a rat stomach preparation, being a potent inhibitor of pentagastrin-stimulated gastric acid secretion and gastric emptying in rodents and humans. The neuroanatomy of endogenous OXM remains an interesting research topic. OXM reportedly mediates its anorexigenic action via direct interaction with the hypothalamus, activating POMC neurones within ARC (Darkin *et al.*, 2004), as well as by suppression of the orexigenic hormone, ghrelin (Coll *et al.*, 2007). It is currently unclear through which receptor OXM mediates its actions. Recently, it was demonstrated that structurally distinct proglucagon-derived peptides regulate food intake and energy expenditure differentially by interacting with a GLP-1 receptor-dependent pathway. Hence, ligand-specific activation of a common GLP-1 receptor increases the complexity of gut–CNS pathways regulating energy homeostasis (Baggio *et al.*, 2004). Little is known about the physiological role of OXM. Central and intra-PVN administration of low doses of OXM caused a robust and substantial inhibition of food intake in rats, a marked reduction in body weight gain and body adiposity. This finding suggests that OXM is a potent regulator of appetite and body weight. OXM offers novel routes for the development of therapeutic agents in the treatment of obesity.

GALP

As described in Chapter 1, galanin-like peptide (GALP) is highly expressed in the ARC. In rats, a biphasic action of GALP on feeding is evident: within 2 h of central administration, GALP stimulates feeding; after 24 h, both feeding and body weight are reduced significantly. In mice, however, GALP elicits a dose-dependent decrease in both feeding and body weight. Further evidence, such as that fasting reduces GALP mRNA levels while leptin induces GALP mRNA expression in the hypothalamus, suggests that GALP is one of the anorexigenic signals in the neural circuitry regulating energy balance (Sahu, 2004).

Prolactin-releasing peptide

PrRP, a ligand for the human orphan GPCR, hGR3/gpr10, originally was reported to cause prolactin secretion from the anterior pituitary cells. Subsequently, a role for PrRP in the regulation of food intake was set forward. Specifically, PrRP and its receptor are localized in the hypothalamus, in particular in DMN, the brainstem NST and ventrolateral medulla (Ellacott *et al.*, 2002). Central infusion of PrRP decreases food intake and body weight gain, causes hyperthermia and increases UCP-1 mRNA expression and oxygen consumption. These data show that the decrease in body weight gain is not due entirely to the reduction in food intake but also to the additional effects on energy expenditure (Lawrence *et al.*, 2002). PrRP mRNA is reduced in situations of negative energy balance and in chronic genetic obesity. CRF and its receptors appear to mediate the anorexigenic and body weight-reducing effects of PrRP. It also interacts with leptin to reduce food intake and body weight, with PrRP neurones expressing LR. Thus, an important role for PrRP in energy homeostasis has been proposed (Ellacott *et al.*, 2002).

Perspectives for Obesity Treatment

Obesity and its co-morbidities have reached epidemic proportions and, consequently, take an increasing toll on life, quality of life and health care resources. Thus, the development of safe and highly efficacious obesity drugs is becoming more of an imperative. Currently available obesity drugs such as orlistat, sibutramine and rimonabant have limited efficacy, as well as safety and/or tolerability concerns which condition prescription to a small percentage of the obese population only. As with other chronic conditions such as hypertension, when effective obesity drugs or other treatments are discontinued, weight gain usually returns. In contrast, gastric and intestinal bypass surgery can result in lasting weight loss. After bariatric surgery, malabsorption is temporary, yet the appetite effects may last many years in relation to the alterations in hormonal signals known to affect appetite (Ramos *et al.*, 2003; Druce *et al.*, 2004).

Primary objectives for antiobesity drugs include reducing food intake, blocking nutrient absorption, increasing thermogenesis, modulating energy storage and influencing the central controller of body weight regulation (Bray and Greenway, 1999). It has to be taken into consideration that redundant pathways control body weight and food intake, which represents a critical aspect for a single agent trying to manipulate energy homeostasis effectively. In this context, as in other diseases, combined therapy may be more effective than monotherapy.

C75, a fatty acid synthase inhibitor

C75 is a synthetic compound that inhibits the fatty acid synthase (FAS), causing anorexia and weight loss in lean and genetically obese mice by both central and peripheral mechanisms (Loftus *et al.*, 2000; Kim *et al.*, 2007; Cheng *et al.*, 2008).

C75 acts in the CNS and in peripheral tissues, in which it increases energy expenditure. In the CNS, C75 injection induces the expression of *c-fos* in hindbrain feeding-related nuclei and in the PVN, the ARC and the central amygdala (Miller *et al.*, 2004). It has been demonstrated that FAS and other enzymes required for the long-chain acid synthesis are highly expressed in neurones in many brain regions, including hypothalamic neurones that regulate feeding behaviour in mice (Cha *et al.*, 2004). C75 administration exerts several effects on neuronal energy metabolism and on neuronal activity. It interferes with the binding of malonyl-CoA to the beta-ketoacyl synthase domain of FAS and affects carnitine palmitoyltransferase-1 in neurones and in peripheral tissues (Bentebibel *et al.*, 2006). C75 action results in the stimulation of fatty acid oxidation and the alteration of glucose metabolism that is modulated by a change in AMPK and in modulation of downstream energy-sensing molecules such as malonyl-CoA (Hu *et al.*, 2003; Landree *et al.*, 2004). This change signals a positive energy balance, resulting in a modified expression of orexigenic and anorexigenic neuropeptides (Cha *et al.*, 2004). Acute and chronic C75 treatments cause reduction of the NPY and AgRP expression in *ob/ob* and diet-induced obesity mice and induction of POMC and CART expression in lean and obese mice. These results show also that C75 exerts its central action independently of leptin (Tu *et al.*, 2005).

Collectively, the results in mice suggest that C75 may play a physiologically relevant role in energy homeostasis, and molecules that act on AMPK are potential therapeutic agents for the treatment of obesity.

Anorexigenic targets

Although CART deficiency does not appear to influence energy balance, there is clear evidence that POMC peptides play a critical role in feeding behaviour, with both POMC-deficient mice and humans developing hyperphagia and obesity (Coll *et al.*, 2007). POMC undergoes extensive posttranslational modification to generate a range of smaller biologically active peptides, the melanocortins, which are agonists for melanocortin receptors. The ultimate pool of bioactive melanocortins released from POMC-expressing neurones is influenced by a number of factors, including the activities of prohormone convertases, carboxypeptidases and both acetylases and deacetylases (Coll *et al.*, 2007). Although α-MSH has always been considered the predominant POMC-derived product controlling energy balance, genetic evidence also strongly implicates β-MSH in appetite regulation, at least in humans (Lee *et al.*, 2006).

There is a great effort to develop agonists and antagonists of peptide receptors that have been associated specifically with energy homeostasis. As described above, the MCR system represents an attractive target. A natural agonist, α-MSH reduces food intake, with mice lacking POMC, the precursor of α-MSH, being obese. This suggests that specific agonists for MC4R might become useful obesity agents (Collins and Kim, 2003). MC3R has been suggested to play a role in nutrient partitioning. Although agonists of the MC3R would not be expected to produce dramatic weight loss, they may favour a more beneficial partitioning of nutrients. The development of dual MC4 and MC3 receptor agonists has been addressed in order to reduce weight dramatically, as well as improve the metabolic co-morbidities of obesity significantly.

Despite the limited effect of leptin in its initial trials as an antiobesity agent, there is still great potential for a leptin-like product as ciliary neurotrophic factor

(CNTF). Very few drugs stimulate robust weight loss, and even fewer can prevent weight gain after termination of therapy. But CNTF, and its analogue, Axokine, are able to do just that (Seeley, 2005). The prolonged effects of CNTF result from the growth of new neurones in the hypothalamus (Kokoeva *et al.*, 2005, 2007). CNTF has been able to correct obesity and diabetes not only in leptin-sensitive obese mice, but also in leptin-resistant mice, including those made obese by a high-fat diet, through activation of the JAK–STAT pathways and inhibition of NPY and AgRP. Moreover, CNTF also acts at the peripheral level. CNTF signals through the CNTFRα-IL-6R-gp130β receptor complex to increase fatty-acid oxidation and reduce insulin resistance in skeletal muscle by activating AMPK, independent of signalling through the brain (Watt *et al.*, 2006). In addition, CNTF suppresses inflammatory signalling cascades associated with lipid accumulation in liver and skeletal muscle (Febbraio, 2007) and reprogrammes adipose tissue to promote mitochondrial biogenesis, enhancing oxidative capacity and reducing lipogenic capacity, thereby resulting in triglyceride loss (Crowe *et al.*, 2008).

Another possibility for obesity treatment that has been explored in experimental animals is gene therapy. For example, the enhanced CNTF gene has been inserted into adenovirus and administrated to rats. After 6 weeks, researchers observed a decrease in body weight and food intake similar to that in rodents treated with the leptin gene. No apparent side effects were evident during the 6 months of treatment. However, potential human application of this drug is still in the distant future. Gene therapy embodies the utopic hope that some day a single drug injection may be a viable option for treating obesity (Kalra and Kalra, 2004).

The resilience of the body's weight-regulatory system to change makes obesity especially difficult to treat. Despite significant advances in the understanding surrounding the regulation of food intake and energy expenditure, a large unmet medical need for effective and safe therapeutics still remains. Neurogenesis and plasticity in these key circuits are being explored to provide alternative therapeutic options. Further research is necessary to understand better the central mechanisms underlying energy homeostasis in order to develop new antiobesity drugs.

References

Abizaid, A., Gao, Q. and Horvath, T.L. (2006) Thoughts for food: brain mechanisms and peripheral energy balance. *Neuron* 51, 691–702.

Adan, R.A., Tiesjema, B., Hillebrand, J.J., la Fleur, S.E., Kas, M.J. and de Krom, M. (2006) The MC4 receptor and control of appetite. *British Journal of Pharmacology* 149, 815–827.

Ahima, R.S. (2006) Adipose tissue as an endocrine organ. *Obesity* 14 (Suppl. 5), 242S–249S.

Ahima, R.S. and Osei, S.Y. (2004) Leptin signaling. *Physiology and Behavior* 81, 223–241.

Ahima, R.S., Qi, Y. and Singhal, N.S. (2006) Adipokines that link obesity and diabetes to the hypothalamus. *Progress in Brain Research* 153, 155–174.

Anderson, U., Filipsson, K., Abbott, C.R., Woods, A., Smith, K., Bloom, S.R., Carling, D. and Small, C.J. (2004) AMPK activated protein kinase play a role in the control of food intake. *Journal of Biological Chemistry* 279, 12005–12008.

Baggio, L.L., Huang, Q., Brown, T.J. and Drucker, D.J. (2004) Oxyntomodulin and glucagon-like peptide-1 differentially regulate murine food intake and energy expenditure. *Gastroenterology* 127, 546–558.

Balthasar, N., Dalgaard, L.T., Lee, C.E., Yu, J., Funahashi, H., Williams, T., Ferreira, M., Tang, V., McGovern, R.A., Kenny, C.D., Christiansen, L.M., Edelstein, E., Choi, B., Boss, O., Aschkenasi, C., Zhang, C.Y., Mountjoy, K., Kishi, T., Elmquist, J.K. and Lowell, B.B. (2005) Divergence of melanocortin pathways in the control of food intake and energy expenditure. *Cell* 123, 493–505.

Banks, W.A. (2003) Is obesity a disease of the blood–brain barrier? Physiological, pathological and evolutionary considerations. *Current Pharmaceutical Design* 9, 801–809.

Banks, W.A. (2004) The many lives of leptin. *Peptides* 25, 331–338.

Banks, W.A., Kastin, A.J., Huang, W., Jaspan, J.B. and Meness, L.M. (1996) Leptin enters brain by a saturable system independent of insulin. *Peptides* 17, 305–311.

Barb, C.R., Robertson, A.S., Barrett, J.B., Kraaling, R.R. and Houseknecht, K.L. (2004) The role of melanocortin-3 and -4 receptor in regulating appetite, energy homeostasis and neuroendocrine function in the pig. *Journal of Endocrinology* 181, 39–52.

Bataille, D., Gespach, C., Tatemoto, K., Marie, J.C., Coudray, A.M., Rosselin, G. and Mutt, V. (1981) Bioactive enteroglucagon (oxyntomodulin): present knowledge on its chemical structure and its biological activities. *Peptides* 2, 41–44.

Bates, S.H., Stearns, W.H., Dundon, T.A., Schuber, T.M., Tso, A.W.K., Wanga, Y., Banks, A.S., Lavery, H.J., Haq, A.K., Flier, E.M., Neela, B.G., Schwartz, M.W. and Myers, M.G. (2003) STAT3 signaling is required for leptin regulation of energy balance but not reproduction. *Nature* 421, 856–859.

Benoit, S.C., Clegg, D.J., Seeley, R.J. and Woods, S.C. (2004) Insulin and leptin as adiposity signals. *Recent Progress in Hormone Research* 59, 267–285.

Bentebibel, A., Sebastián, D., Herrero, L., López-Viñas, E., Serra, D., Asins, G., Gómez-Puertas, P. and Hegardt, F.G. (2006) Novel effect of C75 on carnitine palmitoyltransferase I activity and palmitate oxidation. *Biochemistry* 45, 4339–4350.

Blüher, S. and Mantzoros, C.S. (2007) Leptin in reproduction. *Current Opinion in Endocrinology, Diabetes, and Obesity* 14, 458–464.

Bouret, S.G. and Simerly, R.B. (2004) Minireview: leptin and development of hypothalamic feeding circuits. *Endocrinology* 145, 2621–2626.

Bouret, S.G., Draper, S.J. and Simerly, R.B. (2004) Trophic action of leptin on hypothalamic neurons that regulate feeding. *Nature* 304, 108–110.

Bray, G.A. and Greenway, F.L. (1999) Current and potential drugs for treatment of obesity. *Endocrine Reviews* 20, 805–875.

Caro, J.F., Kolaczynski, J.W., Nyce, M.R., Ohannesian, J.P., Opentanova, I., Goldman, W.R., Lynn, R.B., Zhang, P.L., Sinha, M.K. and Considine, R.V. (1996) Decreased cerebrospinal-fluid/serum leptin ratio in obesity: a possible mechanism for leptin resistance. *The Lancet* 348, 159–161.

Cha, S.H., Hu, Z. and Lane, M.L. (2004) Long-term effects of a fatty acid synthase inhibitor on obese mice: food intake, hypothalamic neuropeptides and UPC3. *Biochemical and Biophysical Research Communications* 317, 301–308.

Chen, C.Y., Inui, A., Asakawa, A., Fujino, K., Kato, I., Chen, C.C., Ueno, N. and Fujimiya, M. (2005) Des-acyl ghrelin acts by CRF type 2 receptors to disrupt fasted stomach motility in conscious rats. *Gastroenterology* 129, 8–25.

Cheng, A., Uetani, N., Simoncic, P.D., Chaubey, V.P., Lee-Loy, A., McGlade, C.J., Kennedy, B.P. and Tremblay, M.L. (2002) Attenuation of leptin action and regulation of obesity by protein tyrosine phosphatase 1B. *Developmental Cell* 2, 497–503.

Cheng, F., Wang, Q., Chen, M., Quiocho, F.A. and Ma, J. (2008) Molecular docking study of the interactions between the thioesterase domain of human fatty acid synthase and its ligands. *Proteins* 70, 1228–1234.

Choi, Y.H., Li, C., Page, K., Westby, A., Della-Fera, M.A., Lin, J., Hartzell, D.L. and Baile, C.A. (2003) Melanocortin receptors mediate leptin effects on feeding and body weight but not adipose apoptosis. *Physiology and Behavior* 79, 795–801.

Coll, A.P., Farooqi, I.S. and O'Rahilly, S. (2007) The hormonal control of food intake. *Cell* 129, 251–262.

Collins, C.A. and Kim, P.R. (2003) Prospects for obesity treatment: MCH receptor antagonists. *Current Opinion in Investigational Drugs* 4, 386–394.

Crowe, S., Turpin, S.M., Ke, F., Kemp, B.E. and Watt, M.J. (2008) Metabolic remodelling in adipocytes promotes CNTF-mediated fat loss in obesity. *Endocrinology* 149, 2546–2556.

Cummings, D.E. and Overduin, J. (2007) Gastrointestinal regulation of food intake. *Journal of Clinical Investigation* 117, 13–23.

Darkin, C.L. (2001) Oxyntomodulin inhibits food intake on the rat. *Endocrinology* 142, 4244–4252.

Darkin, C.L., Small, C.J., Batterham, R.L., Neary, N.M., Cohen, M.A., Patterson, M., Ghatei, M.A. and Bloom, S.R. (2004) Peripheral oxyntomodulin reduces food intake and body weight gain in rats. *Endocrinology* 145, 2687–2695.

Dautzenberg, M.F. and Hauger, R.L. (2002) The CRF peptide family and their receptors: yet more partners discovered. *Trends in Pharmacological Sciences* 23, 71–77.

Dhillon, H., Zigman, J.M., Ye, C., Lee, C.E., McGovern, R.A., Tang, V., Kenny, C.D., Christiansen, L.M., White, R.D., Edelstein, E.A., Coppari, R., Balthasar, N., Cowley, M.A., Chua, S. Jr, Elmquist, J.K. and Lowell, BB. (2006) Leptin directly activates SF1 neurons in the VMH, and this action by leptin is required for normal body-weight homeostasis. *Neuron* 49, 191–203.

Druce, M.R., Small, C.J. and Bloom, S.R. (2004) Minireview: gut peptides regulating satiety. *Endocrinology* 145, 2660–2665.

Drucker, D.J. (2006) The biology of incretin hormones. *Cell Metabolism* 3,153–165.

Ellacott, K.L.J., Lawrence, C.B., Rothwell, N.J. and Luckman, S.M. (2002) PRL-releasing peptide interacts with leptin to reduce food intake and body weight. *Endocrinology* 143, 368–374.

Elmquist, J.K., Ahima, R.S., Elias, C.F., Flier, J.S. and Saper, C.B. (1998) Leptin activates distinct projections from the dorsomedial and ventromedial hypothalamic nuclei. *Neurobiology* 95, 741–746.

Enriori, P.J., Evans, A.E., Sinnayah, P., Jobst, E.E., Tonelli-Lemos, L., Billes, S.K., Glavas, M.M., Grayson, B.E., Perello, M., Nillni, E.A., Grove, K.L. and Cowley, M.A. (2007) Diet-induced obesity causes severe but reversible leptin resistance in arcuate melanocortin neurons. *Cell Metabolism* 5, 181–194.

Farooqi, I.S. (2008) Monogenic human obesity. *Frontiers of Hormone Research* 36, 1–11.

Farooqi, I.S. and O'Rahilly, S. (2007) Genetic factors in human obesity. *Obesity Reviews* 8 (Suppl. 1), 37–40.

Farooqi, I.S., Matarese, G., Lord, G.M., Keogh, J.M., Lawrence, E., Agwu, C., Sanna, V., Jebb, S.A., Perna, F., Fontana, S., Lechler, R.I., DePaoli, A.M. and O'Rahilly, S. (2002) Beneficial effects of leptin on obesity, T cell hyporesponsiveness, and neuroendocrine/metabolic dysfunction of human congenital leptin deficiency. *Journal of Clinical Investigation* 110, 1093–1103.

Farooqi, I.S., Keogh, J.M., Yeo, G.S.H., Lank, E.J., Cheetham, T. and O'Rahilly, S. (2003) Clinical spectrum of obesity and mutations in the melanocortin 4 receptor gene. *New England Journal of Medicine* 348, 1085–1095.

Farooqi, I.S., Wangensteen, T., Collins, S., Kimber, W., Matarese, G., Keogh, J.M., Lank, E., Bottomley, B., Lopez-Fernandez, J., Ferraz-Amaro, I., Dattani, M.T., Ercan, O., Myhre, A.G., Retterstol, L., Stanhope, R., Edge, J.A., McKenzie, S., Lessan, N., Ghodsi, M., De Rosa, V., Perna, F., Fontana, S., Barroso, I., Undlien, D.E. and O'Rahilly, S. (2007) Clinical and molecular genetic spectrum of congenital deficiency of the leptin receptor. *New England Journal of Medicine* 356, 237–247.

Febbraio, M.A. (2007) gp130 receptor ligands as potential therapeutic targets for obesity. *Journal of Clinical Investigation* 117, 841–849.

Flier, J.S. (2006) AgRP in energy balance: will the real AgRP please stand up? *Cell Metabolism* 3, 83–85.

Friedman, J.M. and Halaas, J.L. (1998) Leptin and the regulation of body weight in mammals. *Nature* 395, 763–770.

Frühbeck, G. (2006) Intracellular signalling pathways activated by leptin. *Biochemical Journal* 393, 7–20.

Gao, Q. and Horvath, T.L. (2007) Neurobiology of feeding and energy expenditure. *Annual Reviews in Neuroscience* 30, 367–398.

Gao, Q. and Horvath, T.L. (2008) Neuronal control of energy homeostasis. *FEBS Letters* 582, 132–141.

Gibson, W.T., Farooqi, I.S., Moreau, M., DePaoli, A.M., Lawrence, E., O'Rahilly, S. and Trussell, R.A. (2004) Congenital leptin deficiency due to homozygosity for the Δ133G mutation: report of another case and evaluation of response to four years of leptin therapy. *Journal of Clinical Endocrinology and Metabolism* 89, 4821–4826.

Gray, J., Yeo, G.S., Cox, J.J., Morton, J., Adlam, A.L., Keogh, J.M., Yanovski, J.A., El Gharbawy, A., Han, J.C., Tung, Y.C., Hodges, J.R., Raymond, F.L., O'Rahilly, S. and Farooqi, I.S. (2006) Hyperphagia, severe obesity, impaired cognitive function, and hyperactivity associated with functional loss of one copy of the brain-derived neurotrophic factor (BDNF) gene. *Diabetes* 55, 3366–3371.

Gray, J., Yeo, G.S., Hung, C., Keogh, J., Clayton, P., Banerjee, K., McAulay, A., O'Rahilly, S. and Farooqi, I.S. (2007) Functional characterization of human NTRK2 mutations identified in patients with severe early-onset obesity. *International Journal of Obesity* 31, 359–364.

Halaas, J.L. and Friedman, J.M. (1997) Leptin and its receptor. *Journal of Endocrinology* 155, 215–216.

Han, Z., Yan, J.-Q., Luo, G.-G., Liu, Y. and Wang, Y.-L. (2003) Leptin receptor expression in the basolateral nucleus of amygdala of conditioned taste aversion rats. *World Journal of Gastroenterology* 9, 1034–1037.

Harrold, J.A. (2004) Leptin leads hypothalamic feeding circuits in a new direction. *BioEssays* 26, 1043–1045.

Havel, P.J. (2001) Peripheral signals conveying metabolic information to the brain: short-term and long-term regulation of food intake and energy homeostasis. *Experimental Biology and Medicine* 226, 963–977.

Horvath, T.L. (2005) The hardship of obesity: a soft-wired hypothalamus. *Nature Neuroscience* 8, 561–565.

Hu, Z., Cha, S.H., Chohnan, S. and Lane, M.D. (2003) Hypothalamic malonyl-CoA as a mediator of feeding behavior. *Proceedings of the National Academy of Sciences of the United States of America* 100, 12624–12629.

Hukshorn, C.J. and Saris, W.H. (2004) Leptin and energy expenditure. *Current Opinion in Clinical Nutrition and Metabolic Care* 7, 629–633.

Huo, L., Grill, H.J. and Bjorbaek, C. (2006) Divergent regulation of proopiomelanocortin neurons by leptin in the nucleus of the solitary tract and in the arcuate hypothalamic nucleus. *Diabetes* 55, 567–573.

Huszar, D., Lynch, C.A., Fairchild-Huntress, V., Dunmore, J.H., Fang, Q., Berkemeier, L.R., Gu, W., Kesterson, R.A., Boston, B.A., Cone, R.D., Smith, F.J., Campfield, L.A., Burn, P. and Lee, F. (1997) Targeted disruption of the melanocortin-4 receptor results in obesity in mice. *Cell* 88, 131–141.

Inui, A. (2000) Transgenic approach to the study of body weight regulation. *Pharmacological Reviews* 52, 35–61.

Inui, A. (2001) Ghrelin: an orexigenic and somatotrophic signal from the stomach. *Nature Reviews Neuroscience* 2, 551–560.

Inui, A. (2003) Neuropeptide gene polymorphisms and human behavioural disorders. *Nature Reviews Drug Discovery* 2, 986–998.

Inui, A. (2004) Melanocortin signalling, single nucleotide polymorphism and eating disorders. *International Journal of Psychiatry in Medicine* 34, 397–401.

Jequier, E. (2002) Leptin signaling, adiposity and energy balance. *Annals of the New York Academy of Sciences* 967, 378–388.

Jequier, E. and Tappy, L. (1999) Regulation of body weight in humans. *Physiological Reviews* 79, 451–480.

Kalra, S.P. and Kalra, P.S. (2004) NPY and cohorts in regulating appetite, obesity and metabolic syndrome: beneficial effects of gene therapy. *Neuropeptides* 38, 201–211.

Kamiji, M.M. and Inui, A. (2007) Neuropeptide y receptor selective ligands in the treatment of obesity. *Endocrine Reviews* 28, 664–84.

Kim, E.K., Kleman, A.M. and Ronnett, G.V. (2007) Fatty acid synthase gene regulation in primary hypothalamic neurons. *Neuroscience Letters* 423, 200–204.

Klok, M.D., Jakobsdottir, S. and Drent, M.L. (2007) The role of leptin and ghrelin in the regulation of food intake and body weight in humans: a review. *Obesity Reviews* 8, 21–34.

Knecht, S., Ellger, T. and Levine, J. A. (2008) Obesity in neurobiology. *Progress in Neurobiology* 84, 85–103.

Kokoeva, M.V., Yin, H. and Flier, J.S. (2005) Neurogenesis in the hypothalamus of adult mice: potential role in energy balance. *Science* 310, 679–683.

Kokoeva, M.V., Yin, H. and Flier, J.S. (2007) Evidence for constitutive neural cell proliferation in the adult murine hypothalamus. *Journal of Comparative Neurology* 505, 209–220.

Konturek, S.J., Konturek, J.W., Pawlik, T. and Brzozowki, T. (2004) Brain–gut axis and its role in the control of food intake. *Journal of Physiology and Pharmacology* 55, 137–154.

Korner, J. and Aronne, J.L. (2003) The emerging science of body weight regulation and its impact on obesity treatment. *The Journal of Clinical Investigation* 111, 565–570.

Kristensen, P., Judge, M.E., Thim, L., Ribel, U., Christjansen, K.N., Wulff, B.S., Clausen, J.T., Jensen, P.B., Madsen, O.D., Vrang, N., Larsen, P.J. and Hastrup, S. (1998) Hypothalamic CART is a new anorectic peptide regulated by leptin. *Nature* 393,72–76.

La Cava, A. and Matarese, G. (2004) The weight of leptin in immunity. *Nature Reviews Immunology* 4, 371–379.

Lahlou, N., Issad, T., Lebouc, Y., Carel, J.C., Camoin, L., Roger, M. and Girard, J. (2002) Mutations in the human leptin and leptin receptor genes as models of serum leptin receptor regulation. *Diabetes* 51, 1980–1985.

Landree, L.E., Hanlon, A.L., Strong, D.W., Rumbaugh, G., Miller, I.M., Thupari, J.N., Connolly, E.C., Huganir, R.L., Richardson, C., Witters, L.A., Kuhajda, F.P. and Ronnett, G.V. (2004) C75, a fatty acid synthase inhibitor, modulates AMP-activated protein kinase to alter neuronal energy metabolism. *Journal of Biological Chemistry* 279, 3817–3827.

Lawrence, C.B., Ellacott, K.L.J. and Luckman, S.M. (2002) PRL-releasing peptide reduces food intake and may mediate satiety signaling. *Endocrinology* 143, 360–367.

Lee, Y.S., Challis, B.G., Thompson, D.A., Yeo, G.S., Keogh, J.M., Madonna, M.E., Wraight, V., Sims, M., Vatin, V., Meyre, D., Shield, J., Burren, C., Ibrahim, Z., Cheetham, T., Swift, P., Blackwood, A., Hung, C.C., Wareham, N.J., Froguel, P., Millhauser, G.L., O'Rahilly, S. and Farooqi, I.S. (2006) A POMC variant implicates beta-melanocyte-stimulating hormone in the control of human energy balance. *Cellular Metabolism* 3, 135–140.

Licinio, J., Caglayan, S., Ozata, M., Yildiz, B.O., de Miranda, P.B., O'Kirwan, F., Whitby, R., Liang, L., Cohen, P., Bhasin, S., Krauss, R.M., Veldhuis, J.D., Wagner, A.J., DePaoli, A.M., McCann, S.M. and Wong, ML. (2004) Phenotypic effects of leptin replacement on morbid obesity, diabetes mellitus, hypogonadism, and behavior in leptin-deficient adults. *Proceedings of the National Academy of Sciences of the United States of America* 101, 4531–4536.

Loftus, T.M., Jaworsky, D.E., Frehywot, G.L., Towsend, C.A., Ronnett, G.V., Lane, M.D. and Kuhajda, F.P. (2000) Reduced food intake and body weight in mice treated with fatty acid synthase inhibitors. *Science* 288, 2379–2381.

López, M., Tovar, S., Vázquez, M.J., Williams, L.M. and Diéguez C. (2007) Peripheral tissue–brain interactions in the regulation of food intake. *Proceedings of the Nutrition Society* 66, 131–155.

McDuffie, J.R., Riggs, P.A., Calis, K.A., Freedman, R.J., Oral, E.A., DePaoli, A.M. and Yanovski, J.A. (2004) Effects of exogenous leptin on satiety and satiation in patients with lipodystrophy and leptin insufficiency. *Journal of Clinical Endocrinology and Metabolism* 89, 4258–4263.

Maffei, M., Halaas, J., Ravussin, E., Pratley, R.E., Lee, G.H., Zhang, Y., Fei, H. and Kim, S. (1995) Leptin levels in human and rodent: measurement of plasma leptin and *ob* RNA in obese and weight-reduced subjected. *Nature Medicine* 1, 1155–1161.

Martin, R.L., Perez, E., He, Y.J., Dawson, R. Jr and Millard, W.J. (2000) Leptin resistance is associated with hypothalamic leptin receptor mRNA and protein downregulation. *Metabolism* 49, 1479–1484.

Mergen, M., Mergen, H., Ozata, M., Oner, R. and Oner, C. (2001) A novel melanocortin 4 receptor (MC4R) gene mutation associated with morbid obesity. *Journal of Clinical Endocrinology and Metabolism* 86, 3448–3451.

Miller, I., Ronnett, G.V., Moran, T.H. and Aja, S. (2004) Anorexigenic C75 alters c-Fos in mouse hypothalamic and hindbrain subnuclei. *Neuroreport for Rapid Communication of Neuroscience Research* 15, 925–929.

Miraglia Del Giudice, E., Cirillo, G., Nigro, V., Santoro, N., D'Urso, L., Raimondo, P., Cozzolino, D., Scafato, D. and Perrone L. (2002) Low frequency of melanocortin-4 receptor (MC4R) mutations in a Mediterranean population with early-onset obesity. *International Journal of Obesity* 26, 647–651.

Montague, C.T., Farooqi, I.S., Whitehead, J.P., Soos, M.A., Rau, H. and Wareham, N.J. (1997) Congenital leptin deficiency is associated with severe early-onset obesity in humans. *Nature* 387, 903–908.

Moran, O. and Phillip, M. (2003) Leptin: obesity, diabetes and other peripheral effects – a review. *Pediatric Diabetes* 4, 101–109.

Moran, T.H. (2006) Gut peptide signaling in the controls of food intake. *Obesity* 14 (Suppl. 5), 250S–253S.

Morton, G.J., Cummings, D.E., Baskin, D.G., Barsh, G.S. and Schwartz, M.W. (2006) Central nervous system control of food intake and body weight. *Nature* 443, 289–295.

Niswender, K.D. and Schwartz, M.W. (2003) Insulin and leptin revisited: adiposity signals with overlapping physiological and intracellular signaling capabilities. *Frontiers in Neuroendocrinology* 24, 1–10.

Nogueiras, R., Wiedmer, P., Perez-Tilve, D., Veyrat-Durebex, C., Keogh, J.M., Sutton, G.M., Pfluger, P.T., Castaneda, T.R., Neschen, S., Hofmann, S.M., Howles, P.N., Morgan, D.A., Benoit, S.C., Szanto, I., Schrott, B., Schürmann, A., Joost, H.G., Hammond, C., Hui, D.Y., Woods, S.C., Rahmouni, K., Butler, A.A., Farooqi, I.S., O'Rahilly, S., Rohner-Jeanrenaud, F. and Tschöp, M.H. (2007) The central melanocortin system directly controls peripheral lipid metabolism. *Journal of Clinical Investigation* 117, 3475–3488.

Patel, H.R., Qi, Y., Hawkins, E.J., Hileman, S.M., Elmquist, J.K., Imai, Y. and Ahima, R.S. (2006) Neuropeptide Y deficiency attenuates responses to fasting and high-fat diet in obesity-prone mice. *Diabetes* 55, 3091–3098.

Peters, J.H., McKay, B.M., Simasko, S.M. and Ritter, R.C. (2005) Leptin-induced satiation mediated by abdominal vagal afferents. *American Journal of Physiology – Regulatory, Integrative and Comparative Physiology* 288, R879–R884.

Peters, J.H., Simasko, S.M. and Ritter, R.C. (2007) Leptin analog antagonizes leptin effects on food intake and body weight but mimics leptin-induced vagal afferent activation. *Endocrinology* 148, 2878–2885.

Petrovich, G.D., Setlow, B., Holland, P.C. and Gallagher, M. (2002) Amygdalo–hypothalamic circuit allows learned cues to override satiety and promote eating. *Journal of Neuroscience* 22, 8748–8753.

Pinto, S., Roseberry, A.G., Liu, H., Diano, S., Shanabrough, M., Cai, X., Friedman, J.M. and Horvath, T.L. (2004) Rapid rewiring of arcuate nucleus feeding circuits by leptin. *Science* 304, 110–115.

Pliquett, R.U., Führer, D., Falk, S., Zysset, S., von Cramon, D.Y. and Stumvoll, M. (2006) The effects of insulin on the central nervous system – focus on appetite regulation. *Hormone and Metabolic Research* 38, 442–446.

Prolo, P., Wong, M.L. and Licinio, J. (1998) Leptin. *International Journal of Biochemistry and Cell Biology* 30, 1285–1290.

Ramos, E.J., Xu, Y., Romanova, I., Middleton, F., Chen, C., Quinn, R., Inui, A., Das, U. and Meguid, M.M. (2003) Is obesity an inflammatory disease? *Surgery* 134, 329–335.

Richard, D., Huang, Q. and Timofeeva, E. (2000) The corticotropin-releasing hormone system in the regulation of energy balance in obesity. *International Journal of Obesity* 24, 36–39.

Richard, D., Lin, Q. and Timofeeva, E. (2002) The corticotropin-releasing factor family of peptides and CRF receptors: their roles in the regulation of energy balance. *European Journal of Pharmacology* 440, 189–197.

Rios, M., Fan, G., Fekete, C., Kelly, J., Bates, B., Kuehn, R., Lechan, R.M. and Jaenisch, R. (2001) Conditional deletion of brain-derived neurotrophic factor in the postnatal brain leads to obesity and hyperactivity. *Molecular Endocrinology* 15, 1748–1757.

Sahu, A. (2004) Minireview: a hypothalamic role in energy balance with special emphasis on leptin. *Endocrinology* 145, 2613–2620.

Sarkar, S., Legradi, G. and Lechan, R.M. (2002) Intracerebroventricular administration of α-melanocyte-stimulating hormone increases phosphorylation of CREB in TRH- and CRH-producing neurons of the hypothalamic paraventricular nucleus. *Brain Research* 945, 50–59.

Savontaus, E., Breen, T.L., Kim, A., Yang, L.M., Chua, S.C. and Wardlaw, S.L. (2004) Metabolic effects of transgenic melanocyte-stimulating hormone overexpression in lean and obese mice. *Endocrinology* 145, 3881–3891.

Scarpace, P.J., Nicolson, M. and Matheny, M. (1998) UCP2, UCP3 and leptin gene expression: modulation by food restriction and leptin. *Journal of Endocrinology* 159, 349–357.

Schulz, C. and Lehnert, H. (2004) Central nervous and metabolic effects of intranasally applied leptin. *Endocrinology* 145, 2696–2701.

Schwartz, M.W., Baskin, D.G., Kaiyala, K.J. and Woods, S.C. (1999) Model for regulation of energy balance and adiposity by the central nervous system. *American Journal of Clinical Nutrition* 69, 584–596.

Schwartz, M.W., Woods, S.C., Seeley, R.J., Barsh, G.S., Baskin, D.G. and Leibel, R.L. (2003) Is the energy homeostasis system inherently biased toward weight gain? *Diabetes* 52, 232–238.

Seeley, R.J. (2005) More neurons, less weight. *Nature Medicine* 11, 1276–1278.

Seeley, R.J. and Woods, S.C. (2003) Monitoring of stored and available fuel by the CNS: implications for obesity. *Nature Reviews Neuroscience* 4, 901–909.

Seeley, R.J., Yagaloff, K.A., Fisher, S.L., Burn, P., Thiele, T.E., van Dijk, G., Baskin, D.G. and Schwartz, M.W. (1997) Melanocortin receptors in leptin effects. *Nature* 390, 349.

Shimizu, H., Inoue, K. and Mori, M. (2007) The leptin-dependent and -independent melanocortin signaling system: regulation of feeding and energy expenditure. *Journal of Endocrinology* 193, 1–9.

Tabarin, A., Diz-Chaves, Y., Consoli, D., Monsaingeon, M., Bale, T.L., Culler, M.D., Datta, R., Drago, F., Vale, W.W., Koob, G.F., Zorrilla, E.P. and Contarino, A. (2007) Role of the corticotropin-releasing factor receptor type 2 in the control of food intake in mice: a meal pattern analysis. *European Journal of Neuroscience* 26, 2303–2314.

Tartaglia, L.A., Dembski, M., Weng, X., Deng, N., Culpepper, J., Devos, R., Richards, G.J., Campfield, L.A., Clark, F.T., Deeds, J., Muir, C., Sanker, S., Moriarty, A., Moore, K.J., Smutko, J.S., Mays, G.G., Woolf, E.A., Monroe, C.A. and Tepper, R.I. (1995) Identification and expression cloning of a leptin receptor, OB-R. *Cell* 83, 1263–1271.

Tilg, H. and Moschen, A.R. (2006) Adipocytokines: mediators linking adipose tissue, inflammation and immunity. *Nature Reviews Immunology* 6, 772–783.

Tsuruta, Y., Yoshimatsu, H., Hidaka, S., Kondou, S., Okamoto, K. and Sakata, T. (2002) Hyperleptinemia in *Ay/a* mice upregulates arcuate cocaine- and amphetamine-regulated transcript expression. *American Journal of Physiology – Endocrinology and Metabolism* 282, 967–973.

Tu, Y., Thupari, J.N., Kim, E.-K., Pinn, M.L., Moran, T.H., Ronnett, G.V. and Kuhajda, F.P. (2005) C75 alters central and peripheral gene expression to reduce food intake and increase energy expenditure. *Endocrinology* 146, 486–493.

Unger, T.J., Calderon, G.A., Bradley, L.C., Sena-Esteves, M. and Rios, M. (2007) Selective deletion of Bdnf in the ventromedial and dorsomedial hypothalamus of adult mice results in hyperphagic behavior and obesity. *Journal of Neuroscience* 27, 14265–14274.

Valassi, E., Scacchi, M. and Cavagnini, F. (2008) Neuroendocrine control of food intake. *Nutrition, Metabolism and Cardiovascular Diseases* 18, 158–168.

Watt, M.J., Dzamko, N., Thomas, W.G., Rose-John, S., Ernst, M., Carling, D., Kemp, B.E., Febbraio, M.A. and Steinberg, G.R. (2006) CNTF reverses obesity-induced insulin resistance by activating skeletal muscle AMPK. *Nature Medicine* 12, 541–548.

Woods, S.C. (2005) Signals that influence food intake and body weight. *Physiology and Behavior* 86, 709–716.

Xu, B., Goulding, E.H., Zang, K., Cepoi, D., Cone, R.D., Jones, K.R., Tecott, L.H. and Reichardt, L.F. (2003) Brain-derived neurotrophic factor regulates energy balance downstream of melanocortin-4 receptor. *Nature Neuroscience* 6, 736–742.

Yaswen, L., Diehl, N., Brennan, M.B. and Hochgeschwender, U. (1999) Obesity in the mouse model of pro-opiomelanocortin deficiency responds to peripheral melanocortin. *Nature Medicine* 5, 1066–1070.

Yeo, G.S., Connie Hung, C.C., Rochford, J., Keogh, J., Gray, J., Sivaramakrishnan, S., O'Rahilly, S. and Farooqi, I.S. (2004) A *de novo* mutation affecting human TrkB associated with severe obesity and developmental delay. *Nature Neuroscience* 7, 1187–1189.

Zabolotny, J.M., Bence-Hanulec, K.K., Stricker-Krongrad, A.S., Haj, F., Wang, Y., Minokoshi, Y., Kim, Y.B., Elmquist, J.K., Tartaglia, L.A., Kahn, B.B. and Neel, B.G. (2002) PTP-1B regulates leptin signal transduction *in vivo*. *Developmental Cell* 2, 489–495.

Zorrilla, E.P., Inoue, K., Valdez, G.R., Tabarin, A. and Koob, G.F. (2005) Leptin and postprandial satiety: acute central leptin more potently reduces meal frequency than meal size in the rat. *Psychopharmacology* 177, 324–335.

3 Newcomers and Supporting Actors

JOANNE A. HARROLD[1] AND GARETH WILLIAMS[1,2]

[1]*Neuroendocrine and Obesity Biology Unit, Department of Medicine, University of Liverpool, UK;* [2]*School of Medicine and Dentistry, University of Bristol, UK*

Introduction

For over half a century, scientific research has been motivated by the quest to achieve an understanding of the regulation of energy homeostasis. The identification of various circulating humoral signals, which indicate the body's energy status, and specific neural circuits that can sense and respond appropriately to these signals has improved our level of understanding considerably (Gao and Horvath, 2007, 2008; Harrold and Halford, 2007; Atkinson, 2008). However, some players that appeared to fulfil central roles and provide promising therapeutic targets some years ago – e.g. leptin and NPY Y5 antagonists – now appear to have fallen short of expectations.

As obesity is reaching pandemic proportions, this continues to be one of the most exciting and rapidly advancing topics in biomedical research. The present chapter considers a number of the rising new peptides that have been shown to play a role in the regulation of feeding behaviour, either directly or by influencing the actions of established mediators.

Newcomers

Neuropeptides B and W

Orphan receptors

G protein-coupled receptors (GPCRs) are key regulators of intercellular interactions, participating in almost every physiological response. They exert their effects by being activated by a variety of endogenous ligands. Traditionally, these ligands were identified first, providing tools to characterize the receptors. However, since the late 1980s, homology screening approaches have allowed the GPCRs to be found first and, in turn, used as orphan targets to identify their ligands. Over the past decade, this method has led to the identification of 12 novel neuropeptide families. Interestingly, four of these deorphanized GPCR systems, melanin-concentrating hormone, ghrelin, orexin and neuropeptide B/neuropeptide W, have been found to play a role in the control of energy balance (Harrold and Halford, 2007). Notable examples have included the orexin peptides that regulate energy homeostasis in the short term (Cai *et al.*, 2002). However, in the absence of the appropriate ligands, characterization of the receptors themselves can hint at the physiological roles that the ligands may play.

GPR7 and GPR8 are two orphan GPCRs which were cloned originally from human genomic DNA. They share high sequence identity with each other; the human receptors are 70% homologous but no rodent orthologue of GPR8 has yet been identified (Lee *et al.*, 1999; Singh and Davenport, 2006; Rucinski *et al.*, 2007). The receptors are also significantly similar to opioid and somatostatin receptors – in fact, they have the highest protein sequence identity with these receptors at approximately 40% identity overall (O'Dowd *et al.*, 1995). However, only the GPR7 binds a non-selective opioid ligand with low affinity, and neither binds somatostatin (O'Dowd *et al.*, 1995).

The expression pattern of mRNAs for both GPR7 and GPR8 overlaps with that of opioid and somatostatin receptors in regions including the hypothalamus and hippocampus. However, in strong contrast to the wider distribution of the opioid and somatostatin receptors, GPR7 shows restricted and discrete distribution in the rat. In general, GPR7 displays moderate expression in the hippocampus and dense expression in the hypothalamus (including the suprachiasmatic, paraventricular (PVN), ventromedial (VMH), dorsomedial (DMH) and arcuate (ARC) nuclei), with scattered mRNA also in the olfactory cortex. Some of the amygdaloid nuclei also express GPR7 mRNA (Lee *et al.*, 1999).

Endogenous ligands

The observations that GPR7 and GPR8 are expressed in distinct CNS areas and have marked similarity to opioid and somatostatin receptors provide strong evidence for the existence of endogenous receptor ligands in the CNS that potentially play a role in analgesia and the modulation of neuroendocrine effects, including the regulation of feeding (Kelly *et al.*, 2005). In 2002, a family of endogenous neuropeptide ligands for GPR7 and GPR8 were purified and characterized (Fujii *et al.*, 2002; Shimomura *et al.*, 2002). Neuropeptide B (NPB) was purified from bovine hypothalamic extract. It is a 29-amino acid peptide with a C-6-bromated tryptophan at the N-terminus. The role of bromination is unclear as non-brominated NPB essentially has the same potency and efficacy as the brominated peptide at GPR7 and GPR8 (Tanaka *et al.*, 2003).

In situ hybridization shows discrete localization of the preproNPB mRNA with expression in the PVN, hippocampus and several midbrain and brainstem nuclei (Tanaka *et al.*, 2003). Its expression, along with that of GPR7 and GPR8, in these hypothalamic nuclei, implies a role in the regulation of feeding. Examination of NPB's actions has shown that intracerebroventricular (ICV) administration of the peptide in the dark phase induces short-lasting hyperphagia, followed by a delayed but significantly more pronounced anorexic action. Additionally, NPB has been found to enhance the anorexic action of CRF on coadministration. It has been hypothesized that these late hypophagic effects of NPB arise from its activity in the limbic system, where it may influence the hedonic aspects of eating (Tanaka *et al.*, 2003). Interestingly, mouse knockout models of the gene encoding either NPB or GPR7 have a gender-specific phenotype, with moderate obesity evident in males but not females (Singh and Davenport, 2006).

Expressed sequence tag (EST) database searches identified another isopeptide of the same family, which was named neuropeptide W (NPW; Fujii *et al.*, 2002; Shimomura *et al.*, 2002; Brezillon *et al.*, 2003). As for NPB, NPW's peptide sequence demonstrates no convincing similarity with any known peptide, including opioid and somatostatin. NPW also has an N-terminal tryptophan, and the presence of two arginine residues within the sequence of the 30-amino acid peptide (NPW30) indicates a potential cleavage site, as confirmed by identification of a shorter 23-amino acid product (NPW23).

NPW mRNA was found to show an expression pattern distinct from that of NPB, being confined to midbrain nuclei, and particularly the dorsal raphe nucleus (Tanaka *et al.*, 2003). This, along with the observation that NPB has a slightly higher potency than both NPW sequences at GPR7 (while NPW demonstrates a significantly greater potency than NPB at GPR8), suggests separate roles for the two peptides in the CNS. However, immunohistochemical localization of NPW, using an antibody raised against the sequence of NPW23, identified NPW in cell bodies within the rat hypothalamus (including the PVN, ARC and lateral hypothalamic area (LHA)) and ependymal cells lining the third and lateral ventricles. While NPW-containing processes generally were distributed sparingly throughout the hypothalamus, a dense network was observed in the median eminence (Dun *et al.*, 2003). This distribution is consistent with NPW also playing a role in the regulation of energy homeostasis. In fact, NPW but not NPB reportedly exerts a potent suppressive effect on blood leptin concentrations in the rat, and this mechanism may be related to the involvement of NPW in energy balance regulation (Rucinski *et al.*, 2007).

Central administration of NPW23 and NPW30 to rats has been shown to suppress dark-phase and fasting-induced food intake. Furthermore, these effects are maintained with continuous ICV infusion (Mondal *et al.*, 2003). Conversely, ICV administration of anti-NPW antibodies stimulates food intake (Mondal *et al.*, 2003). In conjunction with the observations that NPW increases body temperature and heat production, this strongly implies that NPW functions as a catabolic signal in the brain.

The precise role of the two orphan GPCRs in the feeding-related actions of NPB and NPW are being unravelled. Mice with a targeted deletion of GPR7 have been reported to develop maturity-onset obesity that is exacerbated by high-fat feeding (Ishii *et al.*, 2003). Obesity develops as a consequence of hyperphagia and reduced energy expenditure. Unlike *ob/ob* mice, neuropeptide Y mRNA levels are decreased in these $GPR7^{-/-}$ animals, while pro-opiomelanocortin (POMC) levels are increased, which is consistent with the observation of elevated plasma glucose, leptin and insulin levels (Ishii *et al.*, 2003). Furthermore, *ob/ob* $GPR7^{-/-}$ and *Ay/a* $GPR7^{-/-}$ double mutants have an increased body weight compared with single mutant animals, implying that obesity occurs independently of leptin and melanocortin signalling. The latter observation was unique to male mice, indicating that GPR7 signalled in a sexually dimorphic manner.

Futher investigation of the biochemical and physiological functions of GPR7, GPR8 and their endogenous ligands will help to elucidate better their precise role in the hypothalamic regulation of energy homeostasis.

Ghrelin

Ghrelin is a hormone secreted predominantly from the stomach, although low levels of ghrelin have also been identified in the brain, particularly within the hypothalamus (Cowley *et al.*, 2003). Its structure is highly conserved among species, suggesting an important physiological role. Ghrelin is known to act as an endogenous ligand for the growth hormone (GH) secretagogue receptor (Kojima *et al.*, 2001), which is expressed in hypothalamic and brainstem nuclei (including the ARC). These distributions of ghrelin and its receptor are consistent with the peptide playing a role in the regulation of energy homeostasis.

Ghrelin and feeding in rodents

In rodents, ghrelin administered either centrally or peripherally stimulates food intake potently (Tshöp *et al.*, 2000; Nakazato *et al.*, 2001; Wren *et al.*, 2001). This effect is blocked by administration of ghrelin antibodies or ghrelin receptor antagonists (Nakazato *et al.*, 2001). Ghrelin's ability to stimulate feeding compares with that of neuropeptide Y (NPY), one of the most potent central appetite stimulants known (Cummings and Schwartz, 2003). Chronic ghrelin administration also enhances weight gain without attenuation of the effects on food intake. Significantly, the weight gain appears to be due to increased fat mass, without changes in lean mass or longitudinal growth (Tschöp *et al.*, 2000). Importantly, the weight gain and adiposity occur independently of ghrelin's ability to modulate GH secretion (Tschöp *et al.*, 2000; Nakazato *et al.*, 2001; Wren *et al.*, 2001).

In contrast to cholecystokinin (CCK) and glucagon-like peptide-1 (GLP-1) (both satiety signals whose release is evoked by eating), plasma ghrelin concentrations rise during fasting and promptly fall after eating. Weight loss and insulin-induced hypoglycaemia also increase peptide secretion and mRNA expression (Tschöp *et al.*, 2000; Toshinai *et al.*, 2001; Cummings *et al.*, 2002). This implies that ghrelin may act as a signal of nutritional status directly from the gut (Horvath *et al.*, 2001). This is supported by the observation that oral administration of the glucose antimetabolite, 2-deoxyglucose, stimulates ghrelin secretion (Cai *et al.*, 2004; Fig. 3.1).

Furthermore, fasting ghrelin levels are suppressed by an oral glucose load but not altered by the same volume of water, demonstrating that nutrient content rather than gastric distension is a crucial signal (Tschöp *et al.*, 2000). Ghrelin administration also results in other local gut effects in addition to its actions on appetite, stimulating gastric emptying and decreasing gastric acid content in rodents (Masuda *et al.*, 2000).

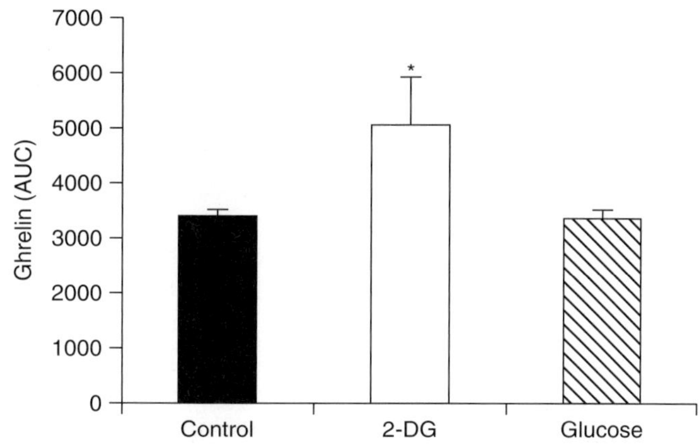

Fig. 3.1. Plasma ghrelin levels in rats 2 h after being dosed with 2 ml of water (control), 2-deoxyglucose (2 mM) or glucose (2 mM) by gavage. Data shown as mean ± SEM from groups of eight rats. * $P < 0.05$ versus controls. 2-DG, 2-deoxyglucose.

Although ghrelin is secreted from the stomach and circulates in the blood, the orexigenic effects of the peptide are proposed to be mediated by hypothalamic circuitry in the CNS, particularly the NPY/agouti-gene related peptide (AgRP) neurones. Ghrelin has been reported to increase the expression of mRNA for NPY and AgRP and to trigger the expression of immediate early genes in the PVN and ARC (Dickson and Luckman, 1997; Hewson and Dickson, 2000). Furthermore, approximately 90% of these Fos-positive neurones in the ARC also express NPY mRNA (Wang *et al.*, 2002). Additionally, ghrelin has been shown to interact directly with NPY neurones of the ARC, inducing Ca^{2+} signalling (Kohno *et al.*, 2003). However, ghrelin-induced food intake is only partly blocked by coadministration of an NPY Y5 (Bagnasco *et al.*, 2003) or Y1 (Chen *et al.*, 2004) antagonist, indicating that additional factors mediate the orexigenic ghrelin signal. The observation that knockout of either NPY or AgRP leads to partial attenuation of ghrelin-induced feeding, whereas combined knockout of both obliterates the peptide's feeding effects completely, suggests that AgRP also partly mediates ghrelin's action (Chen *et al.*, 2004). Interestingly, it has also been shown that leptin suppresses ghrelin-induced activation of NPY neurones in the ARC (Kohno *et al.*, 2007). While ghrelin increases intracellular Ca^{2+} concentrations via mechanisms depending on phospholipase C and adenylate cyclase-PKA pathways in ARC NPY neurones, leptin counteracts the ghrelin responses via a phosphatidylinositol 3-kinase-PDE3 pathway (Kohno *et al.*, 2007). This interaction may play an important role in regulating ARC NPY neurone activity. Moreover, the action of ghrelin has been shown to be attenuated in neural-specific POMC-deficient mice (Tolle and Low, 2008).

The orexin system, arising from cell bodies in the LHA, is another potential mediator (Toshinai *et al.*, 2003) and injection of ghrelin into the LHA is reported to induce c-fos expression in orexin-containing neurones of this nucleus

(Olszewski *et al.*, 2003). Ghrelin enhances feeding via the neuronal pathways of NPY and orexin, which act as orexigenic peptides in the hypothalamus. Ghrelin molecules exist in the stomach and hypothalamus as two major endogenous forms, a form acylated at serine 3 (ghrelin) and a des-acylated form (des-acyl ghrelin). Acylation is indispensable for the binding of ghrelin to the GH secretagogue type 1a receptor (GHS-R1a). Des-acyl ghrelin induces food intake by a mechanism independent of the growth hormone secretagogue receptor, increasing the intracellular calcium concentrations in isolated orexin neurones, perhaps operating in feeding regulation through interactions with a target protein distinct from the GHS-R (Toshinai *et al.*, 2006). Recently, it has been further shown that the orexigenic effect of ghrelin is mediated through central activation of the endogenous cannabinoid system (Kola *et al.*, 2008). An intact cannabinoid signalling pathway is necessary for the stimulatory effects of ghrelin on AMPK activity and food intake, and for the inhibitory effect of ghrelin on PVN neurones.

The activation of central circuitry by circulating ghrelin requires the transfer of the peptide across the blood–brain barrier. However, while ghrelin is transported readily in the brain-to-blood direction in a mouse model, the quantity of transport into the brain appears negligible (Banks *et al.*, 2002). Furthermore, vagotomy has been shown to prevent peripheral ghrelin's influence on the hypothalamus (Date *et al.*, 2002). This suggests that the peptide's effect on the brain is of intrinsic origin. It was reported initially that ghrelin-containing neurones were restricted to the ARC (Lu *et al.*, 2002). However, subsequent observations located ghrelin immunoreactive neurones in a continuum filling the internuclear space between the PVN, VMH and DMH and the ependymal layer of the third ventricle (Cowley *et al.*, 2003). This unique distribution does not overlap with any of the known appetite-regulating hypothalamic cell populations. Ghrelin-binding sites are also present in the hypothalamus (Cowley *et al.*, 2003; Harrold *et al.*, 2004) (Fig. 3.2). This distribution of ghrelin and its receptors argues for a novel central role for the peptide, in addition to its role as a peripheral, gut-secreted hormone.

Knockout experiments, however, indicate that ghrelin is not critical for feeding performance. Ghrelin-deficient mice exhibited normal growth rates, as well as normal spontaneous food intake patterns, memory-related feeding performances, normal basal levels of hypothalamic orexigenic and anorexigenic neuropeptides and no impairment of reflexive hyperphagia after fasting (Wortley *et al.*, 2004; Sato *et al.*, 2008). These results indicate that endogenous ghrelin does not appear to be an essential regulator of food intake, having a redundant role in the regulation of appetite. Double knockout (DBKO) mice simultaneously lacking the ghrelin and ghrelin receptor genes reportedly exhibited decreased body weight, increased energy expenditure and increased motor activity on a standard diet without exposure to a high caloric environment (Pfluger *et al.*, 2008). Mice on the same genetic background lacking either the ghrelin or the ghrelin receptor gene did not exhibit such a phenotype on standard chow, thereby confirming earlier reports. No differences in food intake, meal pattern or lean mass were observed between DBKO, ghrelin-deficient, ghrelin receptor-deficient and wild-type control mice. Only DBKO mutants showed a slight decrease in body length. Thus, simultaneous deletion of ghrelin and its receptor enhances

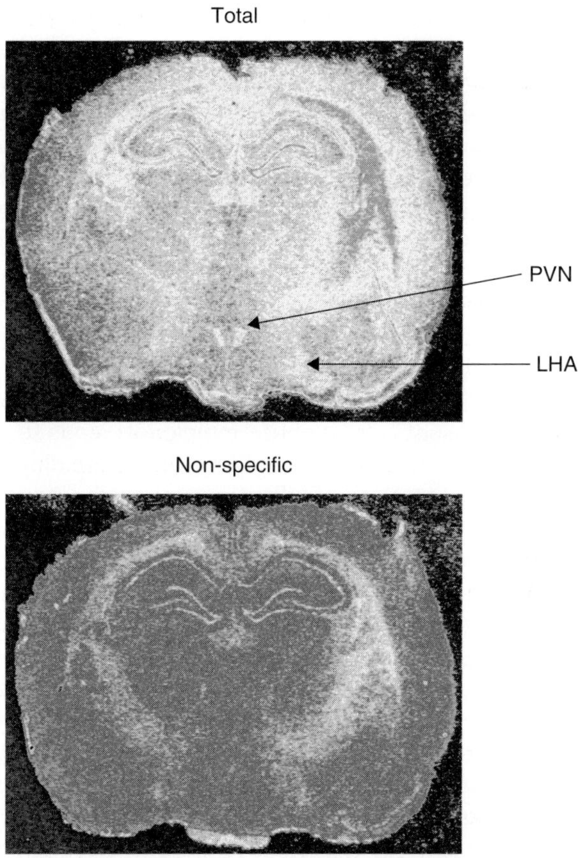

Fig. 3.2. Pseudocolour image of an autoradiogram following exposure to the ligand $[^{125}\text{I-}\textit{His}^{9}]$-ghrelin either in the absence (total) or presence (non-specific) of excess unlabelled ghrelin. PVN, paraventricular nucleus; LHA, lateral hypothalamus.

the metabolic phenotype of single gene-deficient mice compared with wild types, possibly suggesting the existence of additional, as yet unknown, molecular components of the endogenous ghrelin system (Pfluger *et al.*, 2008).

Ghrelin and feeding in humans

Human data also support a role for ghrelin in appetite regulation. As in rodents, plasma ghrelin levels are high in the fasted state, rise progressively between meals and fall in response to feeding in lean individuals (Ariyasu *et al.*, 2001; Cummings *et al.*, 2001; Cummings, 2006; Chaudhri *et al.*, 2008). Ghrelin also enhances appetite in humans, increasing subjective hunger and enhancing food intake (by up to 28%) when administered intravenously to healthy subjects (Wren *et al.*, 2001). Conversely, a fall in ghrelin levels after gastric bypass surgery (levels of ghrelin in gastrectomy patients are around 35% of age-matched controls) may contribute to weight loss in these patients (Ariyasu *et al.*, 2001). Circulating levels decrease with feeding and increase before meals, achieving concentrations sufficient to stimulate hunger and food intake. Preprandial ghrelin surges occur before every meal on various fixed feeding schedules and also among individuals initiating meals voluntarily without time- or food-related cues (Cummings, 2006). Postprandial suppression is not mediated by nutrients in the stomach or duodenum, where most ghrelin is produced. Rather, it results from post-ingestive increases in lower intestinal osmolarity (information probably relayed to the foregut via enteric nervous signalling), as well as from insulin surges (Cummings, 2006).

It has been proposed that ghrelin plays a role in the aetiology of human obesity. The peptide has an inverse relationship with body mass index, with levels being elevated significantly in anorexic individuals (Ariyasu *et al.*, 2001; Shiiya *et al.*, 2002) and lowered in obese subjects (Shiiya *et al.*, 2002; Chaudhri *et al.*, 2008). Weight loss in obese individuals leads to an elevation in ghrelin levels, which may add to the problem of maintaining low body weight after weight loss (Hansen *et al.*, 2002). Conversely, the postprandial suppression of ghrelin, normally observed in lean subjects, appears to be absent in obese individuals, which may contribute to overeating (English *et al.*, 2002). Additionally, individuals with Prader–Willi syndrome also have significantly elevated ghrelin levels and this may contribute to the dramatic hyperphagia that leads to morbid obesity in this condition (DelParigi *et al.*, 2002). Furthermore, polymorphisms of the ghrelin gene may contribute to the genetic predisposition in some cases of human obesity. However, a role for these in weight determination remains controversial (Hinney *et al.*, 2002; Korbonits *et al.*, 2002; Wang *et al.*, 2004; Chaudhri *et al.*, 2008). Considerable evidence implicates ghrelin in both short-term meal initiation and long-term energy homeostasis, thus making it an attractive target for drugs to treat obesity and/or wasting disorders.

Obestatin

On the basis of a bioinformatic prediction, it was discovered that the gene precursor of ghrelin encoded another secreted and bioactive peptide, which was named obestatin, a contraction of obese, from the Latin 'obedere', meaning to devour, and 'statin', denoting suppression (Zhang *et al.*, 2005). First reports appear to demonstrate that this peptide requires an amidation for its biological activity and acts through the orphan receptor, GPR39. Obestatin was shown to have actions opposite to ghrelin on appetite, body weight and gastric emptying, with treatment of rats with obestatin suppressing food intake, inhibiting jejunal contraction and decreasing body weight gain (Zhang *et al.*, 2005). Thus, two peptide hormones with opposed actions seemed to be derived from the same ghrelin gene and acting via distinct receptors (Zhang *et al.*, 2005). However, subsequent studies have failed to observe any effect of obestatin on food intake, body weight, body composition, energy expenditure, locomotor activity, respiratory quotient or hypothalamic neuropeptides involved in energy balance regulation (Nogueiras *et al.*, 2007). Moreover, no mRNA expression of GPR39, the putative obestatin receptor, was observed in the hypothalamus of rats. Therefore, these findings do not support the role of the obestatin/GPR39 system in the regulation of energy balance. The effect of obestatin has been investigated mostly in fasted rodents, a condition associated with high blood levels of ghrelin, which may mask the effect of obestatin. To address the potential interference with ghrelin, the effects of obestatin in ghrelin knockout mice have been studied (Depoortere *et al.*, 2008). Obestatin failed to affect food intake and gastric motility in ghrelin null mice, suggesting that endogenous ghrelin does not mask the effect of obestatin and confirming that obestatin administered peripherally is not a major regulator of satiety signalling or gut motility.

It is noteworthy that GPR39 has been shown to be expressed in human adipose tissue (Catalán *et al.*, 2007). Interestingly, the mRNA expression levels of GPR39 correlated negatively with fasting glucose concentrations, while exhibiting a positive correlation with adiponectin mRNA expression levels. These findings, together with the reduced expression levels of GPR39 observed in omental adipose tissue of obese type 2 diabetic patients, suggest a potential involvement of obestatin signalling in glucose homeostasis (Catalán *et al.*, 2007).

Peptide YY

Peptide YY (PYY), named as a consequence of tyrosine (Y) residues at its amino and carboxy terminals, is a 36-amino acid peptide isolated originally from porcine intestine in 1980 (Tatemoto and Mutt, 1980). It is one of a family of peptides cleaved from the pre-propeptide that includes NPY. PYY is present mainly in L-cells of the gastrointestinal tract and is released in response to feeding in proportion to the caloric content of the meal (Adrian *et al.*, 1985; Pederson-Bjergaard *et al.*, 1996). The release of PYY also reflects the nature of the ingested food, with fat reported to be the most potent nutrient in terms of peptide release. PYY has also been reported to exist in the CNS, although at lower concentrations, with PYY-containing neurones identified in hypothalamic and hindbrain regions (Lundberg *et al.*, 1984).

The 70% structural identity between PYY and NPY, along with their similar CNS distributions, predicts that the two peptides may elicit similar biological responses. This has been corroborated with the observation that PYY produces profound hyperphagia when injected ICV (Morley *et al.*, 1985). Furthermore, multiple injections do not attenuate this effect (Morley *et al.*, 1985). However, unlike NPY, the known functions of PYY are limited to ingestive and gastrointestinal regulation, suggesting that PYY acts primarily to drive feeding.

PYY_{3-36}

PYY exists in two endogenous forms: PYY_{1-36} and PYY_{3-36}. The latter is derived by proteolytic cleavage of PYY through the actions of the enzyme dipeptidyl peptidase-IV (DPP-IV). The percentage of the two forms in the circulation has been reported to differ according to feeding status. In the fasted state in humans, PYY_{1-36} is the dominant form. However, after a meal, PYY_{3-36} is the major circulating form (Grandt *et al.*, 1994).

PYY$_{3-36}$ AND FEEDING IN RODENTS. In striking contrast to the full-length peptide, the N-terminal truncated form of PYY has been shown to decrease appetite when administered either peripherally (at doses that achieve plasma levels within the normal postprandial range) or centrally (Batterham *et al.*, 2002). Additionally, chronic peripheral administration results in reductions in cumulative food intake and body weight, without any apparent attenuation of effects on feeding (Batterham *et al.*, 2002). Furthermore, peripheral administration of PYY$_{3-36}$ induces c-fos expression markedly in the ARC, whereas direct injection of the peptide into this area also inhibits food intake (Batterham *et al.*, 2002). Compared with mice fed a low-fat diet, the high-fat group exhibited lower endogenous plasma PYY and higher tissue PYY but similar PYY mRNA levels, suggesting a possible reduction of PYY release (Le Roux *et al.*, 2006). High-fat-fed mice remained sensitive to the anorectic effects of exogenous intraperitoneal PYY$_{3-36}$. Fasting and postprandial endogenous plasma PYY levels were shown to be attenuated in both obese rodents and humans. The PYY$_{3-36}$ infusion experiments showed that the degree of plasma PYY reduction in obese subjects was likely associated with decreased satiety and relatively increased food intake (Le Roux *et al.*, 2006).

The marked differences in the effects of the two forms of PYY on feeding appear to arise as a consequence of differing affinities of NPY receptors for the peptides. PYY$_{1-36}$ binds to and activates at least three NPY receptor subtypes (Y1, Y2 and Y5), while PYY$_{3-36}$ demonstrates preference for the Y2 receptor. The NPY Y2 receptor mediates the anorectic actions of PYY$_{3-36}$ with rodent studies, implicating the hypothalamus, vagus and brainstem as key target sites. The role of the NPY Y2 receptor in the regulation of food intake has been highlighted through the administration of a selective Y2 agonist. Injection of the agonist into the ARC of rats, at the start of the dark phase, reduces feeding significantly. This effect persists for up to 8 h post-administration (Batterham *et al.*, 2002). Direct administration of the Y2 agonist into the ARC also reduces fasting-induced hyperphagia. However, injection into the PVN has no effect on feeding (Batterham *et al.*, 2002). The role of the Y2 receptor in PYY$_{3-36}$-mediated hypophagia has also been illustrated, with the observation that the effects of the peptide on food intake and body weight are absent in Y2 knockout mice (Batterham *et al.*, 2002).

Y2 receptors are autoreceptors located on NPY/AgRP neurones in the ARC and it has been reported that expression of NPY mRNA is decreased in the ARC by peripheral administration of PYY$_{3-36}$. The addition of the peptide to *ex vivo* hypothalamic explants also inhibits the release of NPY (Batterham *et al.*, 2002). Furthermore, electrophysiological studies suggest that activation by PYY$_{3-36}$ of Y2 receptors on the NPY/AgRP neurones inhibits these cells (Batterham *et al.*, 2002). Both the hypothalamus and medulla oblongata express a high level of Y2 receptors. Diet-induced obese mice have low plasma PYY concentrations, which are accompanied by a compensatory upregulation of PYY and Y2 receptor densities in the medulla (Rahardjo *et al.*, 2007). A low-level response of PYY-medullary regulation to positive energy balance may contribute to the development of diet-induced obesity. Conversely, a normal response of this regulatory axis in obese-resistant mice may contribute to the maintenance of body weight while on a high-fat diet (Rahardjo *et al.*, 2007).

Furthermore, PYY_{3-36}-mediated hypophagia does not arise solely as a consequence of reduced release of these orexigenic peptides. In fact, PYY_{3-36} has a dual action. By inhibiting NPY/AgRP neurones, it also removes the tonic inhibitory tone which these normally exert on the appetite-suppressing effects of POMC/cocaine- and amphetamine-related transcript (CART) neurones of the ARC. This is supported by the observation that addition of PYY_{3-36} to *ex vivo* hypothalamic explants stimulates the release of the anorexigenic peptide α-melanocyte-stimulating hormone (α-MSH) from POMC neurones (Batterham *et al.*, 2002). Yet, an acute anorectic response to peripheral administration of PYY_{3-36} is maintained in *POMC*$^{-/-}$ mice, suggesting that a reduction of NPY/AgRP neuronal activity may play the more prominent role in PYY_{3-36}-induced hypophagia (Challis *et al.*, 2004).

Evidence for a hedonic role for PYY_{3-36} is supported by studies showing that it decreases the motivation to seek high-fat food (Chandarana and Batterham, 2008). Rodent studies using selective Y2 agonists and strategies combining PYY_{3-36}/Y2 agonists with other anorectic agents have revealed increased anorectic and weight-reducing effects. The emerging hedonic effects of PYY_{3-36}, together with the weight-reducing effects observed in obese rodents, suggest that targeting the PYY system may offer a therapeutic strategy for obesity.

PYY_{3-36} AND FEEDING IN HUMANS. PYY_{3-36} also appears to play a physiological role in humans. It has been shown that double-blind intravenous infusion of the peptide, resulting in plasma levels within the normal postprandial range, into healthy fasted non-obese subjects reduced hunger scores significantly for up to 12 h post-infusion (Batterham *et al.*, 2002). Caloric intake during a free-choice meal provided 2 h post-infusion was also reduced by 36%, and the overall reduction in caloric intake over 24 h was 33% (Batterham *et al.*, 2002, 2003). Supraphysiological doses of PYY_{3-36} have no further benefit in food reduction but may cause nausea (Le Roux *et al.*, 2008).

Functional imaging studies in humans have confirmed that PYY_{3-36} activates brainstem and hypothalamic regions (Chandarana and Batterham, 2008). Interestingly, the greatest effects, however, were observed within the orbitofrontal cortex, a brain region involved in reward processing. As mentioned before, a hedonic role for PYY_{3-36} has been supported also by rodent studies, underscoring the relevant role of the peptide in the reward system.

Similarly to ghrelin, obese subjects have lower basal PYY_{3-36} levels and a blunted response to feeding, with a reduced rise in postprandial levels of the peptide (Batterham *et al.*, 2003). This suggests that abnormalities of the PYY system may be involved in the pathogenesis of obesity. However, unlike the situation with the satiety factor leptin, obesity does not appear to be associated with resistance to PYY_{3-36}, as exogenous infusion of the peptide results in a comparable reduction in food intake in both lean and obese groups (Batterham *et al.*, 2003). Nevertheless, the clinical usefulness of PYY_{3-36} or its analogues in human obesity remains to be clarified by long-term studies of its ability to suppress feeding and weight gain.

Endocannabinoids

In the past decade, cannabinoid receptors and their putative ligands have been discovered within the CNS and linked to a number of aspects of feeding behaviour. Recently, interest in the effect on appetite has revived, with research suggesting that endocannabinoids may be key to the hedonic aspects of eating, possibly mediating the craving for and enjoyment of the most palatable and fattening foods (Harrold *et al.*, 2002; Tucci *et al.*, 2006; Harrold and Halford, 2007).

The cannabinoid system consists of two receptors, their endogenous ligands and the uptake mechanisms and hydrolysing enzymes that regulate ligand levels. The two receptor subtypes are classified as the 'central' CB1 receptor, which is distributed widely in the CNS and many peripheral tissues, and the 'peripheral' CB2 receptor, which is not expressed significantly in the CNS (Breivogel and Childers, 1998). The CB1 receptor is the most abundant GPCR expressed in the brain. It is accepted generally that the influences of cannabinoids on feeding behaviour are mediated by the CB1 receptor, which is expressed at particularly high levels in brain regions (including the hippocampus and basal ganglia) that correspond with cannabinoid-mediated behavioural effects (Glass *et al.*, 1997; Harrold *et al.*, 2002; Kirkham, 2005; Harrold and Halford, 2007).

The existence of specific receptor sites indicates the presence of substances, produced within mammalian tissues, for which the cannabinoid receptors are targets. In 1992, the first endocannabinoid was isolated from porcine brain and termed anandamide, from 'ananda' meaning bliss (Devane *et al.*, 1992; Di Marzo *et al.*, 1998). Subsequent searches for additional ligands identified 2-arachidonoylglycerol (2-AG) (Stella *et al.*, 1997). Anandamide and 2-AG are considered to be the primary ligands at CB1 and CB2 receptors. However, other candidate endocannabinoids have been characterized, including nolandin and virodhamine (Hanus *et al.*, 2001; Porter *et al.*, 2002). These ligands have also been identified in various species. The observation that amphibian, mammalian and human CB1 receptors have a high degree of homology suggests that the cannabinoid signalling system plays an important physiological role. Animal models have provided solid evidence that genetically induced obesity leads to long-lasting overstimulation of endocannabinoid system synthesis, resulting in permanent overactivation of CB1, which may then contribute to the maintenance of this disease. The fact that ablation of CB1 receptors results in mice with a lean phenotype, resistance to dietary-induced obesity and enhanced leptin sensitivity suggests that they represent an important orexigenic component of the energy homeostatic circuitry (Ravinet Trillou *et al.*, 2003; Kirkham, 2005). Moreover, CB1 has also been shown to be relevant in key peripheral organs involved in energy balance regulation, such as the adipose tissue and the gastrointestinal system (Pagotto *et al.*, 2006). Importantly, at the peripheral level, CB1 activation has been shown to stimulate lipogenesis in adipocytes (Pagotto *et al.*, 2006). CB1 blockers increase adiponectin production in adipocytes, which leads to increased fatty acid oxidation and free fatty acid clearance. Moreover, CB1 has been shown to be upregulated in adipocytes derived from obese rodents. These results support the role of endocannabinoids in the development and maintenance of obesity, paving the way for taking advantage of both the central and peripheral effects of the CB1 blockers in tackling obesity and its comorbidities (Pagotto *et al.*, 2006).

Endocannabinoids and hyperphagia

Both exogenous and endogenous (anandamide and 2-AG) cannabinoids stimulate feeding (Williams *et al.*, 1998; Williams and Kirkham, 1999; Hao *et al.*, 2000; Harrold and Halford, 2007). The hyperphagia is powerful; peripheral administration of the exogenous cannabinoid Δ^9-tetrahydrocannabinol (Δ^9-THC) stimulates feeding as potently as does central administration of NPY (Corp *et al.*, 1990).

As the orexigenic effect of the cannabinoid agonists is blocked by the CB1 specific antagonist rimonabant, but not by an antagonist of CB2 receptors, this suggests that these actions on feeding are mediated by the central CB1 receptor. Furthermore, as administration of rimonabant alone suppresses food intake in rodents (Colombo *et al.*, 1998; Rowland *et al.*, 2001), this suggests that tonic endocannabinoid activity at these receptors may be a key component of appetite regulation. This tonic activity is supported further by direct measurements of brain endocannabinoid levels in response to fasting and feeding. Fasting increases levels of anandamide and 2-AG in the nucleus accumbens and, to a lesser extent, in the hypothalamus, whereas 2-AG levels decline in the hypothalamus with feeding (Kirkham *et al.*, 2002; Kirkham, 2005). However, levels in the cerebellum, a region not involved directly in the control of feeding, are not influenced by nutritional status (Kirkham *et al.*, 2002).

There is a body of evidence that points towards involvement of established homeostatic pathways, many of which are regulated by the hormone leptin and operate within hypothalamic nuclei. First, leptin administration decreases hypothalamic levels of anandamide and 2-AG, while endocannabinoid levels in the cerebellum are unaffected (Di Marzo *et al.*, 2001). In addition, anandamide increases Fos expression in the PVN of rodents (Wenger *et al.*, 1997; Patel *et al.*, 1998), while administration into the VMH of satiated rats induces significant hyperphagia (Jamshidi and Taylor, 2001). Furthermore, defective leptin signalling in *ob/ob* and *db/db* mice and *fa/fa* Zucker rats is associated with elevated levels of hypothalamic endocannabinoids, and these are reduced in *ob/ob* mice following leptin treatment (Di Marzo *et al.*, 2001). Moreover, CB1 is expressed in a number of leptin-regulated key hypothalamic peptidergic systems of appetite regulation, including those producing CART in the ARC, and melanin-concentrating hormone and orexin in the LHA (Cota *et al.*, 2003). Finally, evidence has been obtained which indicates an interaction between CB1 receptors and the melanocortin system, with the observation that subanorectic doses of rimonabant and the melanocortin receptor agonist α-MSH attenuate baseline feeding synergistically when combined (Verty *et al.*, 2004).

However, hypothalamic 2-AG levels have been found to increase with food deprivation and decline with feeding (Kirkham *et al.*, 2002), suggesting that, once initiated, eating no longer depends on hypothalamic endocannabinoids for maintenance. Furthermore, CB1 receptor binding in the hypothalamus of dietary obese rats is low and unaltered, and no relationship has been identified between CB1 receptor binding density (in any brain region) and leptin levels in these animals (Harrold *et al.*, 2002). These results are notable, drawing attention away from the hypothalamus and leptin-regulated pathways.

In contrast, significant reductions in CB1 receptor binding density, consistent with increased receptor activity, have been identified in the forebrain and hippocampus following 10 weeks of palatable diet feeding (Harrold *et al.*, 2002; Fig. 3.3). These brain areas either are involved directly in the hedonic aspects of eating or are connected to reward-related brain areas (Finkelstein *et al.*, 1996; Gorbachevskaia, 1999; Pecina and Berridge, 2000). Association of the cannabinoid system with reward processes is supported by a number of other lines of evidence. Rimonabant antagonizes the hunger induced by anandamide and 2-AG. However, it also produces changes in ingestive behaviour when administered alone. Rimonabant inhibits consumption of palatable food and drink selectively, with decreased intakes of sucrose, alcohol and a sweet diet observed in rats, mice and marmosets, respectively (Arnone *et al.*, 1997; Simiand *et al.*, 1998). These results suggest that the central cannabinoid system may act to amplify reward indices. In addition, the cannabinoid system appears to interact with known opioidergic reward pathways, indicated by the synergistic actions of rimonabant or the cannabinoid inverse agonist, AM251, with the opioid receptor antagonists, naloxone and nalmefene, on food intake (Welch and Eads, 1999;

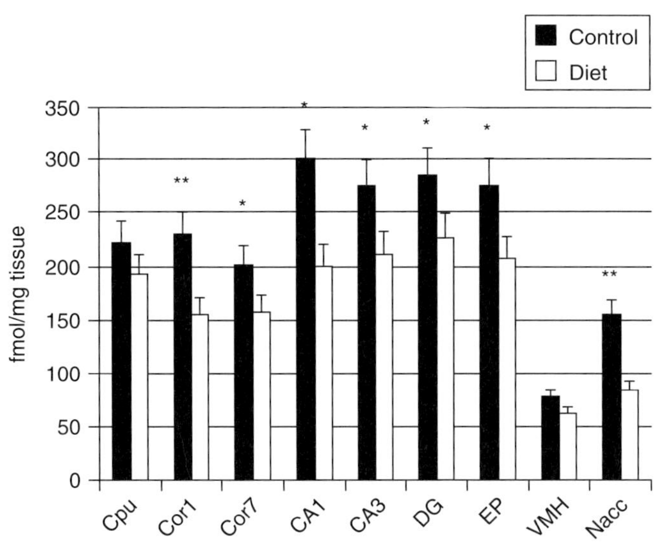

Fig. 3.3. Regional cannabinoid receptor density in rats fed a highly palatable diet and control rats given standard laboratory chow. Data shown as mean ± SEM from groups of 8 rats. * $P < 0.05$ versus controls. Other nuclei examined showed no significant effects of palatable-diet feeding (data not shown). Cpu, caudate-putamen; Cor1, cortex (layer 1); Cor7, cortex (layer 7); CA1 + CA3, hippocampus; DG, dentate gyrus; EP, entopeduncular nucleus; VMH, ventromedial nucleus; Nacc, nucleus accumbens.

Kirkham and Williams, 2001). Furthermore, evidence in humans supports cannabinoid involvement in food reward, with the hyperphagic effects of marijuana in human volunteers being attributed principally to an increase in the consumption of highly palatable sweet foods such as chocolate and biscuits (Iverson, 2000).

Endocannabinoids and obesity

In stark contrast to the cannabinoid system, leptin-regulated hypothalamic orexigenic neuropeptide systems, such as NPY, are reported to be switched off under conditions of excess intake (Widdowson *et al.*, 1997). Additionally, simultaneous deletion of the two most potent orexigenic neuropeptides known to date, NPY and AgRP, fails to produce a lean phenotype, demonstrating the apparent redundancy of neuroendocrine factors that drive food intake. However, even the number of appetite-stimulating peptides identified to date are unable to compensate for the lack of endogenous cannabinoid action, reflecting its crucial role in the regulation of energy balance (Cota *et al.*, 2003).

It is not yet clear to what extent pharmacological agents that act on the cannabinoid system may have sustained actions. For CB1 antagonists to be useful in the treatment of obesity, their effects would have to be sustained over the long term. It has been reported that the anorectic actions of the CB1-specific antagonist rimonabant disappear within 3–6 days of treatment in rodents, which suggests that tolerance develops. However, weight loss is sustained (Colombo *et al.*, 1998). In humans, rimonabant treatment has been shown to promote modest but sustained reductions in weight and waist circumference, together with favourable changes in cardiometabolic risk factors (Després *et al.*, 2005; Van Gaal *et al.*, 2005; Pi-Sunyer *et al.*, 2006). However, the multicentre trials have been limited by a high dropout rate. In addition, recent findings of increased risk of suicide during treatment with rimonabant have led to specific contraindication in patients with depression and severe psychiatric disorders (Christensen *et al.*, 2007).

Histamine

The list of peptides and neurotransmitters known to take part in the complex regulation of body weight is increasing at a phenomenal pace. Among these is histamine, a central neurotransmitter, which has been shown to exert an important role in the regulation of appetite and metabolism (Jørgensen *et al.*, 2007). Studies using both knockout mouse models, as well as pharmacological studies, have revealed that histamine acts as an anorexigenic agent via stimulation of histamine H(1) receptors. One effect of histamine in the regulation of appetite is to act as a mediator of the inhibitory effect of leptin on appetite. It seems that histamine may attenuate and delay the development of leptin resistance in high-fat diet-induced obesity. Furthermore, histamine may also act to accelerate lipolysis. Based on current knowledge, the histaminergic system represents an interesting target for the development of pharmacological agents to control obesity. At present, H(3) receptor antagonists that stimulate the histaminergic system may be the most promising histaminergic drugs for antiobesity therapy (Jørgensen *et al.*, 2007).

Serotonin

More than 35 years of research suggest that endogenous hypothalamic serotonin (5-hydroxytryptamine) plays an important part in within-meal satiation and post-meal satiety processes (Halford *et al.*, 2007; Atkinson, 2008). Thus, the serotonin system provides a viable target for weight control. Numerous serotonin receptor subtypes have been identified; of these, serotonin 5-HT1B and 5-HT2C receptors have been recognized specifically as mediators of serotonin-induced satiety. A number of serotonergic drugs, including selective serotonin reuptake inhibitors (SSRIs), dexfenfluramine and 5-HT2C receptor agonists, have been shown to attenuate rodent body weight gain significantly (Halford *et al.*, 2007). This effect is strongly associated with marked hypophagia and is probably mediated by the hypothalamic melanocortin system. Additionally, sibutramine, dexfenfluramine, fluoxetine and the 5-HT2C receptor agonist, chlorophenylpiperazine, have all been shown to modify appetite in both lean and obese humans, resulting in reduced caloric intake (Halford *et al.*, 2007). Clinical studies demonstrate serotonergic drugs specifically reduce appetite prior to and following the consumption of fixed caloric loads, and cause a reduction in pre-meal appetite and caloric intake at *ad libitum* meals. Weight loss in the obese has also been produced by treatment with both the serotonin precursor, 5-hydroxytryptophan, and the preferential 5-HT2C receptor agonist, chlorophenylpiperazine. A new generation of 5-HT2C receptor selective agonists have been developed and at least one, lorcaserin (APD356), has been shown to be a potent, selective and efficacious agonist of the 5-HT2C receptor, with potential for the treatment of obesity (Thomsen *et al.*, 2008), and is currently undergoing clinical trials. In addition, 5-HT6 receptor antagonists such as PRX-07034 and BVT74316 have been shown to reduce food intake and body weight gain potently in rodent models and have entered clinical trials (Halford *et al.*, 2007). However, the role of the 5-HT6 receptor in the expression of appetite remains to be determined. The hope is that these drugs will not only be free of their predecessors' adverse effect profiles, but will also be equally or more effective at regulating appetite and controlling body weight (Halford *et al.*, 2007).

A possibility receiving little attention is that 5-HT regulates upstream corticotropin-releasing hormone (CRH) signalling systems via activation of 5-HT2C receptors in the PVN. Genetic inactivation of 5-HT2C receptors has been shown to produce a downregulation of CRH mRNA and a blunted CRH and corticosterone release after 5-HT administration (Heisler *et al.*, 2007). These findings thus provide a mechanistic explanation for the longstanding observation of hypothalamic–pituitary–adrenal axis stimulation in response to 5-HT and thereby shed light on the neural circuitry mediating the complex neuroendocrine responses to stress. In addition, genes implicated in serotonergic functioning recently have been shown to predict body mass index (BMI) categories (Fuemmeler *et al.*, 2008).

Dopamine

As explained in detail already, the hypothalamus integrates numerous hormonal and neuronal signals to regulate appetite and metabolism and thereby serves a homeostatic purpose in the regulation of body weight. Additional neural circuits that are superimposed on this system have the potential to override the homeostatic signals, resulting in gluttony. In this context, midbrain dopamine neurones have long been implicated in mediating reward behaviour and the motivational aspects of feeding behaviour.

Mice lacking dopamine D1 receptors are smaller than wild types, but the underlying cause of their growth retardation has not been elucidated. Inactivation of the D2, D3, D4 or D5 receptors has little effect on body weight. While chronic pharmacological blockade of D1 signalling has not been associated with significant effects on appetite or body weight, chronic D2 receptor blockade promotes obesity, with morbidly obese humans exhibiting less D2 receptor availability and individual food reinforcement differences being observed in relation to D2 receptor polymorphisms (Epstein *et al.*, 2007). A decrease in D2-like receptor binding in the striatum has been reported in obese individuals and drug addicts. Although natural and drug rewards share neural substrates, it is not clear whether such effects also contribute to overeating on palatable meals as an antecedent of dietary obesity. Using the D2R agonist quinpirole, an altered D2R signalling in obese Otsuka Long–Evans Tokushima fatty (OLETF) rats similar to drug-induced sensitization has been observed, suggesting a link between this effect and avidity for palatable foods in this model (Hajnal *et al.*, 2008). In addition, along with its involvement in seeking behaviour for drugs of abuse, the D3 receptor may also be involved in seeking behaviour for natural reinforcers such as food (Thanos *et al.*, 2008). Moreover, genes implicated in dopaminergic functioning reportedly predict body mass index categories (Fuemmeler *et al.*, 2008). Recent results further reveal that hormones implicated in regulating the energy balance control system also impinge directly on dopamine neurones (Palmiter, 2007). Thus, leptin and insulin inhibit dopamine neurones directly, whereas ghrelin activates them. Expanding the knowledge on the exact contribution of dopamine signalling on either normal feeding and/or mediating eating for pleasure is warranted.

Other emerging factors

The intense research in the field of body weight and food intake regulation is yielding the identification of a plethora of novel factors with a substantial influence on appetitive behaviour through their actions on the hypothalamus, the brainstem or afferent autonomic nerves.

Brain-specific homeobox factor Bsx

Food intake and activity-induced thermogenesis are important components of energy balance regulation. The molecular mechanism underlying the coordination of food intake with locomotory behaviour to maintain energy homeostasis is unclear. The brain-specific homeobox transcription factor *Bsx* is required for locomotory behaviour, hyperphagia and expression of the hypothalamic neuropeptides, NPY and AgRP, which regulate feeding behaviour and body weight (Sakkou *et al*., 2007). Mice lacking *Bsx* exhibit reduced locomotor activity and lower expression of NPY and AgRP. They also exhibit attenuated physiological responses to fasting, including reduced increase of NPY/AgRP expression, lack of food-seeking behaviour and reduced rebound hyperphagia. Furthermore, *Bsx* gene disruption rescues the obese phenotype of leptin-deficient *ob/ob* mice by reducing their hyperphagia without increasing their locomotor activity. Thus, *Bsx* represents an essential factor for NPY/AgRP neuronal function and locomotory behaviour in the control of energy balance (Sakkou *et al*., 2007).

Prohormone convertase 1/3

Congenital deficiency of the neuroendocrine-specific enzyme prohormone convertase 1/3 has been reported in only a small number of humans in which the disorder leads to a syndrome characterized by obesity, small intestinal dysfunction and dysregulation of glucose homeostasis in humans (Farooqi *et al*., 2007).

Single-minded 1

The hypothalamic transcription factor Single-minded 1 (Sim1) is expressed in a number of regions known to be involved in energy homeostasis, including the PVN and the LH. Haploinsufficiency of Sim1 is associated with hyperphagic obesity and increased linear growth closely resembling the phenotype of agouti yellow (Ay) and MC4R null mice, two classic models of disrupted hypothalamic melanocortin signalling (Coll *et al*., 2007). These similarities may be because MC4 receptors involved in the regulation of food intake signal through Sim1 and/or its transcriptional targets. Sim1 heterozygous mice remain hyperphagic, despite elevated hypothalamic POMC expression, and have an impaired anorectic (appetite-suppressing) response to melanocortins, suggesting that Sim1-expressing neurones in the PVN regulate feeding in response to melanocortin signalling (Coll *et al*., 2007).

Ghrelin O-acyltransferase

Serine-3 of ghrelin is acylated with an eight-carbon fatty acid, octanoate, which is required for its endocrine actions. The recent identification of ghrelin O-acyltransferase (GOAT), a polytopic membrane-bound enzyme that attaches octanoate to serine-3 of ghrelin, may facilitate the search for inhibitors that reduce appetite and diminish obesity in humans (Yang *et al*., 2008). Analysis of the mouse genome has revealed that GOAT belongs to a family of 16 hydrophobic membrane-bound acyltransferases that includes porcupine, which attaches long-chain fatty acids to Wnt proteins. GOAT is the only member of this family that octanoylates ghrelin when coexpressed in cultured endocrine cell lines with prepro-ghrelin. GOAT activity requires catalytic asparagine and histidine residues that are conserved in this family. Consistent with its function, GOAT mRNA is restricted largely to stomach and intestine, the major ghrelin-secreting tissues (Yang *et al.*, 2008).

Supporting Actors

The melanocortin system is one of the critical pathways in the regulation of energy homeostasis (Yang and Harmone, 2003; Harrold and Halford, 2006). This system consists of an array of melanocortin peptides, derived from the common precursor POMC, and the melanocortin receptors through which they signal their effects (see Chapters 1 and 2). A unique feature of the system is the presence of an endogeous antagonist, AgRP, which causes marked and prolonged hyperphagia. AgRP recently has been shown to be able to modulate energy balance via a mechanism independent of MSH and MC3/4-R competitive antagonism, consistent with either inverse agonist activity at MC-R or interaction with a distinct receptor (Tolle and Low, 2008). Additional complexity surrounding the anorectic actions of the melanocortins has emerged with the identification of the peptides, mahogany and syndecan, which are believed to be mediators in the melanocortin pathway for weight control.

Mahogany

Signalling from α-MSH through the melanocortin receptors MC1-R and MC4-R regulates coat colour and body weight, respectively (Huszar *et al.*, 1997; Schallreuter, 1999). The action of α-MSH at these receptors is antagonized by the agouti protein, the expression of which normally is restricted to the skin. Thus, ectopic expression of agouti in A^y mice results in obese animals with yellow fur (Duhl *et al.*, 1994).

The mahogany locus (*mg*), the product of which is a 1428-amino acid protein, was initially identified as a recessive suppressor of agouti at MC1-R (Graham *et al.*, 1997). However, the expression of *mg* was found to be very broad and its location within the hypothalamus, and particularly the VMH, suggested a potential additional role in the regulation of food intake and body weight. More recent work has confirmed that *mg* mutations also suppress obesity in the A^y mouse (Dinulescu *et al.*, 1998). Furthermore, *mg* mutations are able to suppress dietary-induced obesity (Nagle *et al.*, 1999), which is likely to have implications for therapeutic intervention for human obesity. However, *mg* mutations fail to suppress the obese phenotype of mice lacking MC4-R or that of other monogenic obese models (*db/db*, *ob/ob*, *fat/fat*). This suggests that mahogany acts specifically in the agouti pathway either at or upstream from the melanocortin receptors.

In the light of these findings, two potential mechanisms of action for mahogany have been proposed. As the structure of mahogany is similar to that of receptor proteins, it is possible that mahogany acts as an accessory receptor, gathering extracellular molecules to present to high-affinity receptors. In this way, it may present the antagonists, agouti and AgRP, to melanocortin receptors, thereby reducing signalling. Alternatively, mahogany may act to sequester the ligand, effectively increasing the concentration of antagonists in the local environment of the receptor (Nagle *et al.*, 1999).

Intriguingly, it has been demonstrated that although able to suppress obesity in A^y mice, *mg* mutations fail to prevent hyperphagia in these animals. Furthermore, *mg* mutations, in the absence of agouti overexpression, induce hyperphagia and an increase in basal metabolic rate (Nagle *et al.*, 1999). This suggests that although the mahogany protein is required for agouti action, its effects are not limited to agouti or its brain homologue, AgRP, because inhibition of function of these would be expected to lead to hypophagia. The brain uptake of the mahogany peptide has been shown to be higher in young A^y mice, preceding the surge of fat mass and suggesting a role for accelerated blood–brain barrier transport in the epigenetics of these rodents (Pan and Kastin, 2007). Thus, mahogany appears to suppress A^y-induced hyperphagia and concurrently stimulate food intake by interacting with some other pathway. The nature of this additional pathway is unclear. However, the inability of *mg* mutations to suppress obesity in *ob/ob* mice indicates that effectors dependent on mahogany do not include NPY, the expression of which is elevated in these rodents.

Syndecan-3

Syndecans are one of the major cell surface heparin sulphate proteoglycan families. These molecules are known to modulate the activity of a large number of extracellular ligands and thus have the potential to regulate a diversity of biological functions. In recent years, the development of transgenic organisms has allowed a more complete understanding of syndecan function and their putative biological roles. These studies have demonstrated an unforeseen role for the syndecans in the regulation of feeding behaviour.

Transgenic expression of syndecan-1 in the CNS, leading to high levels of expression in the hypothalamic nuclei known to play a role in the regulation of energy balance, results in hyperphagia and maturity-onset obesity (Reizes *et al.*, 2001). This phenotype, which is similar to that observed in animals lacking α-MSH function, and the hypothalamic location of the transgene, which is expressed in a pattern similar to MC4-R and AgRP neurones (Mountjoy *et al.*, 1994; Bagnol *et al.*, 1999), suggests an interaction between syndecan-1 and some component of the α-MSH pathway. Using a ligand-blotting assay, it was shown that syndecan-1 binds AgRP but not α-MSH (Reizes *et al.*, 2001). Additionally, the actions of AgRP are potentiated in HEK 293 cells transfected with MC4-R and syndecan-1 compared to cells transfected with MC4-R alone. Furthermore, transgenically expressed syndecan-1 potentiates the action of AgRP in A^y mice, with animals bearing the syndecan-1 transgene gaining weight more rapidly and to a greater extent that those without CNS expression of syndecan-1. Hypothalamic syndecan generates obesity by binding to and potentiating the inhibitory action of AgRP at melanocortin receptors, most likely MC4-R (Reizes *et al.*, 2001, 2003).

Syndecan-1 causes hyperphagia when misexpressed in the hypothalamus. However, syndecan-3 is the endogenous syndecan in the hypothalamus, being expressed in the PVN, DMH and LHA (Reizes *et al.*, 2001). A physiological role for syndecan-3 in regulating feeding is indicated by the observation that syndecan-3 levels change with nutritional status. Food deprivation increases hypothalamic syndecan-3 levels above those of *ad libitum* fed mice. These elevated levels fall with re-feeding, in a manner similar to AgRP. Furthermore, syndecan-3 knockout mice eat less than their wild-type littermates following a fast and exhibit a reduced adipose content compared with wild-type mice (Strader *et al.*, 2004). On a high-fat diet, syndecan-3-null male and female mice exhibited a partial resistance to obesity due to reduced food intake in males and increased energy expenditure in females relative to that of wild-type mice. As a result, syndecan-3-null mice on a high-fat diet accumulated less adipose mass and showed improved glucose tolerance compared with wild-type controls (Strader *et al.*, 2004). The finding that syndecan-3, an extracellular matrix molecule (ECM), can regulate body weight provides a unique and novel link between the extracellular matrix and body weight regulatory mechanisms. Uniquely, hormones such as leptin, previously thought only to regulate body weight by modulating neuropeptide levels, have now been shown to regulate neuronal plasticity in the hypothalamus (Reizes *et al.*, 2008). Thus, ECMs and syndecans are now being recognized as regulators of plasticity, highlighting the role of syndecans, and in particular, syndecan-3, in neuronal development and synaptic organization in relation to body weight regulation.

Nociceptin

Nociceptin or orphanin FQ (N/OFQ) and its receptor, NOP1, are expressed in hypothalamic nuclei involved in energy homeostasis. N/OFQ administered by ICV or ARC injection increases food intake in satiated rats. The exact mechanisms by which N/OFQ increases food intake are unknown, although it has been proposed that nociceptin via the NOP1 receptor may increase food intake by decreasing the release of the anorectic peptide, CART, and increasing the release of the orexigenic peptide, AgRP (Bewick *et al.*, 2005).

Nesfatin

Nesfatin-1 and its precursor, NUCB2, corresponding to NEFA/nucleobindin2 (NUCB2), were identified by subtraction cloning in cell lines of both neuronal and adipocyte origin (Oh *et al.*, 2006). Nesfatin-1 is a secreted 82-amino acid peptide found to be expressed in the appetite-control hypothalamic nuclei of rats that suppresses food intake dose-dependently after ICV injection. Rat cerebrospinal fluid contains nesfatin-1, and its expression is decreased in the PVN under starved conditions. Whereas injection of an antibody neutralizing nesfatin-1 stimulates appetite, ICV injection of other possible fragments processed from NUCB2 does not promote satiety, with conversion of NUCB2 to nesfatin-1 being necessary to induce feeding suppression (Oh *et al.*, 2006). Nesfatin-1-induced anorexia reportedly takes place in Zucker rats with a leptin receptor mutation, while an anti-nesfatin-1 antibody does not block leptin-induced anorexia. In contrast, central injection of α-MSH elevates NUCB2 gene expression in the PVN, and satiety by nesfatin-1 is abolished by an antagonist of the MC3/4-R. These findings identify nesfatin-1 as another satiety peptide associated with hypothalamic melanocortin signalling (Oh *et al.*, 2006).

Nesfatin-1 has been demonstrated to have either hyperpolarizing or depolarizing effects on a large proportion of different subpopulations of neurones, irrespective of their classification based on electrophysiological fingerprint (magnocellular, neuroendocrine or pre-autonomic) or molecular phenotype (vasopressin, oxytocin, corticotrophin-releasing hormone, thyrotrophin-releasing hormone or vesicular glutamate transporter) (Price *et al.*, 2008). In the PVN, 24% of nesfatin-1 neurones have been shown to overlap with oxytocin, 18% with vasopressin, 13% with CRH and 12% with thyrotropin-releasing hormone (TRH) neurones, while in the supraoptic nucleus (SON), 35% of nesfatin-1 neurones overlapped with oxytocin and 28% with vasopressin (Kohno *et al.*, 2008). After a 48-h fast, re-feeding for 2 h increased dramatically the number of nesfatin-1 neurones expressing c-fos immunoreactivity by approximately 10 times in the PVN and 30 times in the SON, compared with the fasting controls. In the SON, re-feeding also increased significantly the number of nesfatin-1-immunoreactive neurones and NUCB2 mRNA expression, compared with fasting. These results indicate that feeding-activated nesfatin-1 neurones in the PVN and SON may play a role in the postprandial regulation of feeding behaviour and energy homeostasis (Kohno *et al.*, 2008). Additional studies have revealed nesfatin-1-immunoreactive cells in the Edinger–Westphal nucleus, dorsal motor nucleus of vagus and caudal raphe nuclei of rats and that the peptide interacts with GPCRs (Brailoiu *et al.*, 2007). Furthermore, nesfatin was shown to exit the brain by bulk absorption of cerebrospinal fluid without a specific efflux transport system being able to cross the blood–brain barrier, in both the blood-to-brain and brain-to-blood directions, by non-saturable mechanisms (Price *et al.*, 2007). However, the limited penetration of nesfatin under physiological conditions does not limit the pharmacological delivery of this satiety peptide as a potential therapeutic agent.

Conclusions

For a long time, researchers have focused their interest on factors that regulate feeding behaviour. During this time, knowledge of the different pathways influencing feeding has progressed considerably. Consequently, our understanding has moved away from the anatomical concept of feeding and satiety centres towards the presence of specific factors that modulate feeding behaviour by acting on discrete neural pathways. Research efforts continue to expand understanding of the role of signalling molecules between central hypothalamic nuclei and peripheral enteroendocrine cells; and discoveries of novel networks and messengers provide new biological insights on how to manipulate appetite–satiety pathways (Atkinson, 2008). Despite the vast array of peptides that are potentially useful in obesity pharmacotherapy, currently available drugs fall within only four classes of agents: (i) catecholamine stimulants; (ii) serotonin and noradrenaline reuptake inhibitors; (iii) lipase inhibitors; and (iv) more recently, cannabinoid-1 receptor antagonists. The clinical effects of these drugs confer modest improvements and the side effects impact a long-term treatment course negatively.

This chapter adds more molecules to the cast of factors and highlights that their organization for producing an adequate response to the nutritional needs of an individual is getting more and more complex. Brain mechanisms are clearly linked to peripheral systems, while modulatory factors ensure that responses are finely tuned and, finally, the redundancy of different circuits compensates for discrepancies. This multiplicity of regulatory proteins, receptors and modulatory factors reflects the revolutionary importance of body weight maintenance, but also implies that a combinational approach will probably be required for the therapeutic treatment of obesity.

References

Adrian, T.E., Ferri, G.L., Bacarese-Hamilton, A.J., Fuessl, H.S., Polak, J.M. and Bloom, S.R. (1985) Human distribution and release of a putative new gut hormone, peptide YY. *Gastroenterology* 89, 1070–1077.

Ariyasu, H., Takaya, K., Tagami, T., Ogawa, Y., Hosoda, K., Akamizu, T., Suda, M., Koh, T., Natsui, K., Toyooka, S., Shirakami, G., Usui, T., Shimatsu, A., Doi, K., Hosoda, H., Kojima, M., Kangawa, K. and Nakao, K. (2001) Stomach is a major source of circulating ghrelin and feeding state determines plasma ghrelin-like immunoreactivity levels in humans. *Journal of Clinical Endocrinology and Metabolism* 86, 4753–4758.

Arnone, M., Maruani, J., Chaperon, F., Thiebot, M.H., Poncelet, M., Soubrie, P. and Le Fur, G. (1997) Selective inhibition of sucrose and ethanol intake by SR 141716, an antagonist of central cannabinoid (CB1) receptors. *Psychopharmacology* 132, 104–106.

Atkinson, T.J. (2008) Central and peripheral neuroendocrine peptides and signalling in appetite regulation: considerations for obesity pharmacotherapy. *Obesity Reviews* 9, 108–120.

Bagnasco, M., Tulipano, G., Melis, M.R., Argiolas, A., Cocchi, D. and Muller E.E. (2003) *Regulatory Peptides* 111, 161–167.

Bagnol, D., Lu, X.Y., Kaelin, C.B., Day, H.E., Ollmann, M., Gantz, I., Akil, H., Barsh, G.S. and Watson, S.J. (1999) Anatomy of an endogenous antagonist: relationship between agouti-related protein and proopiomelanocortin in brain. *Journal of Neuroscience* 19, 1–7.

Banks, W.A., Tschöp, M., Robinson, S.M. and Heiman, M.L. (2002) Extent and direction of ghrelin transport across the blood–brain barrier is determined by its unique primary structure. *Journal of Pharmacology and Experimental Therapeutics* 302, 822–827.

Batterham, R.L., Cowley, M.A., Small, C.J., Herzog, H., Cohen, M.A., Dakin, C.L., Wren, A.M., Brynes, A.E., Low, M.J., Ghatei, M.A., Cone, R.D. and Bloom, S.R. (2002) Gut hormone PYY(3-36) physiologically inhibits food intake. *Nature* 418, 650–654.

Batterham, R.L., Cohen, M.A., Ellis, S.M., Le Roux, C.W., Withers, D.J., Frost, G.S., Ghatei, M.A. and Bloom, S.R. (2003) Inhibition of food intake in obese subjects by peptide YY3-36. *New England Journal of Medicine* 349, 941–948.

Bewick, G.A., Dhillo, W.S., Darch, S.J., Murphy, K.G., Gardiner, J.V., Jethwa, P.H., Kong, W.M., Ghatei, M.A. and Bloom, S.R. (2005) Hypothalamic cocaine- and amphetamine-regulated transcript (CART) and agouti-related protein (AgRP) neurons coexpress the NOP1 receptor and nociceptin alters CART and AgRP release. *Endocrinology* 146, 3526–3534.

Brailoiu, G.C., Dun, S.L., Brailoiu, E., Inan, S., Yang, J., Chang, J.K. and Dun, N.J. (2007) Nesfatin-1: distribution and interaction with a G protein-coupled receptor in the rat brain. *Endocrinology* 148, 5088–5094.

Breivogel, C.S. and Childers, S.R. (1998) The functional neuroanatomy of brain cannabinoid receptors. *Neurobiological Disorders* 5, 417–431.

Brezillon, S., Lannoy, V., Franssen, J.D., Le Poul, E., Dupriez, V., Lucchetti, J., Detheux, M. and Parmentier, M. (2003) Identification of natural ligands for the orphan G protein-coupled receptors GPR7 and GPR8. *Journal of Biological Chemistry* 278, 776–783.

Cai, X.J., Liu, X.H., Evans, M., Clapham, J.C., Arch, J.R., Morris, R. and Williams, G. (2002) Orexins and feeding: special occasions or everyday occurrence? *Regulatory Peptides* 104, 1–9.

Cai, X.J., Harrold, J.A., Berwaer, M., Williams, G. and Pinkney, J. (2004) Manipulation of luminal glucose availability increases ghrelin concentrations in plasma but not in stomach or duodenum. *Proceedings of the British Pharmacological Society* 2, 10P.

Catalán, V., Gómez-Ambrosi, J., Rotellar, F., Silva, C., Gil, M.J., Rodríguez, A., Cienfuegos, J.A., Salvador, J. and Frühbeck, G. (2007) The obestatin receptor (GPR39) is expressed in human adipose tissue and is down-regulated in obesity-associated type 2 diabetes mellitus. *Clinical Endocrinology* 66, 598–601.

Challis, B.G., Coll, A.P., Yeo, G.S.H., Pinnock, S.B., Dickson, S.L., Thresher, R.R., Dixon, J., Zahn, D., Rochford, J.J., White, A., Oliver, R.L., Millington, G., Aparicio, S.A., Colledge, W.H., Russ, A.P., Carlton, M.B. and O'Rahilly, S. (2004) Mice lacking pro-opiomelanocortin are sensitive to high-fat feeding but response normally to the acute anorectic effects of peptide-$YY_{3\text{-}36}$. *Proceedings of the National Academy of Sciences of the United States of America* 101, 4695–4700.

Chandarana, K. and Batterham, R. (2008) Peptide YY. *Current Opinion in Endocrinology, Diabetes and Obesity* 15, 65–72.

Chaudhri, O.B., Wynne, K. and Bloom, S.R. (2008) Can gut hormones control appetite and prevent obesity? *Diabetes Care* 31 (Suppl. 2), S284–S289.

Chen, H.Y., Trumbauer, M.E., Chen, A.S., Weingarth, D.T., Adams, J.R., Frazier, E.G., Shen, Z., Marsh, D.J., Feighner, S.D., Guan, X.M., Ye, Z., Nargund, R.P., Smith, R.G., Van der Ploeg, L.H., Howard, A.D., MacNeil, D.J. and Oian, S. (2004) Orexigenic action of peripheral ghrelin is mediated by neuropeptide Y and agouti-related protein. *Endocrinology* 145, 2607–2612.

Christensen, R., Kristensen, P.K., Bartels, E.M., Bliddal, H. and Astrup, A. (2007) Efficacy and safety of the weight-loss drug rimonabant: a meta-analysis of randomised trials. *Lancet* 370, 1706–1713.

Coll, A.P., Farooqi, I.S. and O'Rahilly, S. (2007) The hormonal control of food intake. *Cell* 129, 251–262.

Colombo, G., Agabio, R., Diaz, G., Lobina, C., Reali, R. and Gessa, G.L. (1998) Appetite suppression and weight loss after the cannabinoid antagonist SR 141716. *Life Sciences* 63, 113–117.

Corp, E.S., Melville, L.D., Greenberg, D., Gibbs, J. and Smith, G.P. (1990) Effect of fourth ventricle neuropeptide Y and peptide YY on ingestive and other behaviors. *American Journal of Physiology* 259, R317–R323.

Cota, D., Marsicano, G., Tschop, M., Grubler, Y., Flachskamm, C., Schubert, M., Auer, D., Yassouridis, A., Thone-Reineke, C., Ortmann, S., Tomassoni, F., Ceryino, C., Nisoli, E.,

Linthorst, A.C., Pasquali, R., Lutz, B., Stalla, G.K. and Pagotto, U. (2003) The endogenous cannabinoid system affects energy balance via central orexigenic drive and peripheral lipogenesis. *Journal of Clinical Investigation* 112, 423–431.

Cowley, M.A., Smith, R.G., Diano, S., Tschöp, M., Pronchuk, N., Grove, K.L., Strasburger, C.J., Bidlingmaier, M., Esterman, M., Heiman, M.L., Garcia-Segura, L.M., Nillni, E.A., Mendez, P., Low, M.J., Sotonyi, P., Friedman, J.F., Liu, H., Pinto, S., Colmers, W.F., Cone, R.D. and Horvath, T.L. (2003) The distribution and mechanism of action of ghrelin in the CNS demonstrates a novel hypothalamic circuit regulating energy homeostasis. *Neuron* 37, 649–661.

Cummings, D.E. (2006) Ghrelin and the short- and long-term regulation of appetite and body weight. *Physiology and Behaviour* 89, 71–84.

Cummings, D.E. and Schwartz, M.W. (2003) Genetics and pathophysiology of human obesity. *Annual Review of Medicine* 54, 453–471.

Cummings, D.E., Purnell, J.Q., Fravo, R.S., Schmidova, K., Wisee, B.E. and Weigle, D.S. (2001) A preprandial rise in plasma ghrelin suggests a role in meal initiation in humans. *Diabetes* 50, 1714–1719.

Cummings, D.E., Weigle, D.S., Frayo, R.S., Breen, P.A., Ma, P.K., Dellinger, E.P. and Purnell, J.Q. (2002) Plasma ghrelin levels after diet-induced weight loss or gastric bypass surgery. *New England Journal of Medicine* 346, 1623–1630.

Date, Y., Murakami, N., Toshinai, K., Matsukura, S., Niijima, A., Matsuo, H., Kangawa, K. and Nakazato, M. (2002) The role of the gastric afferent vagal nerve in ghrelin-induced feeding and growth hormone secretion in rats. *Gastroenterology* 123, 1120–1128.

DelParigi, A., Tschöp, M., Heiman, M.L., Salbe, A.D., Vozarova, B., Sell, S.M., Bunt, J.C. and Tataranni, P.A. (2002) High circulating ghrelin: a potential cause for hyperphagia and obesity in Prader–Willi syndrome. *Journal of Clinical Endocrinology and Metabolism* 87, 5461–5464.

Depoortere, I., Thijs, T., Moechars, D., De Smet, B., Ver Donck, L. and Peeters, T.L. (2008) Effect of peripheral obestatin on food intake and gastric emptying in ghrelin-knockout mice. *British Journal of Pharmacology* 153, 1550–1557.

Després, J.-P., Golay, A. and Sjöström, L. for the Rimonabant in Obesity–Lipids Study Group (2005) Effects of rimonabant on metabolic risk factors in overweight patients with dyslipidemia. *New England Journal of Medicine* 353, 2121–2134.

Devane, W.A., Hanus, L., Breuer, A., Pertwee, R.G., Stevenson, L.A., Griffin, G., Gibson, D., Mandelbaum, A., Etinger, A. and Mechoulam, R. (1992) Isolation and structure of a brain constituent that binds to the cannabinoid receptor. *Science* 258, 1946–1949.

Dickson, S.L. and Luckman, S.M. (1997) Induction of c-fos messenger ribonucleic acid in neuropeptide Y and growth hormone (GH)-releasing factor neurons in the rat arcuate nucleus following systemic injection of the GH secretagogue, GH-releasing peptide-6. *Endocrinology* 138, 771–777.

Di Marzo, V., Melck, D., Bisogno, T. and De Petrocellis L. (1998) Endocannabinoids: endogenous cannabinoid receptor ligands with neuromodulatory action. *Trends in Neurosciences* 21, 521–528.

Di Marzo, V., Goparaju, S.K., Wang, L., Batkai, S., Jarai, Z., Fezza, F., Miura, G.I., Palmiter, R.D., Sugiura, T. and Kunos, G. (2001) Leptin-regulated endocannabinoids are involved in maintaining food intake. *Nature* 410, 822–825.

Dinulescu, D.M., Fan, W., Boston, B.A., McCall, K., Lamoreux, M.L., Moore, K.J., Montagno, J. and Cone, R.D. (1998) Mahogany (mg) stimulates feeding and increases basal metabolic rate independent of its suppression of agouti. *Proceedings of the National Academy of Sciences of the United States of America* 95, 12707–12712.

Duhl, D.M., Vrieling, H., Miller, K.A., Wolff, G.L. and Barsh, G.S. (1994) Neomorphic agouti mutations in obese yellow mice. *Nature Genetics* 8, 59–65.

Dun, S.L., Brailoiu, C., Yang, J., Kang Chang, J. and Dun, N.J. (2003) Neuropeptide W-immunoreactivity in the hypothalamus and pituitary of the rat. *Neuroscience Letters* 349, 71–74.

English, P.J., Ghatei, M.A., Malik, I.A., Bloom, S.R. and Wilding, J.P. (2002) Food fails to suppress ghrelin levels in obese humans. *Journal of Clinical Endocrinology and Metabolism* 87, 2984–2987.

Epstein, L.H., Temple, J.L., Neaderhiser, B.J., Salis, R.J., Erbe, R.W. and Leddy, J.J. (2007) Food reinforcement, the dopamine D-sub-2 receptor genotype, and energy intake in obese and non-obese humans. *Behavioral Neuroscience* 121, 877–886.

Farooqi, I.S., Volders, K., Stanhope, R., Heuschkel, R., White, A., Lank, E., Keogh, J., O'Rahilly, S. and Creemers, J.W. (2007) Hyperphagia and early-onset obesity due to a novel homozygous missense mutation in prohormone convertase 1/3. *Journal of Clinical Endocrinology and Metabolism* 92, 3369–3373.

Finkelstein, D.I., Reeves, A.K. and Horne, M.K. (1996) An electron microscopic tracer study of projections from the entopeduncular nucleus to the ventrolateral nucleus of the rat. *Neuroscience Letters* 211, 33–36.

Fuemmeler, B.F., Agurs-Collins, T.D., McClernon, F.J., Kollins, S.H., Kail, M.E., Bergen, A.W. and Ashley-Koch, A.E. (2008) Genes implicated in serotonergic and dopaminergic functioning predict BMI categories. *Obesity* 16, 348–355.

Fujii, R., Yoshida, H., Fukusumi, S., Habat, Y., Hosoya, M., Kawamata, Y., Yano, T., Hinuma, S., Kitada, C., Asami, T., Mori, M., Fujisawa, Y. and Fujino. M. (2002) Identification of a neuropeptide modified with bromine as an endogenous ligand for GPR7. *Journal of Biological Chemistry* 277, 34010–34016.

Gao, Q. and Horvath, T.L. (2007) Neurobiology of feeding and energy expenditure. *Annual Reviews in Neuroscience* 30, 367–398.

Gao, Q. and Horvath, T.L. (2008) Neuronal control of energy homeostasis. *FEBS Letters* 582, 132–141.

Glass, M., Dragunow, M. and Faull, R.L.M. (1997) Cannabinoid receptors in the human brain: a detailed anatomical and quantitative autoradiographic study in the fetal, neonatal and adult human brain. *Neuroscience* 77, 299–318.

Gorbachevskaia, A.L. (1999) Projections from the substantia nigra, ventral tegmental area and amygdale to the palladium in dog brain. *Morfologia* 115, 11–14.

Graham, A., Wakamatsu, K., Hunt, G., Ito, S. and Thody, A.J. (1997) Agouti protein inhibits the production of eumelanin and phaeomelanin in the presence and absence of alpha-melanocyte stimulating hormone. *Pigment Cell Research* 10, 298–303.

Grandt, D., Schimiczek, M., Beglinger, C., Layer, P., Goebell, H., Eysselein, V.E. and Reeve, J.R. Jr (1994) Two molecular forms of peptide YY (PYY) are abundant in human blood: characterization of a radioimmunoassay recognizing PYY 1-36 and PYY 3-36. *Regulatory Peptides* 51, 151–159.

Hajnal, A., Margas, W.M. and Covasa, M. (2008) Altered dopamine D2 receptor function and binding in obese OLETF rat. *Brain Research Bulletin* 75, 70–76.

Halford, J.C., Harrold, J.A., Boyland, E.J., Lawton, C.L. and Blundell, J.E. (2007) Serotonergic drugs: effects on appetite expression and use for the treatment of obesity. *Drugs* 67, 27–55.

Hansen, T.K., Dall, R., Hosoda, H., Kojima, M., Kangawa, K., Christiansen, J.S. and Jorgensen, J.O. (2002) Weight loss increases circulating levels of ghrelin in human obesity. *Clinical Endocrinology* 56, 203–206.

Hanus, L., Abu-Lafi, S., Fride, E., Breuer, A., Vogel, Z., Shaley, D.E., Kustanovich, I. and Mechoulam, R. (2001) 2-Arachidonyl glyceryl ether, an endogenous agonist of the

cannabinoid CB1 receptor. *Proceedings of the National Academy of Sciences of the United States of America* 98, 3662–3665.
Hao, S., Ayraham, Y., Mechoulam, R. and Berry, E.M. (2000) Low dose anandamide affects food intake, cognitive function, neurotransmitter and corticosterone levels in diet-restricted mice. *European Journal of Pharmacology* 392, 147–156.
Harrold, J.A. and Halford, J.C. (2006) The hypothalamus and obesity. *Recent Patents in CNS Drug Discovery* 1, 305–314.
Harrold, J.A. and Halford, J.C. (2007) Orphan G-protein-coupled receptors: strategies for identifying ligands and potential for use in eating disorders. *Drugs in R&D* 8, 287–299.
Harrold, J.A. and Williams, G. (2006) Melanocortin-4 receptors, beta-MSH and leptin: key elements in the satiety pathway. *Peptides* 27, 365–371.
Harrold, J.A., Elliott, J.C., King, P.J., Widdowson, P.S. and Williams, G. (2002) Down-regulation of cannabinoid-1 (CB-1) receptors in specific extrahypothalamic regions of rats with dietary obesity: a role for endogenous cannabinoids in driving appetite for palatable food? *Brain Research* 952, 232–238.
Harrold, J.A., Cai, X., Dovey, T., Halford, J. and Pinkney, J. (2004) [^{125}I-His^{9}] ghrelin autoradiography demonstrates ghrelin binding sites in the lateral hypothalamus. *Proceedings of the British Pharmacological Society* 2, 16P.
Heisler, L.K., Pronchuk, N., Nonogaki, K., Zhou, L., Raber, J., Tung, L., Yeo, G.S., O'Rahilly, S., Colmers, W.F., Elmquist, J.K. and Tecott, L.H. (2007) Serotonin activates the hypothalamic–pituitary–adrenal axis via serotonin 2C receptor stimulation. *Journal of Neuroscience* 27, 6956–6964.
Hewson, A.K. and Dickson, S.L. (2000) Systemic administration of ghrelin induces Fos and Egr-1 proteins in the hypothalamic arcuate nucleus of fasted and fed rats. *Journal of Neuroendocrinology* 12, 1047–1049.
Hinney, A., Hoch, A., Geller, F., Schafer, H., Siegfried, W., Goldschmidt, H., Remschmidt, H. and Hebebrand, J. (2002) Ghrelin gene: identification of missense variants and frameshift mutation in extremely obese children and adolescents and healthy normal weight students. *Journal of Clinical Endocrinology and Metabolism* 87, 2716–2719.
Horvath, T.L., Diano, S., Sotonyi, P., Heiman, M. and Tschop, M. (2001) Minireview: ghrelin and the regulation of energy balance – a hypothalamic perspective. *Endocrinology* 142, 4163–4169.
Huszar, D., Lynch, C.A., Fairchild-Huntress, V., Dunmore, J.H., Fang, Q., Berkemeier, L.R., Gu, W., Kesterson, R.A., Boston, B.A., Cone, R.D., Smith, F.J., Campfield, L.A., Burn, P. and Lee, F. (1997) Targeted disruption of the melanocortin-4 receptor results in obesity in mice. *Cell* 88, 131–141.
Ishii, M., Fei, H. and Friedman, J.M. (2003) Targeted disruption of GPR7, the endogenous receptor for neuropeptides B and W, leads to metabolic defects and adult-onset obesity. *Proceedings of the National Academy of Sciences of the United States of America* 100, 10540–10545.
Iverson, L.L. (2000) *The Science of Marijuana*. Oxford University Press, Oxford, UK.
Jamshidi, N. and Taylor, D.A. (2001) Anandamide administration into the ventromedial hypothalamus stimulates appetite in rats. *British Journal of Pharmacology* 134, 1151–1154.
Jørgensen, E.A., Knigge, U., Warberg, J. and Kjaer, A. (2007) Histamine and the regulation of body weight. *Neuroendocrinology* 86, 210–214.
Kelly, M.A., Beuckmann, C.T., Williams, S.C., Sinton, C.M., Motoike, T., Richardson, J.A., Hammer, R.E., Garry, M.G. and Yanagisawa, M. (2005) Neuropeptide B-deficient mice demonstrate hyperalgesia in response to inflammatory pain. *Proceedings of the National Academy of Sciences of the United States of America* 102, 9942–9947.

Kirkham, T.C. (2005) Endocannabinoids in the regulation of appetite and body weight. *Behavioural Pharmacology* 16, 297–313.

Kirkham, T.C. and Williams, C.M. (2001) Synergistic effects of opioid and cannabinoid antagonists on food intake. *Psychopharmacology* 153, 267–270.

Kirkham, T.C., Williams, C.M., Fezza, D. and Di Marzo, V. (2002) Endocannabinoid levels in rat limbic forebrain and hypothalamus in relation to fasting, feeding and satiation: stimulation of eating by 2-arachidonoyl glycerol. *British Journal of Pharmacology* 136, 550–557.

Kohno, D., Gao, H.Z., Muroya, S., Kikuyama, S. and Yada, T. (2003) Ghrelin directly interacts with neuropeptide-Y-containing neurons in the rat arcuate nucleus: Ca^{2+} signalling via protein kinase A and N-type channel-dependent mechanisms and cross-talk with leptin and orexin. *Diabetes* 52, 948–956.

Kohno, D., Nakata, M., Maekawa, F., Fujiwara, K., Maejima, Y., Kuramochi, M., Shimazaki, T., Okano, H., Onaka, T. and Yada, T. (2007) Leptin suppresses ghrelin-induced activation of neuropeptide Y neurons in the arcuate nucleus via phosphatidylinositol 3-kinase- and phosphodiesterase-3-mediated pathway. *Endocrinology* 148, 2251–2263.

Kohno, D., Nakata, M., Maejima, Y., Shimizu, H., Sedbazar, U., Yoshida, N., Dezaki, K., Onaka, T., Mori, M. and Yada, T. (2008) Nesfatin-1 neurons in paraventricular and supraoptic nuclei of the rat hypothalamus coexpress oxytocin and vasopressin and are activated by refeeding. *Endocrinology* 149, 1295–1301.

Kojima, M., Hosoda, H. and Kangawa, K. (2001) Purification and distribution of ghrelin: the natural endogenous ligand for the growth hormone secretagogue receptor. *Hormone Research* 56, 93–97.

Kola, B., Farkas, I., Christ-Crain, M., Wittmann, G., Lolli, F., Amin, F., Harvey-White, J., Liposits, Z., Kunos, G., Grossman, A.B., Fekete, C. and Korbonits, M. (2008) The orexigenic effect of ghrelin is mediated through central activation of the endogenous cannabinoid system. *PLoS ONE* 3, e1797.

Korbonits, M., Gueorguiey, M., O'Grady, E., Lecoeur, C., Swan, D.C., Mein, C.A., Weill, J., Grossman, A.B. and Froguel, P. (2002) A variation in the ghrelin gene increases weight and decreases insulin secretion in tall, obese children. *Journal of Clinical Endocrinology and Metabolism* 87, 4005–4008.

Lee, D.K., Nguyen, T., Porter, C.A., Cheng, R., George, S.R. and O'Dowd, B.F. (1999) Two related G protein-coupled receptors: the distribution of GPR7 in rat brain and the absence of GPR8 in rodents. *Molecular Brain Research* 71, 96–103.

Le Roux, C.W., Batterham, R.L., Aylwin, S.J., Patterson, M., Borg, C.M., Wynne, K.J., Kent, A., Vincent, R.P., Gardiner, J., Ghatei, M.A. and Bloom, S.R. (2006) Attenuated peptide YY release in obese subjects is associated with reduced satiety. *Endocrinology* 147, 3–8.

Le Roux, C.W., Borg, C.M., Murphy, K.G., Vincent, R.P., Ghatei, M.A. and Bloom, S.R. (2008) Supraphysiological doses of intravenous PYY3-36 cause nausea, but no additional reduction in food intake. *Annals of Clinical Biochemistry* 45, Part 1, 93–95.

Lu, S., Guan, J.L., Wang, Q.P., Yamada, S., Goto, N., Date, Y., Nakazato, M., Kojima, M., Kangawa, K. and Shioda, S. (2002) Immunocytochemical observation of ghrelin-containing neurons in the rat arcuate nucleus. *Neuroscience Letters* 321, 157–160.

Lundberg, J.M., Terenius, L., Hokfelt, T. and Tatemoto, K. (1984) Comparative immunohistochemical and biochemical analysis of pancreatic polypeptide-like peptides with special reference to presence of neuropeptide Y in central and peripheral neurons. *Journal of Neuroscience* 4, 2376–2386.

Masuda, Y., Tanaka, T., Inomata, N., Ohnuma, N., Tanaka, S., Itoh, Z., Hosoda, H., Kojima, M. and Kangawa, K. (2000) Ghrelin stimulates gastric acid secretion and motility in rats. *Biochemical and Biophysical Research Communications* 276, 905–908.

Mondal, M.S., Yamaguchi, H., Date, Y., Shimbara, T., Toshinai, K., Shomomura, Y., Mori, M. and Nakazato, M. (2003) A role for neuropeptide W in the regulation of feeding behaviour. *Endocrinology* 144, 4729–4733.

Morley, J.E., Levine, A.S., Grace, M. and Kneip, J. (1985) Peptide YY (PYY), a potent orexigenic agent. *Brain Research* 341, 200–203.

Mountjoy, K.G., Mortrud, M.T., Low, M.J., Simerly, R.B. and Cone, R.D. (1994) Localization of the melanocortin-4 receptor (MC4-R) in neuroendocrine and autonomic control circuits in the brain. *Molecular Endocrinology* 8, 1298–1308.

Nagle, D.L., McGrail, S.H., Vitale, J., Woolf, E.A., Dussault, B.J. Jr, DiRocco, L., Holmgren, L., Montagno, J., Bork, P., Huszar, D., Fairchild-Huntress, V., Ge, P., Keilty, J., Ebeling, C., Baldini, L., Gilchrist, J., Burn, P., Carlson, G.A. and Moore, K.J. (1999) The mahogany protein is a receptor involved in suppression of obesity. *Nature* 398, 148–152.

Nakazato, M., Murakami, N., Date, Y., Kojima, M., Matsuo, H., Kangawa, K. and Matsukura, S. (2001) A role for ghrelin in the central regulation of feeding. *Nature* 409, 194–198.

Nogueiras, R., Pfluger, P., Tovar, S., Arnold, M., Mitchell, S., Morris, A., Perez-Tilve, D., Vázquez, M.J., Wiedmer, P., Castañeda, T.R., DiMarchi, R., Tschöp, M., Schurmann, A., Joost, H.G., Williams, L.M., Langhans, W. and Diéguez, C. (2007) Effects of obestatin on energy balance and growth hormone secretion in rodents. *Endocrinology* 148, 21–26.

O'Dowd, B.F., Scheideler, M.A., Nguyen, T., Cheng, R., Rasmussen, J.S., Marchese, A., Zastawny, R., Heng, H.H., Tsui, L.C., Shi, X., Asa, S., Puy, L. and George, S.R. (1995) The cloning and chromosomal mapping of two novel human opioid- and somatostatin-like receptor genes, GPR7 and GPR8, expressed in distinct areas of the brain. *Genomics* 28, 84–91.

Oh-I., S., Shimizu, H., Satoh, T., Okada, S., Adachi, S., Inoue, K., Eguchi, H., Yamamoto, M., Imaki, T., Hashimoto, K., Tsuchiya, T., Monden, T., Horiguchi, K., Yamada, M. and Mori, M. (2006) Identification of nesfatin-1 as a satiety molecule in the hypothalamus. *Nature* 443, 709–712.

Olszewski, P.K., Li, D., Grace, M.K., Billington, C.J., Kotz, C.M. and Levine, A.S. (2003) Neural basis of orexigenic effects of ghrelin acting within the lateral hypothalamus. *Peptides* 24, 597–602.

Pagotto, U., Marsicano, G., Cota, D., Lutz, B. and Pasquali, R. (2006) The emerging role of the endocannabinoid system in endocrine regulation and energy balance. *Endocrine Reviews* 27, 73–100.

Palmiter, R.D. (2007) Is dopamine a physiologically relevant mediator of feeding behavior? *Trends in Neurosciences* 30, 375–381.

Pan, W. and Kastin, A.J. (2007) Mahogany, blood–brain barrier, and fat mass surge in AVY mice. *International Journal of Obesity* 31, 1030–1032.

Patel, N.A., Moldow, R.L., Patel, J.A., Wu, G. and Chang, S.L. (1998) Arachidonylethanolamide (AEA) activation of FOS proto-oncogene protein immunoreactivity in the rat brain. *Brain Research* 797, 225–233.

Pecina, S. and Berridge, K.C. (2000) Opioid sites in nucleus accumbens shell mediate eating and hedonic 'liking' for food: map based on microinjection fos plumes. *Brain Research* 863, 71–86.

Pederson-Bjergaard, U., Host, U., Kelbaek, H., Schifter, S., Rehfeld, J.F., Faber, J. and Christensen, N.J. (1996) Influence of meal composition on postprandial peripheral plasma concentrations of vasoactive peptides in man. *Scandinavian Journal of Clinical and Laboratory Investigation* 56, 497–503.

Pfluger, P.T., Kirchner, H., Günnel, S., Schrott, B., Perez-Tilve, D., Fu, S., Benoit, S.C., Horvath, T., Joost, H.G., Wortley, K.E., Sleeman, M.W. and Tschöp, M.H. (2008)

Simultaneous deletion of ghrelin and its receptor increases motor activity and energy expenditure. *American Journal of Physiology – Gastrointestinal and Liver Physiology* 294, G610–618.

Pi-Sunyer, F.X., Aronne, L.J., Heshmati, H.M., Devin, J. and Rosenstock, J. for the RIO-North America Study Group (2006) Effect of rimonabant, a cannabinoid-1 receptor blocker, on weight and cardiometabolic risk factors in overweight or obese patients. RIO-North America: a randomized controlled trial. *JAMA* 295, 761–775.

Porter, A.C., Sauer, J.M., Knierman, M.D., Becker, G.W., Berna, M.J., Bao, J., Nomikos, G.G., Carter, P., Bymaster, F.P., Leese, A.B. and Felder, C.C. (2002) Characterization of a novel endocannabinoid, virodhamine, with antagonist activity at the CB1 receptor. *Journal of Pharmacology and Experimental Therapeutics* 301, 1020–1024.

Price, C.J., Hoyda, T.D., Samson, W.K. and Ferguson, A.V. (2008) Nesfatin-1 influences the excitability of paraventricular nucleus neurones. *Journal of Neuroendocrinology* 20, 245–250.

Price, T.O., Samson, W.K., Niehoff, M.L. and Banks, W.A. (2007) Permeability of the blood–brain barrier to a novel satiety molecule nesfatin-1. *Peptides* 28, 2372–2381.

Rahardjo, G.L., Huan, X.F., Tan, Y.Y. and Deng, C. (2007) Decreased plasma peptide YY accompanied by elevated peptide YY and Y2 receptor binding densities in the medulla oblongata of diet-induced obese mice. *Endocrinology* 148, 4704–4710.

Ravinet Trillou, C.R., Arnone, M., Delgorge, C., Gonalons, N., Keane, P., Maffrand, J.P. and Soubrie, P. (2003) Anti-obesity effect of SR 141716, a CB1 receptor antagonist, in diet-induced obese mice. *American Journal of Physiology* 284, R345–R353.

Reizes, O., Lincecum, J., Wang, Z., Goldberger, O., Huang, L., Kaksonen, M., Ahima, R., Hinkes, M.T., Barsh, G.S., Rauvala, H. and Bernfield, M. (2001) Transgenic expression of syndecan-1 uncovers a physiological control of feeding behaviour by syndecan-3. *Cell* 106, 105–116.

Reizes, O., Benoit, S.C., Strader, A.D., Clegg, D.J., Akunuru, S. and Seeley, R.J. (2003) Syndecan-3 modulates food intake by interacting with the melanocortin/AgRP pathway. *Annals of the New York Academy of Sciences* 994, 66–73.

Reizes, O., Benoit, S.C. and Clegg, D.J. (2008) The role of syndecans in the regulation of body weight and plasticity. *International Journal of Biochemistry and Cellular Biology* 40, 28–45.

Rowland, N.E., Mukherjee, M. and Robertson, K. (2001) Effects of the cannabinoid receptor antagonist SR 141716, alone and in combination with dexfenfluramine or naloxone, on food intake in rats. *Psychopharmacology* 159, 111–116.

Rucinski, M., Nowak, K.W., Chmielewska, J., Ziolkowska, A. and Malendowicz, L.K. (2007) Neuropeptide W exerts a potent suppressive effect on blood leptin and insulin concentrations in the rat. *International Journal of Molecular Medicine* 19, 401–405.

Sakkou, M., Wiedmer, P., Anlag, K., Hamm, A., Seuntjens, E., Ettwiller, L., Tschop, M.H. and Treier, M. (2007) A role for brain-specific homeobox factor *Bsx* in the control of hyperphagia and locomotory behavior. *Cell Metabolism* 5, 450–463.

Sato, T., Kurokawa, M., Nakashima, Y., Ida, T., Takahashi, T., Fukue, Y., Ikawa, M., Okabe, M., Kangawa, K. and Kojima, M. (2008) Ghrelin deficiency does not influence feeding performance. *Regulatory Peptides* 145, 7–11.

Schallreuter, K.U. (1999) A review of recent advances on the regulation of pigmentation in the human epidermis. *Cellular and Molecular Biology* 45, 943–949.

Shiiya, T., Nakazato, M., Mizuta, M., Date, Y., Mondal, M.S., Tanaka, M., Nozoe, S., Hosoda, H., Kangawa, K. and Matsukura, S. (2002) Plasma ghrelin levels in lean and obese humans and the effect of glucose on ghrelin secretion. *Journal of Clinical Endocrinology and Metabolism* 87, 240–244.

Shimomura, Y., Harada, M., Goto, M., Sugo, T., Matsumoto, Y., Abe, M., Watanabe, T., Asami, T., Kitada, C., Mori, M., Onda, H. and Fujino, M. (2002) Identification of neuropeptide W as the endogenous ligand for orphan G-protein-coupled receptors GPR7 and GPR8. *Journal of Biological Chemistry* 277, 35826–35832.

Simiand, J., Keane, M., Keane, P.E. and Soubrie, P. (1998) SR 141716, a CB1 cannabinoid receptor antagonist, selectively reduces sweet food intake in marmosets. *Behavioral Pharmacology* 9, 179–181.

Singh, G. and Davenport, A.P. (2006) Neuropeptide B and W: neurotransmitters in an emerging G-protein-coupled receptor system. *British Journal of Pharmacology* 148, 1033–1041.

Stella, N., Schweitzer, P. and Piomelli, D. (1997) A second endogenous cannabinoid that modulates long-term potentiation. *Nature* 388, 773–778.

Strader, A.D., Reizes, O., Woods, S.C., Benoit, S.C. and Seeley, R.J. (2004) Mice lacking the syndecan-3 gene are resistant to diet-induced obesity. *Journal of Clinical Investigation* 114, 1354–1360.

Tanaka, H., Yoshida, T., Miyamoto, N., Motoike, T., Kurosu, H., Shibata, K., Yamanaka, A., Clay Williams, S., Richardson, J.A., Tsujino, N., Garry, M.G., Lerner, M.R., King, D.S., O'Dowd, B.F., Sakurai, T. and Yanagisawa, M. (2003) Characterization of a family of endogenous neuropeptide ligands for the G-protein coupled receptors GPR7 and GPR8. *Proceedings of the National Academy of Sciences of the United States of America* 100, 6251–6256.

Tatemoto, K. and Mutt, V. (1980) Isolation of two novel candidate hormones using a chemical method for finding naturally occurring polypeptides. *Nature* 285, 417–418.

Thanos, P.K., Michaelides, M., Ho, C.W., Wang, G.J., Newman, A.H., Heidbreder, C.A., Ashby, C.R. Jr, Gardner, E.L. and Volkow, N.D. (2008) The effects of two highly selective dopamine D(3) receptor antagonists (SB-277011A and NGB-2904) on food self-administration in a rodent model of obesity. *Pharmacology Biochemistry Behavior* 89, 499–507.

Thomsen, W.J., Grottick, A.J., Menzaghi, F., Reyes-Saldana, H., Espitia, S., Yuskin, D., Whelan, K., Martin, M., Morgan, M., Chen, W., Al-Shama, H., Smith, B., Chalmers, D. and Behan, D. (2008) Lorcaserin, a novel selective human 5-HT2C agonist: *in vitro* and *in vivo* pharmacological characterization. *Journal of Pharmacology and Experimental Therapeutics* 325, 577–587.

Tolle, V. and Low, M.J. (2008) *In vivo* evidence for inverse agonism of agouti-related peptide in the central nervous system of proopiomelanocortin-deficient mice. *Diabetes* 57, 86–94.

Toshinai, K., Mondal, M.S., Nakazato, M., Date, Y., Murakami, N., Kojima, M., Kangawa, K. and Matsukura, S. (2001) Upregulation of ghrelin expression in the stomach upon fasting, insulin-induced hypoglycemia, and leptin administration. *Biochemical and Biophysical Research Communications* 281, 1220–1225.

Toshinai, K., Date, Y., Murakami, N., Shimada, M., Mondal, M.S., Shimbara, T., Guan, J.L., Wang, Q.P., Funahashi, H., Sakurai, T., Shioda, S., Matsukura, S., Kangawa, K. and Nakazato, M. (2003) Ghrelin-induced food intake is mediated via the orexin pathway. *Endocrinology* 144, 1506–1512.

Toshinai, K., Yamaguchi, H., Sun, Y., Smith, R.G., Yamanaka, A., Sakurai, T., Date, Y., Mondal, M.S., Shimbara, T., Kawagoe, T., Murakami, N., Miyazato, M., Kangawa, K. and Nakazato, M. (2006) Des-acyl ghrelin induces food intake by a mechanism independent of the growth hormone secretagogue receptor. *Endocrinology* 147, 2306–2314.

Tschöp, M., Smiley, D.L. and Heiman, M.L. (2000) Ghrelin induces adiposity in rodents. *Nature* 407, 908–913.

Tucci, S.A., Halford, J.C., Harrold, J.A. and Kirkham, T.C. (2006) Therapeutic potential of targeting the endocannabinoids: implications for the treatment of obesity, metabolic syndrome, drug abuse and smoking cessation. *Current Medicinal Chemistry* 13, 2669–2680.

Van Gaal, L.F., Rissanen, A.M., Scheen, A.J., Ziegler, O. and Rossner, S. for the RIO-Europe Study Group (2005) Effects of the cannabinoid-1 receptor blocker rimonabant on weight reduction and cardiovascular risk factors in overweight patients: 1-year experience from the RIO-Europe study. *Lancet* 365, 1389–1397.

Verty, A.N., McFarlane, J.R., McGregor, I.S. and Mallett, P.E. (2004) Evidence for an interaction between CB1 cannabinoid and melanocortin MC4-R receptors in regulating food intake. *Endocrinology* 45, 3224–3231.

Wang, H.J., Geller, F., Dempfle, A., Schauble, N., Friedel, S., Lichtner, P., Fontenla-Horro, F., Wudy, S., Hagemann, S., Gortner, L., Huse, K., Remschmidt, H., Bettecken, T., Meitinger, T., Schafer, H., Hebebrand, J. and Hinney, A. (2004) Ghrelin receptor gene: identification of several sequence variants in extremely obese children and adolescents, healthy normal-weight and underweight students, and children with short stature. *Journal of Clinical Endocrinology and Metabolism* 89, 157–162.

Wang, L., Saint-Pierre, D.H. and Tache, Y. (2002) Peripheral ghrelin selectively increases Fos expression in neuropeptide Y-synthesizing neurons in mouse hypothalamic arcuate nucleus. *Neuroscience Letters* 325, 47–51.

Welch, S.P. and Eads, M. (1999) Synergistic interactions of endogenous opioids and cannabinoid systems. *Brain Research* 848, 183–190.

Wenger, T., Jamali, K.A., Juaneda, C., Leonardelli, J. and Tramu, G. (1997) Arachidonylethanolamide (anandamide) activates the parvocellular part of the hypothalamic paraventricular nucleus. *Biochemical and Biophysical Research Communications* 237, 724–728.

Widdowson, P.S., Upton, R., Henderson, L., Buckingham, R., Wilson, S. and Williams, G. (1997) Reciprocal regional changes in brain NPY receptor density during dietary restriction and dietary-induced obesity in the rat. *Brain Research* 774, 1–10.

Williams, C.M., Rogers, P.J. and Kirkham, T.C. (1998) Hyperphagia in pre-fed rats following oral delta 9-THC. *Physiology and Behavior* 15, 343–346.

Williams, C.M. and Kirkham, T.C. (1999) Anandamide induces over-eating: mediation by central cannabinoid (CB1) receptors. *Psychopharmacology* 143, 315–317.

Wortley, K.E., Anderson, K.D., Garcia, K., Murray, J.D., Malinova, L., Liu, R., Moncrieffe, M., Thabet, K., Cox, H.J., Yancopoulos, G.D., Wiegand, S.J. and Sleeman, M.W. (2004) Genetic deletion of ghrelin does not decrease food intake but influences metabolic fuel preference. *Proceedings of the National Academy of Sciences of the United States of America* 101, 8227–8232.

Wren, A.M., Small, C.J., Abbott, C.R., Dhillo, W.S., Seal, L.J., Cohen, M.A., Batterham, R.L., Taheri, S., Stanley, S.A., Ghatei, M.A. and Bloom, S.R. (2001) Ghrelin causes hyperphagia and obesity in rats. *Diabetes* 50, 2540–2547.

Yang, J., Brown, M.S., Liang, G., Grishin, N.V. and Goldstein, J.L. (2008) Identification of the acyltransferase that octanoylates ghrelin, an appetite-stimulating peptide hormone. *Cell* 132, 387–396.

Yang, Y.K. and Harmone, C.M. (2003) Recent developments in our understanding of the melanocortin system in the regulation of food intake. *Obesity Reviews* 4, 239–248.

Zhang, J.V., Ren, P.G., Avsian-Kretchmer, O., Luo, C.W., Rauch, R., Klein, C. and Hsueh, A.J. (2005) Obestatin, a peptide encoded by the ghrelin gene, opposes ghrelin's effects on food intake. *Science* 310, 996–999.

4 The Gut as a Second Brain

Caroline J. Small, Katie Wynne and Stephen R. Bloom

Department of Metabolic Medicine, Division of Investigative Science, Imperial College, London, UK

Introduction

Food intake and body weight are tightly regulated by the brainstem, hypothalamus (see Chapters 1–3) and reward circuits (see Chapter 11). These centres integrate diverse cognitive inputs with humoral and neuronal signals of nutritional status. Many peptides are synthesized and released from the digestive tract. While their roles in the regulation of gastrointestinal function have been known for decades, it is now evident that they also influence eating behaviour. Our knowledge of the role of gut hormones in this complex homeostatic system has expanded enormously in recent years (Cummings and Overduin, 2007; Näslund and Hellström, 2007; Wren and Bloom, 2007; Chaudhri *et al*., 2008; Vincent *et al*., 2008). The chapter reviews the participation of gastrointestinal hormones in the control of appetite and body weight and, at the same time, considers the therapeutic potential of some promising gut hormones and how they alter the central nervous system (CNS) to control appetite (Field *et al*., 2008).

The Brain–Gut Axis

At the heart of appetite regulation lies the brain–gut axis. Peripheral signals of energy status (such as leptin) and satiety (the gut hormones) alter neuronal activity within the hypothalamus, brainstem and other deep brain structures, and thus influence feeding and energy intake. Once a meal is ingested, satiety hormones are released from the gut in a coordinated manner (Coll *et al*., 2007; Cummings and Overduin, 2007). These hormones act to optimize the digestive process as well as modulating appetite, energy expenditure and therefore behaviour. Previous work on cholecystokinin (CCK) and ghrelin has suggested that there is a gut nutrient sensor that signals to brain appetite centres in order to reduce appetite after meals (Moran and Schwartz, 1994; Kojima and Kangawa, 2002; Cummings and Overduin, 2007). It has become increasingly apparent that other components of the gut endocrine system play an important physiological role in postprandial satiety. Recent work has identified the gut hormone, peptide YY (PYY) (Batterham *et al.*, 2002, 2003a), oxyntomodulin (Oxm) (Cohen *et al.*, 2003; Dakin *et al.*, 2004) and pancreatic polypeptide (PP) (Batterham *et al.*, 2003b), which inhibit appetite, and ghrelin (Wren *et al.*, 2001a,b), which stimulates food intake. These hormones are active within the plasma range observed in humans and their manipulation represents a novel approach to the treatment of obesity. Below, their influence on appetite, their potential contribution to obesity and their changes following weight loss are described.

Cholecystokinin

CCK is released rapidly from the gastrointestinal tract postprandially and plasma levels remain elevated for up to 5 h (Liddle *et al.*, 1985). Although it is distributed widely within the gastrointestinal tract, the majority is found in the upper small intestine (Larsson and Rehfeld, 1978). The stimulatory effect of CCK on gallbladder contraction, pancreatic enzyme release and intestinal motility, along with its inhibitory effect on gastric emptying, are well established. CCK also has an inhibitory effect on food intake and was the first gut hormone to be implicated in satiety (Gibbs *et al.*, 1973). Although, there was some controversy over whether CCK affected the taste of food adversely, rather than increasing satiety, inhibition of food intake occurs in rodents at low doses, without any adverse effects (West *et al.*, 1984, 1987). Its effects on energy intake in humans are similar to rodents; it has been demonstrated that an intravenous infusion of the terminal octapeptide of CCK reduces both meal size and meal duration (Kissileff *et al.*, 2003).

The usefulness of CCK as a therapeutic agent for obesity is limited because of its short half-life; it is ineffective when given more than 30 min before a meal (West *et al.*, 1984, 1987). Neither does it appear to be effective when administered chronically, as repeated administration does not alter body weight in rats, although food intake is reduced, meal frequency increases, so overall intake is unchanged (West *et al.*, 1984, 1987). In fact, when given to rodents as a continuous intraperitoneal infusion, the anorectic effect is lost after 24 h (Crawley *et al.*, 1984). There is, however, some evidence that CCK could influence body weight by interacting with other signals of adiposity. Studies of specific receptor agonists suggest that the CCK_A receptor mediates the effects of CCK on appetite (Asin *et al.*, 1992) and chronic administration of CCK_A receptor antagonists or anti-CCK antibodies accelerates weight gain in rodents, though without evidence of significant hyperphagia (McLaughlin and Baile, 1981; Meereis-Schwanke *et al.*, 1998). The Otsuka Long–Evans Tokushima fatty (OLETF) rat, which lacks CCK_A receptors, is hyperphagic and obese (Moran *et al.*, 1998; Schwartz *et al.*, 1999). These effects of CCK on body weight may be the result of interaction with leptin, as peripheral administration of CCK is able to potentiate the central effect of leptin on body weight (Matson and Ritter, 1999). Recently, CCK has been shown to have a negative impact on energy balance, facilitating the access of leptin to hypothalamic areas, thus allowing leptin to act on hypothalamic targets involved in body weight control (Merino *et al.*, 2008). Furthermore, a peripheral synergistic interaction between CCK and urocortins to suppress food intake and gastric emptying through corticotropin-releasing factor receptor-2 in lean but not obese mice has been reported (Gourcerol *et al.*, 2007).

The CCK_A receptor is present on the vagus nerve, enteric neurones, brainstem and the dorsomedial nucleus of the hypothalamus (DMH), and there is evidence that CCK may signal appetite via both brainstem and hypothalamic regions (Moran, 2006; Coll *et al.*, 2007; Cummings and Overduin, 2007). Peripheral administration of CCK induces localized synthesis of c-fos, a marker of neuronal activation, in brainstem areas and release of CCK at low concentrations from the gut modulates vagus nerve activity, which then relays satiety signals to the brainstem. CCK may signal nutritional status via the hypothalamus by crossing the blood–brain barrier and acting on receptors expressed on the DMH, where it reduces the level of a potent orexigenic peptide, NPY (Moran, 2006; Coll *et al.*, 2007; Cummings and Overduin, 2007; Morris *et al.*, 2008).

Peptide YY

PYY was first isolated from porcine intestine in 1980 (Tatemoto and Mutt, 1980). PYY belongs to the PP-fold peptide family that also includes NPY and PP. These peptides have other common features: all have 36 amino acids containing several tyrosine residues and all undergo C-terminal amidation, which is necessary for biological activity. The PP-fold configuration consists of a polyproline helix and α-helix connected by a β turn, resulting in a characteristic U-shaped peptide. There is also marked evolutionary conservation of the amino acid sequence between the peptides, with 42% homology between rat PP and PYY.

Five cloned receptors for the PP-fold peptide family have been described, Y1–Y5 (Larhammar, 1996). They are all seven transmembrane domain receptors coupled to inhibitory G proteins, resulting in inhibition of adenylate cyclase. However, Y1 also increases intracellular calcium and Y2 regulates both calcium and potassium channels. The receptors are classified according to their affinity for PYY, PP and NPY fragments and analogues having diverse distributions and functions. While PYY binds with high affinity to all Y receptors, PYY_{3-36}, the active fragment, shows selectivity for Y2 and Y5 receptors.

PYY is expressed widely in endocrine L-cells throughout the gastrointestinal tract, where it is co-localized with glucagon-like peptide-1 (GLP-1) and Oxm (Chandarana and Batterham, 2008). PYY immunoreactive cells are almost absent in the stomach, there are relatively few in the duodenum and jejunum, but they increase dramatically in frequency in the ileum and colon and are at very high levels in the rectum. PYY has been described in the myenteric plexus and endocrine pancreas of many species. Immunoreactivity for PYY has also been reported in human adrenal medulla. In the CNS, PYY immunoreactive nerve terminals are present in the hypothalamus, medulla, pons and spinal cord (Ekblad and Sundler, 2002).

PYY is released into the circulation in response to food intake, rising to a plateau after 1–2 h. Release is partly proportional to calorie intake, but it is also influenced by meal composition (Batterham *et al.*, 2006). Higher plasma PYY concentrations are seen following isocaloric meals of fat compared to intake of protein or carbohydrate (Chandarana and Batterham, 2008). An intraduodenal meal increases plasma PYY even before nutrients have reached the PYY-containing cells of the ileum. This suggests release through a neural reflex, possibly via the vagus. In addition to nutrients, PYY release is also stimulated by gastric acid, CCK and infusion of bile acids into the ileum or colon in animal studies. PYY is not released by gastric distension. Other factors also alter circulating PYY; plasma PYY concentrations are increased by insulin-like growth factor-1, bombesin and calcitonin-gene related peptide, and decreased by GLP-1.

Early studies on the actions of peripherally administered PYY demonstrated numerous effects on the gastrointestinal tract. PYY administration significantly delays gastric emptying, gastric and pancreatic secretion and the cephalic phase of gallbladder emptying, but increases ileal postprandial fluid and electrolyte absorption. Peripheral administration of PYY was also shown to decrease appetite, with PYY_{3-36} administered peripherally to mouse, rat or human, inhibiting food intake markedly (Batterham *et al.*, 2002; Batterham and Bloom, 2003). The pattern of c-fos expression in the brain after peripheral administration of PYY_{3-36} shows a marked induction of c-fos in the arcuate nucleus (ARC). Injection of PYY_{3-36} directly into the ARC inhibits food intake, and chronic administration of PYY_{3-36} leads to a decrease in food intake and body weight. Addition of PYY_{3-36} to *ex vivo* hypothalamic explants inhibits the release of NPY and stimulates that of α-melanocyte-stimulating hormone (α-MSH). Peripheral administration of PYY_{3-36} in rats causes a decrease in expression of ARC NPY mRNA. PYY_{3-36} has a high affinity for the Y2 receptor (Y2R). Inhibition of appetite is seen with a Y2R specific agonist and is absent in the Y2R knockout mouse (Batterham *et al.*, 2002; Batterham and Bloom, 2003). It appears that circulating PYY_{3-36} inhibits appetite by acting directly on the ARC via the Y2R, a pre-synaptic inhibitory autoreceptor (Batterham *et al.*, 2002; Batterham and Bloom, 2003).

These results were reproduced by Halatchev *et al.* (2004), who confirmed that intraperitoneal administration of PYY_{3-36} inhibited food intake in rodents dose-dependently in both dark-phase food intake and following a fast. However, this effect was seen only in animals acclimatized to handling and intraperitoneal injections. Challis *et al.* (2003) showed that intraperitoneal injections of PYY_{3-36} reduced food intake at 6 and 24 h post-injection following a 24-h fast in mice, but not in non-fasted, freely feeding animals, although the first measurement was taken 6 h post-injection, when PYY_{3-36} may already have had its effect. Some investigators have not been able to reproduce the original observation that peripheral administration of PYY_{3-36} inhibited food intake in rodents (Tschöp *et al.*, 2004). However, many other investigators have confirmed the original findings (Challis *et al.*, 2003, 2004; Cox and Randich, 2004; Halatchev *et al.*, 2004; Riediger *et al.*, 2004). It is likely that the anorectic effects of PYY_{3-36} are influenced by stress (Halatchev *et al.*, 2004). In order to overcome this influence, extensive handling and habituation of the animals to the experimental procedures is necessary prior to PYY administration.

PYY is hypothesized to mediate its anorectic actions by switching off the ARC NPY neurones, decreasing hypothalamic NPY and thus food intake. This is supported by the observation that PYY_{3-36} decreases NPY mRNA and NPY release from *ex vivo* hypothalamic explants (Batterham *et al.*, 2002). This switching off of the NPY neurones leads to an activation of the ARC pro-opiomelanocortin (POMC)/cocaine- and amphetamine-regulated transcript (CART) neurones. This is supported by the observations that PYY_{3-36} increases POMC mRNA, α-MSH release from hypothalamic explants and increases the electrical activity of the POMC neurones (Batterham *et al.*, 2002; Challis *et al.*, 2003). However, subsequent evidence has shown that its mechanism of action may be more complex. *POMC*$^{-/-}$ mice have been found to retain a normal, acute anorectic response to peripherally administered PYY_{3-36} (Halatchev *et al.*, 2004; Martin *et al.*, 2004), suggesting that melanocortin peptides may not be required for the actions of PYY_{3-36}. In addition, it has been shown further that melanocortin-4 receptor knockout mice (MC4-R$^{-/-}$) are responsive to the anorexigenic effects of PYY_{3-36} (Halatchev *et al.*, 2004; Martin *et al.*, 2004). Together, these results suggest that the melanocortin system is not essential for the anorectic actions of PYY_{3-36}.

Administration of PYY_{3-36} into the CNS has strikingly opposing actions to those seen peripherally. Injections of PYY_{3-36} into the third, lateral or fourth cerebral ventricles (Corp *et al.*, 1990; O'Shea *et al.*, 1997), the paraventricular nucleus (PVN) (Stanley and Leibowitz, 1985) or the hippocampus (Hagan *et al.*, 1998) in rodents stimulate food intake potently. ICV injections of PYY_{3-36} also increase food intake, but this orexigenic action is reduced in both Y1 mice and Y5 knockout mice (Kanatani *et al.*, 2000), suggesting that these receptors may play a role in the CNS feeding effects of PYY_{3-36}. Batterham *et al.* (2002) proposed that the discrepancy between the CNS and peripheral effects of PYY_{3-36} could be explained by peripheral injections specifically activating the Y2 receptor in the hypothalamic ARC, where the blood–brain barrier was relatively permeable. This anatomical specificity was confirmed by showing that injections of a Y2 agonist administered directly into the ARC inhibited food intake in a dose-dependent fashion. This effect was not reproducible with injections of the Y2 agonist into the hypothalamic PVN, which was not directly in communication with circulating hormones. Thus, the Y2 receptor mediates the anorectic actions of PYY_{3-36} with rodent studies, implicating the hypothalamus, vagus and brainstem as key target sites. Functional imaging in humans has confirmed that PYY_{3-36} activates brainstem and hypothalamic regions. The greatest effects, however, were observed within the orbitofrontal cortex, a brain region involved in reward processing (Chandarana and Batterham, 2008).

Both the hypothalamus and medulla oblongata express a high level of Y2 receptors. Diet-induced obese mice exhibit low plasma PYY, which may cause a compensatory upregulation of PYY and Y2 receptor densities in the medulla (Rahardjo *et al.*, 2007). A low-level response of PYY-medullary regulation to positive energy balance may contribute to the development of high-fat diet-induced obesity; conversely, a normal response of this regulatory axis in the obese-resistant mice may be responsible for the maintenance of body weight while on a high-fat diet (Rahardjo *et al.*, 2007).

Early light-phase injection of PYY_{3-36} to mice fed *ad libitum* results in a trend toward increased levels of hypothalamic NPY and agouti-related peptide mRNA and a decrease in POMC mRNA at the beginning of the dark phase (Parkinson *et al.*, 2008). Furthermore, plasma levels of ghrelin were increased significantly and there was a trend toward decreased plasma PYY_{3-36} levels at the beginning of the dark phase, indicating that PYY_{3-36} injection resulted in an acute anorexigenic effect followed by a delayed orexigenic effect.

In humans, food intake in a free-choice meal is reduced by 30% following an intravenous infusion of PYY_{3-36}, which results in plasma levels similar to those achieved physiologically after a meal (Batterham *et al.*, 2002). It has been investigated if obese subjects are resistant to the anorectic effect of PYY_{3-36} infusion (Batterham and Bloom, 2003). Caloric intake during a buffet lunch offered 2 h after the infusion of PYY_{3-36} was decreased by 30% in obese subjects and by 31% in lean subjects. Overall, PYY reduced 24-h caloric intake significantly in both the obese (16.5%) and lean groups (23.5%). This is in contrast to the marked resistance to the action of leptin in the obese, greatly limiting its therapeutic effectiveness. In this study, PYY_{3-36} infusion also caused a reduction in the fasting preprandial concentrations of the hunger hormone, ghrelin, suggesting an interaction between these two gut hormones and a possible mechanism by which PYY reduced hunger in humans. In addition, they also showed that endogenous fasting and postprandial PYY levels were significantly lower in obese subjects and plasma PYY levels correlated negatively with body mass index (BMI). This suggests that PYY deficiency may contribute to the pathogenesis of this condition. Increasing plasma PYY levels, either by exogenous administration or by stimulating endogenous release, is therefore an attractive strategy for the treatment of obesity. However, it should be taken into consideration that supraphysiological doses of intravenous PYY_{3-36} cause nausea, but no additional reduction in food intake (Le Roux *et al.*, 2008).

Further investigation is needed to clarify whether PYY actually causes reduced calorie intake or whether the rate of food delivery to the ileo-colonic segment influences PYY levels, thus affecting satiation (Grudell and Camilleri, 2007). Animal studies have revealed brain regions that control homeostatic feeding, but the rampant overeating contributing to the obesity epidemic suggests the participation of 'non-homeostatic' control centres. Under conditions of high plasma PYY concentrations, mimicking the fed state, changes in neural activity within the caudolateral orbital frontal cortex predict feeding behaviour independently of meal-related sensory experiences (Batterham *et al.*, 2007). In contrast, in conditions of low levels of PYY, hypothalamic activation predicts food intake. Thus, the presence of a postprandial satiety factor switches food intake regulation from a homeostatic to a hedonic, corticolimbic area. These findings provide insight into the neural networks in humans that respond to a specific satiety signal to regulate food intake. An increased understanding of how such homeostatic and higher brain functions are integrated may pave the way for the development of new treatment strategies for obesity.

Pancreatic polypeptide

Pancreatic polypeptide (PP) is secreted by cells situated at the periphery of the pancreatic islets and, to a lesser extent, by the exocrine pancreas and distal gut

(Chaudhri *et al.*, 2008). Postprandially, PP is released in a biphasic manner in proportion to food intake and remains elevated for up to 6 h. Peripheral PP administration reduces food intake in lean and genetically obese mice and reduces food intake in normal-weight humans when given as an intravenous infusion (Batterham *et al.*, 2003b).

There is some evidence that artificially increasing PP levels could reduce body weight. PP has high affinity for Y4 and Y5 receptors (Larhammar, 1996), which are present in both the ARC and brainstem. However, evidence suggests several types of Y receptors may be involved in the feeding response to PP (Kanatani *et al.*, 2000). Part of the influence of circulating PP on appetite may be mediated via the vagal pathway to the brainstem (Asakawa *et al.*, 2003). Mice overexpressing PP to supraphysiological levels are lean and hypophagic, with reduced gastric emptying (Ueno *et al.*, 1999). Repeated administration of PP to genetically obese mice results in reduced insulin resistance and hyperlipidaemia, increased energy expenditure, hypophagia and reduced weight gain (Asakawa *et al.*, 2003).

Although PP could represent a possible therapeutic target for the treatment of obesity, its effect on appetite and body weight in obese humans is still unclear. Obese subjects demonstrate low basal circulating PP levels and a reduced second phase release after a meal, suggesting that decreased PP signalling could contribute to a lack of satiety and the development of obesity (Chaudhri *et al.*, 2008). As might be expected, circulating PP levels are higher in very lean individuals, such as anorexic subjects (Uhe *et al.*, 1992; Fujimoto *et al.*, 1997). However, this apparent relationship of PP with body weight in humans remains controversial; some investigators have shown similar levels in lean and obese patients with stable body weight (Jorde and Burhol, 1984; Meryn *et al.*, 1986; Uhe *et al.*, 1992; Koska *et al.*, 2004), while others have observed that low-dose PP inhibits food intake in humans (Jesudason *et al.*, 2007). A prospective study in Pima Indians, an ethnic group with a high prevalence of obesity, demonstrated that high fasting baseline levels of PP were correlated positively with future weight change, whereas a high postprandial PP release was correlated negatively with weight change over the subsequent years of follow-up (Koska *et al.*, 2004). PP has been administered as a twice-daily infusion in subjects with obesity and hyperphagia secondary to Prader–Willi syndrome. These patients exhibit a reduced basal and blunted postprandial PP release, which may contribute to their hyperphagia (Zipf *et al.*, 1981, 1983). PP replacement results in a reduction in food intake in these patients (Berntson *et al.*, 1993), but its effect in common polygenic obesity remains to be elucidated fully.

Glucagon-like peptide-1 and oxyntomodulin

GLP-1 and Oxm are products of the preproglucagon gene, which is expressed in the CNS, the L-cells of the small intestine and the pancreas (Wynne and Bloom, 2006; Holst, 2007). Preproglucagon is cleaved, by prohormone convertase 1 and 2, into different products, depending on the tissue. In the pancreas, the glucagon sequence is cleaved out, whereas the part containing the GLP-1 and

GLP-2 is secreted as a single, large inactive peptide. The posttranslational processing in the gut and brain are similar. The glucagon sequence remains in a larger peptide, glicentin, thought to be inactive. The two glucagon-like peptides are cleaved out and secreted separately. Glicentin is later cleaved into GRPP (inactive N-terminal fragment) and Oxm.

Oxm and GLP-1 are released from the L-cells of the distal small intestine, 5–30 min after food ingestion and in proportion to meal calorie intake. The secretion of Oxm may be in response to fat which has undergone hydrolysis to fatty acids within the gut. Oxm increases energy expenditure, while reducing energy intake, resulting in negative energy balance (Wynne *et al.*, 2005, 2006). As discussed above, L-cells also coexpress other anorexigenic peptides, such as PYY. Increasing plasma levels of GLP-1 and Oxm results in postprandial satiety via activation of the GLP-1 receptor (Chaudhri *et al.*, 2006). The raised plasma levels of these two gut hormones also inhibit gastric acid secretion and motility.

GLP-1 receptors are found in the brainstem, ARC and PVN. Intracerebroventricular (ICV) or direct administration into the PVN of GLP-1 to rats inhibits food intake potently, while the specific GLP-1 receptor antagonist, exendin 9-39, causes an increase in food intake (Coll *et al.*, 2007; Holst, 2007). In addition, chronic ICV administration of GLP-1 decreases body weight and chronic administration of exendin 9-39 increases body weight. ICV administration of Oxm inhibits food intake in the rat with greater potency than does GLP-1. Oxm appears to act via a GLP-1-like receptor, because its anorectic actions are blocked by coadministration of the GLP-1 receptor antagonist, exendin 9-39 (Wynne and Bloom, 2006). However, the affinity of Oxm for GLP-1R is approximately two orders of magnitude weaker than that of GLP-1, even though Oxm exerts a comparable effect on food intake. It is therefore possible there may be a separate Oxm receptor, which has not yet been cloned.

Peripheral administration of GLP-1 in humans and rats inhibits food intake and, in rats, also results in c-fos expression in the brainstem. These and other findings suggest that the main site for appetite inhibition by peripheral GLP-1 is the dorsal vagal complex, acting, in part, directly through the area postrema. The role for GLP-1 in physiological control of human appetite is not clear (Coll *et al.*, 2007; Holst, 2007). GLP-1 does decrease gastric emptying dependently and this may have effects on food intake. Peripheral administration inhibits food intake in normal individuals, diabetics and non-diabetic obese men. A meta-analysis of the effect of GLP-1 infusion has shown an average reduction in calorie intake of 11.7%, which is dose dependent and does not differ between obese and lean individuals (Verdich *et al.*, 2001a).

Some reports have suggested that GLP-1 secretion is reduced in obese subjects, with weight loss normalizing the levels (Verdich *et al.*, 2001b). However, other findings have not supported these observations (Feinle *et al.*, 2002; Velasquez-Mieyer *et al.*, 2003; Vilsboll *et al.*, 2003a,b). Interestingly, the anorectic effects of GLP-1 are preserved in obesity; prandial subcutaneous GLP-1 given for 5 days to obese but otherwise healthy human subjects resulted in a reduction of calorie intake of 15% and a weight loss of 0.5 kg (Näslund *et al.*, 2004). A reduced secretion of GLP-1 could therefore contribute to the pathogenesis of obesity, with GLP-1 receptor agonists being potential targets for treatment.

The therapeutic potential of GLP-1 is limited by its rapid breakdown. GLP-1 is deactivated by dipeptidyl peptidase IV (DPP-IV), which cleaves off the two N-terminal amino acid residues. Inhibition of DPP-IV has been shown to be an effective treatment for type 2 diabetes mellitus, even without an effect on body weight (Meier *et al.*, 2002). Various resistant analogues such as exendin 4 (exenatide and albumin-based forms such as liraglutide) improve glycaemic control and reduce body weight (Holst, 2007).

Oxm is also a potent inhibitor of food intake when administered intraperitoneally to rats (Dakin *et al.*, 2004), resulting in c-fos expression in the ARC, a region partially outside the blood–brain barrier, while producing little activation of neurones in the nucleus of the solitary tract (NTS) in the brainstem. These experiments demonstrate that Oxm has a very different pattern of neuronal activation from that of GLP-1. When the antagonist exendin 9-39 was injected into the ARC, circulating Oxm no longer inhibited food intake, suggesting an arcuate site of action. By contrast, the effect of circulating GLP-1, acting via the brainstem, was unaffected (Dakin *et al.*, 2004).

In human subjects, peripheral administration of GLP-1 via an IV route results in satiety. Intravenous infusion of Oxm in 13 healthy subjects in a randomized double-blind placebo-controlled crossover study reduced *ad libitum* calorie intake of a free-choice buffet meal significantly (mean decrease of 19.3 ± 5.6%, $p < 0.01$) and caused a significant reduction in hunger scores. In addition, cumulative 12 h caloric intake was reduced significantly by infusion of Oxm (mean decrease of 11.3 ± 6.2%, $p < 0.05$). Fasting levels of ghrelin were suppressed significantly by Oxm (44 ± 10% mean reduction of postprandial decrease, $p < 0.01$) (Cohen *et al.*, 2003).

The feeling of satiety produced by Oxm and GLP-1 is likely to be due to their effects on the CNS, as well as their effect on gastric emptying. The GLP-1 receptor is present in the NTS and ARC. The NTS receives afferent input from the vagal and glossopharyngeal nerves and integrates both neuronal and humoral factors. This area is also able to synthesize GLP-1, thus the GLP-1-containing neurones may influence their own activity.

GLP-1-containing neurones of the NTS project to the ARC and hypothalamic nuclei, such as the dorsal medial nucleus, and PVN, which are involved in appetite control. There are two populations of neuronal circuits within the ARC. One circuit inhibits food intake and consists of neurones which coexpress POMC and CART (see Part I). The other circuit, coexpressing NPY and agouti-related peptide (AgRP), stimulates food intake. The GLP-1 projections may act on these neurones in order to inhibit appetite. There is also evidence that the ARC is influenced by GLP-1 from the periphery via the area postrema and subfornical organ (Tang-Christensen *et al.*, 2001). In contrast, circulating Oxm may act directly from the periphery on the ARC feeding circuits (Dakin *et al.*, 2004). Thus, there are several routes by which the gut can communicate with appetite circuits.

Oxm may also exert its effects on appetite via suppression of ghrelin, an orexigenic peptide produced by endocrine cells in the oxyntic glands of the stomach. Oxm administration, producing plasma concentrations comparable to postprandial levels, reduces preprandial ghrelin by around 44% in human subjects

(Cohen *et al.*, 2003). These effects are also observed in rodent studies (Dakin *et al.*, 2004).

In addition to their effects on satiety, Oxm and GLP-1 also promote meal-induced insulin secretion. GLP-1 has been found to upregulate insulin gene expression and potentiate all steps of insulin biosynthesis. An intravenous infusion of GLP-1 is capable of normalizing blood glucose levels completely in patients with long-standing type 2 diabetes who cannot be controlled by sulphonylurea therapy. Furthermore, a 6-week subcutaneous infusion of GLP-1 to type 2 diabetics normalizes glycosylated fructosamine and reduces HbA1c by 1.3%. This infusion of GLP-1 was also found to reduce body weight by 2 kg (Zander *et al.*, 2002), an effect which might be particularly useful in type 2 diabetes, where obesity is commonly a significant issue.

Ghrelin

Many cues for meal initiation are learned by association, but signals from the gut endocrine system are also involved. Ghrelin is produced and released primarily by gastric oxyntic cells and both the expression and circulating levels are upregulated by fasting (Kojima *et al.*, 1999; Wren *et al.*, 2001a). Total gastrectomy reduces plasma ghrelin by about 60%, as the remaining circulating ghrelin is released by duodenum, ileum, caecum and colon. Ghrelin is a 28-amino acid peptide with addition of an acyl side chain, *n*-octanoic acid, to the third serine residue, which is necessary for binding to the GHS-R type 1a and for ghrelin's effects on food intake (López *et al.*, 2007). The GHS-R is expressed in hypothalamic and brainstem nuclei, including the ARC.

Plasma ghrelin levels are regulated by food intake; in human subjects with a regular meal schedule, plasma ghrelin rises during fasting and falls postprandially (Ariyasu *et al.*, 2001; Cummings *et al.*, 2001; Tschöp *et al.*, 2001a). This reduction in circulating ghrelin is regulated by both energy intake and circulating nutritional signals, such as glucose (Tschöp *et al.*, 2000; Sakata *et al.*, 2002), but not mechanic stimuli such as gastric distension (Tschöp *et al.*, 2000). In addition, ghrelin levels demonstrate a diurnal variation: in humans, basal ghrelin levels are high in the morning and low at night (Cummings *et al.*, 2001), whereas in rodents, ghrelin peaks at the end of light and dark periods (Murakami *et al.*, 2002).

In rodents, exogenous ghrelin administration is a potent stimulus to feeding, with maximum effects observed within an hour of peripheral administration (Wren *et al.*, 2000, 2001a) in response to plasma levels comparable to those observed after a 24-h fast (Wren *et al.*, 2001a). Conversely, blockade of the endogenous action of ghrelin by central infusion of anti-ghrelin antibodies attenuates fasting-induced re-feeding. Chronic ghrelin administration induces adiposity (Tschöp *et al.*, 2000; Wren *et al.*, 2001a) without attenuation of the effects on food intake (Wren *et al.*, 2000, 2001a). Ghrelin has local gut effects in addition to its effects on appetite, stimulating gastric emptying and decreasing gastric acid secretion in rodents (Masuda *et al.*, 2000). It has also been reported to reverse the temporary paralysis of the ileus observed after abdominal surgery (Trudel *et al.*, 2002).

Exogenous infusion of ghrelin increases food intake at a buffet meal by 28% in human subjects, compared to a saline control administration (Wren *et al.*, 2001b). Despite the increased food intake following the ghrelin infusion, there is no increase in satiety after the meal and total food intake remains higher compared to the control group (Wren *et al.*, 2001b). Ghrelin may initiate feeding by producing a feeling of hunger, as rising preprandial plasma ghrelin levels correlate with hunger scores in humans eating spontaneously (Cummings *et al.*, 2004).

A role for ghrelin in the aetiology of human obesity has been proposed. Ghrelin has an inverse relationship with BMI and is significantly lower in obese subjects compared to lean individuals (Shiiya *et al.*, 2002). Ghrelin is the 'hormone of hunger' and this picture would fit with the notion of its role in the homeostatic control of body weight – high circulating ghrelin in thin individuals would favour increased food intake and positive energy balance. Weight loss in obese people results in an elevation in ghrelin level (Hansen *et al.*, 2002), which may contribute to the difficulty seen in maintaining body weight after weight loss. Food fails to suppress ghrelin levels to the same extent in obese humans (English *et al.*, 2002), which again could impair postprandial satiety and contribute to overeating. Individuals with Prader–Willi syndrome have grossly elevated ghrelin levels and this could be a cause of their hyperphagia (Cummings *et al.*, 2002a). Mutations in the ghrelin gene have been identified in humans, but a role for these in weight determination remains controversial (Hinney *et al.*, 2002; Wang *et al.*, 2004).

Leptin

Circulating leptin levels reflect both long-term energy stores and short-term changes in energy balance. Plasma leptin levels are correlated highly with adipose tissue mass (see Chapter 5), but short-term food restriction can also suppress circulating leptin acutely, which can be reversed by re-feeding. Exogenous leptin administration to rodents, both centrally and peripherally, reduces spontaneous and fasting-induced hyperphagia, while chronic peripheral administration reduces food intake, resulting in loss of fat mass and body weight (Frühbeck *et al.*, 1998). Although it is expressed predominantly by adipocytes, leptin is also produced at lower levels by chief and endocrine P-cells in the gastric epithelium (Bado *et al.*, 1998). Gastric leptin release also contributes to circulating leptin, as shown from experiments in which feeding fasted animals substantially depleted gastric leptin and increased serum leptin (Bado *et al.*, 1998). Furthermore, gastric leptin regulates intestinal nutrient absorption, delays gastric emptying and signals short-term satiety via vagal afferent nerves (Frühbeck, 2002). Leptin, infused into the upper gastrointestinal tract arterial supply reduces meal size and enhances satiation evoked by CCK. Leptin activates vagal afferent neurones, and this activation is likely to participate in meal termination by enhancing vagal sensitivity to CCK. These findings are consistent with the view that leptin and CCK exert an influence on food intake synergistically by accessing multiple neural systems (viscerosensory, motivational, affective and motor) at multiple points along the neuroaxis (Peters *et al.*, 2006).

Gut Hormones in Obesity and Following Weight Loss

Obesity may be thought of as a state of chronic adaptation to the hormonal changes of increased fat mass. Results to date suggest that increased BMI is associated with increased plasma leptin and decreased plasma ghrelin, adiponectin, PP and PYY (Tschöp *et al.*, 2001b; Näslund and Hellström, 2007). Obese people have a similar sensitivity to the appetite inhibitory effects of exogenous PYY_{3-36} infusion as lean people (Batterham *et al.*, 2003a), i.e no 'PYY resistance' takes place in obesity. However, the sensitivity to ghrelin infusion remains to be established. Although the gastric oxyntic glands appear to be capable of producing ever-increasing concentrations of ghrelin in states of undernutrition, it might be possible that the low levels of PYY in obesity are due to L-cell failure. In this setting, PYY replacement might provide a realistic therapeutic antiobesity treatment.

The most successful treatment so far to achieve lasting weight reduction is gastric and intestinal bypass surgery (Le Roux *et al.*, 2006, 2007; Chaudhri *et al.*, 2008). In one series, mean weight loss at 15 years post-bypass surgery was 29.5 kg (Mitchell *et al.*, 2001). However, the morbidity and mortality associated with bypass surgery, in addition to practical and financial constraints, usually limit this approach to the severely obese patient (BMI $\geq$ 40 kg/m^2 or BMI $\geq$ 35 kg/m^2 with co-morbidities). While surgery reduces calorie absorption initially, its sustained effect is due to reduction in calorie intake. The levels of PP are increased after jejunoileal bypass surgery for obesity (Jorde and Burhol, 1982). Thus, PP could contribute to the loss in appetite and weight which occurs after bypass of the small intestine. Bypass surgery also alters circulating gut hormones such as Oxm, PYY (Sarson *et al.*, 1981; Wynne and Bloom, 2006), ghrelin (Cummings *et al.*, 2002b; Cummings and Shannon, 2003) and GLP-1 (Borg *et al.*, 2007; Rodieux *et al.*, 2008), suggesting its long-term action on appetite might be secondary to these alterations in circulating hormones (Le Roux *et al.*, 2006, 2007; Chaudhri *et al.*, 2008). Thus, the success of bypass surgery is as much hormonal as mechanical. Induced malabsorption post-surgery is usually only temporary, whereas the reduction of appetite is permanent. Plasma ghrelin levels were measured recently in patients who had undergone gastric bypass. Bariatric surgery reduces plasma ghrelin despite weight loss and this may contribute to appetite suppression, reduced food intake and the sustained weight loss which occurs post-surgery (Cummings and Shannon, 2003). Circulating ghrelin levels were 77% lower in the bypass group compared with BMI-matched controls, and the usual pre-meal peaks were lost (Cummings *et al.*, 2002b). However, other reports suggest that the effects are more complex and that the changes depend on the extent to which bariatric surgery affects fundus functionality (Frühbeck *et al.*, 2004a,b,c). After bypass surgery, there is significant elevation in plasma PYY and Oxm. In rats, bypass surgery results in a three-fold increase in circulating PYY, with a 21% reduction in body weight after 28 days (Le Roux *et al.*, 2006). Therefore, the success of bypass surgery may be due, in part, to a decrease in circulating ghrelin and an increase in circulating PYY and Oxm. These changes in gut hormones act on the ARC of the hypothalamus, either directly or via the brainstem. Thus, altered gut signals following bypass surgery could explain why bypass patients frequently describe amazingly reduced hunger and permanent weight loss following surgery.

Summary and Conclusions

PYY is released postprandially from the gastrointestinal L-cells with other anorectic peptides, GLP-1 and Oxm. Following peripheral administration of PYY_{3-36}, the circulating form of PYY, to mouse, rat or human, there is marked inhibition of food intake. Obese subjects have lower basal fasting PYY levels and have a smaller postprandial rise. Furthermore, obesity does not appear to be associated with resistance to PYY (as it is with leptin) and exogenous infusion of PYY_{3-36} also results in a reduction in food intake in obese individuals. GLP-1 and Oxm, products of the preproglucagon gene, decrease food intake and body weight in rodents when administered either peripherally or directly into the CNS. In addition, both have been shown to decrease food intake in humans. Both GLP-1 and Oxm are thought to mediate their effects through the GLP-1 receptor. Ghrelin, an anorexigenic hormone produced by the stomach, increases in the circulation following a period of fasting. Administration of ghrelin either peripherally or directly into the CNS increases food intake in rodents and chronic administration leads to obesity. Ghrelin is thought to act through the growth hormone secretagogue receptor. Further infusion into normal healthy volunteers increases both food intake and appetite.

Several new, centrally acting chemical entities are being developed by pharmaceutical companies to target receptor systems involved in appetite regulation. However, these same systems also affect many other CNS functions, using the same receptors; for example, the serotonin system. The peripheral administration of natural gut hormones as therapeutic agents has the advantage of targeting only the relevant brain appetite systems. Gut hormones are released every day after meals without side effects and continue to exert their effect without escape. Thus, the administration of naturally occurring gut hormones may offer a long-term therapeutic approach to weight control without deleterious side effects.

References

Ariyasu, H., Takaya, K., Tagami, T., Ogawa, Y., Hosoda, K., Akamizu, T., Suda, M., Koh, T., Natsui, K., Toyooka, S., Shirakami, G., Usui, T., Shimatsu, A., Doi, K., Hosoda, H., Kojima, M., Kangawa, K. and Nakao, K. (2001) Stomach is a major source of circulating ghrelin, and feeding state determines plasma ghrelin-like immunoreactivity levels in humans. *Journal of Clinical Endocrinology and Metabolism* 86, 4753–4758.

Asakawa, A., Inui, A., Yuzuriha, H., Ueno, N., Katsuura, G., Fujimiya, M., Fujino, M.A., Niijima, A., Meguid, M.M. and Kasuga, M. (2003) Characterization of the effects of pancreatic polypeptide in the regulation of energy balance. *Gastroenterology* 124, 1325–1336.

Asin, K.E., Gore, P.A. Jr, Bednarz, L., Holladay, M. and Nadzan, A.M. (1992) Effects of selective CCK receptor agonists on food intake after central or peripheral administration in rats. *Brain Research* 571, 169–174.

Bado, A., Levasseur, S., Attoub, S., Kermorgant, S., Laigneau, J.P., Bortoluzzi, M.N., Moizo, L., Lehy, T., Guerre-Millo, M., Le Marchand-Brustel, Y. and Lewin, M.J. (1998) The stomach is a source of leptin. *Nature* 394, 790–793.

Batterham, R.L. and Bloom, S.R. (2003) The gut hormone peptide YY regulates appetite. *Annals of the New York Academy of Sciences* 994, 162–168.

Batterham, R.L., Cowley, M.A., Small, C.J., Herzog, H., Cohen, M.A., Dakin, C.L., Wren, A.M., Brynes, A.E., Low, M.J., Ghatei, M.A., Cone, R.D. and Bloom, S.R. (2002) Gut hormone PYY(3-36) physiologically inhibits food intake. *Nature* 418, 650–654.

Batterham, R.L., Cohen, M.A., Ellis, S.M., Le Roux, C.W., Withers, D.J., Frost, G.S., Ghatei, M.A. and Bloom, S.R. (2003a) Inhibition of food intake in obese subjects by peptide YY3-36. *New England Journal of Medicine* 349, 941–948.

Batterham, R.L., Le Roux, C.W., Cohen, M.A., Park, A.J., Ellis, S.M., Patterson, M., Frost, G.S., Ghatei, M.A. and Bloom, S.R. (2003b) Pancreatic polypeptide reduces appetite and food intake in humans. *Journal of Clinical Endocrinology and Metabolism* 88, 3989–3992.

Batterham, R.L., Heffron, H., Kapoor, S., Chivers, J.E., Chandarana, K., Herzog, H., Le Roux, C.W., Thomas, E.L., Bell, J.D. and Withers, D.J. (2006) Critical role for peptide YY in protein-mediated satiation and body-weight regulation. *Cell Metabolism* 4, 223–233.

Batterham, R.L., Ffytche, D.H., Rosenthal, J.M., Zelaya, F.O., Barker, G.J., Withers, D.J. and Williams, S.C. (2007) PYY modulation of cortical and hypothalamic brain areas predicts feeding behaviour in humans. *Nature* 450, 106–109.

Berntson, G.G., Zipf, W.B., O''Dorisio, T.M., Hoffman, J.A. and Chance, R.E. (1993) Pancreatic polypeptide infusions reduce food intake in Prader-Willi syndrome. *Peptides* 14, 497–503.

Borg, C.M., Le Roux, C.W., Ghatei, M.A., Bloom, S.R. and Patel, A.G. (2007) Biliopancreatic diversion in rats is associated with intestinal hypertrophy and with increased GLP-1, GLP-2 and PYY levels. *Obesity Surgery* 17, 1193–1198.

Challis, B.G., Pinnock, S.B., Coll, A.P., Carter, R.N., Dickson, S.L. and O'Rahilly, S. (2003) Acute effects of PYY3-36 on food intake and hypothalamic neuropeptide expression in the mouse. *Biochemical Biophysical Research Communications* 311, 915–919.

Challis, B.G., Coll, A.P., Yeo, G.S., Pinnock, S.B., Dickson, S.L., Thresher, R.R., Dixon, J., Zahn, D., Rochford, J.J., White, A., Oliver, R.L., Millington, G., Aparicio, S.A., Colledge, W.H., Russ, A.P., Carlton, M.B. and O'Rahilly, S. (2004) Mice lacking pro-opiomelanocortin are sensitive to high-fat feeding but respond normally to the acute anorectic effects of peptide-YY(3-36). *Proceedings of the National Academy of Sciences of the United States of America* 101, 4695–4700.

Chandarana, K. and Batterham, R. (2008) Peptide YY. *Current Opinion in Endocrinology, Diabetes and Obesity* 15, 65–72.

Chaudhri, O.B., Parkinson, J.R., Kuo, Y.T., Druce, M.R., Herlihy, A.H., Bell, J.D., Dhillo, W.S., Stanley, S.A., Ghatei, M.A. and Bloom, S.R. (2006) Differential hypothalamic neuronal activation following peripheral injection of GLP-1 and oxyntomodulin in mice detected by manganese-enhanced magnetic resonance imaging. *Biochemical Biophysical Research Communications* 350, 298–306.

Chaudhri, O.B., Wynne, K. and Bloom, S.R. (2008) Can gut hormones control appetite and prevent obesity? *Diabetes Care* 31 (Suppl. 2), S284–S289.

Cohen, M.A., Ellis, S.M., Le Roux, C.W., Batterham, R.L., Park, A., Patterson, M., Frost, G.S., Ghatei, M.A. and Bloom, S.R. (2003) Oxyntomodulin suppresses appetite and reduces food intake in humans. *Journal of Clinical Endocrinology and Metabolism* 88, 4696–4701.

Coll, A.P., Farooqi, I.S. and O'Rahilly, S. (2007) The hormonal control of food intake. *Cell* 129, 251–262.

Corp, E.S., Melville, L.D., Greenberg, D., Gibbs, J. and Smith, G.P. (1990) Effect of fourth ventricular neuropeptide Y and peptide YY on ingestive and other behaviors. *American Journal of Physiology – Regulatory Integrative and Comparative Physiology* 259, R317–R323.

Cox, J.E. and Randich, A. (2004) Enhancement of feeding suppression by PYY(3-36) in rats with area postrema ablations. *Peptides* 25, 985–989.

Crawley, J.N., Kiss, J.Z. and Mezey, E. (1984) Bilateral midbrain transections block the behavioral effects of cholecystokinin on feeding and exploration in rats. *Brain Research* 322, 316–321.

Cummings, D.E. and Overduin, J. (2007) Gastrointestinal regulation of food intake. *Journal of Clinical Investigation* 117, 13–23.

Cummings, D.E. and Shannon, M.H. (2003) Ghrelin and gastric bypass: is there a hormonal contribution to surgical weight loss? *Journal of Clinical Endocrinology and Metabolism* 88, 2999–3002.

Cummings, D.E., Purnell, J.Q., Frayo, R.S., Schmidova, K., Wisse, B.E. and Weigle, D.S. (2001) A preprandial rise in plasma ghrelin levels suggests a role in meal initiation in humans. *Diabetes* 50, 1714–1719.

Cummings, D.E., Clement, K., Purnell, J.Q., Vaisse, C., Foster, K.E., Frayo, R.S., Schwartz, M.W., Basdevant, A. and Weigle, D.S. (2002a) Elevated plasma ghrelin levels in Prader Willi syndrome. *Nature Medicine* 8, 643–644.

Cummings, D.E., Weigle, D.S., Frayo, R.S., Breen, P.A., Ma, M.K., Dellinger, E.P. and Purnell, J.Q. (2002b) Plasma ghrelin levels after diet-induced weight loss or gastric bypass surgery. *New England Journal of Medicine* 346, 1623–1630.

Cummings, D.E., Frayo, R.S., Marmonier, C., Aubert, R. and Chapelot, D. (2004) Plasma ghrelin levels and hunger scores in humans initiating meals voluntarily without time- and food-related cues. *American Journal of Physiology – Endocrinology and Metabolism* 287, E297–E304.

Dakin, C.L., Small, C.J., Batterham, R.L., Neary, N.M., Cohen, M.A., Patterson, M., Ghatei, M.A. and Bloom, S.R. (2004) Peripheral oxyntomodulin reduces food intake and body weight gain in rats. *Endocrinology* 145, 2687–2695.

Ekblad, E. and Sundler, F. (2002) Distribution of pancreatic polypeptide and peptide YY. *Peptides* 23, 251–261.

English, P.J., Ghatei, M.A., Malik, I.A., Bloom, S.R. and Wilding, J.P. (2002) Food fails to suppress ghrelin levels in obese humans. *Journal of Clinical Endocrinology and Metabolism* 87, 2984.

Feinle, C., Chapman, I.M., Wishart, J. and Horowitz, M. (2002) Plasma glucagon-like peptide-1 (GLP-1) responses to duodenal fat and glucose infusions in lean and obese men. *Peptides* 23, 1491–1495.

Field, B.C., Wren, A.M., Cooke, D. and Bloom, S.R. (2008) Gut hormones as potential new targets for appetite regulation and the treatment of obesity. *Drugs* 68, 147–163.

Frühbeck, G. (2002) Peripheral actions of leptin and its involvement in disease. *Nutrition Reviews* 60, S47–S55.

Frühbeck, G., Jebb, S.A. and Prentice, A.M. (1998) Leptin: physiology and pathophysiology. *Clinical Physiology* 18, 399–419.

Frühbeck, G., Diez-Caballero, A. and Gil, M.J. (2004a) Fundus functionality and ghrelin concentrations after bariatric surgery. *New England Journal of Medicine* 350, 308–309.

Frühbeck, G., Diez-Caballero, A., Gil, M.J., Montero, I., Gómez-Ambrosi, J., Salvador, J. and Cienfuegos, J.A. (2004b) The decrease in plasma ghrelin concentrations following bariatric surgery depends on the functional integrity of the fundus. *Obesity Surgery* 14, 606–612.

Frühbeck, G., Rotellar, F., Hernández-Lizoain, J.L., Gil, M.J., Gómez-Ambrosi, J., Salvador, J. and Cienfuegos, J.A. (2004c) Fasting plasma ghrelin concentrations 6 months after gastric bypass are not determined by weight loss or changes in insulinemia. *Obesity Surgery* 14, 1208–1215.

Fujimoto, S., Inui, A., Kiyota, N., Seki, W., Koide, K., Takamiya, S., Uemoto, M., Nakajima, Y., Baba, S. and Kasuga, M. (1997) Increased cholecystokinin and pancreatic polypeptide responses to a fat-rich meal in patients with restrictive but not bulimic anorexia nervosa. *Biological Psychiatry* 41, 1068–1070.

Gibbs, J., Young, R.C. and Smith, G.P. (1973) Cholecystokinin decreases food intake in rats. *Journal of Comparative Physiology and Psychology* 84, 488–495.

Gourcerol, G., Wang, L., Wang, Y.H., Million, M. and Taché, Y. (2007) Urocortins and cholecystokinin-8 act synergistically to increase satiation in lean but not obese mice: involvement of corticotropin-releasing factor receptor-2 pathway. *Endocrinology* 148, 6115–6123.

Grudell, A.B. and Camilleri, M. (2007) The role of peptide YY in integrative gut physiology and potential role in obesity. *Current Opinion in Endocrinology, Diabetes and Obesity* 14, 52–57.

Hagan, M.M., Castaneda, E., Sumaya, I.C., Fleming, S.M., Galloway, J. and Moss, D.E. (1998) The effect of hypothalamic peptide YY on hippocampal acetylcholine release *in vivo*: implications for limbic function in binge-eating behavior. *Brain Research* 805, 20–28.

Halatchev, I.G., Ellacott, K.L., Fan, W. and Cone, R.D. (2004) Peptide YY3-36 inhibits food intake in mice through a melanocortin-4 receptor-independent mechanism. *Endocrinology* 145, 2585–2590.

Hansen, T.K., Dall, R., Hosoda, H., Kojima, M., Kangawa, K., Christiansen, J.S. and Jorgensen, J.O. (2002) Weight loss increases circulating levels of ghrelin in human obesity. *Clinical Endocrinology* 56, 203–206.

Hinney, A., Hoch, A., Geller, F., Schafer, H., Siegfried, W., Goldschmidt, H., Remschmidt, H. and Hebebrand, J. (2002) Ghrelin gene: identification of missense variants and a frameshift mutation in extremely obese children and adolescents and healthy normal weight students. *Journal of Clinical Endocrinology and Metabolism* 87, 2716.

Holst, J.J. (2007) The physiology of glucagon-like peptide 1. *Physiological Reviews* 87, 1409–1439.

Jesudason, D.R., Monteiro, M.P., McGowan, B.M., Neary, N.M., Park, A.J., Philippou, E., Small, C.J., Frost, G.S., Ghatei, M.A. and Bloom, S.R. (2007) Low-dose pancreatic polypeptide inhibits food intake in man. *British Journal of Nutrition* 97, 426–429.

Jorde, R. and Burhol, P.G. (1982) Effect of jejunoileal bypass operation and Billroth II resection on postprandial plasma pancreatic polypeptide release. *Scandinavian Journal of Gastroenterology* 17, 613–617.

Jorde, R. and Burhol, P.G. (1984) Fasting and postprandial plasma pancreatic polypeptide (PP) levels in obesity. *International Journal of Obesity* 8, 393–397.

Kanatani, A., Mashiko, S., Murai, N., Sugimoto, N., Ito, J., Fukuroda, T., Fukami, T., Morin, N., MacNeil, D.J., Van der Ploeg, L.H., Saga, Y., Nishimura, S. and Ihara, M. (2000) Role of the Y1 receptor in the regulation of neuropeptide Y-mediated feeding: comparison of wild-type, Y1 receptor-deficient, and Y5 receptor-deficient mice. *Endocrinology* 141, 1011–1016.

Kissileff, H.R., Carretta, J.C., Geliebter, A. and Pi-Sunyer, F.X. (2003) Cholecystokinin and stomach distension combine to reduce food intake in humans. *American Journal of Physiology – Regulatory Integrative and Comparative Physiology* 285, R992–R998.

Kojima, M. and Kangawa, K. (2002) Ghrelin, an orexigenic signaling molecule from the gastrointestinal tract. *Current Opinion in Pharmacology* 2, 665–668.

Kojima, M., Hosoda, H., Date, Y., Nakazato, M., Matsuo, H. and Kangawa, K. (1999) Ghrelin is a growth-hormone-releasing acylated peptide from stomach. *Nature* 402, 656–660.

Koska, J., DelParigi, A., de Courten, B., Weyer, C. and Tataranni, P.A. (2004) Pancreatic polypeptide is involved in the regulation of body weight in pima Indian male subjects. *Diabetes* 53, 3091–3096.

Larhammar, D. (1996) Structural diversity of receptors for neuropeptide Y, peptide YY and pancreatic polypeptide. *Regulatory Peptides* 65, 165–174.

Larsson, L.I. and Rehfeld, J.F. (1978) Distribution of gastrin and CCK cells in the rat gastrointestinal tract. Evidence for the occurrence of three distinct cell types storing COOH-terminal gastrin immunoreactivity. *Histochemistry* 58, 23–31.

Le Roux, C.W., Aylwin, S.J., Batterham, R.L., Borg, C.M., Coyle, F., Prasad, V., Shurey, S., Ghatei, M.A., Patel, A.G. and Bloom, S.R. (2006) Gut hormone profiles following bariatric surgery favor an anorectic state, facilitate weight loss, and improve metabolic parameters. *Annals of Surgery* 243, 108–114.

Le Roux, C.W., Welbourn, R., Werling, M., Osborne, A., Kokkinos, A., Laurenius, A., Lönroth, H., Fändriks, L., Ghatei, M.A., Bloom, S.R. and Olbers, T. (2007) Gut hormones as mediators of appetite and weight loss after Roux-en-Y gastric bypass. *Annals of Surgery* 246, 780–785.

Le Roux, C.W., Borg, C.M., Murphy, K.G., Vincent, R.P., Ghatei, M.A. and Bloom, S.R. (2008) Supraphysiological doses of intravenous PYY3-36 cause nausea, but no additional reduction in food intake. *Annals of Clinical Biochemistry* 45, 93–95.

Liddle, R.A., Goldfine, I.D., Rosen, M.S., Taplitz, R.A. and Williams, J.A. (1985) Cholecystokinin bioactivity in human plasma. Molecular forms, responses to feeding, and relationship to gallbladder contraction. *Journal of Clinical Investigation* 75, 1144–1152.

López, M., Tovar, S., Vázquez, M.J., Williams, L.M. and Diéguez, C. (2007) Peripheral tissue–brain interactions in the regulation of food intake. *Proceedings of the Nutrition Society* 66, 131–155.

McLaughlin, C.L. and Baile, C.A. (1981) Obese mice and the satiety effects of cholecystokinin, bombesin and pancreatic polypeptide. *Physiology and Behaviour* 26, 433–437.

Martin, N.M., Small, C.J., Sajedi, A., Patterson, M., Ghatei, M.A. and Bloom, S.R. (2004) Pre-obese and obese agouti mice are sensitive to the anorectic effects of peptide YY(3-36) but resistant to ghrelin. *International Journal of Obesity* 28, 886–893.

Masuda, Y., Tanaka, T., Inomata, N., Ohnuma, N., Tanaka, S., Itoh, Z., Hosoda, H., Kojima, M. and Kangawa, K. (2000) Ghrelin stimulates gastric acid secretion and motility in rats. *Biochemical Biophysical Research Communications* 276, 905–908.

Matson, C.A. and Ritter, R.C. (1999) Long-term CCK-leptin synergy suggests a role for CCK in the regulation of body weight. *American Journal of Physiology – Regulatory Integrative and Comparative Physiology* 276, R1038–R1045.

Meereis-Schwanke, K., Klonowski-Stumpe, H., Herberg, L. and Niederau, C. (1998) Long-term effects of CCK-agonist and -antagonist on food intake and body weight in Zucker lean and obese rats. *Peptides* 19, 291–299.

Meier, J.J., Gallwitz, B., Schmidt, W.E. and Nauck, M.A. (2002) Glucagon-like peptide 1 as a regulator of food intake and body weight: therapeutic perspectives. *European Journal of Pharmacology* 440, 269–279.

Merino, B., Cano, V., Guzmán, R., Somoza, B. and Ruiz-Gayo, M. (2008) Leptin-mediated hypothalamic pathway of cholecystokinin (CCK-8) to regulate body weight in free-feeding rats. *Endocrinology* 149, 1994–2000.

Meryn, S., Stein, D. and Straus, E.W. (1986) Fasting- and meal-stimulated peptide hormone concentrations before and after gastric surgery for morbid obesity. *Metabolism* 35, 798–802.

Mitchell, J.E., Lancaster, K.L., Burgard, M.A., Howell, L.M., Krahn, D.D., Crosby, R.D., Wonderlich, S.A. and Gosnell, B.A. (2001) Long-term follow-up of patients' status after gastric bypass. *Obesity Surgery* 11, 464–468.

Moran, T.H. (2006) Gut peptide signaling in the controls of food intake. *Obesity* 14, 250S–253S.

Moran, T.H. and Schwartz, G.J. (1994) Neurobiology of cholecystokinin. *Critical Reviews in Neurobiology* 9, 1–28.

Moran, T.H., Katz, L.F., Plata-Salaman, C.R. and Schwartz, G.J. (1998) Disordered food intake and obesity in rats lacking cholecystokinin A receptors. *American Journal of Physiology – Regulatory Integrative and Comparative Physiology* 274, R618–R625.

Morris, M.J., Chen, H., Watts, R., Shulkes, A. and Cameron-Smith, D. (2008) Brain neuropeptide Y and CCK and peripheral adipokine receptors: temporal response in obesity induced by palatable diet. *International Journal of Obesity* 32, 249–258.

Murakami, N., Hayashida, T., Kuroiwa, T., Nakahara, K., Ida, T., Mondal, M.S., Nakazato, M., Kojima, M. and Kangawa, K. (2002) Role for central ghrelin in food intake and secretion profile of stomach ghrelin in rats. *Journal of Endocrinology* 174, 283–288.

Näslund, E. and Hellström, P.M. (2007) Appetite signaling: from gut peptides and enteric nerves to brain. *Physiology and Behavior* 92, 256–262.

Näslund, E., King, N., Mansten, S., Adner, N., Holst, J.J., Gutniak, M. and Hellström, P.M. (2004) Prandial subcutaneous injections of glucagon-like peptide-1 cause weight loss in obese human subjects. *British Journal of Nutrition* 91, 439–446.

O'Shea, D., Morgan, D.G., Meeran, K., Edwards, C.M., Turton, M.D., Choi, S.J., Heath, M.M., Gunn, I., Taylor, G.M., Howard, J.K., Bloom, C.I., Small, C.J., Haddo, O., Ma, J.J., Callinan, W., Smith, D.M., Ghatei, M.A. and Bloom, S.R. (1997) Neuropeptide Y induced feeding in the rat is mediated by a novel receptor. *Endocrinology* 138, 196–202.

Parkinson, J.R., Dhillo, W.S., Small, C.J., Chaudhri, O.B., Bewick, G.A., Pritchard, I., Moore, S., Ghatei, M.A. and Bloom, S.R. (2008) PYY3-36 injection in mice produces an acute anorexigenic effect followed by a delayed orexigenic effect not observed with other anorexigenic gut hormones. *American Journal of Physiology – Endocrinology and Metabolism* 294, E698–708.

Peters, J.H., Simasko, S.M. and Ritter, R.C. (2006) Modulation of vagal afferent excitation and reduction of food intake by leptin and cholecystokinin. *Physiology and Behavior* 89, 477–485.

Rahardjo, G.L., Huang, X.F., Tan, Y.Y. and Deng, C. (2007) Decreased plasma peptide YY accompanied by elevated peptide YY and Y2 receptor binding densities in the medulla oblongata of diet-induced obese mice. *Endocrinology* 148, 4704–4710.

Riediger, T., Bothe, C., Becskei, C. and Lutz, T.A. (2004) Peptide YY directly inhibits ghrelin-activated neurons of the arcuate nucleus and reverses fasting-induced c-Fos expression. *Neuroendocrinology* 79, 317–326.

Rodieux, F., Giusti, V., D'Alessio, D.A., Suter, M. and Tappy, L. (2008) Effects of gastric bypass and gastric banding on glucose kinetics and gut hormone release. *Obesity* 16, 298–305.

Sakata, I., Nakamura, K., Yamazaki, M., Matsubara, M., Hayashi, Y., Kangawa, K. and Sakai, T. (2002) Ghrelin-producing cells exist as two types of cells, closed- and opened-type cells, in the rat gastrointestinal tract. *Peptides* 23, 531–536.

Sarson, D.L., Scopinaro, N. and Bloom, S.R. (1981) Gut hormone changes after jejunoileal (JIB) or biliopancreatic (BPB) bypass surgery for morbid obesity. *International Journal of Obesity* 5, 471–480.

Schwartz, G.J., Whitney, A., Skoglund, C., Castonguay, T.W. and Moran, T.H. (1999) Decreased responsiveness to dietary fat in Otsuka Long-Evans Tokushima fatty rats lacking CCK-A receptors. *American Journal of Physiology – Regulatory Integrative and Comparative Physiology* 277, R1144–R1151.

Shiiya, T., Nakazato, M., Mizuta, M., Date, Y., Mondal, M.S., Tanaka, M., Nozoe, S., Hosoda, H., Kangawa, K. and Matsukura, S. (2002) Plasma ghrelin levels in lean and obese humans and the effect of glucose on ghrelin secretion. *Journal of Clinical Endocrinology and Metabolism* 87, 240–244.

Stanley, B.G. and Leibowitz, S.F. (1985) Neuropeptide Y injected in the paraventricular hypothalamus: a powerful stimulant of feeding behavior. *Proceedings of the National Academy of Sciences of the United States of America* 82, 3940–3943.

Tang-Christensen, M., Vrang, N. and Larsen, P.J. (2001) Glucagon-like peptide containing pathways in the regulation of feeding behaviour. *International Journal of Obesity* 25 (Suppl. 5), S42–S47.

Tatemoto, K. and Mutt, V. (1980) Isolation of two novel candidate hormones using a chemical method for finding naturally occurring polypeptides. *Nature* 285, 417–418.

Trudel, L., Tomasetto, C., Rio, M.C., Bouin, M., Plourde, V., Eberling, P. and Poitras, P. (2002) Ghrelin/motilin-related peptide is a potent prokinetic to reverse gastric postoperative ileus in rat. *American Journal of Physiology – Gastrointestinal and Liver Physiology* 282, G948–G952.

Tschöp, M., Smiley, D.L. and Heiman, M.L. (2000) Ghrelin induces adiposity in rodents. *Nature* 407, 908–913.

Tschöp, M., Wawarta, R., Riepl, R.L., Friedrich, S., Bidlingmaier, M., Landgraf, R. and Folwaczny, C. (2001a) Postprandial decrease of circulating human ghrelin levels. *Journal of Endocrinological Investigation* 24, RC19–RC21.

Tschöp, M., Weyer, C., Tataranni, P.A., Devanarayan, V., Ravussin, E. and Heiman, M.L. (2001b) Circulating ghrelin levels are decreased in human obesity. *Diabetes* 50, 707–709.

Tschöp, M., Castañeda, T.R., Joost, H.G., Thone-Reineke, C., Ortmann, S., Klaus, S., Hagan, M.M., Chandler, P.C., Oswald, K.D., Benoit, S.C., Seeley, R.J., Kinzig, K.P., Moran, T.H., Beck-Sickinger, A.G., Koglin, N., Rodgers, R.J., Blundell, J.E., Ishii, Y., Beattie, A.H., Holch, P., Allison, D.B., Raun, K., Madsen, K., Wulff, B.S., Stidsen, C.E., Birringer, M., Kreuzer. O.J., Schindler, M., Arndt, K., Rudolf, K., Mark, M., Deng, X.Y., Whitcomb, D.C., Halem, H., Taylor, J., Dong, J., Datta, R., Culler, M., Craney, S., Flora, D., Smiley, D. and Heiman, M.L. (2004) Physiology: does gut hormone PYY3-36 decrease food intake in rodents? *Nature* 430, 1.

Ueno, N., Inui, A., Iwamoto, M., Kaga, T., Asakawa, A., Okita, M., Fujimiya, M., Nakajima, Y., Ohmoto, Y., Ohnaka, M., Nakaya, Y., Miyazaki, J.I. and Kasuga, M. (1999) Decreased food intake and body weight in pancreatic polypeptide-overexpressing mice. *Gastroenterology* 117, 1427–1432.

Uhe, A.M., Szmukler, G.I., Collier, G.R., Hansky, J., O'Dea, K. and Young, G.P. (1992) Potential regulators of feeding behavior in anorexia nervosa. *American Journal of Clinical Nutrition* 55, 28–32.

Velasquez-Mieyer, P.A., Cowan, P.A., Umpierrez, G.E., Lustig, R.H., Cashion, A.K. and Burghen, G.A. (2003) Racial differences in glucagon-like peptide-1 (GLP-1) concentrations and insulin dynamics during oral glucose tolerance test in obese subjects. *International Journal of Obesity* 27, 1359–1364.

Verdich, C., Flint, A., Gutzwiller, J.P., Näslund, E., Beglinger, C., Hellstrom, P.M., Long, S.J., Morgan, L.M., Holst, J.J. and Astrup, A. (2001a) A meta-analysis of the effect of glucagon-like peptide-1 (7-36) amide on *ad libitum* energy intake in humans. *Journal of Clinical Endocrinology and Metabolism* 86, 4382–4389.

Verdich, C., Toubro, S., Buemann, B., Lysgard, M.J., Juul, H.J. and Astrup, A. (2001b) The role of postprandial releases of insulin and incretin hormones in meal-induced satiety – effect of obesity and weight reduction. *International Journal of Obesity* 25, 1206–1214.

Vilsboll, T., Krarup, T., Sonne, J., Madsbad, S., Volund, A., Juul, A.G. and Holst, J.J. (2003a) Incretin secretion in relation to meal size and body weight in healthy subjects and people with type 1 and type 2 diabetes mellitus. *Journal of Clinical Endocrinology and Metabolism* 88, 2706–2713.

Vilsboll, T., Agerso, H., Krarup, T. and Holst, J.J. (2003b) Similar elimination rates of glucagon-like peptide-1 in obese type 2 diabetic patients and healthy subjects. *Journal of Clinical Endocrinology and Metabolism* 88, 220–224.

Vincent, R.P., Ashrafian, H. and Le Roux, C.W. (2008) Mechanisms of disease: the role of gastrointestinal hormones in appetite and obesity. *Nature Clinical Practice Gastroenterology and Hepatology* 5, 268–277.

Wang, H.J., Geller, F., Dempfle, A., Schauble, N., Friedel, S., Lichtner, P., Fontenla-Horro, F., Wudy, S., Hagemann, S., Gortner, L., Huse, K., Remschmidt, H., Bettecken, T., Meitinger, T., Schafer, H., Hebebrand, J. and Hinney, A. (2004) Ghrelin receptor gene: identification of several sequence variants in extremely obese children and adolescents, healthy normal-weight and underweight students, and children with short normal stature. *Journal of Clinical Endocrinology and Metabolism* 89, 157–162.

West, D.B., Fey, D. and Woods, S.C. (1984) Cholecystokinin persistently suppresses meal size but not food intake in free-feeding rats. *American Journal of Physiology – Regulatory Integrative and Comparative Physiology* 246, R776–R787.

West, D.B., Greenwood, M.R., Sullivan, A.C., Prescod, L., Marzullo, L.R. and Triscari, J. (1987) Infusion of cholecystokinin between meals into free-feeding rats fails to prolong the intermeal interval. *Physiology and Behaviour* 39, 111–115.

Wren, A.M. and Bloom, S.R. (2007) Gut hormones and appetite control. *Gastroenterology* 132, 2116–2130.

Wren, A.M., Small, C.J., Ward, H.L., Murphy, K.G., Dakin, C.L., Taheri, S., Kennedy, A.R., Roberts, G.H., Morgan, D.G., Ghatei. M.A. and Bloom, S.R. (2000) The novel hypothalamic peptide ghrelin stimulates food intake and growth hormone secretion. *Endocrinology* 141, 4325–4328.

Wren, A.M., Small, C.J., Abbott, C.R., Dhillo, W.S., Seal, L.J., Cohen, M.A., Batterham, R.L., Taheri, S., Stanley, S.A., Ghatei, M.A. and Bloom, S.R. (2001a) Ghrelin causes hyperphagia and obesity in rats. *Diabetes* 50, 2540–2547.

Wren, A.M., Seal, L.J., Cohen, M.A., Brynes, A.E., Frost, G.S., Murphy, K.G., Dhillo, W.S., Ghatei, M.A. and Bloom, S.R. (2001b) Ghrelin enhances appetite and increases food intake in humans. *Journal of Clinical Endocrinology and Metabolism* 86, 5992.

Wynne, K. and Bloom, S.R. (2006) The role of oxyntomodulin and peptide tyrosine-tyrosine (PYY) in appetite control. *Nature Clinical Practice Endocrinology and Metabolism* 2, 612–620.

Wynne, K., Park, A.J., Small, C.J., Patterson, M., Ellis, S.M., Murphy, K.G., Wren, A.M., Frost, G.S., Meeran, K., Ghatei, M.A. and Bloom, S.R. (2005) Subcutaneous oxyntomodulin reduces body weight in overweight and obese subjects: a double-blind, randomized, controlled trial. *Diabetes* 54, 2390–2395.

Wynne, K., Park, A.J., Small, C.J., Meeran, K., Ghatei, M.A., Frost, G.S. and Bloom, S.R. (2006) Oxyntomodulin increases energy expenditure in addition to decreasing energy intake in overweight and obese humans: a randomised controlled trial. *International Journal of Obesity* 30, 1729–1736.

Zander, M., Madsbad, S., Madsen, J.L. and Holst, J.J. (2002) Effect of 6-week course of glucagon-like peptide 1 on glycaemic control, insulin sensitivity, and beta-cell function in type 2 diabetes: a parallel-group study. *Lancet* 359, 824–830.

Zipf, W.B., O'Dorisio, T.M., Cataland, S. and Sotos, J. (1981) Blunted pancreatic polypeptide responses in children with obesity of Prader–Willi syndrome. *Journal of Clinical Endocrinology and Metabolism* 52, 1264–1266.

Zipf, W.B., O'Dorisio, T.M., Cataland, S. and Dixon, K. (1983) Pancreatic polypeptide responses to protein meal challenges in obese but otherwise normal children and obese children with Prader–Willi syndrome. *Journal of Clinical Endocrinology and Metabolism* 57, 1074–1080.

5 Elements of the Adipostat

HANS HAUNER

Else Kröner-Fresenius Centre for Nutritional Medicine, Technical University of Munich, Germany

Introduction

Obesity is defined as a state of increased body weight, in particular increased body fat mass that is associated with an elevated risk of adverse health effects. It represents a disorder of energy homeostasis which develops when energy intake exceeds energy expenditure. Genetic, environmental, behavioural and psychosocial factors are known to contribute to the wide variation of body weight among individuals. Long-term changes in the energy balance are reflected by adaptations in the adipose tissue mass. This requires an enormous plasticity of the adipose organ, as changes in energy balance vary substantially according to external conditions. In fact, studies in rodents have demonstrated clearly that adipocytes can change their volume markedly and are highly flexible in their response to broad variations in nutrient supply.

Nevertheless, it is apparent from many studies in rodents and humans that the individual's body weight is controlled by internal physiological feedback systems, as proposed by the 'set point' theory (Cabanac, 2001). When body weight is increased or decreased, e.g. by dietary interventions, adaptive changes of energy intake and/or expenditure will occur subsequently to regain the original set point. In this line, obesity can be considered as a result of alterations in the physiological mechanisms that control body weight (Cabanac, 2001; Jebb *et al.*, 2006; Siervo *et al.*, 2008). Advances in the genetics of body weight regulation indicate that genetic factors are powerful determinants of body weight (Bell *et al.*, 2005). However, the individual weight is not fixed for the whole lifetime but can change, at least over longer time periods, which is what takes place more likely in the majority of the population and is known as the 'settling point' theory.

As our genetic background has remained unchanged during past centuries, the current epidemic of obesity is primarily a consequence of recent changes in our environment and lifestyle. It is widely accepted that the ability to store fat in times of nutritional abundance is an advantage during the evolution of mankind.

Likewise, subjects with a low energy expenditure and/or the ability to reduce their energy requirements in times of food scarcity may have had a greater chance for survival. The 'thrifty gene' hypothesis originally proposed by Neel (1962) is based on the idea of genetic selection in periods of famine or shortness of food. The concept that particular genotypes predispose individuals to obesity and related diseases is supported by data from adoption and twin studies, as well as from obesity-prone populations like the Pima Indians (e.g. Ravussin *et al.*, 1988).

Stability of Body Weight

It is long known that animals that are force-fed to become obese reduce food intake when allowed to have free access to food and body weight will return to baseline levels. In contrast, when animals have restricted food intake they will lose weight. When access to food is restored, they increase food intake to bring their body weight back to normal. Similar observations were made in controlled small-scale human studies (Sims and Danforth, 1987). The major volume changes in both settings occur in adipose tissue. Thus, the fat organ has an extreme capacity to adapt to changes in food supply or energy requirements. The hypothesis that arises from these observations is that factors related to body fat mass may be responsible for, or at least may be involved in the adaptive changes in food intake and energy expenditure, or both processes.

The idea that body fat regulates food intake was introduced originally by Kennedy (1952) as a result of his studies on weight change after lesions of the ventromedial hypothalamus. This so-called 'lipostatic' theory claimed that the amount of body fat was involved in the control of daily food intake. However, for many years, no information on the nature of the signals that controlled the lipostat was available. Until recently, metabolites such as fatty acids or glycerol frequently were considered to be among the postulated candidates responsible for controlling the size of the adipose tissue mass. Both fatty acids and glycerol are released in substantial amounts from adipose tissue, especially in the obese state, during basal lipolysis or when lipolysis is stimulated by catecholamines. In this context, a series of studies has dealt with the possible differential effects of fatty acids on satiety and body weight and adipose tissue mass, respectively. Feeding experiments in rodents have shown that a diet rich in saturated fat is more adipogenic than a diet rich in unsaturated fat. Likewise, a diet rich in arachidonic acid was found to promote the new formation of fat cells as compared to a diet rich in n-3 fatty acids, which was associated with a much lower rate of fat cell formation and adipose tissue expansion. There is also a number of studies showing that an increased fatty acid oxidation results in a reduction in food intake, and vice versa. Stimulation of hepatic fatty acid oxidation via increased expression of CPT-1α may enhance satiety (Leonhardt and Langhans, 2004). Other animal experiments suggest that long-chain fatty acids stimulate the release of cholecystokinin (CCK), peptide YY and glucagon-like peptide-1 in the small intestine, providing another mechanism of satiety signalling. There is also evidence from human studies that the composition of ingested fatty acids has some effect on the response of gut hormones (see Chapter 4). However, the published data are not consistent and it is questionable if usual variations in fatty acid composition, apart from energy balance, are of physiological significance for the control of adipose tissue mass.

Even 30 years ago, it had been shown already that subcutaneous or intracerebral injection of glycerol in mice caused a reduction of food intake and, subsequently, a decrease in body weight (Wirtshafter and Davis, 1977). In another study in rodents, Glick (1980) found that intra-arterial infusion of glycerol produced a reduction of daily energy intake that was three times the amount of calories infused. The release of glycerol from adipose tissue was also reported to be proportional to the amount of body fat and to adipocyte size (Björntorp *et al.*, 1969; Goldrick and McLoughlin, 1970). Although such data are in favour of a possible negative feedback mechanism between glycerol release from adipose tissue and energy intake, the physiological role of glycerol in the regulation of energy balance is far from being understood fully.

During the past decade, two notable events have stimulated an explosion in the research of body weight regulation. One was the discovery of leptin as the signal of adipose tissue to communicate the extent of its energy reservoir to the integrating hypothalamic areas of the central nervous system (Zhang *et al.*, 1994; Frühbeck and Gómez-Ambrosi, 2001). The other cause of rising concern and interest in obesity research is the worldwide obesity epidemic, together with its accompanying adverse health consequences, which exert an important impact on health care systems (Van Gaal *et al.*, 2006; Allender and Rayner, 2007).

Our understanding of the factors and elements that sense and control body fat mass (adipostat) has grown substantially. It is obvious that these determinants are heterogeneous, including factors that are produced by cellular components of adipose tissue, but also by other organs. As adipose tissue no longer can be considered as an inert organ for fat storage, it is important to review the current body of knowledge on adipose tissue biology to obtain a better understanding of how internal and external factors may interact in their control of body fat mass.

Adipose Tissue as the Major Energy Depot in the Body

Converting and packing calories in the essentially unhydrous form of triglycerides is the most efficient way in mammals to store excess energy. In contrast, glycogen is highly hydrated and exhibits a more decreased fuel capacity. Even a lean adult has a total fat mass between 10 and 20 kg, which is equivalent to a total energy reserve of between 70,000 and 140,000 kcal. This is sufficient to survive a period of between 50 and 100 days of total fasting. In obese subjects, these energy reserves are much greater and warrant an even greater resistance to periods of undernutrition. Recently, it has been observed that the number of fat cells stays constant in adulthood in both lean and obese individuals, even after marked weight loss, indicating that the number of adipocytes is set during childhood and adolescence (Spalding *et al.*, 2008). In view of the high uncertainty of food supply in most periods of human history, the ability to store and mobilize fat energy became extremely important. For this reason, rather efficient mechanisms to maintain these abilities were developed and are still operating.

However, under conditions of continuous overwhelming supply of energy and – at the same time – decreased necessity for physical activity, this genetic and physiologic make-up turns out to promote obesity and other chronic diseases of welfare.

Adipose Tissue as an Endocrine Organ

It was believed until recently that adipose tissue was only a passive storage organ which served mainly to store excess energy or to provide energy in case of increased demand or food shortage. During the past decade, the concept that adipose tissue is far more than a lipid-storing organ has emerged (Frühbeck *et al.*, 2001; Gimeno and Klaman, 2005; Trayhurn, 2005; Kahn *et al.*, 2006; Matsuzawa, 2006; Trayhurn *et al.*, 2006; Lago *et al.*, 2007). A growing number of factors were found to be produced and released from adipose tissue. These secreted products were termed adipokines. Today, it is evident that adipose tissue is a multifunctional organ which maintains an intensive crosstalk with many other organs. From recent studies, it is also becoming more and more obvious that adipose tissue is integrated fully in the complex network of energy homeostasis and body weight regulation, as well as nutrient partitioning.

To date, more than one hundred factors have been identified as being produced and released by adipose tissue. Interestingly, adipocyte size has been shown to be an important determinant of adipokine secretion, with a preferential expression of proinflammatory factors with increasing adipocyte size (Skurk *et al.*, 2007). The shift towards a dominance of proinflammatory adipokine secretion results largely from a dysregulation of hypertrophic, very large cells. In addition to adipocytes, by far the largest cell fraction, other cell types are also resident in adipose tissue, such as stromal cells or preadipocytes and macrophages, as well as other cells possibly contributing to the secretory function (Hauner, 2005; Fain, 2006). Interestingly, proinflammatory T-lymphocytes are present in visceral adipose tissue and may contribute to local inflammatory cell activation before the appearance of macrophages, suggesting that these cells may play an important role in the initiation and perpetuation of adipose tissue inflammation, as well as in the development of insulin resistance (Kintscher *et al.*, 2008). However, the specific role of the single cellular fractions awaits to be determined. The secreted products belong to different types/families of molecules and may exert a variety of actions. Figure 5.1 gives an overview of the most important and best-studied secretory products. However, the complete picture about the physiological functions of these products released from adipose tissue remains to be disentangled fully.

When the possible biological functions of the secreted products are discussed, it is important to distinguish between a local paracrine action and a possible systemic role. Only a limited number of fat cell products is released into the blood stream in detectable or significant amounts. Weight gain or obesity in humans results in increased circulating levels of some of these factors, with the exception of only a few adipokines which decrease with increasing fat mass. Table 5.1 comprises a list of products secreted from human adipose tissue reflecting

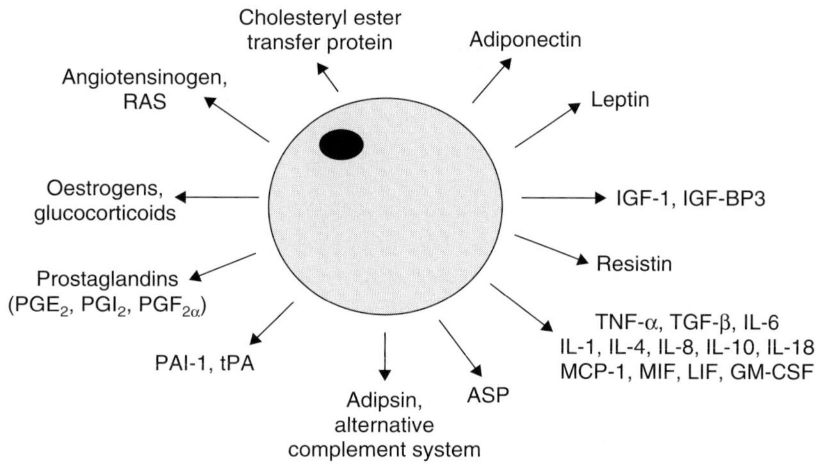

Fig. 5.1. Schematic representation of main secretory products released by human adipose tissue. ASP, acylation-stimulating protein; GM-CSF, granulocyte-macrophage colony-stimulating factor; IGF, insulin-like growth factor; IGF-BP, insulin-like growth factor-binding protein; IL, interleukin; LIF, leukaemia inhibitory factor; MCP-1, monocyte chemoattractant protein-1; MIF, macrophage migration inhibitory factor; PAI-1, plasminogen activator inhibitor-1; PGE_2, prostaglandin E_2; $PGF_{2\alpha}$, prostaglandin $F_{2\alpha}$; PGI_2, prostacyclin; RAS, renin-angiotensin system; TGF-β, transforming growth factor-β; TNF-α, tumour necrosis factor-α; tPA, tissue-type plasminogen activator.

Table 5.1. Main factors secreted from human adipose tissue and their circulating concentrations in obese patients compared to normal-weight individuals.

Adipokine	Plasma levels
Elevated in obesity:	
Leptin	+ + +
PAI-1	+ +
Angiotensin II	+
Interleukin-6	+
Interleukin-8	+
Interleukin-18	+
MIF	+
Soluble TNF receptors	+
TNF-α	(+)
Decreased in obesity:	
Adiponectin	– –
Omentin	–
Interleukin-10	(–)

Note: PAI-1, plasminogen activator inhibitor-1; MIF, macrophage migration inhibitory factor; TNF, tumour necrosis factor.

the circulating profile observed in obesity. Interestingly, many of the secreted factors may be acting only locally, with some factors even exerting a dual action, both on surrounding cells and on distant organs. Another possibility is that some factors have a primary local action, but subsequent alterations at the local level may induce secondary metabolic effects on other organs, e.g. via induction of lipolysis or insulin resistance.

Despite a shift in the research focus to the peripheral function of adipose tissue, substantial progress has also been made concerning the regulation of appetite and satiety in the central nervous system. From an integrative point of view, it is obvious that the central regulation of body weight or energy balance depends on the input from peripheral organs, with the list of candidates whose circulating concentrations are proportional to the adipose tissue mass increasing at a phenomenal pace during the past few years. At present, insulin and leptin are considered to be the best characterized signals from the periphery that exert critical functions in the matching between energy intake and expenditure (Benoit *et al.*, 2004). However, one has to keep in mind that food intake and energy expenditure are two complex processes that are determined by many central and peripheral neural, hormonal and neurochemical signals (see Part I). Below, some of the well-established and more recently identified peripheral signals involved in the maintenance of body fat mass and energy balance are described in more detail.

Leptin

For many years, the hypothesis that elevated fatty acids which were released from an enlarged fat mass might play a central role, at least in the development of some metabolic disturbances such as dyslipidaemia, impaired glucose metabolism and insulin resistance, dominated scientific discussion (Boden, 1997). This situation changed completely with the discovery of leptin in 1994. Leptin was identified as a fat cell-derived protein of cytokine-like structure that signals the size of energy stores to the brain (Zhang *et al.*, 1994; Ahima, 2006). In addition, leptin was found to act as a satiety hormone by interfering with hypothalamic regulatory systems in the control of food intake. Leptin deficiency was identified as the cause of the phenotypic alterations in *ob/ob* mice, including hyperphagia, low body temperature and low metabolic rate. Injection of leptin in these animals not only reduced food intake, but also increased energy expenditure. Both effects were reported to contribute to the reduction of elevated body weight in this and other rodent models of obesity (Ahima, 2006).

Numerous subsequent studies have shown consistently that circulating leptin concentrations are correlated closely with body fat mass and fat cell size in both animal models and humans (Ahima, 2006). Although circulating leptin levels increase with body fat mass and are proportional to the size of energy depots, leptin is also subject to short-term regulation and variation. For example, fasting and/or caloric restriction result in a rapid fall in leptin not proportional to fat mass loss (Considine *et al.*, 1996). Leptin concentrations also display a circadian secretion pattern rising from minimum levels in the morning to a peak in the evening (Frühbeck *et al.*, 1998).

The primary functional role of leptin is, however, not to reduce but to defend body fat. The decline of leptin after weight loss decreases energy expenditure and increases a food-seeking behaviour to conserve fat stores. However, the threshold for leptin action is influenced by developmental processes also involving the hypothalamus and chronic changes in fat stores. The physiological responses to leptin are asymmetrical: decreased concentrations of leptin provoke strong counterregulatory responses such as triggering food intake and reducing energy expenditure to improve the chance for survival, whereas high or elevated levels of leptin produce minimal effects such as moderate suppression of food intake, thereby allowing the deposition of additional fat stores, particularly under environmental conditions when energy supply is warranted and plentiful (Leibel, 2002).

Studies aiming to identify and understand the underlying mechanisms which link leptin to energy metabolism have shown the peripheral actions of leptin to be involved in the control of energy metabolism, in addition to its central action on the appetite and sympathetic nerve activity (Frühbeck, 2001, 2006). Leptin has also been found to stimulate the oxidation of fatty acids by activation of AMP-activated protein kinase, an enzyme that stimulates fatty acid oxidation in muscle potently by inhibiting acetyl coenzyme A carboxylase (Minokoshi *et al.*, 2002).

Another metabolic effect of leptin, which may also affect energy balance, is the specific suppression of stearoyl-CoA desaturase-1 (SCD-1). This enzyme is expressed in the liver and catalyses the biosynthesis of monounsaturated fatty acids. Mice lacking SCD-1 are lean and hypermetabolic. In addition, *ob/ob* mice with a mutation in the SCD-1 gene are less obese and have an increased energy expenditure, indicating that this specific effect of leptin is also associated with an increased energy dissipation (Cohen *et al.*, 2002). However, it remains to be examined fully if such mechanisms might contribute significantly to energy homeostasis in humans.

Leptin is considered to be one of the major adipocyte-derived factors that controls energy balance, and thereby the adipose tissue mass, by acting on diverse brain structures. The crosstalk between adipose tissue and the brain via leptin is complex as the leptin signal to the central nervous system may be modified by binding proteins in the plasma, the passage across the blood–brain barrier, as well as by brain sensitivity to leptin. A number of elegant studies have shown that leptin increases the hypothalamic expression of pro-opiomelanocortin and its cleavage product, α-melanocyte-stimulating hormone, which is an agonist of the melanocortin-4 receptor. At the same time, leptin suppresses the expression of neuropeptide Y and agouti-related peptide, with the latter acting as an antagonist to the melanocortin-4 receptor (Morton *et al.*, 2006; Coll *et al.*, 2007; Gao and Horvath, 2008).

Human obesity is not characterized by leptin deficiency but by elevated circulating concentrations, which are several-fold higher in obese as compared to lean subjects (Considine *et al.*, 1996). It was concluded from these findings that a decreased responsiveness of obese humans to leptin might exist in the target structures of this satiety hormone. The molecular mechanisms underlying this resistance are still far from being understood completely. Interestingly, there is only a close correlation between circulating leptin and body mass index (BMI) up to a range of 30–35 kg/m^2. At higher levels of BMI, there is no further linear increase in leptin. Even in the BMI range between 25 and 35 kg/m^2, there is a high biological variation of circulating leptin concentrations whose determinants are poorly understood.

Interestingly, studies have further indicated that leptin plays a key role in the restoration of body weight after diet-induced weight loss. Maintenance of body weight reduction is accompanied not only by decreased leptin concentrations but also by decreased energy expenditure, sympathetic nervous system tone and lower circulating concentrations of thyroxine and triiodothyronine. All these changes may favour weight regain. Administration of low doses of leptin to weight-reduced subjects resulted in the normalization of energy expenditure, skeletal muscle work efficiency, sympathetic activity and thyroid hormone levels to pre-weight-loss levels, indicating that relative leptin insufficiency contributed to weight regain after dietary weight reduction (Rosenbaum *et al.*, 2005).

In addition to leptin, many other candidates released by adipose tissue also contribute to the control of energy metabolism and need to be taken under consideration.

Classic cytokines

Adipose tissue is known to be an important production site for a variety of cytokines, originating not only from fat cells but also from adipocyte precursor cells and immune cells, particularly macrophages. Immune cells accumulate in adipose tissue, along with expansion of the fat mass (Wellen and Hotamisligil, 2005; Hotamisligil, 2006; Lago *et al.*, 2007). Obesity has been described as a state of chronic low-level inflammation, which may mediate the metabolic disturbances characteristic of an enlarged adipose tissue mass (see Chapter 8).

Additional candidates to participate in the regulation of energy metabolism originating from expanded adipose tissue are the classic cytokines, tumour necrosis factor-alpha (TNF-α) and interleukin-6 (IL-6) (Frühbeck *et al.*, 2001). It is well known that these and other cytokines, including interleukin-1 and -8, are able to inhibit food intake. These factors cause anorexia if elevated levels have access to the brain or are produced centrally. These cytokines may act via specific receptors in the central nervous system and exert multiple effects, apart from those related to control of body weight and energy balance. Furthermore, cytokines are known to increase the metabolic rate, at least under certain conditions, such as acute stress situations. It is also noteworthy that cytokines usually act in concert and are able to stimulate the release of other cytokines and chemokines. In this respect, some cytokines can enhance the effect of others in producing anorexia and weight loss. It remains to be clarified fully as to what extent cytokines really are involved in the physiological control of energy balance beyond severe diseases such as acute infection, wasting disorders, etc. Irrespective of this complexity, studies during the past years have elucidated that TNF-α and IL-6 may be of particular importance with regard to energy metabolism and adipose tissue mass.

TNF-α

It was reported originally by Hotamisligil *et al.* (1993, 1995) that adipose tissue from rodent models of obesity expresses high levels of TNF-α. These authors also showed elevated serum concentrations of this cytokine and an association with impaired insulin action in obese animals, as infusion of a soluble TNF-α antibody improved insulin sensitivity (Hotamisligil *et al.*, 1993). Subsequent studies revealed that human obesity was also associated with an increased adipose expression of TNF-α and its two receptor subtypes, indicating a general upregulation of the TNF system (Hotamisligil *et al.*, 1995; Kern *et al.*, 1995; Hube *et al.*, 1999). The increased production of TNF-α by adipose tissue may have a variety of consequences, as demonstrated thereafter in a number of studies. Elevated levels of TNF-α in adipose tissue are associated with increased lipolytic activity, impaired insulin action on glucose transport, reduced expression of lipoprotein lipase, inhibition of fat cell recruitment and, possibly, induction of fat cell apoptosis (Hube and Hauner, 1999). One of the main physiological purposes of increased TNF-α expression in obesity might be to limit adipose tissue growth. The price to pay may be the induction of insulin resistance and subsequent disturbances of glucose and lipid metabolism. At present, it is still unclear if and to what extent this local production contributes to energy homeostasis.

In contrast to rodents, it is an unsettled question as to whether elevated adipose expression of TNF-α is a significant determinant of insulin resistance in humans. Circulating concentrations of TNF-α are elevated only modestly in obese as compared to lean subjects (Hauner *et al.*, 1998) and some groups have failed to unravel an association between the TNF-α system and insulin sensitivity (Kellerer *et al.*, 1996). Infusion of a neutralizing TNF-α antibody under experimental conditions was not found to affect insulin sensitivity in type 2 diabetic subjects (Ofei *et al.*, 1996). Interestingly, data from the clinical use of neutralizing TNF-α antibodies for the treatment of autoimmune diseases are now available. For example, in patients with spondylarthropathy, treatment with either infliximab or etanercept for 1 year was associated with a significant increase in body weight by 2.2 kg (Briot *et al.*, 2005). However, the weight gain was due to an increase in lean mass, rather than in body fat. Likewise, administration of infliximab in obese subjects did not change insulin sensitivity (Di Rocco *et al.*, 2004). Thus, it appears that TNF-α, independent of its site of production, may exert only a minor role in the control of energy homeostasis, with the exception of a general activation of the immune and stress response in severe diseases such as acute sepsis, when massively elevated concentrations not only induce anorexia but also increase energy expenditure.

IL-6

IL-6 originally was described as a cytokine produced mainly by immune cells, but also by other cell types (Frühbeck *et al.*, 2001). Numerous studies suggest that this cytokine is involved in the regulation of many metabolic and endocrine processes. Several groups reported that IL-6 was also produced by adipose tissue. Fried *et al.* (1998) showed that omental adipose tissue secreted significantly more IL-6 than subcutaneous abdominal fat. Interestingly, the secretion of IL-6 from isolated adipocytes was much lower than that from whole adipose tissue pieces, indicating that cells other than adipocytes were the major source of this cytokine in adipose tissue. Clinical data also suggest a positive association between body fat mass and circulating IL-6, with the elevated concentrations found in obese subjects being due, at least partially, to increased release of IL-6 from adipose tissue (Mohamed-Ali *et al.*, 1997). Our group was able to demonstrate by immunohistochemistry that IL-6 was produced by cultured adipocytes (Päth *et al.*, 2001). In addition, the complete IL-6 receptor system was found to be expressed in adipose tissue. We saw a very weak stimulatory effect on lipolysis and a modest inhibitory effect on adipogenesis.

More important, with respect to the control of energy balance, is to answer the question whether or not peripheral IL-6 is involved in food intake or energy expenditure. It is known from rodent studies that IL-6, like other interleukins and TNF-α, is able to enter the brain, but there are also data that IL-6 is produced in various brain structures. Thus, IL-6 may reduce food intake by a direct effect on the hypothalamic areas controlling appetite. It is also noteworthy that mice deficient in IL-6 develop maturity-onset obesity, which is reversed by replacement of IL-6 (Wallenius *et al.*, 2002). It was also reported that intracerebroventricular administration of IL-6 increased energy expenditure and resulted in the loss of body fat following prolonged exposure (Wallenius *et al.*, 2003). These findings provide evidence for a role of IL-6 in the regulation of energy homeostasis in rodents.

In humans, IL-6 levels in the cerebrospinal fluid have been shown to be correlated inversely with BMI, suggesting that severe obesity is coupled to a relative central IL-6 deficiency (Stenlöf *et al.*, 2003). This finding is in contrast to the BMI-dependent elevated plasma concentrations of IL-6 in the periphery (Mohamed-Ali *et al.*, 1997). This discrepancy remains to be explored further, but the higher concentrations of IL-6 in the cerebrospinal fluid as compared to the peripheral serum levels may indicate that most of the IL-6 probably is produced in the central nervous system. Taken together, the evidence for IL-6 exerting a significant role in energy metabolism and adipose tissue mass control in humans is so far insufficient.

Adiponectin

Adiponectin is a 30-kDa adipose-specific secretory protein that consists of an amino-terminal collagen-like domain and a carboxy-terminal head domain with structural similarities with complement factor C1q (Berg *et al.*, 2002). The protein is expressed abundantly in adipose tissue but, in contrast to other adipokines, expression and circulating levels are decreased with increasing BMI (Tilg and Moschen, 2008). Adiponectin levels have been found to be correlated negatively with insulin resistance and type 2 diabetes mellitus, with weight loss resulting in a significant increase in circulating levels, indicating that the decrease in adiponectin production in obesity is reversible (Yang *et al.*, 2001).

In addition to its multiple antidiabetic and antiatherosclerotic properties, adiponectin has been found to have a central action that results in a loss of body weight in mice (Qi *et al.*, 2004). This study also showed that adiponectin was transported into the cerebrospinal fluid after intravenous injection. In contrast to leptin, adiponectin decreased body weight mainly by stimulating energy expenditure. Adiponectin has been shown further to potentiate the effect of leptin on thermogenesis and lipid levels. Similar to leptin, adiponectin increased the expression of hypothalamic corticotropin-releasing hormone (Qi *et al.*, 2004). Interestingly, agouti mice that overproduce the agouti protein, which antagonizes the melanocortin receptor signalling system, are insensitive to the action of adiponectin in the brain, suggesting that leptin and adiponectin may use common pathways in the central nervous system (Qi *et al.*, 2004).

Another interesting finding in this context is that peripheral infusion of the globular domain of adiponectin enhances lipid oxidation in muscle (Yamauchi *et al.*, 2001) and decreases body weight without inhibiting food intake in mice fed a sucrose- and fat-rich diet (Fruebis *et al.*, 2001). Available information suggests that adiponectin is another member of the coordinated system that links adipose tissue function to the central control of energy balance and glucose homeostasis, stimulating AMP-activated protein kinase in the hypothalamus and increasing food intake (Kubota *et al.*, 2007). It is noteworthy that adiponectin exhibits particularly beneficial antidiabetic, antiatherosclerotic and cardioprotective properties (Tilg and Moschen, 2008).

Plasminogen activator inhibitor-1 (PAI-1)

Plasminogen activator inhibitor-1 (PAI-1), the most important endogenous inhibitor of fibrinolysis, has also been reported as being produced by adipose tissue, both in rodents (Sawdey and Loskutoff, 1991) and humans (Alessi *et al.*, 1997; Eriksson *et al.*, 1998; Gottschling-Zeller *et al.*, 2000). Adipose tissue expression of PAI-1 has been detected both in preadipose and in fully developed fat cells. PAI-1 expression is promoted by cytokines (transforming growth factor-beta, TNF-α, interleukin-1β) and angiotensin II, with a clearly higher expression in omental than subcutaneous adipose tissue (Alessi *et al.*, 1997; Gottschling-Zeller *et al.*, 2000; Skurk *et al.*, 2001; Skurk and Hauner, 2004).

Increased levels of PAI-1 are linked not only to thrombosis but also to insulin resistance and body weight control. Ma *et al.* (2004) studied the relationship between PAI-1 and obesity and insulin resistance, respectively, using a gene knockout approach. Surprisingly, the authors found that PAI-1-deficient mice did not develop obesity and insulin resistance on a high-fat diet. The *PAI-1*$^{-/-}$ mice exhibited an increased metabolic rate and total energy expenditure compared with wild-type mice, along with a marked increase in uncoupling protein 3 mRNA expression in skeletal muscle, likely mechanisms contributing to the prevention of obesity. The transgenic animals also showed higher levels of peroxisome proliferator-activator receptor-γ and adiponectin mRNA. Finally, the transgenic mice were characterized by increased insulin sensitivity and improved glucose tolerance (Ma *et al.*, 2004). These findings suggest that PAI-1 represents another secretory product from adipose tissue involved directly in the control of adipose tissue mass and energy metabolism, although additional studies are needed to corroborate this possibility.

Insulin

The pancreatic hormone insulin is also well known to enter the brain from the circulation and to act as a central satiety signal similar to leptin (Benoit *et al.*, 2004; Morton *et al.*, 2006). Insulin receptors are expressed by brain neurones involved in energy intake, with administration of insulin directly into the brain reducing food intake. It has also been shown that insulin circulates at levels proportional to body fat content. The receptor-mediated transport of insulin across the blood–brain barrier into the brain is also proportional to plasma concentrations. Insulin exerts its anorectic action by decreasing the expression of the orexigenic neuropeptide Y and by increasing the expression of the anorexigenic α-melanocyte-stimulating hormone (Plum *et al.*, 2005). Although there is growing evidence that insulin participates in the central nervous system regulation of energy homeostasis, an estimation of its physiological significance is complicated by its profound multiple peripheral actions. Available data also indicate that leptin plays a more critical role compared to insulin, despite similar mechanisms of action on brain structures (Benoit *et al.*, 2004; Morton *et al.*, 2006).

However, in contrast to leptin, the increase in circulating insulin, along with the increase in body fat mass, is secondary to a decrease in insulin sensitivity. In obesity, insulin secretion from the pancreas increases in both the basal state and in response to a meal to compensate for insulin resistance. In contrast, weight loss is associated with an increase in insulin sensitivity, which may also support weight regain of adipose tissue by mechanisms such as increased expression and activation of lipoprotein lipase, which acts as gatekeeper for fatty acid transport into adipose tissue. Although insulin has a clear anabolic action in peripheral organs such as muscle and adipose tissue, an increased delivery of insulin to the brain may help to protect the body from further detrimental weight gain. In this sense, insulin is also an important element of the adipostat.

Other emerging candidates

The intense research efforts in adipocyte biology gradually are revealing the intricate adipokine-mediated interplay among white adipose tissue, metabolic disorders, appetite and energy balance. Undoubtedly, in the coming years, unravelling the exact contribution as elements of the adipostat of the more recently identified adipokines, such as resistin, visfatin, vaspin, apelin, interferon-gamma-inducible protein-10, omentin and chemerin, among others (Fukuhara *et al.*, 2005, 2007; Arner, 2006; Herder *et al.*, 2006; Lago *et al.*, 2007; Wada, 2008), together with the likely identification of novel factors, is warranted.

Intervention Strategies

In view of the growing evidence of a functional role of adipokine secretion, the potential modification of the secretory profile of adipose tissue represents an interesting therapeutic alternative. In principle, two intervention strategies can be applied, namely a dietary and a pharmacological approach. The treatment of choice is weight loss by dietary measures, as this approach has been shown repeatedly to be associated with restoration of normal secretion of the adipokines under investigation. The improvement in the secretory pattern is usually proportional to the extent of weight reduction. It remains to be outlined better if these changes are the result of stable weight loss or are due to calorie restriction, as most measurements are usually performed immediately after completing a weight loss programme.

In addition to the negative energy balance attained through diet, frequently used drugs with anti-inflammatory activity, such as thiazolidinediones, metformin and AT_1-receptor antagonists, have been shown to influence the secretory function of human adipose tissue. One example of such effects is the downregulation of adipose PAI-1 production by drugs such as troglitazone, metformin or candesartan (Skurk and Hauner, 2004). Even statins have been found to restore IL-6 secretion in human adipose cells. As mentioned before, subcutaneous administration of low doses of leptin after dietary weight loss has been reported to normalize energy expenditure, sympathetic nervous activity and thyroid function to baseline levels, thereby preventing the adaptive changes which serve to restore the original adipose tissue depots (Rosenbaum *et al*., 2005).

Integrated View and Conclusions

In conclusion, the current body of knowledge indicates convincingly that adipose tissue represents an active organ which secretes a variety of factors into the local environment, as well as into the circulation, and maintains a complex communication with other organs. The interaction between adipose tissue and hypothalamic centres is of particular significance, but the liver and muscle may also represent important elements of the adipostat. In addition, mediators of the immune system should be also viewed as relevant determinants of the adipostat, at least under defined conditions. Undoubtedly, further research on the relationship between adipose tissue function and energy balance is extremely important to understand better how to achieve and maintain a healthy body weight. This research is promising, not only with regard to the control of adipose tissue size, but also with respect to preventing the other unfavourable consequences of an expanded fat mass.

References

Ahima, R.S. (2006) Adipose tissue as an endocrine organ. *Obesity* 14 (Suppl. 5), 242S–249S.

Alessi, M.C., Pirelli, F., Morange, P., Henry, M., Nalbone, G. and Juhan-Vague, I. (1997) Production of plasminogen activator inhibitor-1 by adipose tissue: possible link between visceral fat accumulation and vascular disease. *Diabetes* 46, 860–867.

Allender, S. and Rayner, M. (2007) The burden of overweight and obesity-related ill health in the UK. *Obesity Reviews* 8, 467–473.

Arner, P. (2006) Visfatin – a true or false trail to type 2 diabetes mellitus. *Journal of Clinical Endocrinology and Metabolism* 91, 28–30.

Bell, C.G., Walley, A.J. and Froguel, P. (2005) The genetics of human obesity. *Nature Reviews Genetics* 6, 221–234.

Benoit, S.C., Clegg, D.J., Seeley, R.J. and Woods, S.C. (2004) Insulin and leptin as adiposity signals. *Recent Progress in Hormone Research* 59, 267–285.

Berg, A.H., Combs, T.P. and Scherer, P.E. (2002) ACRP30/adiponectin: an adipokine regulating glucose and lipid metabolism. *Trends in Endocrinology and Metabolism* 13, 84–89.

Björntorp, P., Bergman, H., Varnas-Kas, E. and Lindholm, B. (1969) Lipid mobilization in relation to body composition in man. *Metabolism* 18, 112–117.

Boden, G. (1997) Role of fatty acids in the pathogenesis of insulin resistance and NIDDM. *Diabetes* 46, 3–10.

Briot, K., Garnero, P., Le Henanff, A., Dougados, M. and Roux, C. (2005) Body weight, body composition, and bone turnover changes in patients with spondyloarthropathy receiving anti-tumour necrosis factor (alpha) treatment. *Annals of Rheumatic Diseases* 64, 1137–1140.

Cabanac, M. (2001) Regulation and the ponderostat. *International Journal of Obesity* 25 (Suppl. 5), S7–12.

Cohen, P., Miyazaki, M., Socci, N.D., Hagge-Greenberg, A., Liedtke, W., Soukas, A.A., Sharma, R., Hudgins, L.C., Ntambi, J.M. and Friedman, J.M. (2002) Role of stearoyl-CoA desaturase-1 in leptin-mediated weight loss. *Science* 297, 240–243.

Coll, A.P., Farooqi, I.S. and O'Rahilly, S. (2007) The hormonal control of food intake. *Cell* 129, 251–262.

Considine, R.V., Sinha, M.K., Heiman, M.L., Kriauciunas, A., Stephens, T.W., Nyce, M.R., Ohannesian, J.P., Marco, C.C., McKee, L.J., Bauer, T.L. and Caro, J.F. (1996) Serum immunoreactive-leptin concentrations in normal-weight and obese humans. *New England Journal of Medicine* 334, 292–295.

Di Rocco, P., Manco, M., Rosa, G., Greco, A.V. and Mingrone, G. (2004) Lowered tumor necrosis factor receptors, but not increased insulin sensitivity, with infliximab. *Obesity Research* 12, 734–739.

Eriksson, P., Reynisdottir, S., Lönnqvist, F., Stemme, V., Hamsten, A. and Arner, P. (1998) Adipose secretion of plasminogen activator inhibitor-1 in non-obese and obese individuals. *Diabetologia* 41, 65–71.

Fain, J.N. (2006) Release of interleukins and other inflammatory cytokines by human adipose tissue is enhanced in obesity and primarily due to the non-fat cells. *Vitamins and Hormones* 74, 443–477.

Fried, S.K., Bunkin, D.A. and Greenberg, A.S. (1998) Omental and subcutaneous adipose tissue of obese subjects release interleukin-6: depot differences and regulation by glucocorticoids. *Journal of Clinical Endocrinology and Metabolism* 83, 847–850.

Fruebis, J., Tsao, T.S., Javorschi, S., Ebbets-Reed, D., Erickson, M.R., Yen, F.T., Bihain, B.E. and Lodish, H.F. (2001) Proteolytic cleavage product of 30-kDa adipocyte complement-related protein increases fatty acid oxidation in muscle and causes weight loss in mice. *Proceedings of the National Academy of Sciences of the United States of America* 98, 2005–2010.

Frühbeck, G. (2001) A heliocentric view of leptin. *Proceedings of the Nutrition Society* 60, 301–318.

Frühbeck, G. (2006) Hunting for new pieces to the complex puzzle of obesity. *Proceedings of the Nutrition Society* 65, 329–347.

Frühbeck, G. and Gómez-Ambrosi, J. (2001) Rationale for the existence of additional adipostatic hormones. *FASEB Journal* 15, 1996–2006.

Frühbeck, G., Jebb, S.A. and Prentice, A.M. (1998) Leptin: physiology and pathophysiology. *Clinical Physiology* 18, 399–419.

Frühbeck, G., Gómez-Ambrosi, J., Muruzábal, F.J. and Burrell, M.A. (2001) The adipocyte: a model for integration of endocrine and metabolic signaling in energy metabolism regulation. *American Journal of Physiology – Endocrinology and Metabolism* 280, E827–E847.

Fukuhara, A., Matsuda, M., Nishizawa, M., Segawa, K., Tanaka, M., Kishimoto, K., Matsuki, Y., Murakami, M., Ichisaka, T., Murakami, H., Watanabe, E., Takagi, T., Akiyoshi, M., Ohtsubo, T., Kihara, S., Yamashita, S., Makishima, M., Funahashi, T., Yamanaka, S., Hiramatsu, R., Matsuzawa, Y. and Shimomura, I. (2005) Visfatin: a protein secreted by visceral fat that mimics the effects of insulin. *Science* 307, 426–430. Retraction in: Fukuhara, A., Matsuda, M., Nishizawa, M., Segawa, K., Tanaka, M., Kishimoto, K., Matsuki, Y., Murakami, M., Ichisaka, T., Murakami, H., Watanabe, E., Takagi, T., Akiyoshi, M., Ohtsubo, T., Kihara, S., Yamashita, S., Makishima, M., Funahashi, T., Yamanaka, S., Hiramatsu, R., Matsuzawa, Y. and Shimomura, I. (2007) *Science* 318, 565.

Gao, Q. and Horvath, T.L. (2008) Neuronal control of energy homeostasis. *FEBS Letters* 582, 132–141.

Gimeno, R.E. and Klaman, L.D. (2005) Adipose tissue as an active endocrine organ: recent advances. *Current Opinion in Pharmacology* 5, 122–128.

Glick, Z. (1980) Food intake of rats administered with glycerol. *Physiology and Behavior* 25, 621–626.

Goldrick, R.B. and McLoughlin, G.M. (1970) Lipolysis and lipogenesis from glucose in human fat cells of different sizes. *Journal of Clinical Investigation* 49, 1213–1223.

Gottschling-Zeller, H., Birgel, M., Röhrig, K. and Hauner, H. (2000) Effect of tumor necrosis factor alpha and transforming growth factor beta 1 on plasminogen activator inhibitor-1 secretion from subcutaneous and omental human fat cells in suspension culture. *Metabolism* 49, 666–671.

Hauner, H. (2005) Secretory factors from human adipose tissue and their functional role. *Proceedings of the Nutrition Society* 64, 163–169.

Hauner, H., Bender, M., Haastert, B. and Hube, F. (1998) Plasma concentrations of soluble TNF-alpha receptors in obese subjects. *International Journal of Obesity* 22, 1239–1243.

Herder, C., Hauner, H., Kempf, K., Kolb, H. and Skurk, T. (2006) Constitutive and regulated expression and secretion of interferon-c-inducible protein 10 (IP-10/CXCL10) in human adipocytes. *International Journal of Obesity* 31, 403–410.

Hotamisligil, G.S. (2006) Inflammation and metabolic disorders. *Nature* 444, 860–867.

Hotamisligil, G.S., Shargill, N.S. and Spiegelman, B.M. (1993) Adipose expression of tumor necrosis factor-alpha: direct role in obesity-linked insulin resistance. *Science* 257, 87–91.

Hotamisligil, G.S., Arner, P., Caro, J.F., Atkinson, R.L. and Spiegelman, B.M. (1995) Increased adipose tissue expression of tumor necrosis factor-alpha in human obesity and insulin resistance. *Journal of Clinical Investigation* 95, 2409–2415.

Hube, F. and Hauner, H. (1999) The role of TNF-α in human adipose tissue: prevention of weight gain at the expense of insulin resistance. *Hormone and Metabolic Research* 31, 626–631.

Hube, F., Birgel, M., Lee, Y.-M. and Hauner, H. (1999) Expression pattern of tumour necrosis factor receptors in subcutaneous and omental human adipose tissue: role of obesity and non-insulin-dependent diabetes mellitus. *European Journal of Clinical Investigation* 29, 672–678.

Jebb, S.A., Siervo, M., Frühbeck, G., Goldberg, G.R., Murgatroyd, P.R. and Prentice, A.M. (2006) Variability of appetite control mechanisms in response to 9 weeks of progressive overfeeding in humans. *International Journal of Obesity* 30, 1160–1162.

Kahn, S.E., Hull, R.L. and Utzschneider, K.M. (2006) Mechanisms linking obesity to insulin resistance and type 2 diabetes. *Nature* 444, 840–846.

Kellerer, M., Rett, K., Renn, W., Groop, L. and Häring, H.U. (1996) Circulating TNF-alpha and leptin levels in offspring of NIDDM patients do not correlate to individual insulin sensitivity. *Hormone and Metabolic Research* 28, 737–743.

Kennedy, G.C. (1952) The role of depot fat in the hypothalamic control of food intake in the rat. *Proceedings of the Royal Society B* 140, 578–592.

Kern, P.A., Saghizadeh, M., Ong, J.M., Bosch, R.J., Deem, R. and Simsolo, R.B. (1995) The expression of tumor necrosis factor in human adipose tissue. *Journal of Clinical Investigation* 95, 2111–2119.

Kintscher, U., Hartge, M., Hess, K., Foryst-Ludwig, A., Clemenz, M., Wabitsch, M., Fischer-Posovszky, P., Barth, T.F., Dragun, D., Skurk, T., Hauner, H., Blüher, M., Unger, T., Wolf, A.M., Knippschild, U., Hombach, V. and Marx, N. (2008) T-lymphocyte infiltration in visceral adipose tissue. A primary event in adipose tissue inflammation and the development of obesity-mediated insulin resistance. *Arteriosclerosis, Thrombosis, and Vascular Biology* 28, 1304–1310.

Kubota, N., Yano, W., Kubota, T., Yamauchi, T., Itoh, S., Kumagai, H., Kozono, H., Takamoto, I., Okamoto, S., Shiuchi, T., Suzuki, R., Satoh, H., Tsuchida, A., Moroi, M., Sugi, K., Noda, T., Ebinuma, H., Ueta, Y., Kondo, T., Araki, E., Ezaki, O., Nagai, R., Tobe, K., Terauchi, Y., Ueki, K., Minokoshi, Y. and Kadowaki, T. (2007) Adiponectin stimulates AMP-activated protein kinase in the hypothalamus and increases food intake. *Cell Metabolism* 6, 55–68.

Lago, F., Diéguez, C., Gómez-Reino, J. and Gualillo, O. (2007) The emerging role of adipokines as mediators of inflammation and immune responses. *Cytokine and Growth Factor Reviews* 18, 313–325.

Leibel, R. (2002) The role of leptin in the control of body weight. *Nutrition Reviews* 60, S15–S19.

Leonhardt, M. and Langhans, W. (2004) Fatty acid oxidation and control of food intake. *Physiology and Behaviour* 83, 645–651.

Ma, L.-J., Mao, S.-L., Taylor, K.L., Kanjanabuch, T., Guan, Y., Zhang, Y., Brown, N.J., Swift, L.L., McGuinness, O.P., Wasserman, D.H., Vaughan, D.E. and Fogo, A.B. (2004) Prevention of obesity and insulin resistance in mice lacking plasminogen activator inhibitor-1. *Diabetes* 53, 336–346.

Matsuzawa, Y. (2006) Therapy insight: adipocytokines in metabolic syndrome and related cardiovascular disease. *Nature Clinical Practice Cardiovascular Medicine* 3, 35–42.

Minokoshi, Y., Kim, Y.-B., Peroni, O.B., Fryer, L.G., Müller, C., Carling, D. and Kahn, B.B. (2002) Leptin stimulates fatty-acid oxidation by activating AMP-activated protein kinase. *Nature* 415, 339–343.

Mohamed-Ali, V., Goodbrick, S., Rawesh, A., Katz, D.R., Miles, J.M., Yudkin, J.S., Klein, S. and Coppack, S.W. (1997) Subcutaneous adipose tissue releases interleukin-6, but not tumor necrosis factor-α *in vivo*. *Journal of Clinical Endocrinology and Metabolism* 82, 4196–4200.

Morton, G.J., Cummings, D.E., Baskin, D.G., Barsh, G.S. and Schwartz, M.W. (2006) Central nervous system control of food intake and body weight. *Nature* 443, 289–295.

Neel, J. (1962) Diabetes mellitus: a 'thrifty' genotype rendered detrimental by 'progress'? *American Journal of Human Genetics* 14, 353–362.

Ofei, F., Hurel, S., Newkirk, J., Sopwith, M. and Taylor, R. (1996) Effects of an engineered human anti-TNF-alpha antibody (CDP571) on insulin sensitivity and glycemic control in patients with NIDDM. *Diabetes* 45, 881–885.

Päth, G., Bornstein, S.R., Gurniak, M., Chrousos, G.P., Scherbaum, W.A. and Hauner, H. (2001) Human breast adipocytes express interleukin-6 (IL-6) and its receptor system: increased IL-6 production by β-adrenergic activation and effects of IL-6 on adipocyte function. *Journal of Clinical Endocrinology and Metabolism* 86, 2281–2288.

Plum, L., Schubert, M. and Brüning, J.G. (2005) The role of insulin receptor signaling in the brain. *Trends in Endocrinology and Metabolism* 16, 59–65.

Qi, Y., Takahashi, N., Hileman, S.M., Berg, A.H., Pajvani, U.B., Scherer, P.E. and Ahima, R.S. (2004) Adiponectin acts in the brain to decrease body weight. *Nature Medicine* 10, 524–529.

Ravussin, E., Lillioja, S., Knowler, W.C., Christin, L., Freymond, D., Abbott, W.G., Boyce, V., Howard, B.V. and Bogardus, C. (1988) Reduced rate of energy expenditure as a risk factor for body weight gain. *New England Journal of Medicine* 318, 467–474.

Rosenbaum, M., Goldsmith, R., Bloomfield, D., Magnano, A., Weimer, L., Heymsfield, S., Gallagher, D., Mayer, L., Murphy, E. and Leibel, R.L. (2005) Low-dose leptin reverses skeletal muscle, autonomic, and neuroendocrine adaptations to maintenance of reduced weight. *Journal of Clinical Investigation* 115, 3579–3586.

Sawdey, M.S. and Loskutoff, D.J. (1991) Regulation of murine type 1 plasminogen activator inhibitor gene expression *in vivo*: tissue specificity and induction by lipopolysaccharide, tumor necrosis factor-alpha, and transforming growth factor-beta. *Journal of Clinical Investigation* 88, 1346–1353.

Siervo, M., Frühbeck, G., Dixon, A., Goldberg, G.R., Coward, W.A., Murgatroyd, P.R., Prentice, A.M. and Jebb, S.A. (2008) Efficiency of autoregulatory homeostatic responses to imposed caloric excess in lean men. *American Journal of Physiology – Endocrinology and Metabolism* 294, E416–424.

Sims, E.A. and Danforth, E. (1987) Expenditure and storage of energy in man. *Journal of Clinical Investigation* 79, 1019–1025.

Skurk, T. and Hauner, H. (2004) Obesity and impaired fibrinolysis: role of adipose production of plasminogen activator inhibitor-1. *International Journal of Obesity* 28, 1357–1364.

Skurk, T., Lee, Y.-M. and Hauner, H. (2001) Angiotensin II and its metabolites stimulate PAI-1 protein release from human adipocytes in primary culture. *Hypertension* 37, 1336–1340.

Skurk, T., Alberti-Huber, C., Herder, C. and Hauner, H. (2007) Relationship between adipocyte size and adipokine expression and secretion. *Journal of Clinical Endocrinology and Metabolism* 92, 1023–1033.

Spalding, K.L., Arner, E., Westermark, P.O., Bernard, S., Buchholz, B.A., Bergmann, O., Blomqvist, L., Hoffstedt, J., Näslund, E., Britton, T., Concha, H., Hassan, M., Rydén, M., Frisén, J. and Arner, P. (2008) Dynamics of fat cell turnover in humans. *Nature* 453, 783–787.

Stenlöf, K., Wernstedt, I., Fjällman, T., Wallenius, V., Wallenius, K. and Jansson, J.O. (2003) Interleukin-6 levels in the central nervous system are negatively correlated with fat mass in overweight/obese subjects. *Journal of Clinical Endocrinology and Metabolism* 88, 4379–4383.

Tilg, H. and Moschen, A.R. (2008) Role of adiponectin and PBEF/visfatin as regulators of inflammation: involvement in obesity-associated diseases. *Clinical Science* 114, 275–288.

Trayhurn, P. (2005) Endocrine and signalling role of adipose tissue: new perspectives on fat. *Acta Physiologica Scandinava* 184, 285–293.

Trayhurn, P., Bing, C. and Wood, I.S. (2006) Adipose tissue and adipokines – energy regulation from the human perspective. *Journal of Nutrition* 136, 1935S–1939S.

Van Gaal, L.F., Mertens, I.L. and De Block, C.E. (2006) Mechanisms linking obesity with cardiovascular disease. *Nature* 444, 875–880.

Wada, J. (2008) Vaspin: a novel serpin with insulin-sensitizing effects. *Expert Opinion on Investigational Drugs* 17, 327–333.

Wallenius, V., Wallenius, K., Ahren, B., Rudling, M., Carlsten, H., Dickson, S.L., Ohlsson, C. and Jansson, J.O. (2002) Interleukin-6-deficient mice develop maturity-onset obesity. *Nature Medicine* 8, 75–79.

Wallenius, K., Jansson, J.O. and Wallenius, V. (2003) The therapeutic potential of interleukin-6 in treating obesity. *Expert Opinion in Biological Therapy* 3, 1061–1070.

Wellen, K.E. and Hotamisligil, G.S. (2005) Inflammation, stress, and diabetes. *Journal of Clinical Investigation* 115, 1111–1119.

Wirtshafter, D. and Davis, J.D. (1977) Body weight: reduction by long-term glycerol treatment. *Science* 198, 1271–1273.

Yamauchi, T., Kamon, J., Minokoshi, Y., Ito, Y., Waki, H., Uchida, S., Yamashita, S., Noda, M., Kita, S., Ueki, K., Eto, K., Akanuma, Y., Froguel, P., Foufelle, F., Ferre, P., Carling, D., Kimura, S., Nagai, R., Kahn, B.B. and Kadowaki, T. (2001) Adiponectin stimulates glucose utilization and fatty-acid oxidation by activating AMP-activated protein kinase. *Nature Medicine* 8, 1288–1295.

Yang, W.S., Lee, W.J., Funahashi, T., Tanaka, S., Matsuzawa, Y., Chao, C.L., Chen, C.L., Tai, T.Y. and Chuang, L.M. (2001) Weight reduction increases plasma levels of an adipose tissue derived anti-inflammatory protein, adiponectin. *Journal of Clinical Endocrinology and Metabolism* 86, 3815–3819.

Zhang, Y., Proenca, R., Maffei, M., Barone, M., Leopold, L. and Friedman, J.M. (1994) Positional cloning of the mouse obese gene and its human homologue. *Nature* 372, 425–432.

6 Natriuretic Peptides and Other Lipolytic Peptides Involved in the Control of Lipid Mobilization in Humans

Max Lafontan, Coralie Sengenes, Cédric Moro, Jean Galitzky and Michel Berlan

Unité de Recherches sur les obésités, Université Paul Sabatier, Institut Louis Bugnard, Hôpital Rangueil, France

Background

Adipose tissue (AT) lipolysis, i.e. the catabolic process leading to the breakdown of triacylglycerols (TAG) into non-esterified fatty acids (NEFAs) and glycerol, is often considered as a well-established metabolic pathway. However, it is not understood fully what governs the basal rate of AT lipolysis. During this process, intracellular TAG undergoes hydrolysis through the action of neutral lipases located inside the fat cell, hormone-sensitive lipase (HSL) and adipose triglyceride lipase (ATGL). After TAG hydrolysis, NEFA and glycerol leave the fat cells and are transported by the bloodstream to other tissues (mainly the liver for glycerol and the liver, skeletal muscle and heart for NEFA). NEFAs act as metabolic substrates, as well as signalling molecules (Duplus and Forest, 2002; Arner, 2005). In addition to their role in AT metabolism, they can regulate glucose utilization in muscle and are important signals to the liver and pancreas beta cell as well. However, some NEFAs that appear during lipolysis do not leave the fat cell and can be re-esterified into intracellular TAG in a futile cycle. The amount of NEFA released into the blood is the result of a balance between TAG breakdown and resynthesis. The glycerol formed during lipolysis is not reutilized to any major extent by fat cells because, under normal conditions, they contain only minimal amounts of the enzyme, glycerol kinase. However, glyceroneogenesis occurs in adipocytes and could provide the glycerol-3-phosphate required for NEFA re-esterification. This is a phenomenon which could be of importance in some physiological conditions in the adipocyte (Beale *et al.*, 2003, 2004). In a normal-weight human, the mean turnover rate of TAG in the total fat mass is about 100–300 g/day. An imbalance between hydrolysis and synthesis of TAG can be

 Peptides in Energy Balance and Obesity (ed. G. Frühbeck)

important in the development of obesity. Altered lipolysis could be an element leading to obesity. Inter-individual variations in AT lipolysis are of importance in rates of weight loss. Conversely, excessive lipolytic rates, in conjunction with impairment of NEFA utilization by muscle and liver, may be a major contributor to the metabolic abnormalities found in persons with android or upper-body obesity and lead to metabolic disorders (insulin resistance, hyperglycaemia, hyperlipidaemia and type 2 diabetes) (Langin, 2006).

NEFA levels have been proposed as a major link between obesity and insulin resistance/type 2 diabetes (Boden, 2002; Bays *et al.*, 2004) and they are also a predictive risk factor for sudden death in the population (Jouven *et al.*, 2001). Elevated NEFA concentrations have been shown to reproduce some of the metabolic abnormalities of obesity. It has been hypothesized that visceral AT lipolysis releases excess NEFA into the portal vein, exposing the liver to higher NEFA concentrations. Plasma NEFA levels are approximately 20% greater in obese men and women. The contribution of visceral AT lipolysis to hepatic NEFA delivery increases with increasing visceral fat in humans and this effect is greater in women than in men (Nielsen *et al.*, 2004). Acute and chronic elevations in plasma NEFA produce peripheral (skeletal muscle and hepatic) insulin resistance (Bergman and Ader, 2000; McGarry, 2002; Bays *et al.*, 2004); they also modulate vascular tone and tissue blood flow (Steinberg and Baron, 2002). NEFAs reportedly interfere with skeletal muscle insulin signalling pathways via protein kinase C-induced phosphorylation of IRS-1 and the reduction of IRS-1-mediated actions (Yu *et al.*, 2002). However, NEFA availability differs between lean, overweight and obese subjects. A blunted increase in the lipolytic rate in overweight and obese men compared with lean individuals limits the availability of plasma NEFA as a fuel during exercise. Nevertheless, the rate of total fat oxidation was found to be similar in lean, overweight and obese subjects because of a compensatory increase in the oxidation of non-systemic fatty acids (Kanaley *et al.*, 1993; Horowitz and Klein, 2000; Goodpaster *et al.*, 2002; Mittendorfer *et al.*, 2004). Through its NEFA storing capacity, AT makes a major contribution to the control of daily lipid flux in the body (Frayn, 2002).

Catecholamines and insulin are the major regulators of lipolysis and lipid mobilization in humans (Lafontan and Berlan, 1993; Arner, 1999; Langin and Lafontan, 2004) (Fig. 6.1). This review focuses on various peptide hormones such as parathyroid hormone (PTH), growth hormone (GH), tumour necrosis factor-α (TNF-α), interleukin-6 (IL-6) and some other recently discovered agents that exhibit lipolytic activity and which also may contribute to the regulation of lipid mobilization in humans. The well-studied action of insulin will be considered briefly, while special attention will be paid to the role of cardiac hormones, the natriuretic peptides (NPs), which recently have been shown by our laboratory to exert potent lipolytic and lipid mobilizing effects in humans (Lafontan *et al.*, 2005, 2008). They represent a new and promising pathway for a better understanding of the control of exercise-induced mobilization in humans under physiological and pathological conditions. Moreover, it will be important to determine their contribution to metabolic regulation when their production and hence the circulating plasma levels are altered.

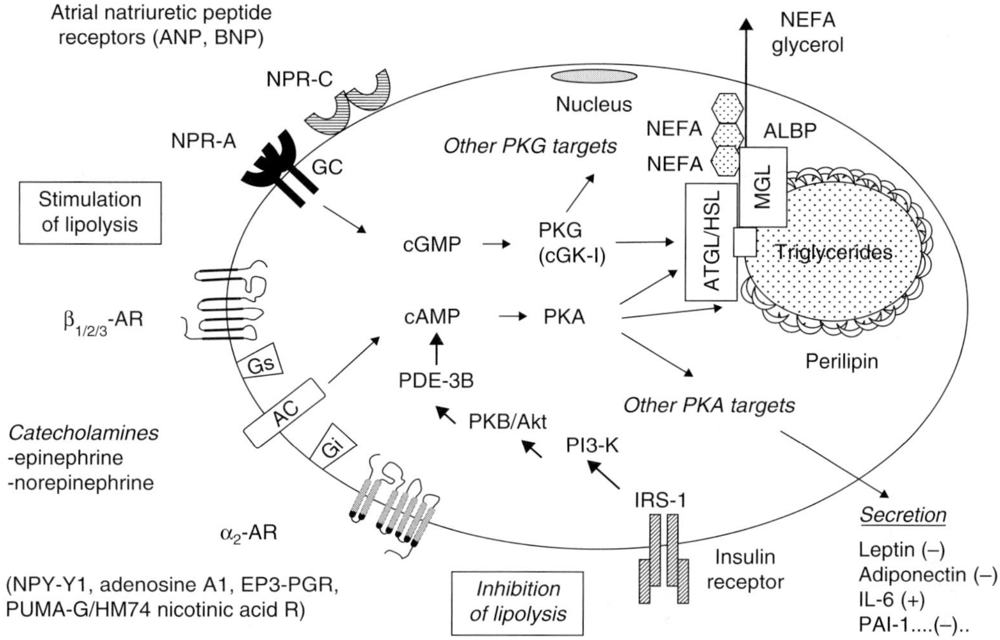

Fig. 6.1. Major pathways involved in the control of human fat cell lipolysis. Signal transduction pathways for catecholamines via adrenergic receptors (AR), atrial natriuretic peptide via type A receptor (NPR-A) and insulin. Protein kinases (PKA and PKG (cGK-I)) are involved in target protein phosphorylation. HSL phosphorylation promotes its translocation from the cytosol to the surface of the lipid droplet. Perilipin phosphorylation induces an important physical alteration of the droplet surface that facilitates the action of ATGL and HSL and initiation of lipolysis. Docking of adipocyte lipid-binding protein (ALBP) to HSL favours the fat cell outflow of NEFAs released by the hydrolysis of triglycerides. PKA and PKG (cGK-I) phosphorylate a number of other substrates that are shown in the diagram and can also influence the secretion of various adipocyte productions.

AC, adenylyl cyclase; ALBP, adipocyte lipid-binding protein; AR, adrenergic receptor; ATGL, adipose triglyceride lipase; GC, guanylyl cyclase; Gi, inhibitory GTP-binding protein; Gs, stimulatory GTP-binding protein; HSL, hormone-sensitive lipase; IRS-1, insulin receptor substrate-1; IL-6, interleukin-6; MGL, monoglyceride lipase; NEFA, non-esterified fatty acid; PAI-1, plasminogen activator inhibitor-1; PDE-3B, phosphodiesterase-3B; PI3-K, phosphatidylinositol-3-phosphate kinase; PKA, protein kinase A; PKB/Akt, protein kinase B; PKG (cGK-1), protein kinase G; (+) stimulation; (–) inhibition.

Insulin: A Major Antilipolytic Agent

Insulin plays a major role in the control of NEFA disposal. It regulates the rate of lipolysis and NEFA efflux (i.e. inhibits lipolysis) but also fat storage (i.e. elevates the rate of resynthesis of TAG from the NEFA; the re-esterification effect) and glucose uptake by fat cells. Studies over the past 20 years have elucidated the main features of how insulin acts at the molecular level as a major regulator of lipolysis (Smith and Manganiello, 1989; Degerman *et al.*, 1990; Kido *et al.*, 2001;) (Fig. 6.1). The effects of insulin on lipolysis and lipid mobilization occur very rapidly (within minutes). In the postprandial situation, or when insulin is infused intravenously using the euglycaemic hyperinsulinaemic clamp technique, lipolysis is suppressed rapidly and strikingly. Reduction of plasma insulin levels, as observed during fasting, physical exercise or even after acute somatostatin administration, leads to a sharp increase in the lipolytic rate. Overnight postabsorptive plasma insulin concentrations suppress NEFA release (Jensen, 1997). A number of circulating factors (such as TNF-α, interleukins, insulin itself, fatty acids and glycation products) have been shown to influence the effects of insulin at the target cell level and may lead to hyperglycaemia and type 2 diabetes on dysregulation of their action (Pirola *et al.*, 2004).

It seems reasonable to propose that the well-known upper-body obesity-related events are linked to regional variations in lipolysis regulation and NEFA production. There has been considerable disagreement in previous *in vitro* studies regarding regional differences in the sensitivity to insulin's antilipolytic effects (Smith *et al.*, 1979; Rebuffe-Scrive *et al.*, 1987; Dowling *et al.*, 1995). Nevertheless, it is clear that moderate changes in fasting insulin levels or insulin efficacy noticeably alter fat cell lipolysis and fasting plasma NEFA concentrations. Striking depot-specific differences, modulated by obesity and AT distribution, have been found in fat cell responsiveness to insulin. Insulin-induced antilipolysis and activation of NEFA re-esterification are blunted in omental compared with subcutaneous fat cells. Various functional differences have been identified at the insulin receptor level and the postreceptor insulin signalling cascade (Lefebvre *et al.*, 1998; Zierath *et al.*, 1998). Other partners in the insulin signalling cascade, such as type 3B phosphodiesterase (PDE-3B), which is responsible for the antilipolytic action of insulin in fat cells, and protein tyrosine phosphatases (PTPase), involved in the dephosphorylation of the insulin receptor, also contribute to the modulation of the action of insulin. Endogenous PTPase activity, including PTPase1B, is increased in omental AT and this may contribute to the relative insulin resistance of this fat depot (Wu *et al.*, 2001).

Increases in baseline systemic NEFA flux have been reported in upper-body obese women. They have been attributed partly to a decreased sensitivity to the antilipolytic effect of insulin, independent of fat cell size, and to increased lipolytic rates associated with subcutaneous fat cell hypertrophy (Jensen *et al.*, 1989). Irrespective of fat cell size, subcutaneous abdominal adipocytes are more resistant to the antilipolytic effect of insulin than gluteal adipocytes (Johnson *et al.*, 2001). To assess whether the differences reported in isolated fat cells exist *in vivo*, measurement of dose–response characteristics of systemic, splanchnic and leg NEFA release have been performed in normal humans to evaluate whether upper-body and splanchnic AT respond differently from leg AT to exogenously administered insulin. The results have confirmed the regional heterogeneity of insulin-regulated NEFA release *in vivo*. Visceral AT is more resistant to insulin's antilipolytic effects than is leg and non-splanchnic body fat (Meek *et al.*, 1999). Nevertheless, visceral fat may be a marker for, but not the source of, excess postprandial NEFA in obesity, since the increased postprandial NEFAs release observed in upper-body obese women and type 2 diabetics originates from non-splanchnic upper-body fat, not visceral fat (Guo *et al.*, 1999; Basu *et al.*, 2001).

Growth Hormone

The effect of GH on fuel metabolism includes direct stimulation of lipolysis and protein synthesis and indirect inhibition of proteolysis via insulin-like growth factor-1 (IGF-1). Exposure to GH leads to increased plasma NEFA, ketone bodies, IGF-1, insulin and glucose. Fasting and stress contribute to GH secretion, while food intake inhibits GH release. GH deficiency in humans is associated with reduced lean body mass and increased fat mass, which are normalized by GH replacement (Jorgensen, 1991). Although GH treatment in adults reduces abdominal obesity and improves insulin sensitivity as well as the blood lipid profile, the physiological contribution of GH to the control of human AT lipid mobilization has remained elusive and is not yet elucidated fully. In humans, pulsatile and continuous GH administration leads to an increase in plasma concentration, seen during exercise in humans. This promotes a significant increase in NEFA after 2–3 h, reflecting stimulation of lipolysis and ketogenesis (Moller *et al.*, 1990, 1992). Small physiological GH pulses increase interstitial glycerol concentrations in both femoral and abdominal AT (Gravholt *et al.*, 1999). Moreover, the normal nocturnal rise in plasma GH concentrations also leads to site-specific regulation of lipolysis in AT (Boyle *et al.*, 1992; Samra *et al.*, 1999). The lipolytic sensitivity to GH shows increases during fasting (Moller *et al.*, 1993). The nocturnal mean peak of GH preceded that of NEFA by 2 h; this is a lag-time which fits with that found after a GH bolus administration (Rosenthal and Woodside, 1988). GH has been suspected of playing a role in the deterioration of the metabolic control of type 1 diabetics (Press *et al.*, 1984). *In vitro* studies have shown that GH stimulates lipolysis in human adipocytes (Harant *et al.*, 1994); the effect is delayed (2–3 h) when compared with that of catecholamines and the exact mechanism of action is not fully established. The transducing pathways are suspected to involve those used by catecholamines (i.e. cAMP- and PKA-dependent pathways). GH-dependent modifications of the interactions between adenylyl cyclase and Giα_2 have been reported. Consequently, the GH-related relief of Gi-dependent inhibition of cAMP production increases lipolysis (Doris *et al.*, 1994; Yip and Goodman, 1999) (Fig. 6.1). GH-deficient patients exhibit a reduction of lipolysis and plasma NEFA concentrations (Boyle *et al.*, 1992).

GH-dependent stimulation of lipolysis probably represents a physiological adaptation to stress (maintenance of basal lipolysis during fasting and exercise). When the capacity of GH to increase lipolysis is blocked, the protein-retaining and insulin-antagonistic effects of GH on glucose metabolism are either abolished or weakened dramatically. This is an observation compatible with a key role for lipolysis in orchestrating the actions of GH (Moller *et al.*, 2003). GH and cortisol are co-secreted during stress conditions. It is probable that both hormones are involved in the regulation of AT metabolism during fasting and stress. Acute lipolytic effects of cortisol have been reported (Divertie *et al.*, 1991; Djurhuus *et al.*, 2002). GH and cortisol stimulate systemic and regional lipolysis independently and in an additive manner when coadministered (Ottoson *et al.*, 2000; Djuurhus *et al.*, 2003). A small synthetic peptide sequence of human GH (AOD-9041) has been shown to increase human and rodent fat cell lipolysis *in vitro*. Its efficiency in lipid mobilization has been observed after chronic oral administration in rodents, with the mechanisms of action remaining to be clarified in humans (Hefferman *et al.*, 2000). GH (which varies in plasma concentration range between 1 and 30 μg/l) exhibits some delayed lipid mobilizing properties when compared with fast-acting lipolytic agents and can be considered as a weaker regulator of lipolysis than catecholamines and insulin. Endogenous GH plays a very limited metabolic role during the daily feed/fast cycle but is essential for the increased lipolytic rate found with more prolonged fasting (Sakharova *et al.*, 2008).

Other Peptides with Lipolytic and/or Antilipolytic Activity

Adrenocorticotropic hormone (ACTH), α-melanocyte-stimulating hormone (α-MSH) and lipotropin (beta-LPH) exert potent lipolytic effects in rodent fat cells via melanocortin 2 receptors (Boston, 1999). These peptides have no effect in human fat cells. Glucagon and glucagon-like peptide-1 (GLP-1), which also act in rodent fat cells, do not stimulate lipolysis in human isolated subcutaneous fat cells. Moreover, no significant effect of either GLP-1 or glucagon on either lipolysis rate or blood flow were detected in muscle or AT during local or experimental intravenous (IV) hyperglucagonaemia (Bertin *et al.*, 2001; Gravholt *et al.*, 2001).

Cachexia-inducing tumours release a lipid-mobilizing factor (LMF) that promotes release of glycerol when incubated with murine adipocytes. Induction of lipolysis by LMF was associated with an increase in intracellular cyclic AMP levels. The lipolytic activity of LMF was attenuated by eicosapentaenoic acid (Tisdale, 2002). The serum and urine of cachectic cancer patients contain LMF, the activity of which is correlated with the extent of weight loss (Groundwater *et al.*, 1990). Zinc-α_2-glycoprotein (ZAG), a protein of 43 kDa, acts as a lipid-mobilizing factor to stimulate lipolysis in adipocytes, leading to cachexia in mice implanted with ZAG-producing tumours. ZAG was detected in the major fat depots of mice and in 3T3-L1 adipocytes. Both dexamethasone and a β_3-agonist increased ZAG mRNA levels in 3T3-L1 cells. ZAG gene expression and protein were also found in human fat cells (visceral and subcutaneous AT). Murine and human ZAG share up to 100% identity in specific regions hypothesized to be important in lipid metabolism (Sanchez *et al.*, 1999). ZAG is a new AT protein factor that may be involved in the modulation of lipolysis in adipocytes (Bing *et al.*, 2004). Its physiological role remains to be established (Agustsson *et al.*, 2007; Bing and Trayhurn, 2008).

Parathyroid hormone

PTH stimulates lipolysis in human fat cells (Sinha *et al.*, 1976; Taniguchi *et al.*, 1987; Bousquet-Melou *et al.*, 1995). The N-terminal 1–34 peptide portion of the hormone is responsible for adenylyl cyclase stimulation and cAMP production leading to stimulation of lipolysis. Furthermore, in humans, PTH stimulates lipid mobilization *in vivo*. There appears to be no defect in the adenylate cyclase system in the fat cell in response to PTH in patients with pseudohypoparathyroidism. The effect of PTH appears at rather high PTH concentrations, which are clearly extra-physiological. There are no data concerning a physiological contribution of PTH in the control of lipid mobilization in humans.

Interleukin-6

Subcutaneous AT was shown to secrete large amounts of IL-6 and this secretion was correlated with the BMI of the subjects (Mohamed-Ali *et al.*, 1997). Circulating IL-6 levels may reflect, at least in part, AT IL-6 production (Bastard *et al.*, 2000) and it has been proposed that locally secreted IL-6 could act on adipocytes by a paracrine/autocrine mechanism (see Chapter 5). Human adipocytes express both IL-6 and its receptor system consisting of the IL-6 receptor and the signal transducing protein, gp130 (Bastard *et al.*, 2002). IL-6 stimulates lipolysis in human adipocytes (Päth *et al.*, 2001) and exerts anti-insulin actions. It was found that IL-6-treated adipocytes exhibit reduced insulin-stimulated lipogenesis and glucose transport and fail to maintain their adipocyte phenotype (e.g. downregulation of various adipogenic markers). Likewise, expression of insulin receptor-β (IR-β) and IRS-1 is reduced, as is insulin-induced activation of IR-β, Akt/PKB and ERK1/2. Expression of suppressor of cytokine signalling 3 (SOCS3), a potential inhibitor of insulin signalling (Lagathu *et al.*, 2003; Rotter *et al.*, 2003) is also induced by IL-6. Recombinant human IL-6 (rhIL-6) infusion, leading to plasma IL-6 concentrations of about 140 pg/ml in healthy volunteers, resulted in an increment of plasma NEFA and glycerol concentrations and an increased rate of appearance of NEFA and glycerol measured by isotope dilution techniques (van Hall *et al.*, 2003). Plasma cortisol concentrations were increased by 50%, with transient changes in epinephrine also observed during IL-6 infusion, while putative concomitant changes in GH levels were not determined. Inclusion of IL-6 inside a physiological loop of lipolytic process regulation remains to be established (Jensen, 2003). A recent study suggests that higher circulating IL-6 concentrations are associated with increased isoproterenol-stimulated lipolysis, especially in omental adipocytes in women (Morisset *et al.*, 2008).

Tumour necrosis factor-*α*

TNF-α is a macrophage-secreted product that is also released by fat cells. Macrophages release cytokines in response to lipopolysaccharide that stimulate lipolysis in freshly isolated rat adipocytes. Interestingly, TNF-α can account for most of the action on adipocytes. Stimulation of lipolysis by TNF-α is not direct, since it becomes apparent only after long-lasting exposure of human and rodent adipocytes to the cytokine (Hauner *et al.*, 1995). TNF-α mechanisms of action have been explored in rodent fat cells with altered TNF-α receptor expression (TNFR1 and TNFR2). Experiments were performed on preadipocyte cell lines established from wild-type mice ($TNFR1^{+/+}$–$TNFR2^{+/+}$) and from mice lacking TNFR1 ($TNFR1^{-/-}$), TNFR2 ($TNFR2^{-/-}$) or both ($TNFR1^{-/-}$–$TNFR2^{-/-}$) to demonstrate the role of the different TNF-α receptors in the induction of the lipolytic effects. TNF-α-induced lipolysis, as well as inhibition of insulin-stimulated glucose transport, are mediated predominantly by TNFR1 (Sethi *et al.*, 2000; Xu and Hotamisligil, 2001). Experiments have demonstrated that TNF-α regulates lipolysis partly by decreasing perilipin protein levels at the lipid droplet surface and activating the extracellular signal-related kinase (ERK) pathway (Souza *et al.*, 2003; Langin and Arner, 2006). Blunting the endogenous inhibition of lipolysis through down-regulation of Gi protein represents another possible mechanism (Gasic *et al.*, 1999). In human fat cells, TNF-α activates the three mammalian mitogen-activated protein kinases (MAPK) in a distinct time- and concentration-dependent manner. TNF-α-induced lipolysis is mediated only by p44/42 ERK and Jun-kinase, but not by p38-kinase (Ryden *et al.*, 2002).

Leptin

Leptin is produced essentially by adipocytes and secreted into the bloodstream. Mutations of the leptin gene in humans are associated with severe obesity, glucose intolerance and insulin resistance, which are reversed by leptin therapy (Clément *et al.*, 1998; Farooqi *et al.*, 1999). Leptin can act directly on tissues independently of hypothalamic mediation and plays a potential role in inflammation. It acts directly on macrophages to increase their recruitment in AT (Curat *et al.*, 2004), their production of cytokines and their phagocytic activity (Gainsford *et al.*, 1996; Pickup and Crook, 1998; Bouloumié *et al.*, 1999). In rodents, leptin has been shown *in vitro* to reduce the expression of lipogenic enzymes in preadipocytes (Bai *et al.*, 1996) and to increase glycerol release from mature adipocytes (Siegrest-Kaiser *et al.*, 1997). Leptin causes concentration-dependent stimulation of lipolysis in rat fat cells and has no effect in fat cells of the *fa/fa* rat, which has a defective leptin Ob-R_L receptor. Apparently, the lipolytic effect of leptin occurs at the adenylyl cyclase/Gi protein level and leptin-induced lipolysis opposes the tonic inhibition of endogenous adenosine in white adipocytes (Frühbeck *et al.*, 1997, 2001; Siegrest-Kaiser *et al.*, 1997). Nevertheless, leptin (within a concentration range from 25 to 250 ng/ml) has no direct lipolytic effect in human adipocytes, either in children or adults (Eliman *et al.*, 2002).

Leptin administration has been shown previously to stimulate the sympathetic nervous system in rodents (Collins *et al.*, 1996; Haynes *et al.*, 1997). Transgenic mice overexpressing leptin (Lep-Tg) exhibit substantial reductions of adipose mass. However, the absence of β_3-adrenergic receptor has virtually no effect on the phenotype of the Lep-Tg mice. It does not result in a chronically elevated lipolytic state but, instead, in low basal lipolysis characterized by a decrease in perilipin and PKA activity in white fat (Ke *et al.*, 2003). Nevertheless, leptin, when overexpressed ectopically through adenovirus administration to rodents, has been shown to exert a novel form of lipolysis in which glycerol is released without proportional release of NEFA. Upregulation of peroxisome proliferator activated receptor-α (PPARα), acyl CoA oxidase (ACO), carnitine palmitoyl transferase-1 (CPT-1) and NEFA oxidation was observed in rat fat cells after leptin treatment. The fat cells undergo striking modifications, have an increased number of mitochondria and are transformed into fat-oxidizing machines (Shimabukuro *et al.*, 1997; Wang *et al.*, 1999; Orci *et al.*, 2004). It is unknown if the same leptin-dependent action, which is usually observed at supra-physiological doses of leptin, solely modifies leptin effects (e.g. other cytokine receptors could be activated by such concentrations of leptin) and operates in human fat cells.

Angiopoietin-like protein 3

Angiopoietin-like protein 3 (ANGPTL3) is a secretory protein of 70 kDa, the mRNA of which is expressed in the liver of human, rat and mouse. This protein contains a coiled-coil region and fibrinogen-like motif and belongs to the angiopoietin-like family, having a similar amino acid sequence and structure. ANGPTL3 also acts on endothelial cells and vascular neogenesis to a lesser degree than angiopoietins (Camenisch *et al.*, 2002). It inhibits the activity of lipoprotein lipase that probably accounts for an increase in plasma TAG. Human ANGPTL3 stimulates the release of NEFA and glycerol from 3T3-L1 adipocytes. Specific binding of ANGPTL3 to AT has been shown using fluorescence-labelled protein and ^{125}I-labelled protein by binding analysis in rodent and human AT. ANGPTL3 is a liver-specific factor targeting the adipocyte but its lipolytic action, if any, on human fat cells remains to be established.

Adenylyl cyclase inhibitors – antilipolytic agents

Various hormones and autacoid agents are known to control adenylyl cyclase activity negatively and inhibit cAMP production and lipolysis by their interaction with plasma membrane receptors. So far, two peptides with well-established antilipolytic actions have been described. Neuropeptide Y (NPY) and peptide YY (PYY) exert antilipolytic effects in human fat cells (Valet *et al.*, 1990). NPY/PYY-receptor stimulation, via Gi-protein coupling, inhibits adenylyl cyclase activity and cAMP production (Fig. 6.1). Differences exist in human NPY/PYY receptor distribution according to the anatomical location of AT; the highest level of expression being found in subcutaneous adipocytes (Castan *et al.*, 1993). The human fat cell NPY/PYY receptor is an NPY-Y_1-receptor subtype which, when stimulated, sustains a strong antilipolytic effect. In addition, its stimulation is also associated with a positive action on leptin secretion by human fat cells (Serradeil-Le Gal *et al.*, 2000). In the absence of suitable pharmacological tools for *in vivo* assays, the physiological relevance of this pathway has not been established in humans.

Natriuretic peptides contribute to the control of lipolysis and lipid mobilization in humans

Natriuretic peptides

The first member of the natriuretic peptide (NP) family, atrial natriuretic peptide (ANP), was discovered when de Bold *et al.* (1981) showed for the first time that infusion of atrial tissue extracts in rats promoted strong natriuresis and diuresis. Subsequently, two other peptides with natriuretic properties were isolated: brain natriuretic peptide (BNP) and the C-type natriuretic peptide. ANP and BNP essentially are synthesized by cardiomyocytes, but the thymus and macrophages have also been identified as sites of synthesis of the hormone (Kiemer and Vollmar, 2001). ANP and BNP regulate a variety of physiological events; they have natriuretic, vasodilating and lusitropic properties. They also influence sympathetic nervous system activity and the renin–angiotensin system. ANP and BNP, which could be considered as endocrine hormones, apparently are antagonists to vasopressin, endothelins and the renin–vasopressin–aldosterone system. Most of the effects are mediated by the stimulation of guanosine 3′,5′-cyclic monophosphate (cGMP) production by target cells (Silberbach and Roberts, 2001). Further studies have also focused on a role in the control of myocardium function and heart remodelling. *In vitro* studies suggest a role for BNP as a regulator of myocardial structure via control of cardiac fibroblast function (Tsuruda *et al.*, 2002). The role of CNP *in vivo* is less well defined. Although CNP might not be a modulator of diuresis and natriuresis, it is a vasodilator expressed by endothelial cells. The functions of ANP are not restricted simply to blood pressure homeostasis, they also seem to play an important role in the immune system (Kiemer and Vollmar, 2001). Production of ANP and BNP is stimulated in pathological conditions, with plasma levels of these peptides being increased in subjects with left ventricular hypertrophy (Schirmer and Omland, 1999), asymptomatic ventricular dysfunction (Lerman *et al.*, 1993; McDonagh *et al.*, 1998) and overt heart failure (Cowie *et al.*, 1997). Considerable attention has been focused on the potential utility of plasma levels of these peptides to establish a diagnosis of heart failure or for identifying left ventricular dysfunction (Maisel *et al.*, 2002). Strenuous endurance exercise is followed by increases in plasma ANP and BNP (Ohba *et al.*, 2001; Köning *et al.*, 2003).

Molecular cloning techniques have revealed the existence of three subtypes of natriuretic peptide receptors (NPR). These receptors either present a guanylyl cyclase activity (NPR-A and NPR-B) or not (NPR-C). NPR-A and NPR-B are the active forms of the receptors since their stimulation activates the guanylylcyclase (GC) function of the receptor protein and the production of the second messenger, cGMP. The increment of cGMP promoted by NPR-A and NPR-B activation has also been reported to inhibit phospholipase C (PLC) signalling in smooth muscle cells (Kuhn, 2003). NPR-C, which has a short cytoplasmic domain without GC activity, influences plasma NP levels, ANP half-life and systemic effects such as natriuresis and blood pressure. Although some mechanisms of action mediated by NPR-C still remain unclear, it is recognized that NPR-C acts as a clearance receptor which allows the activity of the NP system to be tailored to specific local needs

(Matsukawa *et al.*, 1999). Some studies have shown that the NPR-C may be coupled to adenylyl cyclase inhibition via a heterotrimeric Gi protein. The majority of the receptors in vascular smooth muscle cells are of the NPR-C subtype.

Lipolytic effect of natriuretic peptides

ANP receptors have been identified in various tissues, including rat fat cells, and it was shown that their stimulation increased cGMP (Göke *et al.*, 1989; de Leon *et al.*, 1995; Jeandel *et al.*, 1988). Human AT expresses NPR-A and NPR-C mRNA (Sarzani *et al.*, 1996; Garruti *et al.*, 2007). The original finding that led to a number of subsequent studies was the discovery of the lipolytic action of NPs (Lafontan *et al.*, 2005, 2008). NPs exert potent lipolytic effects similar to those induced by the β-adrenergic receptor agonist, isoproterenol. The relative order of lipolytic potency of the peptides (ANP>BNP>>CNP) corroborates the presence of NPR-A in fat cells (Fig. 6.2a). This point was confirmed by binding studies performed on human fat cell membranes using [^{125}I]ANP as a radioligand (Sengenes *et al.*, 2000).

Generally, lipolysis in adipocytes is stimulated by hormones that activate adenylyl cyclase, elevate intracellular cAMP levels and activate PKA, which phosphorylates perilipin and HSL (Langin and Lafontan, 2004). The main enzyme that is also involved in the degradation of cAMP in the adipocyte, PDE-3B, also modulates lipolysis when activated by insulin or inhibited by methylxanthine drugs (Fig. 6.1). NPs promote a strong and sustained increment of intracellular cGMP in human fat cells (Moro *et al.*, 2004b) and this effect is unrelated to an inhibition of the cGMP-inhibitable phosphodiesterase PDE-3B of the adipocyte (Sengenes *et al.*, 2000). In fact, although PDE3-3B catalytic sites have similar high affinity for cAMP and cGMP, with the V_{max} for cAMP being much higher (4–10 times) for cGMP, it is considered that cGMP could inhibit PDE3-3B catalytic activity transiently through competition with cAMP at the catalytic site. ANP-induced lipolysis is associated with an increase in the serine phosphorylation of HSL in mature human adipocytes, as well as in adipocyte precursors differentiated into adipocytes. The effect was not shown in rat adipocytes, a non-responsive species (Sengenes *et al.*, 2002). The signal transduction pathway stimulated by ANP to promote lipolysis in human fat cells is strictly connected to an increase in intracellular cGMP concentrations. The non-hydrolysable analogue of cGMP, 8-bromo-cGMP, mimicked the lipolytic effects of ANP. Since PKA may be activated by cGMP, it was verified that ANP did not stimulate PKA activity and that inhibition of PKA by H-89 did not affect ANP-induced lipolysis. ANP-mediated lipolysis did not involve crosstalk between cGMP and PKA. It is a PKG, identified as cGK-I, which promotes perilipin and HSL phosphorylation and which is at the origin of ANP-induced lipolysis. The cGMP analogue inhibitor of cGK-I, 8-pCPT-cGMPS, inhibited HSL activation and lipolysis (Sengenes *et al.*, 2003). This finding in isolated human fat cells confirmed early data in a rat cell-free system where phosphorylation and activation of HSL was shown (Khoo *et al.*, 1977; Strålfors and Belfrage, 1985). Although subsequent studies have shown a role for the MAP kinases in the control of lipolysis induced by TNF-α in

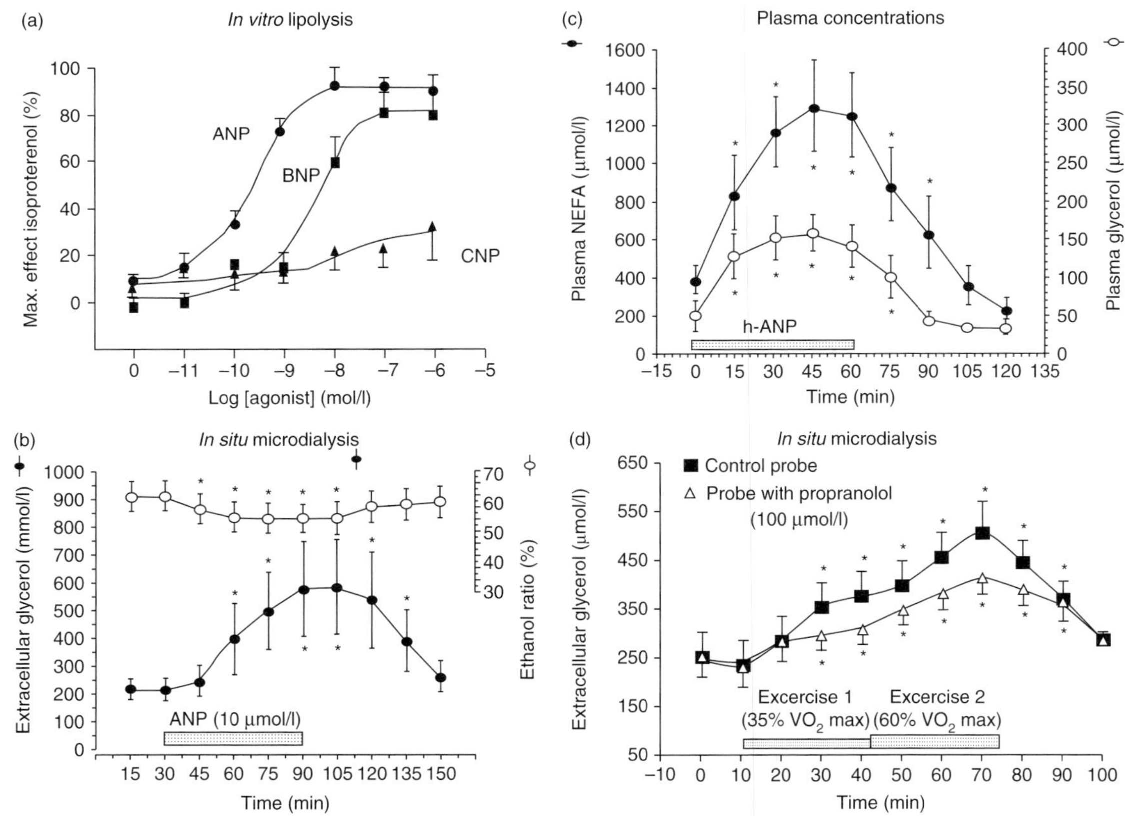

Fig. 6.2. *In vitro* and *in vivo* effects on natriuretic peptides on isolated fat cell lipolysis and lipid mobilization in humans. Values are means ± SEM of 6–8 subjects. *Significantly different from basal value. (a) Comparison of the lipolytic effects of natriuretic peptides in isolated human fat cells (Sengenes *et al.*, 2000). (b) Changes in extracellular glycerol concentrations (EGC) and in ethanol ratio (ethanol dialysate concentration/ethanol perfusate levels × 100), which reveals changes in local blood flow, during the infusion of human ANP (10 µmol/l) in a microdialysis probe implanted in human subcutaneous adipose tissue (Sengenes *et al.*, 2000). (c) Effect of IV infusion of human ANP (50 ng/min/kg during 60 min) on plasma NEFA and glycerol levels (Galitzky *et al.*, 2001). (d) Comparison of the mean changes in EGC values in abdominal subcutaneous adipose tissue during two successive exercise bouts performed at 35% (exercise 1) and 60% (exercise 2) of VO_2 max and during recovery. Control microdialysis probe was perfused with Krebs–Ringer buffer. For the study of local β-adrenergic receptor blockade, a microdialysis probe was supplemented with propranolol (100 µmol/l) (Moro *et al.*, 2004a).

human fat cells (Ryden *et al.*, 2002; Zhang *et al.*, 2002), neither ERK nor p38 MAP-kinase were involved in the ANP-mediated HSL phosphorylation. Moreover, the MAP kinase inhibition did not affect the ANP-induced HSL phosphorylation. Importantly, it was observed that the lipolytic effect of NPs was not controlled by insulin. Insulin treatment of human fat cells has no effect on the ANP-induced lipolytic response (Sengenes *et al.*, 2000; Moro *et al.*, 2004b). The occurrence of NP-induced lipolysis is a primate fat cell specificity. NPs do not stimulate lipolysis in various species. One of the major explanations of such species-related differences is that adipocytes from ANP-non-responsive species present a predominance of the ANP clearance receptor (NPR-C) and a very low expression of biologically active NPR-A. Stimulation of the low population of NPR-A is not sufficient to promote an increment in cGMP levels required for HSL activation (Sengenes *et al.*, 2002).

Induction of lipid mobilization by administration of pharmacological doses of ANP

The lipolytic action revealed *in vitro* in isolated human subcutaneous fat cells was confirmed *in vivo* after the administration of ANP in a microdialysis probe implanted in human subcutaneous abdominal AT (SCAAT). ANP infusion in the microdialysis probe promoted an increment in the extracellular concentration of glycerol in AT, as well as vasodilatation of the vessels draining the fat depots (Sengenes *et al.*, 2000) (Fig. 6.2). Both events contribute to a coordinated enhancement of lipid mobilization in SCAAT.

The lipid-mobilizing effect of an IV infusion of h-ANP, as well as various metabolic and cardiovascular variables, were studied in young obese and normal-weight subjects. Microdialysis probes were inserted into SCAAT to measure changes in the extracellular glycerol concentrations during IV h-ANP administration. Spectral analysis of blood pressure and heart rate oscillations, recorded using digital photoplethysmography, were also used to assess changes in autonomic nervous system activity. Intravenous infusion of h-ANP (50 ng/min/kg) for 60 min stimulated a marked increase in plasma glycerol and NEFA (Fig. 6.2) and a slight increase in insulin plasma levels in both lean and obese men. Plasma norepinephrine concentrations rose weakly and similarly during h-ANP infusion in lean and obese men. The effects of h-ANP infusion on the autonomic nervous system were similar in both groups, with an increase in the spectral energy of the low-frequency band of systolic blood pressure variability and a decrease in the spectral energy of the high-frequency band of heart rate. In SCAAT, h-ANP IV infusion increased extracellular glycerol concentration and blood flow similarly in both groups. The increase in extracellular glycerol observed during h-ANP infusion was not modified when 0.1 mmol/l propranolol was added to the microdialysis probe perfusate to prevent β-adrenoceptor (β-AR) activation. These *in vivo* studies show that ANP is a potent lipid-mobilizing hormone that acts independently of the activation of the sympathetic nervous system. Apparently, obesity did not modify the lipid-mobilizing effect of ANP in young obese subjects (Galitzky *et al.*, 2001). Before giving final answers concerning the role of the ANP lipid-mobilizing pathway in obese subjects, it is necessary to expand this kind of approach to a larger number of obese patients and to pay special attention to sex, age, the duration of the obese state and the existence or not of metabolic disorders related to the obese state (Moro *et al.*, 2007a; Chatzinikolaou *et al.*, 2008).

Contribution of ANP to the physiological control of lipid mobilization in humans

Exercise-induced lipid mobilization in humans was considered to depend mainly on sympathetic nervous system activation and catecholamine action. Exercise-induced inhibition of insulin release also contributes to the lipid-mobilizing effect during exercise. Examining the results showing that ANP/BNP exerted lipolytic and lipid-mobilizing effects in humans, a putative physiological contribution of ANP/BNP to exercise-induced lipid mobilization was hypothesized. Strenuous endurance exercise is followed by increases in plasma BNP (Köning *et al.*, 2003). During exercise, the sympathetic nervous system is activated and, concomitantly, ANP is released from the exercising heart. Lipid mobilization assays were performed in healthy young men using *in situ* microdialysis in SCAAT during two successive exercise bouts performed at 35% and 60% VO_2 max after placebo or oral non-selective β-antagonist (tertatolol) treatment. In placebo-treated subjects, as expected, exercise promoted an increase in the extracellular glycerol concentration (EGC). The infusion of 0.1 mmol/l propranolol (non-selective β-antagonist) in the microdialysis probe only reduced the EGC increase promoted by exercise partially (Fig. 6.2). This suggested a possible contribution of another lipid-mobilizing pathway, which was confirmed by an additional study. Indeed, oral β-AR blockade (given 1 h before the beginning of exercise) did not prevent exercise-induced lipid mobilization in SCAAT. Since blockade of fat-cell β-ARs was verified, the existence of another partner and the possible contribution of ANP was further strengthened. The exercise-induced increase in plasma ANP was magnified greatly by administration of oral tertatolol. A positive correlation has been found between EGC and plasma ANP levels, but also between extracellular cGMP and EGC (Moro *et al.*, 2004a) (Fig. 6.3).

These studies show that exercise-induced lipid mobilization, resistant to local propranolol, is related partly to the action of ANP. Oral β-AR blockade promotes strong exercise-related ANP release by the heart, which explains lipid mobilization in SCAAT. The physiological contribution of ANP, concomitantly with the SNS, to the control of lipid mobilization is supported by these results. Due to the lack of a suitable NPR-A antagonist, usable in clinical studies (Moro *et al.*, 2004b), it is difficult to determine the relative contribution of both pathways to the physiological control of lipid mobilization. Putative release, during exercise, of other biological compounds known to exert lipolytic effects in human fat cells also should be considered. Moderate exercise had either no effect (Henderson *et al.*, 1989) or slightly increased plasma PTH concentrations (Tsai *et al.*, 1997). GH is known to be secreted during the type of exercise where ANP action is proposed (Stich *et al.*, 2000a; Moro *et al.*, 2004a). GH is a less potent lipolytic agent than

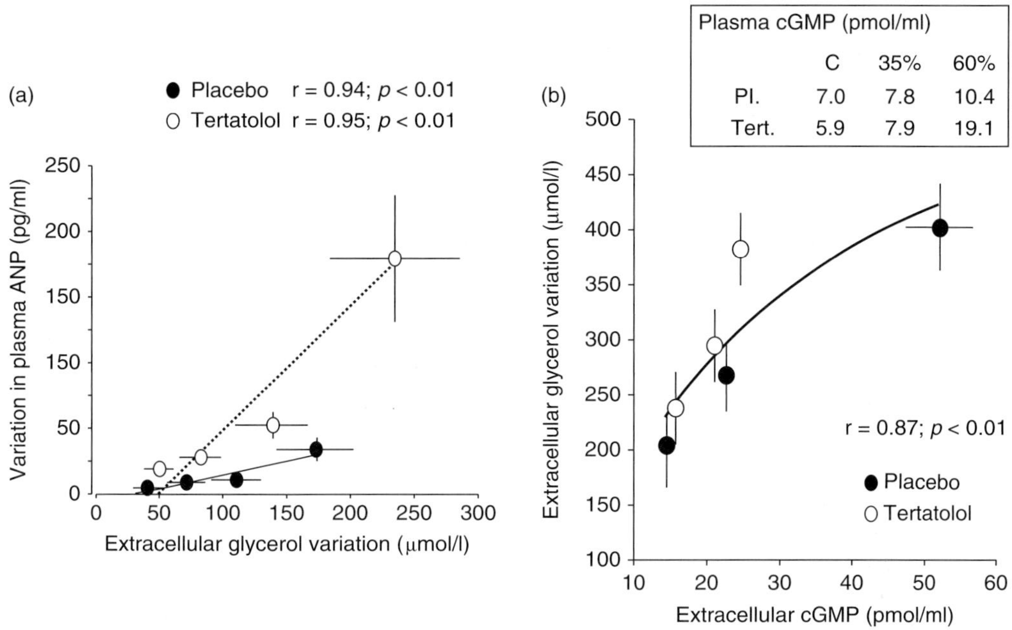

Fig. 6.3. Comparison of the mean changes in plasma ANP and cGMP levels, as well as extracellular glycerol (EGC) and cGMP concentration values during two successive exercise bouts performed at 35% (exercise 1) and 60% (exercise 2) of VO_2 max in placebo and tertatolol (β-adrenergic receptor antagonist)-treated patients (Moro *et al.*, 2004a). (a) Correlation between plasma atrial natriuretic peptide (ANP) concentrations and EGC increase from the probes perfused with propranolol in placebo- and tertalolol-treated subjects. ANP and EGC values were determined at the 10th and the 20th min of each exercise bout. (b) Correlation between EGC and extracellular cGMP concentrations. Plasma cGMP, EGC and extracellular cGMP concentrations were determined at rest and during each exercise bout.

catecholamines, with the ANP- and GH-related lipolytic effects not fully explaining the acute and short-term lipid mobilization promoted by exercise (Hansen, 2002). High concentrations of IL-6 appear to be necessary to stimulate lipolysis, and a lipolytic effect of this agent is seen only after several hours (van Hall *et al.*, 2003). Neither glucagon nor ACTH are active in human fat cells. Other putative agents, such as prostaglandins, adenosine or NPY, the release of which could be affected by exercise (Valet *et al.*, 1990), are known to exert antilipolytic effects on human fat cells and do not represent potential partners to explain the lipid-mobilizing action observed in our studies.

Conclusions and Future Trends

This review discusses how various peptides act as important hormonal regulators of lipolysis. Insulin is the most important and widely recognized factor operating through its antilipolytic activity. Modulations of insulin release and action have a direct effect on lipolysis and lipid mobilization. It is less certain whether some of the peptides exerting lipolytic effects *in vitro* contribute significantly to the control of lipid mobilization. Before adding a lipolytic agent to the list of the proven lipolytic hormones, it will be necessary to determine whether the circulating or local concentration (i.e. at the adipocyte level) of the lipolytic agent is able to initiate lipid mobilization *in vivo*. The time-course and magnitude of the changes in lipid mobilization promoted by the various putative factors require attention when physiological and/or pathological relevance is investigated. During exercise, plasma insulin concentrations fall, whereas the levels of catecholamines, ANP, GH and IL-6 increase. Changes in hormone concentrations and lipolytic responses are rapid during exercise and depend largely on the intensity and duration of the exercise. The functional hierarchy of modulators of lipolysis merits discussion. Catecholamines, ANP and insulin are probably the factors which could operate initially, whereas the more gradual increase of GH and IL-6 support the idea that they could contribute to a more delayed lipolytic action. The variability in lipolytic rate in different AT beds is a well-established phenomenon existing in humans and rodents. So far, regional differences usually have been defined based on adrenergic and insulin receptor density and function (Horowitz, 2003; Lafontan and Berlan, 2003). At the present time, the contribution of other lipolytic and antilipolytic systems has not been investigated so thoroughly. Differences in local AT insulin, α_2- and β-adrenergic receptor affinity, density and function have been considered to be responsible for the regional heterogeneity in exercise-induced lipid mobilization (Wahrenberg *et al.*, 1989; Stich *et al.*, 2000b).

ANP and BNP have been described by our group as being potent lipolytic agents in isolated human fat cells, whereas the action of CNP is very weak. The lipolytic effect is mediated by the stimulation of fat cell plasma membrane NPR-A and operates through an original cGMP-dependent pathway leading to HSL phosphorylation/activation. NPR-C is also expressed in human fat cells. Moreover, neprilysin (neutral endopeptidase 24.11, CD10, NEP), the enzyme that is involved in NP degradation, has been identified in human fat cell membranes (Fig. 6.4).

Physiological or pharmacological modulations of its activity could interfere with effects of NPs on fat cells. The administration of pharmacological doses of h-ANP has demonstrated clearly the lipid-mobilizing effect of the hormone and its ability to reach fat cells by IV administration.

The putative relevance of circulating NPs in physiological conditions remains to be outlined completely. The rise in plasma ANP during head-down bed rest has been associated with increased lipolysis in SCAAT and whole-body lipid oxidation in healthy young men, independently of catecholamine elevation (Moro *et al.*, 2007b). During exercise, it has been shown that, in addition to catecholamines, ANP plays a noticeable role in the control of lipid mobilization in SCAAT (Moro *et al.*, 2004a, 2006). The action of ANP on diverse AT depots will be addressed further in the future to verify if fat cell sensitivity to ANP/BNP differs according to the anatomical fat location as reported for other regulatory pathways (Lafontan and Berlan, 2003). Elevated plasma levels of ANP were found after prolonged exercise (e.g. long-distance running) (Niessner *et al.*, 2003)

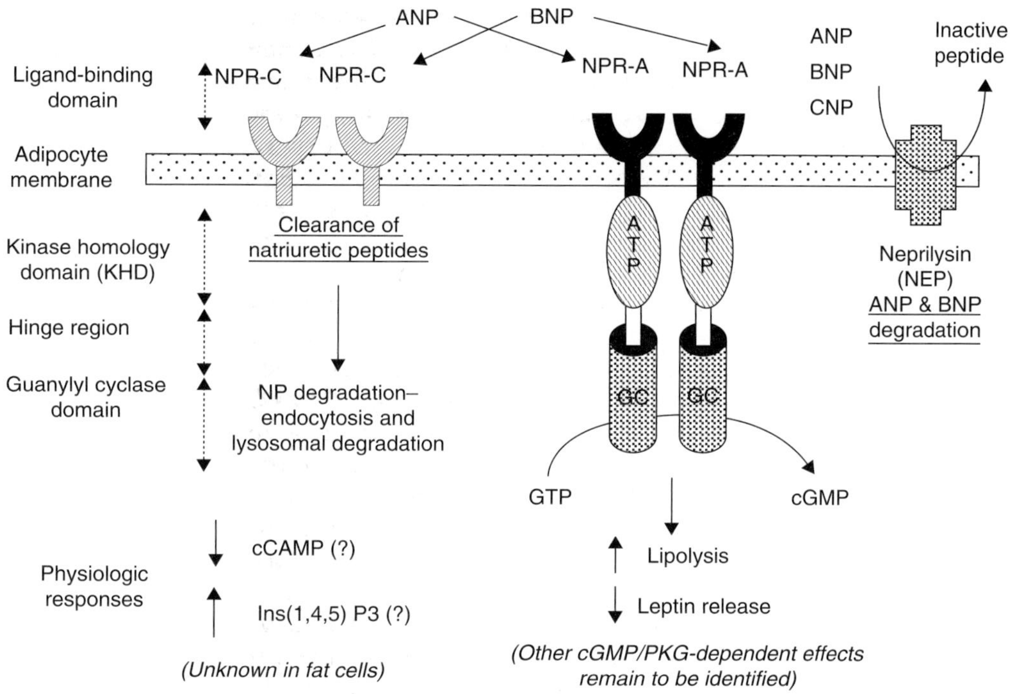

Fig. 6.4. Functional domains of the ANP-dependent receptor system in human adipocyte. Note that neprilysin (neutral endopeptidase 24.11, CD10, NEP), which is involved in natriuretic peptide degradation, has been identified in the human fat cell membrane.

and an exaggerated release of ANP has been reported in older individuals (Poveda *et al.*, 1998). One of the positive and specific effects of ANP during prolonged exercise could be to sustain lipid mobilization from AT. ANP may contribute, by elevating circulating NEFA levels, to the huge energy demand during long-distance running.

Changes in the lipid-mobilizing potencies of NPs have been reported in obese women submitted to a 28-day low-calorie diet associated with significant weight loss. The lipid-mobilizing effects of ANP were enhanced, basal lipolysis was increased, fasting insulin levels were reduced but NPR-A and NPR-C changes were not reported (Lafontan *et al.*, 2008). In previous studies, expression of NPR-C (clearance receptor) was shown to be reduced by fasting in rats (Sarzani *et al.*, 1995). Other ANP-related functions were improved by food restriction, such as short-term calorie restriction followed by body weight reduction, enhanced ANP-induced diuresis and natriuresis and decreased blood pressure in obese hypertensive subjects (Dessi-Fulgheri *et al.*, 1999). It might be speculated that during periods of low energy supply, a reduction in NPR-C receptor density takes place. According to the classical paradigm, clearance of ANP by NPR-C uptake could be reduced, while interactions with NPR-A would be facilitated.

A variation in the promoter region of the *Npr3* gene (encoding for NPR-C) has been associated with lower plasma ANP and higher blood pressure in obese hypertensive patients who had a reduced NPR-A-to-NPR-C expression ratio (based on mRNA determinations) in AT (Sarzani *et al.*, 1999). An epidemiological study has further shown that homozygosity for the A(-55) allelic variant of the *Npr3* gene is associated with lower body mass index (BMI). Male subjects carrying the A(-55)A NPR-C genotype exhibited a significantly lower prevalence of overweight, obesity and abdominal obesity. These individuals also presented a lower 20-year rate of being overweight compared with CC+CA individuals (Sarzani *et al.*, 2004). Preliminary studies have not revealed major differences between lean and obese subjects in the response to ANP administration (Galitzky *et al.*, 2001). However, experiments should be promoted to evaluate the impact of the NPs in older obese subjects of both sexes.

An epidemiological study on the Framingham Heart Study offspring cohort has shown that plasma ANP and BNP levels are reduced in obese subjects (Wang *et al.*, 2004). In addition, reduced BNP levels occur in obese individuals with heart failure (Mehra *et al.*, 2004). Although hypertension is probably associated with elevated plasma NP levels, it is noticeable that obese and overweight subjects exhibit low plasma NP levels, which may exacerbate salt and fluid retention and aggravate hypertension. Obesity is a major factor of hypertension and hypertension-related disorders such as left ventricular hypertrophy. Occurrence of an altered distribution of fat cell NPR distribution and altered NP responsiveness has been proposed in the obese (Dessi-Fulgheri *et al.*, 1998).

The role of the ANP-dependent lipid-mobilizing pathway becomes of major importance when subjects are submitted to treatment with β-AR antagonists, which expectedly block the lipolytic effects. This is the situation for millions of people taking β-AR antagonist treatment for cardiovascular diseases and heart failure (Giesler *et al.*, 2004). Through an ANP release, the possibility of fat cell TAG mobilization in the setting of β-AR blockade remains open. The lipid-mobilizing activity remaining, when performing exercise, in the subjects treated with a β-AR antagonist is due essentially to the ANP-dependent pathway. Administration of a β-AR antagonist along with exercising represents a prerequisite to promote enhanced ANP release (Berlin *et al.*, 1993). Subjects receiving β-AR antagonist medication (i.e. metoprolol, bisoprolol, atenolol, propranolol and sotalol) are characterized by substantially elevated ANP, BNP and cGMP plasma concentrations (Luchner *et al.*, 1998). Patients with chronic coronary artery disease exhibit much higher exercise releases of ANP and BNP when they are treated with β-blockers (Marie *et al.*, 2004). These results suggest that the cardiac ANP system may contribute to the therapeutic mechanisms of β-AR antagonists, generating an attractive therapeutic mechanism for further protecting the diseased heart against stress. Possible relationships between ANP/BNP increase and alterations of plasma NEFA levels have never been considered in detail, although cardiac heart failure has been described as a ketone-prone state (Lommi *et al.*, 1997). Thus, the efficiency of this ANP-dependent lipid-mobilizing loop becomes of particular interest in β-blocker-treated patients to maintain some lipid-mobilizing possibilities and avoid accumulation of fat. Regular practice of exercise will be especially beneficial in such patients to limit the deleterious actions of fat cell β-AR blockade on fat accumulation.

Circulating concentrations of NPs (especially BNP) are very high in patients with congestive heart failure and/or ventricle hypertrophy (Lerman *et al.*, 1993; Cowie *et al.*, 1997; McDonagh *et al.*, 1998; Schirmer and Omland, 1999; Gardner, 2003). Weight loss in heart failure reportedly is associated with a poor prognosis. Heart failure is associated with a number of metabolic and neurohormonal dysfunctions, with the real contribution of NPs to the disease remaining to be established (Kalra and Tigas, 2002). Importantly, congestive heart failure is associated with a higher prevalence or risk of developing type 2 diabetes. Elevated NEFA concentrations are suspected to play a pivotal role as the majority of patients with type 2 diabetes are overweight or obese and are characterized by day-long elevations in plasma NEFA concentrations, which are not suppressed normally by meals or glucose load (Reaven *et al.*, 1988; Amato *et al.*, 1997; Mokdad *et al.*, 2003). Elevated circulating NEFA can cause/aggravate insulin resistance in muscle and the liver (Bays *et al.*, 2004) and represents a predictive risk factor for sudden death in the population (Jouven *et al.*, 2001). It is therefore important to know if NPs, in addition to insulin resistance, operate as partners in disorders related to the metabolic syndrome. Controversies exist concerning plasma levels of ANP/BNP in diabetic patients (Wang *et al.*, 2004).

In addition to the putative hormonal factors proposed previously (i.e. catecholamines, IL-6 and TNF-α), increased endogenous ANP/BNP concentrations in congestive heart failure may promote lipid mobilization and could contribute to cardiac cachexia. Nevertheless, it is also possible that the body may adapt to the raised levels of endogenous BNP through downregulation of NP signalling pathways or upregulation of the clearing pathways (i.e. increased clearance NPR-C expression and/or enhanced activity of neprilysine, the neutral endopeptidase enzyme, which plays a role in NP degradation).

BNP has now been approved for treatments in acute heart failure since it has beneficial effects on central haemodynamics and urinary excretion of Na^+ (Richards *et al.*, 2002; Gardner, 2003; Maisel, 2003). It will be essential to verify if infusion of BNP promotes, like the IV administration of ANP (Galitzky *et al.*, 2001), a potent and sustained lipid-mobilizing effect and increased plasma NEFA levels, which could alter heart function and counteract some of the positive actions of the compound at other sites. A number of additional studies will be required to answer the numerous questions raised by the discovery of the metabolic effects of ANP in order to integrate the impact of this endocrine pathway in the broader perspective of physiopathology.

References

Agustsson, T., Rydén, M., Hoffstedt, J., van Harmelen, V., Dicker, A., Laurencikiene, J., Isaksson, B., Permert, J. and Arner, P. (2007) Mechanism of increased lipolysis in cancer cachexia. *Cancer Research* 67, 5531–5537.

Amato, L., Paolisso, G., Cacciatore, F., Ferrara, N., Ferrara, P., Canonico, S., Varricchio, M. and Rengo, F. (1997) Congestive heart failure predicts the development of non-insulin-dependent diabetes mellitus in the elderly. The Osservatorio Geriatrico Regione Campania Group. *Diabetes and Metabolism* 23, 213–218.

Arner, P. (1999) Catecholamine-induced lipolysis in obesity. *International Journal of Obesity and Metabolic Disorders* 23 (Suppl. 1), 10–13.

Arner, P. (2005) Human fat cell lipolysis: biochemistry, regulation and clinical role. *Best Practice in Research in Clinical Endocrinology and Metabolism* 19, 471–482.

Bai, Y., Zhang, S., Kim, K.-S., Lee, J.-K. and Kim, K.-H. (1996) Obese gene expression alters the ability of 30A5 preadipocytes to respond to lipogenic hormones. *Journal of Biological Chemistry* 271, 13939–13942.

Bastard, J.-P., Jardel, C., Bruckert, E., Blondy, P., Capeau, J., Laville, M., Vidal, H. and Hainque, B. (2000) Elevated levels of interleukin-6 are reduced in serum and subcutaneous adipose tissue of obese women after weight loss. *Journal of Clinical Endocrinology and Metabolism* 85, 3338–3342.

Bastard, J.-P., Maachi, M., Nhieu, J.T.V., Jardel, C., Bruckert, E., Grimaldi, A., Robert, J.-J., Capeau, J. and Hainque, B. (2002) Adipose tissue IL-6 content correlates with resistance to insulin activation of glucose uptake both *in vivo* and *in vitro*. *Journal of Clinical Endocrinology and Metabolism* 87, 2084–2089.

Basu, A., Basu, R., Shah, P., Vella, A., Rizza, R.A. and Jensen, M.D. (2001) Systemic and regional free fatty acid metabolism in type 2 diabetes. *American Journal of Physiology* 280, E1000–E1006.

Bays, H., Mandarino, L. and DeFronzo, R.A. (2004) Role of the adipocyte, free fatty acids and ectopic fat in pathogenesis of type 2 diabetes mellitus: peroxisomal proliferator-activated receptor agonists provide a rational therapeutic approach. *Journal of Clinical Endocrinology and Metabolism* 89, 463–478.

Beale, E.G., Antoine, B. and Forest, C. (2003) Glyceroneogenesis in adipocytes: another textbook case. *Trends in Biochemical Sciences* 28, 402–403.

Beale, E.G., Hammer, R.E., Antoine, B. and Forest, C. (2004) Disregulated glyceroneogenesis: PCK1 as a candidate diabetes and obesity gene. *Trends in Endocrinology and Metabolism* 15, 129–135.

Bergman, R.N. and Ader, M. (2000) Free fatty acids and pathogenesis of type 2 diabetes mellitus. *Trends in Endocrinology and Metabolism* 11, 351–356.

Berlin, L., Lechat, P., Deray, G., Landault, C., Maistre, G., Chermat, V., Brouart, R., Ressayre, C. and Puech, A.J. (1993) Beta-adrenoceptor blockade potentiates acute exercise-induced release of atrial natriuretic peptide by increasing atrial diameter in normotensive healthy subjects. *European Journal of Pharmacology* 44, 127–133.

Bertin, E.P.A., Bolinder, J. and Hagstrom-Toft, E. (2001) Action of glucagon and glucagon-like peptide-1-(7-36) amide on lipolysis in human subcutaneous adipose tissue and skeletal muscle *in vivo*. *Journal of Clinical Endocrinology and Metabolism* 86, 1229–1234.

Bing, C. and Trayhurn, P. (2008) Regulation of adipose tissue metabolism in cancer cachexia. *Current Opinion in Clinical Nutrition and Metabolic Care* 11, 201–207.

Bing, C., Bao, Y., Sanders, P., Manieri, M., Cinti, S., Tisdale, M.J. and Trayhurn, P. (2004) Zinc-α_2-glycoprotein, a lipid mobilizing factor, is expressed in adipocytes and is upregulated in mice with cancer cachexia. *Proceedings of the National Academy of Sciences of the United States of America* 101, 2500–2505.

Boden, G. (2002) Interaction between free fatty acids and glucose metabolism. *Current Opinion in Clinical Nutrition and Metabolic Care* 5, 545–549.

Bold, A. de, Bornstein, J., Veress, H.B. and Sonnenberg, A.T. (1981) A rapid and potent natriuretic response to intravenous injection of atrial myocardial extracts in rats. *Life Sciences* 28, 89–94.

Boston, B.A. (1999) The role of melanocortins in adipocyte function. *Annals of the New York Academy of Sciences* 885, 75–84.

Bouloumié, A., Marumo, T., Lafontan, M. and Busse, R. (1999) Leptin induces oxidative stress in human endothelial cells. *FASEB Journal* 13, 1231–1238.

Bousquet-Melou, A., Galitzky, J., Lafontan, M. and Berlan, M. (1995) Control of lipolysis in intra-abdominal fat cells of non-human primates: comparison with humans. *Journal of Lipid Research* 36, 451–461.

Boyle, P.J., Avogaro, A., Smith, L., Bier, D.M., Pappu, A.S., Illingworth, D.R. and Cryer, P.E. (1992) Role of GH in regulating nocturnal rate of lipolysis and plasma mevalonate levels in normal and diabetic humans. *American Journal of Physiology* 263, E168–E172.

Camenisch, G., Pisabarro, M.T., Sherman, D., Kowalski, J., Nagel, M., Hass, P., Xie, M.-H., Gurney, A., Bodary, S., Liang, X.H., Clark, K., Beresini, M., Ferrara, N. and Gerber, H.-P. (2002) ANGPTL3 stimulates endothelial cell adhesion and migration via integrin $\alpha_v\beta_3$ and induces blood vessel formation *in vivo*. *Journal of Biological Chemistry* 277, 17281–17290.

Castan, I., Valet, P., Larrouy, D., Voisin, T., Remaury, A., Daviaud, D., Laburthe, M. and Lafontan, M. (1993) Distribution of PYY receptors in human fat cells: an antilipolytic system alongside the α2-adrenergic one. *American Journal of Physiology* 265, E74–E80.

Chatzinikolaou, A., Fatouros, I., Petridou, A., Jamourtas, A., Avloniti, A., Douroudos, I., Mastorakos, G., Lazaropoulou, C., Papassotiriou, I., Tournis, S., Mitrakou, A. and Mougios, V. (2008) Adipose tissue lipolysis is upregulated in lean and obese men during acute resistance exercise. *Diabetes Care* 31, 1397–1399.

Clément, K., Vaisse, C., Lahlou, N., Cabrol, S., Pelloux, V., Cassuto, D., Gourmelen, M., Dina, C., Chambaz, J., Lacorte, J.-M., Basdevant, A., Bougnères, P., Lebouc, Y., Froguel, P. and Guy-Grand, B. (1998) A mutation in the human leptin receptor gene causes obesity and pituitary dysfunction. *Nature* 392, 398–401.

Collins, S., Kuhn, C.M., Petro, A.M., Swick, A.G., Chrunyk, B.A. and Surwit, R.S. (1996) Role of leptin in fat regulation. *Nature* 380, 677.

Cowie, M.R., Struthers, A.D., Wood, D.A., Coats, A.J., Thompson, S.G., Poole-Wilson, P.A. and Sutton, G.C. (1997) Value of natriuretic peptides in assessment of patients with possible new heart failure in primary care. *Lancet* 350, 1349–1353.

Curat, C.A., Miranville, A., Sengenes, C., Diehl, M., Tonus, C., Busse, R. and Bouloumié, A. (2004) From blood monocytes to adipose-tissue-resident macrophages: induction of diapedesis by human mature adipocytes. *Diabetes* 53, 1285–1292.

Degerman, E., Smith, C.J., Tornquist, H., Vasta, V., Belfrage, P. and Manganiello, V.C. (1990) Evidence that insulin and isoprenaline activate the cGMP-inhibited low Km cAMP-phosphodiesterase in rat fat cells by phosphorylation. *Proceedings of the National Academy of Sciences of the United States of America* 87, 533–537.

Dessi-Fulgheri, P., Sarzani, R. and Rappelli, A. (1998) The natriuretic peptide system in obesity-related hypertension: new pathophysiological aspects. *Journal of Nephrology* 11, 296–299.

Dessi-Fulgheri, P., Sarzani, R., Serenelli, M., Tamburini, P., Spagnolo, D., Giantomassi, L., Espinosa, E. and Rappelli, A. (1999) Low calorie diet enhances renal, hemodynamic and human humoral effects of exogenous atrial natriuretic peptide in obese hypertensives. *Hypertension* 33, 658–662.

Divertie, G.D., Jensen, M.D. and Miles, J.M. (1991) Stimulation of lipolysis in humans by physiological hypercortisolemia. *Diabetes* 40, 1228–1232.

Djurhuus, C.B., Gravholt, C.H., Nielsen, S., Mengel, A., Christiansen, J.S., Schmitz, O.E. and Moller, N. (2002) Effects of cortisol and regional interstitial glycerol levels in humans. *Journal of Clinical Endocrinology and Metabolism* 283, E172–E177.

Djurhuus, C.B., Gravholt, C.H., Nielsen, S., Pedersen, S.B., Moller, N. and Schmitz, O. (2003) Additive effects of cortisol and growth hormone on regional and systemic lipolysis in humans. *American Journal of Physiology* 286, E488–E494.

Doris, R., Vernon, R.G., Houslay, M.D. and Kilgour, E. (1994) Growth hormone decreases the response to anti-lipolytic agonists and decreases the levels of Gi2 in rat adipocytes. *Biochemical Journal* 297, 41–45.

Dowling, H.J., Fried, S.K. and Pi-Sunyer, F.X. (1995) Insulin resistance in adipocytes of obese women: effect of body fat distribution and race. *Metabolism* 44, 987–995.

Duplus, E. and Forest, C. (2002) Is there a single mechanism for fatty acid regulation of the gene transcription. *Biochemical Pharmacology* 64, 893–901.

Eliman, A., Kamel, A. and Marcus, C. (2002) *In vitro* effects of leptin on human adipocyte metabolism. *Hormone Research* 58, 88–93.

Farooqi, I.S., Jebb, S.A., Langmack, G., Lawrence, E., Cheetham, C.H., Prentice, A.M., Hughes, I.A., McCamish, M.A. and O'Rahilly, S. (1999) Effects of recombinant leptin therapy in a child with congenital leptin deficiency. *New England Journal of Medicine* 341, 879–882.

Frayn, K.N. (2002) Adipose tissue as a buffer for daily lipid flux. *Diabetologia* 45, 1201–1210.

Frühbeck, G., Aguado, M. and Martinez, J.A. (1997) *In vitro* lipolytic effect of leptin on mouse adipocytes: evidence for a possible autocrine/paracrine role of leptin. *Biochemical and Biophysical Research Communications* 240, 590–594.

Frühbeck, G., Gomez-Ambrosi, J. and Salvador, J. (2001) Leptin-induced lipolysis opposes the tonic inhibition of endogenous adenosine in white adipocytes. *FASEB Journal* 15, 333–340.

Gainsford, T., Willson, T.A., Metcalf, D., Handman, E., McFarlane, C., Ng, A., Nicola, N.A., Alexander, W.S. and Hilton, D.J. (1996) Leptin can induce proliferation, differentiation and functional activation of hematopoietic cells. *Proceedings of the National Academy of Sciences of the United States of America* 93, 14564–14568.

Galitzky, J., Sengenes, C., Thalamas, C., Marques, M.-A., Senard, J.-M., Lafontan, M. and Berlan, M. (2001) The lipid mobilizing effect of atrial natriuretic peptides is unrelated to sympathetic nervous system activation or obesity in young men. *Journal of Lipid Research* 42, 536–544.

Gardner, D.G. (2003) Natriuretic peptides: markers or modulators of cardiac hypertrophy. *Trends in Endocrinology and Metabolism* 14, 411–416.

Garruti, G., Giusti, V., Nussberger, J., Darimont, C., Verdumo, C., Amstutz, C., Puglisi, F., Giorgino, F., Giorgino, R. and Cotecchia, S. (2007) Expression and secretion of the atrial natriuretic peptide in human adipose tissue and preadipocytes. *Obesity* 15, 2181–2189.

Gasic, S., Tian, B. and Green, A. (1999) Tumor necrosis factor alpha stimulates lipolysis in adipocytes by decreasing Gi protein concentrations. *Journal of Biological Chemistry* 274, 6770–6775.

Giesler, G., Lenihan, D.J. and Durand, J.-B. (2004) The update on the rationale, use and selection of β-blockers in heart failure. *Current Opinion in Cardiology.* 19, 250–253.

Göke, R., Göke, B., Noll, B., Richter, G., Christoph-Fehmann, H. and Arnold, R. (1989) Receptors for atrial natriuretic peptide on isolated rat adipocytes. *Biochemical Research* 10, 463–467.

Goodpaster, B.H., Wolfe, R.R. and Kelley, D.E. (2002) Effects of obesity on substrate utilization during exercise. *Obesity Research* 10, 575–584.

Gravholt, C., Schmitz, O., Simonsen, L., Bulow, J., Christiansen, J. and Moller, N. (1999) Effects of a physiological GH pulse on interstitial glycerol in abdominal and femoral adipose tissue. *American Journal of Physiology* 277, E848–E854.

Gravholt, C., Moller, N., Jensen, M., Christiansen, J. and Schmitz, O. (2001) Physiological levels of glucagon do not influence lipolysis in abdominal adipose tissue as assessed by microdialysis. *Journal of Clinical Endocrinology and Metabolism* 86, 2085–2089.

Groundwater, M., Bulcavage, L., Barton, C., Adamson, C., Ferrier, I. and Tisdale, M. (1990) Alteration of serum and urinary lipolytic activity with weight loss in cachectic cancer patients. *British Journal of Cancer* 62, 816–821.

Guo, Z., Hensrud, D.D., Johnson, C.M. and Jensen, M.D. (1999) Regional postprandial fatty acid metabolism in different obesity phenotypes. *Diabetes* 48, 1586–1592.

Hall, G. van, Steenberg, A., Sacchetti, M., Fisher, C., Keller, C., Schjerling, P., Hiscock, N., Moller, H., Saltin, B., Febbraio, M.A. and Pedersen, B.K. (2003) Interleukin-6 stimulates lipolysis and fat oxidation in humans. *Journal of Clinical Endocrinology and Metabolism* 88, 3005–3010.

Hansen, T.K. (2002) Pharmacokinetics and acute lipolytic actions of growth hormone. Impact of age, body composition, binding proteins and other hormones. *Growth Hormone and IGF Research* 12, 342–358.

Harant, I., Beauville, M., Crampes, F., Rivière, D., Tauber, M.-T., Tauber, J.-P. and Garrigues, M. (1994) Response of fat cells to gowth hormone: effect of long-term treatment with recombinant human growth hormone in GH-deficient adults. *Journal of Clinical Endocrinology and Metabolism* 78, 1392–1395.

Hauner, H., Petruschke, T., Russ, M., Röhrig, K. and Eckel, J. (1995) Effects of tumor necrosis factor alpha (TNFα) on glucose transport and lipid metabolism of newly differentiated human fat cells in culture. *Diabetologia* 38, 764–771.

Haynes, W.G., Morgan, D.A., Walsh, S.A., Mark, A.L. and Sivitz, W.I. (1997) Receptor-mediated regional sympathetic nerve activation by leptin. *Journal of Clinical Investigation* 100, 270–278.

Hefferman, M.A., Jiang, W.J., Thorburn, A.W. and Ng, F.M. (2000) Effects of oral administration of a synthetic fragment of human growth hormone on lipid metabolism. *American Journal of Physiology* 279, E501–507.

Henderson, S.A., Graham, H.K., Mollan, R.A., Riddoch, C., Sheridan, B. and Johnston, H. (1989) Calcium homeostasis and exercise. *International Orthopaedics* 13, 69–73.

Horowitz, J.F. (2003) Fatty acid mobilization from adipose tissue during exercise. *Trends in Endocrinology and Metabolism* 14, 386–392.

Horowitz, J.F. and Klein, S. (2000) Oxidation of non-plasma fatty acids during exercise is increased in women with abdominal obesity. *Journal of Applied Physiology* 89, 2276–2282.

Jeandel, L., Okamura, H., Belles-Isles, M., Chabot, J.-G., Dihl, F., Morel, G., Kelly, P.-A. and Heisler, S. (1988) Immunocytochemical localization, binding and effects of atrial natriuretic peptide in rat adipocytes. *Molecular and Cellular Endocrinology* 62, 69–78.

Jensen, M., Haymond, M., Rizza, R., Cryer, P. and Miles, J. (1989) Influence of body fat distribution on free fatty acid metabolism in obesity. *Journal of Clinical Investigation* 83, 1168–1173.

Jensen, M.D. (1997) Lipolysis: contribution from regional fat. *Annual Review of Nutrition* 17, 127–139.

Jensen, M.D. (2003) Cytokine regulation of lipolysis in humans? *Journal of Clinical Endocrinology and Metabolism* 88, 3003–3004.

Johnson, J.A., Fried, S.K., Pi-Sunyer, F.X. and Albu, J.B. (2001) Impaired insulin action in subcutaneous adipocytes from women with visceral obesity. *American Journal of Physiology* 280, E40–E49.

Jorgensen, J.O. (1991) Human growth hormone replacement therapy: pharmacological and clinical aspects. *Endocrine Reviews* 12, 189–207.

Jouven, X., Charles, M.-A., Desnos, M. and Ducimetière, P. (2001) Circulating non-esterified fatty acid level as a predictive risk factor for sudden death in population. *Circulation* 104, 756–761.

Kalra, P.R. and Tigas, S. (2002) Regulation of lipolysis: natriuretic peptides and the development of cachexia. *International Journal of Cardiology* 85, 125–132.

Kanaley, J.A., Cryer, P.E. and Jensen, M.D. (1993) Fatty acid kinetic responses to exercise. Effects of obesity, body fat distribution and energy restricted diet. *Journal of Clinical Investigation* 92, 255–261.

Ke, Y., Qiu, J., Ogus, S., Shen, W.-J., Kraemer, F.B. and Chehab, F.F. (2003) Overexpression of leptin in transgenic mice leads to decreased basal lipolysis, PKA activity and perilipin levels. *Biochemical and Biophysical Research Communications* 312, 1165–1170.

Khoo, J.C., Sperry, P.J., Gill, G.N. and Steinberg, D. (1977) Activation of hormone-sensitive lipase and phosphorylase kinase by purified cyclic GMP-dependent protein kinase. *Proceedings of the National Academy of Sciences of the United States of America* 74, 4843–4847.

Kido, Y., Nakae, J. and Accili, D. (2001) The insulin receptor and its cellular targets. *Journal of Clinical Endocrinology and Metabolism* 86, 972–979.

Kiemer, A.K. and Vollmar, A.M. (2001) The atrial natriuretic peptide regulates the production of inflammatory mediators in macrophages. *Annals of Rheumatic Diseases* 60, 68–70.

Köning, D., Schumacher, O., Heinrich, L., Schmid, A., Berg, A. and Dickhuth, H.H. (2003) Myocardial stress after competitive exercise in professional road cyclists. *Medicine and Science in Sports and Exercise* 35, 1679–1683.

Kuhn, M. (2003) Structure, regulation and function of mammalian membrane guanylyl cyclase receptors, with a focus on guanylyl cyclase-A. *Circulation Research* 93, 700–709.

Lafontan, M. and Berlan, M. (1993) Fat cell adrenergic receptors and the control of white and brown fat cell function. *Journal of Lipid Research* 34, 1057–1091.

Lafontan, M. and Berlan, M. (2003) Do regional differences in adipocyte biology provide new pathophysiological insights? *Trends in Pharmacological Sciences* 24, 276–283.

Lafontan, M., Moro, C., Sengenes, C., Galitzky, J., Crampes, F. and Berlan, M. (2005) An unsuspected metabolic role for atrial natriuretic peptides: the control of lipolysis, lipid mobilization, and systemic non-esterified fatty acids levels in humans. *Arteriosclerosis, Thrombosis and Vascular Biology* 25, 2032–2042.

Lafontan, M., Moro, C., Berlan, M., Crampes, F., Sengenes, C. and Galitzky, J. (2008) Control of lipolysis by natriuretic peptides and cyclic GMP. *Trends in Endocrinology and Metabolism* 19, 130–137.

Lagathu, C., Bastard, J.-P., Auclair, M., Maachi, M., Capeau, J. and Caron, M. (2003) Chronic interleukin-6 (IL-6) treatment increased IL-6 secretion and induced insulin resistance in adipocyte: prevention by rosiglitazone. *Biochemical and Biophysical Research Communications* 311, 372–379.

Langin, D. (2006) Adipose tissue lipolysis as a metabolic pathway to define pharmacological strategies against obesity and the metabolic syndrome. *Pharmacological Research* 53, 482–491.

Langin, D. and Arner, P. (2006) Importance of TNFalpha and neutral lipases in human adipose tissue lipolysis. *Trends in Endocrinology and Metabolism* 17, 314–320.

Langin, D. and Lafontan, M. (2004) Lipolysis and lipid mobilization in human adipose tissue. In: Bray, G.A. and Bouchard, C. (eds) *Handbook of Obesity. Etiology and Pathophysiology*. Marcel Dekker, Inc, New York, Basel, pp. 515–532.

Lefebvre, A.-M., Laville, M., Vega, N., Riou, J.P., Gaal, L.V., Auwerx, J. and Vidal, H. (1998) Depot-specific differences in adipose tissue gene expression in lean and obese subjects. *Diabetes* 47, 98–103.

Leon, H. de, Bonhomme, M.-C., Thibault, G. and Garcia, R. (1995) Localization of atrial natriuretic factor receptors in the mesenteric arterial bed. Comparison with angiotensin II and endothelin receptors. *Circulation Research* 77, 64–72.

Lerman, A., Gibbons, R.J., Rodeheffer, R.J., Bailey, K.R., McKinley, L.J., Heublein, D.M. and Burnett, J.C. Jr (1993) Circulating N-terminal atrial natriuretic peptide as a marker for symptomless left ventricular dysfunction. *Lancet* 341, 1105–1109.

Lommi, J., Koskinen, P., Naveri, H., Harkonen, M. and Kupari, M. (1997) Heart failure ketosis. *Journal of Internal Medicine* 242, 231–238.

Luchner, A., Jr, J.C.B., Jougasaki, M., Hense, H.-W., Riegger, G.A.J. and Schunkert, H. (1998) Augmentation of the cardiac natriuretic peptides by beta-receptor antagonism: evidence from a population-based study. *Journal of the American College of Cardiology* 32, 1839–1844.

McDonagh, T.A., Robb, S.D., Murdoch, D.R., Morton, J.J., Ford, I., Morrisson, C.E., Tunsdall-Pedoe, H., McMurray, J.J. and Dargie, H.J. (1998) Biochemical detection of left-ventricular systolic dysfunction. *Lancet* 351, 9–13.

McGarry, J.D. (2002) Dysregulation of fatty acid metabolism in the etiology of type 2 diabetes. *Diabetes* 51, 7–18.

Maisel, A.S. (2003) Nesiritide: a new therapy for the treatment of heart failure. *Cardiovasular Toxicology* 3, 37–42.

Maisel, A.S., Krishnaswamy, P., Nowak, R.M., McCord, J., Hollander, J.E., Duc, P., Omland, T., Storrow, A.B., Abraham, W.T., Wu, A.H., Clopton, P., Steg, P., Westheim, A., Knudsen, C.W., Perez, A., Kazanegra, R., Hermann, H.C., P.A. McCullough and Investigators, B.N.P.M.S. (2002) Rapid measurement of B-type natriuretic peptide in the emergency diagnosis of heart failure. *New England Journal of Medicine* 347, 161–167.

Marie, P.-Y., Mertes, P.M., Hassan-Sebag, N., Talence, N.D., Djaballah, K., Djaballah, W., Friberg, J., Olivier, P., Karcher, G., Zannad, F. and Bertrand, A. (2004) Exercise release of cardiac natriuretic peptides is markedly enhanced when patients with coronary artery disease are treated medically by beta-blockers. *Journal of the American College of Cardiology* 43, 353–359.

Matsukawa, N., Grzesik, W.J., Takahashi, N., Pandey, K.N., Pang, S., Yamauchi, M. and Smithies, O. (1999) The natriuretic peptide clearance receptor locally modulates the physiological effects of the natriuretic peptide system. *Proceedings of the National Academy of Sciences of the United States of America* 96, 7403–7408.

Meek, S.E., Nair, K.S. and Jensen, M.D. (1999) Insulin regulation of regional free fatty acid metabolism. *Diabetes* 48, 10–14.

Mehra, M.R., Uber, P.A., Park, M.H., Scott, R.L., Ventura, H.O., Harris, B.C. and Frohlich, E.D. (2004) Obesity and suppressed B-type natriuretic peptide levels in heart failure. *Journal of the American College of Cardiology* 43, 1590–1595.

Mittendorfer, B., Fields, D.A. and Klein, S. (2004) Excess body fat in men decreases plasma fatty acid availability and oxidation during endurance exercise. *American Journal of Physiology* 286, E354–E362.

Mohamed-Ali, V., Goodrick, S., Rawesh, A., Katz, D.R., Miles, J.M., Yudkin, J.S., Klein, S. and Coppack, S.W. (1997) Subcutaneous adipose tissue releases interleukin-6 but not tumor necrosis factor-α, *in vivo*. *Journal of Clinical Endocrinology and Metabolism* 82, 4196–4200.

Mokdad, A., Ford, E.S., Bowman, B.A., Dietz, W.H., Vinicor, F., Bales, V.S. and Marks, J.S. (2003) Prevalence of obesity, diabetes and obesity-related health risk factor 2001. *Journal of the American Medical Association* 289, 76–79.

Moller, N., Jorgensen, J.O.L., Schmitz, O., Moller, J.S., Christiansen, J.S., Alberti, K.G.M.M. and Orskov, H. (1990) Effect of growth hormone pulse on total and forearm substrate fluxes in humans. *American Journal of Physiology* 258, E86–E91.

Moller, N., Schmitz, O., Porksen, J., Moller, J. and Jorgensen, J.O. (1992) Dose–response studies on the metabolic effects of a growth hormone pulse in humans. *Metabolism* 41, 172–175.

Moller, N., Porksen, N., Ovesen, P. and Alberti, K.G.M.M. (1993) Evidence for increased sensitivity of fuel mobilization to growth hormone during short-term fasting in humans. *Hormone and Metabolic Research* 25, 175–179.

Moller, N., Gjedsted, J., Gormsen, L., Fuglsang, J. and Djurhuus, C. (2003) Effects of growth hormone on lipid metabolism in humans. *Growth Hormone and IGF Research* 13 (Suppl. A), S18–S21.

Morisset, A.S., Huot, C., Légaré, D. and Tchernof, A. (2008) Circulating IL-6 concentrations and abdominal adipocyte isoproterenol-stimulated lipolysis in women. *Obesity* 16, 1487–1492.

Moro, C., Crampes, F., Sengenes, C., Glisezinski, I. de, Galitzky, J., Thalamas, C., Lafontan, M. and Berlan, M. (2004a) Atrial natriuretic peptide contributes to the physiological control of lipid mobilization in humans. *FASEB Journal* 18, 908–910.

Moro, C., Galitzky, J., Sengenes, C., Crampes, F., Lafontan, M. and Berlan, M. (2004b) Functional and pharmacological characterization of the natriuretic peptide-dependent lipolytic pathway in human fat cells. *Journal of Pharmacology and Experimental Therapeutics* 308, 984–992.

Moro, C., Polak, J., Hejnova, J., Klimcakova, E., Crampes, F., Stich, V., Lafontan, M. and Berlan, M. (2006) Atrial natriuretic peptide stimulates lipid mobilization during repeated bouts of endurance exercise. *American Journal of Physiology Endocrinology and Physiology* 290, E864–E869.

Moro, C., Pillard, F., Glisezinski, I. de, Crampes, F., Thalamas, C., Harant, I., Marques, M.A., Lafontan, M. and Berlan, M. (2007a) Sex differences in lipolysis-regulating mechanisms in overweight subjects: effect of exercise intensity. *Obesity* 15, 2245–2255.

Moro, C., Pillard, F., Glisezinski, I. de, Crampes, F., Thalamas, C., Harant, I., Marques, M.A., Lafontan, M. and Berlan, M. (2007b) Atrial natriuretic peptide contribution to lipid mobilization and utilization during head-down bed rest in humans. *American Journal of Physiology – Regulatory and Integrative Comparative Physiology* 293, R612–R617.

Nielsen, S., Guo, Z., Johnson, M., Hensrud, D.D. and Jensen, M.D. (2004) Splanchnic lipolysis in human obesity. *Journal of Clinical Investigation* 113, 1582–1588.

Niessner, A., Ziegler, S., Slany, J., Billensteiner, E., Woloszczuk, W. and Geyer, G. (2003) Increases in plasma levels of atrial and brain natriuretic peptides after running a marathon: are their effects partly counterbalanced by adrenocortical steroids? *European Journal of Endocrinology* 149, 555–559.

Ohba, H., Takada, H., Musha, H., Nagashima, J., Mori, N., Awaya, T., Omiya, K. and Maruyama, M. (2001) Effects of prolonged strenuous exercise on plasma levels of brain atrial natriuretic peptide and brain natriuretic peptide in healthy men. *American Heart Journal* 141, 751–758.

Orci, L., Cook, W.S., Ravazzola, M., Wang, M.-Y., Park, B.-H., Montesano, R. and Unger, R.H. (2004) Rapid transformation of white adipocytes into fat oxidizing machines. *Proceedings of the National Academy of Sciences of the United States of America* 101, 2058–2063.

Ottoson, M., Lönnroth, P., Björntorp, P. and Eden, S. (2000) Effects of cortisol and growth hormone on lipolysis in human adipose tissue. *Journal of Clinical Endocrinology and Metabolism* 85, 799–803.

Päth, G., Bornstein, S.R., Gurniak, M., Chrousos, G.P., Scherbaum, W.A. and Hauner, H. (2001) Human breast adipocytes express interleukin-6 (IL-6) and its receptor system: increased IL-6 production by beta-adrenergic activation and effects of IL-6 on adipocyte function. *Journal of Clinical Endocrinology and Metabolism* 86, 2281–2288.

Pickup, J.C. and Crook, M.A. (1998) Is type 2 diabetes mellitus a disease of the innate immune system. *Diabetologia* 41, 1241–1248.

Pirola, L., Johnston, A.M. and Van Obberghen, E. (2004) Modulation of insulin action. *Diabetologia* 47, 170–184.

Poveda, J.J., Berrazueta, J.R., Ochoteco, A., Montalban, C., Garcia-Unzueta, M.T., Fernandez, C., Pena, N. and Amado, J.A. (1998) Age-related responses of vasoactive factors during acute exercise. *Hormone and Metabolic Research* 30, 668–672.

Press, M., Tamborlane, W.V. and Sherwin, R.S. (1984) Importance of raised growth hormone levels in mediating the metabolic derangements of diabetes. *New England Journal of Medicine* 310, 810–815.

Reaven, G.M., Hollenbeck, C., Jeng, C.-Y., Wu, M.S. and Chen, Y.-D. (1988) Measurement of plasma glucose, free fatty acid, lactate and insulin for 24 hours in patients with NIDDM. *Diabetes* 37, 1020–1024.

Rebuffe-Scrive, M., Lönnroth, P., Marin, P., Wesslau, C., Björntorp, P. and Smith, U. (1987) Regional adipose tissue metabolism in men and postmenopausal women. *International Journal of Obesity* 11, 347–355.

Richards, M.A., Lainchbury, J.G., Nicholls, M.G., Troughton, R.W. and Yandle, T.G. (2002) BNP in hormone-guided treatment of heart failure. *Trends in Endocrinology and Metabolism* 13, 151–155.

Rosenthal, M.J. and Woodside, W.F. (1988) Nocturnal regulation of free fatty acids in healthy young and elderly men. *Metabolism* 37, 645–648.

Rotter, V., Nagaev, I. and Smith, U. (2003) Interleukin-6 (IL-6) induces insulin resistance in 3T3-L1 adipocytes and is, like IL-8 and tumor necrosis factor-α, overexpressed in human fat cells from insulin-resistant subjects. *Journal of Biological Chemistry* 278, 45777–45784.

Ryden, M., Dicker, A., Harmelen, V.V., Hauner, H., Brunnberg, M., Perbeck, L., Lönnqvist, F. and Arner, P. (2002) Mapping of early signaling events in tumor necrosis-alpha-mediated lipolysis in human fat cells. *Journal of Biological Chemistry* 277, 1085–1091.

Sakharova, A.A., Horowitz, J.F., Surya, S., Goldenberg, N., Harber, M.P., Symons, K. and Barkan, A. (2008) Role of growth hormone in regulating lipolysis, proteolysis, and hepatic glucose production during fasting. *Journal of Clinical Endocrinology and Metabolism* 93, 2755–2759.

Samra, J.S., Clark, M.L., Humphreys, S.M., MacDonald, I.A., Bannister, P.A., Matthews, D.R. and Frayn, K.N. (1999) Suppression of the nocturnal rise in growth hormone reduces subsequent lipolysis in subcutaneous adipose tissue. *European Journal of Clinical Investigation* 29, 1045–1052.

Sanchez, L.M., Chirino, A.J. and Bjorkman, P.J. (1999) Crystal structure of human ZAG, a fat-depleting factor related to MHC molecules. *Science* 283, 1914–1919.

Sarzani, R., Paci, M.V., Zingaretti, C.M., Pierleoni, C., Cinti, S., Cola, G., Rappelli, A. and Dessi-Fulgheri, P. (1995) Fasting inhibits natriuretic peptides clearance receptor expression in rat adipose tissue. *Journal of Hypertension* 13, 1241–1246.

Sarzani, R., Dessi-Fulgheri, P., Paci, V.M., Espinosa, E. and Rappelli, A. (1996) Expression of natriuretic peptide receptors in human adipose and other tissues. *Journal of Endocrinological Investigation* 1996, 581–585.

Sarzani, R., Dessi-Fulgheri, P. and Salvi, F. (1999) A novel promoter variant of the natriuretic peptide clearance receptor gene is associated with lower atrial natriuretic peptide and higher blood pressure in obese hypertensives. *Journal of Hypertension* 17, 1301–1305.

Sarzani, R., Strazzullo, P., Salvi, F., Iacone, R., Pietrucci, F., Siani, A., Barba, G., Gerardi, M.C., Dessi-Fulgheri, P. and Rappelli, A. (2004) Natriuretic peptide clearance receptor alleles and susceptibility to abdominal obesity. *Obesity Research* 12, 351–356.

Schirmer, H. and Omland, T. (1999) Circulating N-terminal pro-atrial natriuretic peptide is an independent predictor of left ventricular hypertrophy in the general population. The Tromso Study. *European Heart Journal* 20, 755–763.

Sengenes, C., Berlan, M., Glisezinski, I. de, Lafontan, M. and Galitzky, J. (2000) Natriuretic peptides: a new lipolytic pathway in human adipocytes. *FASEB Journal* 14, 1345–1351.

Sengenes, C., Zakaroff-Girard, A., Moulin, A., Berlan, M., Bouloumié, A., Lafontan, M. and Galitzky, J. (2002) Natriuretic peptide-dependent lipolysis in fat cells is a primate specificity. *American Journal of Physiology* 283, R257–R265.

Sengenes, C., Bouloumié, A., Hauner, H., Berlan, M., Busse, R. and Lafontan, M. (2003) Involvement of a cGMP-dependent pathway in natriuretic peptide-mediated hormone-sensitive lipase phosphorylation in human adipocytes. *Journal of Biological Chemistry* 278, 48617–48626.

Serradeil-Le Gal, C., Lafontan, M., Raufaste, D., Marchand, J., Pouzet, B., Casellas, P., Pascal, M., Maffrand, J.-P. and Fur, G.L. (2000) Characterization of NPY receptors controlling lipolysis and leptin secretion in human adipocytes. *FEBS Letters* 475, 150–156.

Sethi, J., Xu, H., Uysal, K., Wiesbrock, S., Scheja, L. and Hotamisligil, G. (2000) Characterisation of receptor-specific TNFα functions in adipocyte cell lines lacking type 1 and 2 TNF receptors. *FEBS Letters* 469, 77–82.

Shimabukuro, M., Koyama, K., Chen, G., Wang, M.-Y., Trieu, F., Lee, Y., Newgard, C.B. and Unger, R.H. (1997) Direct antidiabetic effect of leptin through triglyceride depletion of tissues. *Proceedings of the National Academy of Sciences of the United States of America* 94, 4637–4641.

Siegrest-Kaiser, C.A., Pauli, V., Juge-Aubry, C.E., Boss, O., Pernin, A., Chin, W.W., Cusin, I., Rohner-Jeanrenaud, F., Burger, A.G., Zapf, J. and Meier, C.A. (1997) Direct effects of leptin on brown and white adipose tissues. *Journal of Clinical Investigation* 100, 2858–2864.

Silberbach, M. and Roberts, C.T. (2001) Natriuretic peptide signalling. Molecular and cellular pathways to growth regulation. *Cellular Signalling* 13, 221–231.

Sinha, T.K., Thajchayapong, P., Queener, S.F., Allen, D.O. and Bell, N.H. (1976) On the lipolytic action of parathyroid hormone in man. *Metabolism* 25, 251–260.

Smith, C.J. and Manganiello, V.C. (1989) Role of hormone-sensitive low Km cAMP phosphodiesterase in regulation of cAMP-dependent protein kinase and lipolysis in rat adipocytes. *Molecular Pharmacology* 35, 381–386.

Smith, U., Hammersten, J., Björntorp, P. and Kral, J.G. (1979) Regional differences and effect of weight reduction on human fat cell metabolism. *European Journal of Clinical Investigation* 9, 327–332.

Souza, S.C., Palmer, H.J., Kang, Y.-H., Yamamoto, M.T., Muliro, K.V., Paulson, K.E. and Greenberg, A.S. (2003) TNF-α induction of lipolysis is mediated through activation of the extracellular signal related kinase pathway in 3T3-L1 adipocytes. *Journal of Cellular Biochemistry* 89, 1077–1086.

Steinberg, H.O. and Baron, A.D. (2002) Vascular function, insulin resistance and fatty acids. *Diabetologia* 45, 623–634.

Stich, V., Glisezinski, I. de, Berlan, M., Bulow, J., Galitzky, J., Harant, I., Suljkovicova, H., Lafontan, M., Rivière, D. and Crampes, F. (2000a) Adipose tissue lipolysis is increased during a repeated bout of aerobic exercise. *Journal of Applied Physiology* 88, 1277–1283.

Stich, V., Glisezinski, I. de, Crampes, F., Hejnova, J., Cottet-Emard, J.-M., Galitzky, J., Lafontan, M., Rivière, D. and Berlan, M. (2000b) Activation of alpha2-adrenergic receptors impairs exercise-induced lipolysis in SCAT of obese subjects. *American Journal of Physiology* 279, R499–R504.

Strålfors, P. and Belfrage, P. (1985) Phosphorylation of hormone-sensitive lipase by cyclic GMP-dependent protein kinase. *FEBS Letters* 180, 280–284.

Taniguchi, A., Kataoka, K., Kono, T., Oseko, F., Okuda, H., Nagata, I. and Imura, H. (1987) Parathyroid hormone-induced lipolysis in human adipose tissue. *Journal of Lipid Research* 28, 490–494.

Tisdale, M.J. (2002) Cachexia in cancer patients. *Nature Reviews on Cancer* 2, 862–871.

Tsai, K.S., Lin, J.C., Chen, C.K., Cheng, W.C. and Yang, C.H. (1997) Effect of exercise and exogenous glucocorticoid on serum level of intact parathyroid hormone. *International Journal of Sports Medicine* 18, 583–587.

Tsuruda, T., Boerritger, G., Huntley, B.K., Nosher, J.A., Cataliotti, A., Costello-Boerritger, L.C., Chen, H.H. and Burnett, J.C. Jr (2002) Brain natriuretic peptide is produced in cardiac fibroblasts and induces matrix metalloproteinases. *Circulation Research* 91, 1127–1134.

Valet, P., Berlan, M., Beauville, M., Crampes, F., Montastruc, J.-L. and Lafontan, M. (1990) Neuropeptide Y and peptide YY inhibit lipolysis in human and dog fat cells through a pertussis toxin-sensitive G protein. *Journal of Clinical Investigation* 85, 291–295.

Wahrenberg, H., Lönnqvist, F. and Arner, P. (1989) Mechanisms underlying regional differences in lipolysis in human adipose tissue. *Journal of Clinical Investigation* 84, 458–467.

Wang, M.Y., Lee, Y. and Unger, R.H. (1999) Novel form of lipolysis induced by leptin. *Journal of Biological Chemistry* 274, 17541–17544.

Wang, T.J., Larson, M.G., Levy, D., Benjamin, E.J., Leip, E.P., Wilson, P.W.F. and Vasan, R.S. (2004) Impact of obesity on plasma natriuretic peptide levels. *Circulation* 109, 594–600.

Wu, X., Hoffstedt, J., Deeb, W., Singh, R., Sedkova, N., Zilbering, A., Zhu, L., Park, P.K., Arner, P. and Goldstein, B. (2001) Depot-specific variation in protein–tyrosine phosphatase activities in human omental and subcutaneous adipose tissue: a potential contribution to differential insulin sensitivity. *Journal of Clinical Endocrinology and Metabolism* 86, 5973–5980.

Xu, H. and Hotamisligil, G.S. (2001) Signaling pathways utilized by tumor necrosis factor receptor-1 in adipocytes to suppress differentiation. *FEBS Letters* 506, 97–102.

Yip, R.G.-C. and Goodman, H.M. (1999) Growth hormone and dexamethasone stimulate lipolysis and activate adenylyl cyclase in rat adipocytes by selectively shifting $Gi\alpha_2$ to lower density membrane fractions. *Endocrinology* 140, 1219–1227.

Yu, C., Chen, Y., Cline, G.W., Zhang, D., Zong, H., Wang, Y., Bergeron, R., Kim, J.K., Cushman, S.W., Cooney, G.J., Atcheson, B., White, M.F., Kraegen, E.W. and Shulman, G.I. (2002) Mechanisms by which fatty acids inhibit insulin activation of insulin receptor substrate-1 (IRS-1)-associated phosphatidylinositol 3-kinase activity in muscle. *Journal of Biological Chemistry* 277, 50230–50236.

Zhang, H.H., Halbleib, M., Ahmad, F., Manganiello, V.C. and Greenberg, A.S. (2002) Tumor necrosis factor-α stimulates lipolysis in differentiated human adipocytes through activation of extracellular signal-related kinase and elevation of intracellular cAMP. *Diabetes* 51, 2929–2935.

Zierath, J., Livingston, J., Thorne, A., Bolinder, J., Reynisdottir, S., Lonnqvist, F. and Arner, P. (1998) Regional difference in insulin inhibition of non-esterified fatty acid release from human adipocytes: relation to insulin receptor phosphorylation and intracellular signalling through the insulin receptor substrate-1 pathway. *Diabetologia* 41, 1343–1354.

7 The Adipo–Hepato–Insular Axis in Glucose Homeostasis

Javier Gómez-Ambrosi,[1] Victoria Catalán[1] and Gema Frühbeck[1,2]

[1]*Metabolic Research Laboratory, Clínica Universitaria de Navarra, University of Navarra, Spain;* [2]*Department of Endocrinology, Clínica Universitaria de Navarra, University of Navarra, Spain*

Introduction

Glucose is the most important metabolic fuel for living organisms. In contrast to other organs that are able to oxidize fatty acids, glucose is the only fuel the brain uses under physiological conditions, although it can also use ketone bodies during prolonged fasting. In healthy adult humans, postabsorptive serum glucose concentrations range between 72 and 108 mg/dl, (4.0 and 6.0 mmol/l) with a mean value around 90 mg/dl (5.0 mmol/l) (Cryer, 2003). Glucose homeostasis, i.e. the maintenance of blood glucose levels within this narrow range, is a critical physiological function involving several organs and cell types.

Glucose is obtained from the digestion of dietary carbohydrates and can be stored as glycogen, mainly in the liver and muscles, to ensure continuous supply over longer periods. Glucose release into the circulation can therefore follow glycogenolysis, the breakdown of glycogen, and can be metabolized from other precursors, such as lactate, some amino acids and glycerol, in a process known as gluconeogenesis. Although several tissues have the ability to synthesize and hydrolyse glycogen, only the liver and kidneys express the enzyme necessary for the release of glucose to the blood (glucose-6-phosphatase, G6Pase). The liver and kidneys are also the only organs that express the enzymes necessary for gluconeogenesis, with the hepatic contribution predominating (≈80%) (Cryer, 2003).

Plasma glucose concentrations are the net integrated balance of glucose influx to the circulation, as described above, and glucose utilization by tissues (Fig. 7.1). Most circulating glucose is used by the brain (45–60%), skeletal muscle (15–20%), kidneys (10–15%), blood cells (5–10%), splanchnic organs (3–6%) and adipose tissue (2–4%) (Cryer, 2003). Systemic glucose balance is orchestrated mainly by insulin, glucagon and epinephrine, but also by many other hormones produced by the pancreas, liver and adipose tissue. In addition, the role of the central nervous system in the control of glucose levels has been

 Peptides in Energy Balance and Obesity (ed. G. Frühbeck)

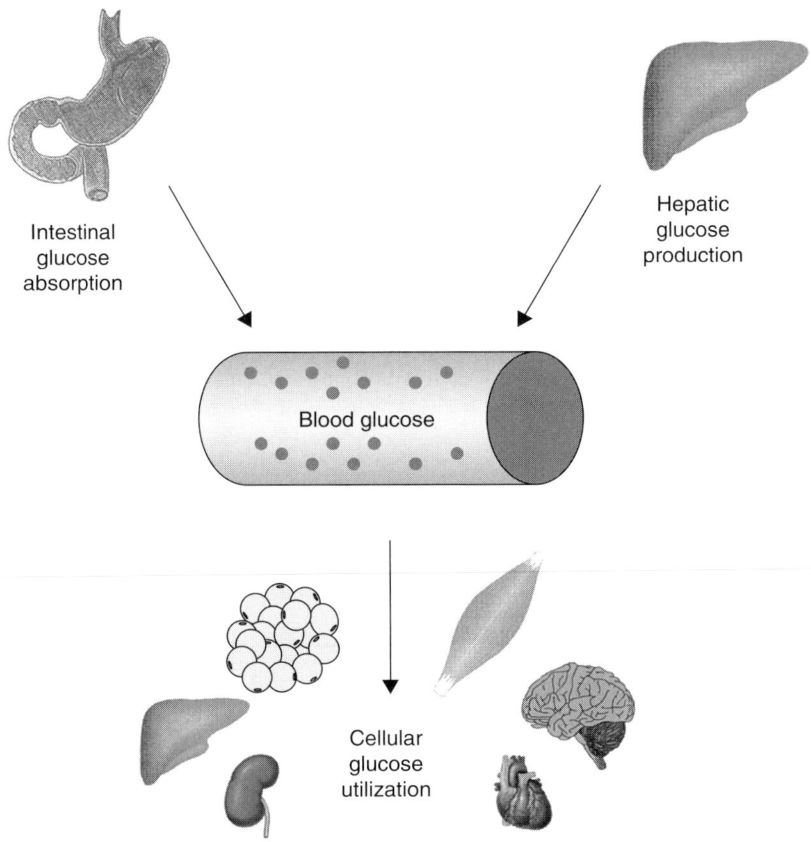

Fig. 7.1. A simplified model of the main factors affecting glucose homeostasis.

recognized recently (Cota *et al.*, 2007; Rother *et al.*, 2008). The crosstalk between these peripheral organs in glucose homeostasis is the focus of this chapter.

Glucose homeostasis is a critical physiological process that becomes deranged in situations of insulin resistance, which may be overcome through hyperinsulinaemia, and finally drive to pancreatic β-cell failure and the development of type 2 diabetes mellitus (T2DM) (Fig. 7.2).

Diabetes mellitus is considered an epidemic in many developed and newly industrialized countries, affecting more than 170 million people worldwide (Kahn *et al.*, 2006). T2DM accounts for around 90% of all cases of diabetes mellitus and is characterized by a decreased inhibition of liver glucose production and a decreased stimulatory effect of insulin on peripheral glucose uptake. Obesity is the major risk factor associated with the rise in T2DM incidence. However, the link between obesity and T2DM, i.e. the mechanisms by which increased fat mass lead to insulin resistance, have yet to be fully unravelled (Kahn *et al.*, 2006).

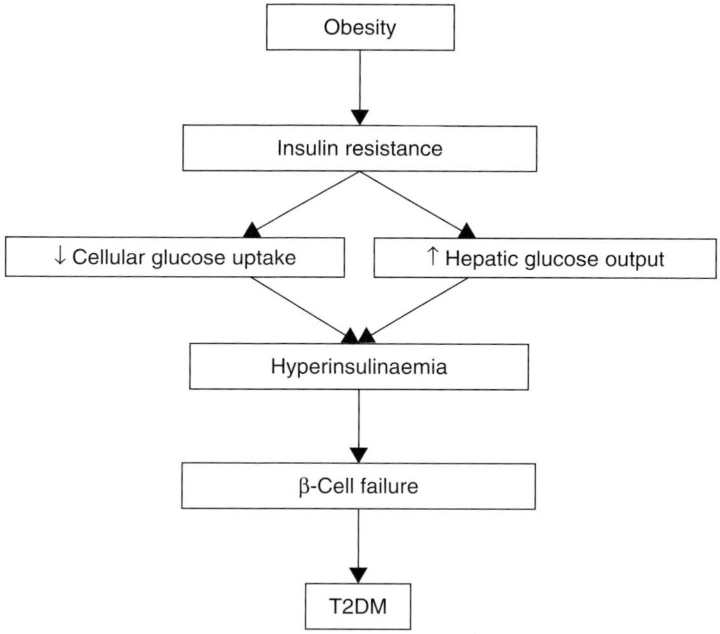

Fig. 7.2. Steps leading from obesity to the development of type 2 diabetes mellitus (T2DM).

Role of the Endocrine Pancreas in Glucose Homeostasis

The pancreas is an organ with endocrine and exocrine functions. The latter relates mainly to the regulation of gastrointestinal physiology. The major physiological function of the endocrine pancreas is the maintenance of glucose homeostasis. The pancreas senses the concentration of glucose in blood and, through the release of insulin and glucagon, regulates glucose utilization by peripheral tissues. Insulin and glucagon are produced in the islets of Langerhans, which are the endocrine units of the pancreas comprising 1–2% of the total organ weight.

Insulin is secreted from the β-cell of the islets, has anabolic properties and lowers blood glucose by suppression of endogenous glucose production and promoting glucose uptake by insulin-sensitive tissues (Cryer, 2003; Kahn *et al.*, 2006). Although insulin shares many features with many other peptides, it plays a unique role in body physiology. Its absence, or the absence of its receptor, causes major abnormalities in multiple metabolic pathways and is lethal (Accili *et al.*, 1996; Duvillié *et al.*, 1997). Glucose homeostasis is maintained by the fine regulation of insulin secretion, with insulin action promoting glucose transport into muscle and adipocytes and inhibiting hepatic glucose output (Fig. 7.3). In the liver, insulin stimulates both glycolysis and glycogen synthesis and inhibits glycogenolysis and gluconeogenesis, as well as ketogenesis. In addition, insulin

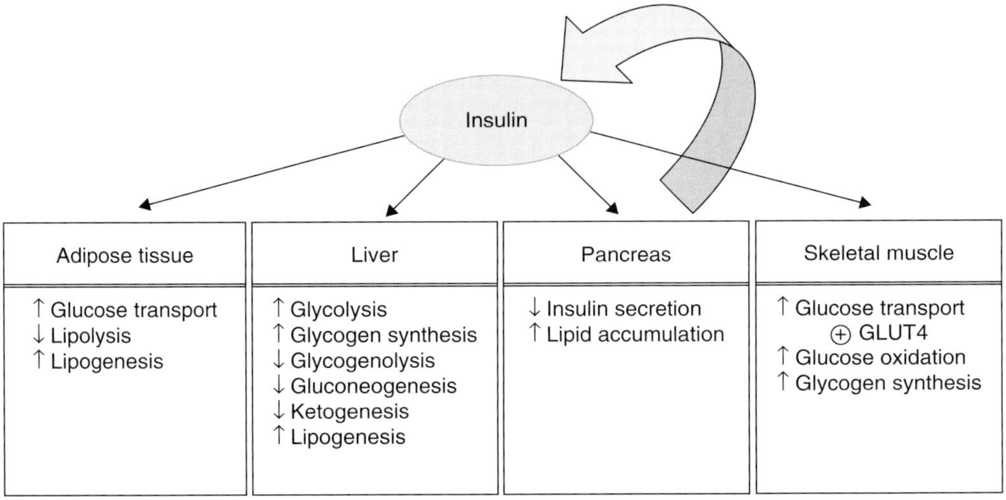

Fig. 7.3. Main actions of insulin in glucose homeostasis. GLUT4, glucose transporter 4.

suppresses lipolysis and induces lipogenesis in adipose tissue. Insulin also stimulates lipoprotein lipase (LPL) in peripheral tissues.

Insulin signalling takes place through phosphorylation of multiple proteins such as insulin receptor substrate (IRS)-1 to -4, with IRS-1 and IRS-2 exhibiting a predominant role. Phosphorylation of tyrosine residues in these substrates leads to the activation of downstream signalling pathways, including phosphoinositide 3-kinase (PI3K), Akt/PKB and mitogen-activated protein kinase (MAPK). These pathways act in a coordinated way to integrate the regulation of fuel metabolism, gene expression, cell growth and differentiation (Kahn *et al.*, 2006; Muoio and Newgard, 2008).

Glucagon is secreted by the α-cell of the islets when circulating blood glucose falls. It has catabolic properties functioning as a counter-regulatory hormone opposing the actions of insulin. Glucagon maintains blood glucose levels during fasting by promoting glycogenolysis and gluconeogenesis, as well as by inhibiting glycogenesis and glycolysis in the liver, therefore preventing hypoglycaemia (Jiang and Zhang, 2003). In addition, glucagon has many extrahepatic effects, including increased lipolysis in adipose tissue and anorexigenic effects acting as a satiety factor in the brain. The increased secretion of glucagon that takes place in diabetes, together with the failure of meal-associated glucagon suppression, worsens hyperglycaemia by increasing hepatic glucose output (Jiang and Zhang, 2003).

Activation of the glucagon receptors leads to the increase in intracellular levels of cAMP and subsequent activation of protein kinase A (PKA). In addition, glucagon also activates phospholipase C (PLC), which subsequently produces the release of intracellular calcium, through inositol 1,4,5-triphosphate, and protein kinase C (PKC) activation, via 1,2-diacylglycerol (DAG). These pathways act coordinately to regulate the activity of key enzymes involved in carbohydrate and lipid metabolism (Jiang and Zhang, 2003; Drucker, 2005).

Obese patients exhibit increased fasting insulin concentrations and glucose-stimulated insulin secretion. In addition, glucagon concentrations may also be elevated in obesity (Basu *et al.*, 2005). Inappropriately elevated concentrations of insulin and glucagon, together with insulin resistance, contribute to obesity-associated impaired glucose homeostasis (Koeslag *et al.*, 2003).

Role of the Liver in Glucose Homeostasis

The liver plays a critical role in maintaining glucose homeostasis, being the major source of glucose production in the postabsorptive state. The rate of hepatic glucose production is adjusted to the rate of glucose uptake by peripheral tissues, thus allowing the maintenance of normoglycaemia. This fine regulation results from an optimal insulin:glucagon ratio that controls the activity of gluconeogenic and glycogenolytic enzymes.

In the postprandial state, insulin suppresses hepatic glycogenolysis and gluconeogenesis and induces hepatic glycogen synthesis (Fig. 7.3). In the fasting state, hepatic glucose output is activated by glucagon through the stimulation of gluconeogenesis and glycogenolysis. The genes encoding the rate-limiting enzymes controlling hepatic glucose output, i.e. phosphoenolpyruvate carboxykinase (PEPCK) and G6Pase, are controlled transcriptionally by the peroxisome proliferator activated receptor (PPAR) γ coactivator-1α (PGC-1α) (Lin *et al.*, 2005). PGC-1α is strongly induced by fasting through the transducer of regulated CREB activity 2 (TORC2)-mediated activation of the cAMP response element binding protein (CREB), which coactivates key hepatic transcription factors such as hepatocyte nuclear factor 4 α (HNF4α), PPARα and forkhead box O1 (FOXO1), among others. PGC-1α deficiency clearly impairs gluconeogenic gene expression and hepatic glucose production, leading to fasting hypoglycaemia (Lin *et al.*, 2005).

The liver has been implicated as a primary site of obesity-associated insulin resistance (Stumvoll *et al.*, 2005). In this sense, the onset of hepatic insulin resistance frequently precedes the appearance of extra-hepatic insulin resistance in humans. Obesity is related to increased hepatic glucose production, which correlates with fasting hyperglycaemia. Whether the obesity-associated increase in hepatic glucose production is due to a direct consequence of impaired insulin suppression of hepatic gluconeogenesis and glycogenolysis (Basu *et al.*, 2005) or whether it is due to circulatory factors, mainly secreted by the adipose tissue, has not been fully clarified (Barzilai *et al.*, 1999).

Role of the Adipose Tissue in Glucose Homeostasis

Adipose tissue is an essential organ for the maintenance of glucose homeostasis (Frühbeck and Gómez-Ambrosi, 2005; Rosen and Spiegelman, 2006). It is well known that excess of fat produces insulin resistance (Kahn *et al.*, 2006; Guilherme *et al.*, 2008). However, contrarily, situations in which the amount of body fat is scarce, such as patients with lipodystrophy or genetically modified mice

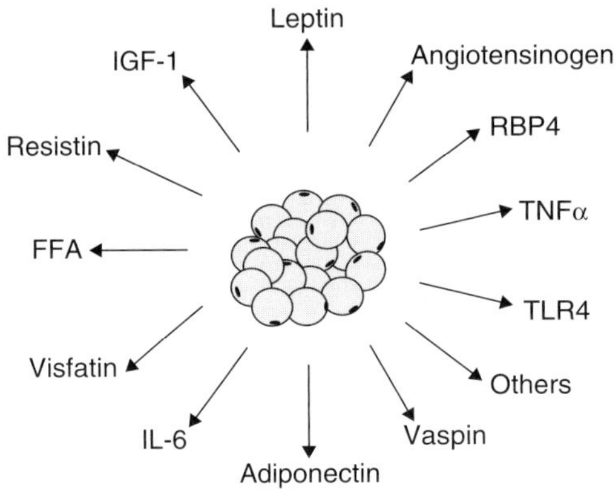

Fig. 7.4. Principal elements secreted/expressed by adipose tissue and involved in glucose homeostasis.

with a minor amount of fat, are also associated with insulin resistance and diabetes mellitus (Frühbeck and Gómez-Ambrosi, 2003; Garg, 2004; Sell *et al.*, 2006). This observation has been explained by the inability of adipose tissue to store excess energy, which has to be accommodated in the liver and muscles, inducing insulin resistance in these organs (Danforth, 2000). This theory is confirmed by looking at transgenic mice overexpressing the leptin receptor specifically in adipocytes, which are unable to store fat in adipose tissue. They show protection against diet-induced obesity but develop insulin resistance due to fat accumulation in the liver, skeletal muscle and heart (Wang *et al.*, 2005b, 2008a). Thus, an adequate body fat amount and functionality are essential for maintaining normal blood glucose concentrations.

For a long time, adipose tissue has been considered to be a passive tissue for the storage of excess energy in the form of fat. However, adipose tissue also secretes a wide variety of biologically active molecules, collectively called adipokines, thus representing an extremely active endocrine organ (Fig. 7.4). These adipokines, which include leptin, adiponectin, tumour necrosis factor (TNF)-α, interleukin-6 (IL-6), resistin, visfatin and retinol binding protein (RBP4), among others, play a key role in glucose homeostasis (Fig. 7.3) (Frühbeck *et al.*, 2001; Frühbeck and Gómez-Ambrosi, 2003; Ahima *et al.*, 2006; Rosen and Spiegelman, 2006; Trujillo and Scherer, 2006; Guilherme *et al.*, 2008). In addition to adipokines, adipose tissue also regulates the release of free fatty acids (FFA) that impair insulin sensitivity in muscle and the liver (Kahn *et al.*, 2006; Guilherme *et al.*, 2008).

Leptin

Leptin is a 16-kDa hormone produced mainly by adipocytes in proportion to fat size stores (Zhang *et al.*, 1994). It was thought originally to be involved only in food intake and body weight regulation acting at its hypothalamic receptors. There are at least five isoforms of the leptin receptor (OB-R) produced by alternative splicing. The full-length isoform, OB-Rb, contains intracellular motifs required for activation of the Janus kinase/signal transducer and activator of transcription (JAK/STAT) signal transduction pathway and is considered to be the functional receptor. Leptin receptors are expressed in almost all tissues, underlining a high functional pleiotropism involving energy homeostasis, reproduction, angiogenesis, immunity, wound healing, bone remodelling and cardiovascular function (Tartaglia *et al.*, 1995; Frühbeck *et al.*, 1998b; Myers *et al.*, 2008). After binding of OB-Rb, leptin signals through the main signalling pathways, including those involving IRS, PI3K, protein kinase B (PKB), PKC, MAPKs, PLC and nitric oxide (Frühbeck, 2006; Myers *et al.*, 2008). Plasma leptin concentrations are increased in obese patients, being strongly correlated with the body mass index (BMI) and the percentage of body fat, as well as with the leptin mRNA expression in adipose tissue.

Leptin-deficient *ob/ob* and leptin receptor-deficient *db/db* mice show early-onset obesity and diabetes. Peripheral administration of leptin to *ob/ob* mice reverses hyperglycaemia and hyperinsulinaemia even before weight loss takes place, indicating that leptin plays a role in glucose homeostasis (Frühbeck and Salvador, 2000; Ceddia *et al.*, 2002). However, the true contribution of leptin in the maintenance of glucose homeostasis in humans remains controversial. Although leptin concentrations are increased in insulin-resistant as compared to adiposity-matched insulin-sensitive individuals (Segal *et al.*, 1996), leptin replacement improves glycaemic control greatly in leptin-deficient individuals (Farooqi *et al.*, 1999; Licinio *et al.*, 2004) and also in leptin-deficient women with lipodystrophy (Oral *et al.*, 2002). In addition, leptin also reverses insulin resistance and diabetes mellitus in mice with congenital lipodystrophy (Shimomura *et al.*, 1999). This apparent insulin-sensitizing effect seen at the whole-body level is not always observed in the individual tissues (Ceddia *et al.*, 2002).

Leptin and the pancreas

In addition to regulating insulin sensitivity, leptin also modulates insulin secretion (Fig. 7.5). Functional OB-R have been detected in pancreatic β-cells (Kieffer *et al.*, 1996), with compelling *in vitro* and *in vivo* evidence showing that leptin inhibits both basal and glucose-stimulated insulin secretion (Frühbeck and Salvador, 2000; Kieffer and Habener, 2000; Ceddia *et al.*, 2002; Seufert, 2004; Covey *et al.*, 2006). In this sense, the existence of a negative feedback signal from adipose tissue to the pancreatic β-cell has been established (Kieffer and Habener, 2000). Leptin inhibition of insulin secretion involves the activation of ATP-sensitive K^+ channels (Kieffer *et al.*, 1997), interfering with the PLC–PKC signalling system (Chen *et al.*, 1997), and the suppression of the second phase of insulin secretion by reducing the activity of the Ca^{2+}-dependent PKC isoform (Ookuma *et al.*, 1998). Leptin also inhibits insulin secretion antagonizing cAMP signalling through the activation of phosphodiesterase 3B (Frühbeck, 2006). Moreover, leptin also decreases glucose-mediated insulin secretion through a hypothalamic effect involving the melanocortinergic system (Muzumdar *et al.*, 2003). Furthermore, leptin downregulates preproinsulin mRNA expression acutely in *ob/ob* mice and humans (Murakami *et al.*, 2001). This effect apparently takes place through the repression of protein phosphatase 1 (PP1) and the induction of suppressor of cytokine signalling 3 (SOCS3) (Seufert, 2004) (Fig. 7.6).

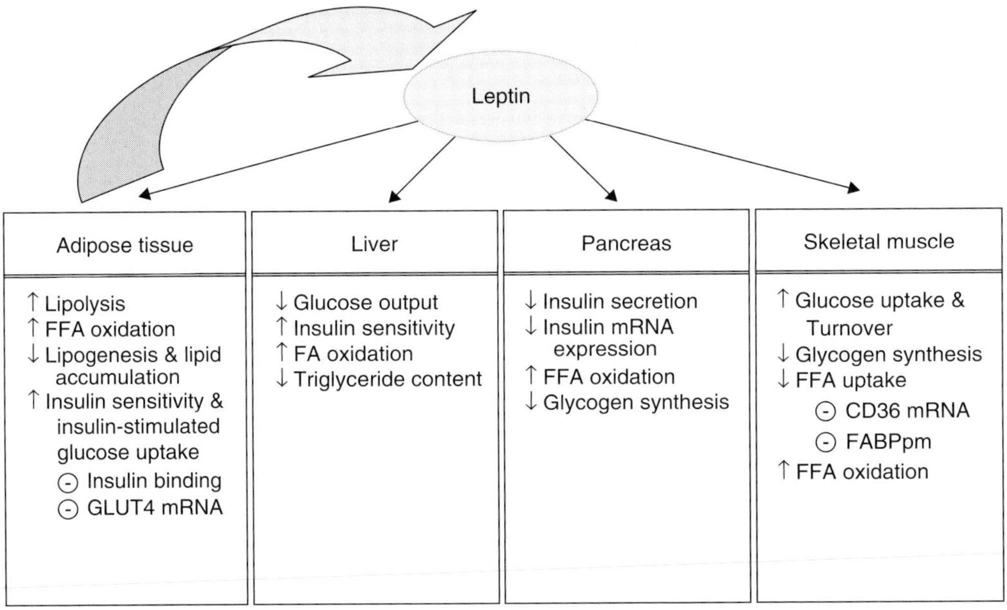

Fig. 7.5. Overview of the main effects of leptin on glucose homeostasis. FFA, free fatty acids; GLUT4, glucose transporter 4; FAT/CD36, fatty acid translocase; FABP, fatty acid binding protein.

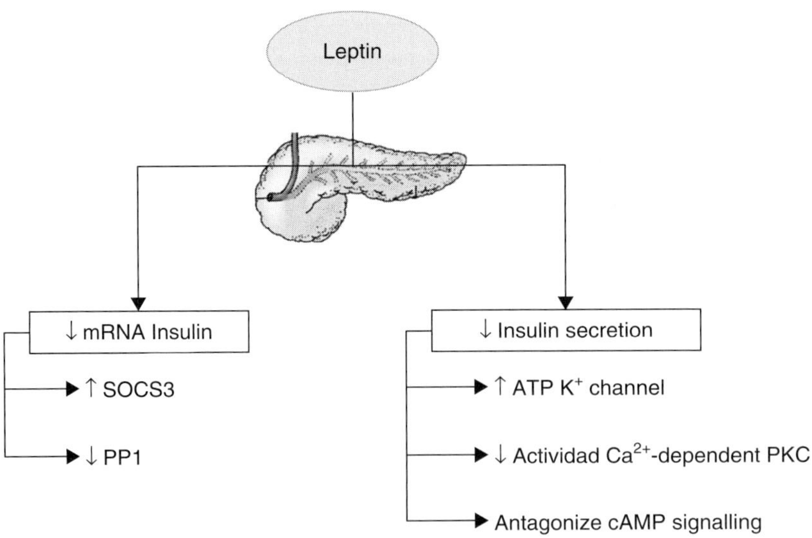

Fig. 7.6. Effects of leptin on the pancreas. SOCS3, suppressor of cytokine signalling 3; PP1, protein phosphatase 1; PKC, protein kinase C.

Besides modulating insulin expression and secretion, there is a complex crosstalk between the leptin and insulin signalling pathways that leads to differential modification of the metabolic effects of insulin exerted via IRS-1 and IRS-2 and which can modify insulin-induced changes in gene expression (Frühbeck, 2006). Leptin also regulates other signalling pathways activated by insulin such as Akt, glycogen synthase kinase and PKC (Frühbeck and Salvador, 2000; Kieffer and Habener, 2000; Frühbeck *et al.*, 2001; Ceddia *et al.*, 2002; Seufert, 2004). The importance of leptin in the regulation of β-cell physiology and islet biology is demonstrated in mice with pancreas-specific leptin receptor deficiency (Morioka *et al.*, 2007). On the other hand, insulin increases the expression and secretion of leptin from adipocytes in rodents and humans through the regulation of glucose transport and metabolism and the involvement of adipocyte determination and differentiation-dependent factor 1/sterol regulatory element binding protein 1 (ADD1/SREBP1) (Kim *et al.*, 1998; Mueller *et al.*, 1998).

Finally, it has also been proposed that leptin acts as a protecting factor against the lipid accumulation-induced damage in the β-cell (lipotoxicity) that leads to β-cell failure and T2DM (Unger, 2003). In conclusion, leptin serves as an inhibitory signal from adipose tissue to prevent overproduction of insulin and hyperinsulinaemia. Moreover, insulin increases the expression and secretion of leptin from adipocytes establishing a classic feedback loop, which influences glucose homeostasis and the development of obesity-associated insulin resistance (Frühbeck and Salvador, 2000; Kieffer and Habener, 2000).

Leptin and the liver

Leptin receptors are expressed in the liver, with leptin exerting an important role in the regulation of glucose and lipid metabolism in this organ (Fig. 7.5) (Frühbeck and Salvador, 2000; Ceddia *et al.*, 2002). Overall, leptin improves insulin sensitivity through complex effects on hepatic gene expression of key metabolic enzymes and on intrahepatic partitioning of metabolic fluxes. Leptin administration in rodents enhances the inhibition of hepatic glucose production exerted by insulin through a marked decrease in hepatic glycogenolysis and/or directly increasing glycogen synthesis via PI3K-dependent activation of phosphodiesterase 3B and a subsequent decrease in cAMP.

At the same time, leptin increases gluconeogenesis, upregulating the mRNA expression of PEPCK and G6Pase, through centrally mediated effects. It has been shown that leptin and insulin exhibit complex interactions in their signalling pathways (Szanto and Kahn, 2000). In this sense, leptin may antagonize some functions of insulin via IRS-1 dephosphorylation. Moreover, a leptin-dependent short-term inhibition of gluconeogenesis, mediated through IRS-2, and a positive crosstalk between the insulin and leptin cascades at the level of JAK2 and STAT5b have also been described (Anderwald *et al.*, 2002; Ceddia *et al.*, 2002; Carvalheira *et al.*, 2003). Therefore, although it has been proposed that leptin regulates hepatic glucose metabolism by an insulin-mimicking effect on glycogenolysis and a glucagon-like effect on gluconeogenesis (Nemecz *et al.*, 1999), the complex regulatory effect of leptin on the liver needs to be elucidated fully.

Analogously to what has been described in β-cells, leptin induces changes in hepatic gene expression consistent with a switch in energy utilization from glucose use and fatty acid synthesis to fatty acid oxidation. In line with this, leptin replacement in patients with severe lipodystrophy improves insulin action significantly and reduces hepatic triglycerides markedly (Petersen *et al.*, 2002), underlining that failure of leptin action represents a major physiological mechanism for hepatic steatosis (Fishman *et al.*, 2007).

Leptin and the skeletal muscle

The functional isoform of the OB-R is readily detectable in skeletal muscles of different muscle fibre type and composition (Frühbeck *et al.*, 1999), indicating that skeletal muscle is a clear target for the direct metabolic effects of leptin (Ceddia, 2005; Dube *et al.*, 2007).

Intravenous administration of leptin into mice increases glucose turnover and glucose uptake (Fig. 7.5). Similar effects are observed after both intravenous and intracerebroventricular infusion, evidencing that the effects of leptin on glucose metabolism are, at least in part, centrally mediated. Furthermore, leptin modulates muscular glucose metabolism directly, increasing glucose transport and utilization (Wang *et al.*, 1999a; Ceddia, 2005), apparently with no changes in glucose transporter 4 (GLUT4) expression levels (Wang *et al.*, 1999a). In addition, leptin is able to inhibit glycogen synthesis in the soleus muscle of leptin-deficient *ob/ob* mice, suggesting that leptin may attenuate insulin action on glucose storage in muscle. Moreover, chronic leptin administration enhances insulin-stimulated glucose disposal in skeletal muscle of normal and high-fat diet-fed rodents (Ceddia *et al.*, 2002; Yaspelkis *et al.*, 2004; Ceddia, 2005).

In addition to the described effects on glucose metabolism, leptin promotes fatty acid utilization in skeletal muscle showing insulin antagonizing effects (Muoio and Newgard, 2008). Moreover, chronic leptin administration decreases fatty acid uptake and mRNA expression of the fatty acid transporters, fatty acid translocase (FAT/CD36) and plasma membrane-associated fatty acid binding protein (FABPpm) (Steinberg *et al.*, 2002a; Dube *et al.*, 2007). Leptin increased fatty acid oxidation in skeletal muscle of lean women, while this effect was absent in the muscle of obese women (Steinberg *et al.*, 2002b). Enhanced glucose and lipid metabolism exerted by leptin in skeletal muscle takes place through PI3K and IRS-2, extracellular signal-regulated kinase 2 (ERK2) and AMP-activated protein kinase (AMPK) phosphorylation, a fuel gauge critically involved in energy metabolism (Kellerer *et al.*, 1997; Frühbeck and Salvador, 2000; Ceddia *et al.*, 2002; Minokoshi *et al.*, 2002; Ceddia, 2005; Kahn *et al.*, 2005).

In summary, published evidence suggests that leptin increases glucose and fatty acid oxidation in skeletal muscle, therefore driving metabolism towards fuel utilization rather than storage. Leptin plays a determining role on fuel homeostasis, reducing lipotoxicity and enhancing skeletal muscle and whole-body insulin sensitivity (Frühbeck and Salvador, 2000; Ceddia *et al.*, 2002; Ceddia, 2005).

Leptin effect on adipose tissue

Leptin administration induces a dramatic loss of adipose mass in rodents. This effect is not only mediated by a reduction in food intake, but also by a direct effect on adipose tissue (Fig. 7.5). Leptin inhibits lipogenesis and stimulates lipolysis in adipocytes through a direct effect (Siegrist-Kaiser *et al.*, 1997; Frühbeck *et al.*, 1998a) or a centrally mediated influence (Gallardo *et al.*, 2007) without the release of FFAs, which are oxidized intracellularly (Wang *et al.*, 1999b). In addition to its effects on lipid mobilization, leptin decreases insulin sensitivity and insulin-stimulated glucose uptake and incorporation into lipids in rodent adipocytes (Ceddia, 2005). It has been reported that this effect occurs through the inhibition of insulin binding to its receptor (Walder *et al.*, 1997), the decrease of GLUT4 mRNA expression (Wang *et al.*, 1999a) and the impairment of insulin signalling (Pérez *et al.*, 2004). Therefore, high concentrations of leptin induce lipid utilization in adipocytes, rather than carbohydrates, transforming the adipocytes into fat-oxidizing machines in order to avoid excessive fat accumulation (Orci *et al.*, 2004). Interestingly, obese individuals exhibit increased concentrations of leptin at the same time as their adipocytes accumulate high amounts of fat, leading to the observation that fat storage in adipocytes requires inactivation of leptin's paracrine activity (Wang *et al.*, 2005a).

The importance of leptin action on adipocytes is evidenced further in a mice model in which the expression of OB-R is reduced specifically in white adipose tissue (Huan *et al.*, 2003). These mice show increased adiposity and insulin resistance, indicating that a dysfunctional adipocyte leptin response has a great impact on whole-body glucose homeostasis.

Adiponectin

Adiponectin, also known as Acrp30, adipoQ, apM1 and GBP28, is another adipokine highly expressed in adipose tissue (Kadowaki and Yamauchi, 2005; Kadowaki *et al.*, 2006). It is secreted as a 244-amino acid protein accounting for approximately 0.01% of total serum protein (Trujillo and Scherer, 2005). Adiponectin consists of an amino-terminal collagen-like domain and a carboxy-terminal globular domain belonging structurally to the complement factor C1q family (Kadowaki and Yamauchi, 2005; Kadowaki *et al.*, 2006). The basic form of adiponectin in serum consists of a homotrimer of three 30-kDa subunits. Trimers associate to form low molecular weight (LMW) hexamers of 180 kDa and high molecular weight (HMW) multimers of up to 18-mers of over 400 kDa (Kadowaki and Yamauchi, 2005; Trujillo and Scherer, 2005; Kadowaki *et al.*, 2006). Adiponectin exerts a wide variety of physiological effects through binding of at least three different receptors. The first two isoforms described were called AdipoR1 and AdipoR2 and exhibited a specific tisular expression pattern. AdipoR1 and AdipoR2 bind globular and recombinant adiponectin, although their pathophysiological relevance remains to be determined fully (Yamauchi *et al.*, 2003). Both the LMW and HMW forms, but not the globular form, bind T-cadherin, the third receptor described. T-cadherin is a glycosylphosphatidylinositol-anchored protein that may act as a co-receptor for a signalling receptor through which adiponectin transmits metabolic signals, but its functional implications remain to be disentangled (Hug *et al.*, 2004).

Serum concentrations of adiponectin are decreased in obese individuals (Arita *et al.*, 1999; Weyer *et al.*, 2001) and increase after weight loss (Yang *et al.*, 2001). Adiponectin concentrations are also lower in people with cardiovascular disease (Ouchi *et al.*, 1999), dyslipidaemia (Matsubara *et al.*, 2002b), hypertension (Kazumi *et al.*, 2002; Mallamaci *et al.*, 2002) and lipodystrophy (Haque *et al.*, 2002). Hypoadiponectinaemia is associated with insulin resistance, and patients with T2DM are reported to have decreased concentrations of adiponectin (Hotta *et al.*, 2000; Weyer *et al.*, 2001; Spranger *et al.*, 2003). In addition, adiponectin is associated independently with a reduced risk of T2DM in apparently healthy individuals (Spranger *et al.*, 2003). Moreover, administration of adiponectin, either in the full length or in the globular form, induces glucose lowering effects and ameliorates insulin resistance in rodent models of diabetes and obesity (Berg *et al.*, 2001; Combs *et al.*, 2002), as well as having antiatherogenic properties (Matsuzawa *et al.*, 2004). Adiponectin-deficient mice, as well as mice lacking adiponectin receptors, confirm the protective effects of this adipokine in the development of insulin resistance and atherosclerosis (Kubota *et al.*, 2002; Maeda *et al.*, 2002; Yamauchi *et al.*, 2007).

Adiponectin and the pancreas

Administration of globular adiponectin has no effect on insulin or glucagon secretion in mice (Combs *et al.*, 2001; Fruebis *et al.*, 2001), although some authors have described an increase in glucagon secretion following adiponectin injection, probably due to a fall in glucose concentrations (Berg *et al.*, 2001). AdipoR1 and AdipoR2 are expressed in human and rat pancreatic β-cells, at levels similar to the liver and higher than in muscle. Globular adiponectin increases expression of LPL, an enzyme contributing to the delivery of FFA to tissues, while FFA upregulates adiponectin receptors in β-cells. These observations suggest that adiponectin may be involved in the supply of nutrients to β-cells during periods of caloric restriction (Kharroubi *et al.*, 2003). In normal islets, adiponectin does not have an effect on insulin secretion (Winzell *et al.*, 2004). However, in islets from high-fat fed induced insulin-resistant mice, adiponectin exhibits a glucose-dependent dual effect on insulin secretion, decreasing basal insulin secretion but potentiating glucose-stimulated insulin secretion (Fig. 7.7). Therefore, adiponectin seems to be of importance in preventing the deterioration of glucose homeostasis in insulin resistance by both increasing insulin sensitivity and elevating glucose-stimulated insulin secretion (Winzell *et al.*, 2004). Moreover, a protective effect of adiponectin against lipotoxicity-induced β-cell apoptosis has been reported (Rakatzi *et al.*, 2004). Metabolic effects of adiponectin in β-cells seem to be mediated through induction of AMPK (Huypens *et al.*, 2005).

The regulatory effect of insulin on adiponectin gene expression and protein secretion has not been clarified fully. Chronic exposure to insulin reduces adiponectin mRNA expression in murine adipocytes *in vitro* (Fasshauer *et al.*, 2002). These findings are in agreement with *in vivo* studies in humans showing that

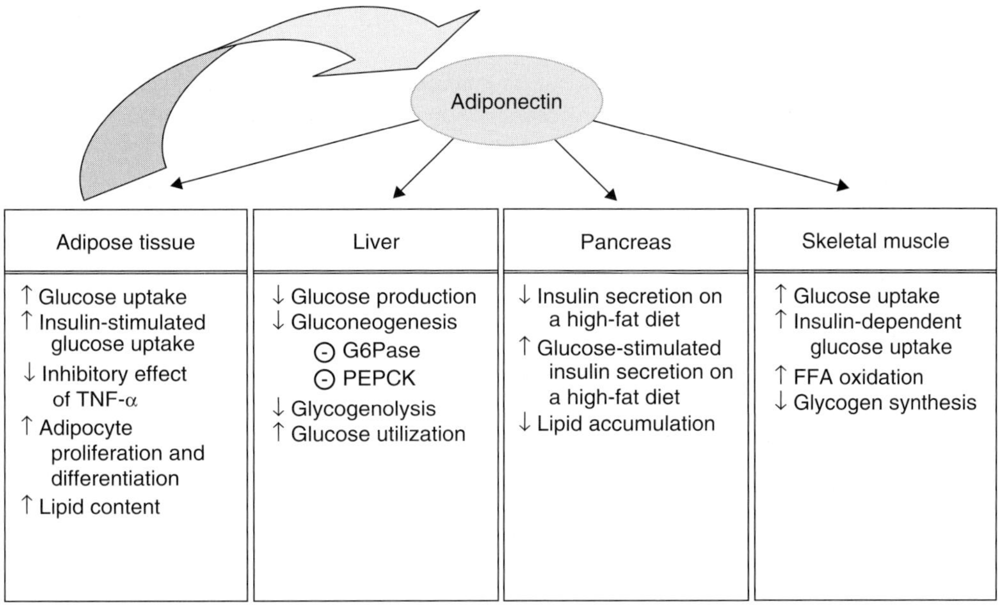

Fig. 7.7. Main actions of adiponectin in glucose homeostasis. TNF-α, tumour necrosis factor-α; G6Pase, glucose-6-phosphatase; PEPCK, phosphoenolpyruvate carboxykinase; FFA, free fatty acids.

fasting insulin is correlated negatively with adiponectin serum concentrations (Matsubara *et al.*, 2002a; Gavrila *et al.*, 2003). In contrast, other studies suggest that adiponectin synthesis is increased by insulin in human (Halleux *et al.*, 2001) and rat (Cong *et al.*, 2007) adipocytes.

Adiponectin and the liver

Adiponectin does not affect the rates of glucose uptake, glycolysis or glycogen synthesis in the liver. However, adiponectin inhibits hepatic glucose production through a reduction in the expression of the gluconeogenic enzymes, PEPCK and G6Pase (Combs *et al.*, 2001; Yamauchi *et al.*, 2002) independently of the presence of insulin (Zhou *et al.*, 2005) and enhances insulin action (Fig. 7.7) (Berg *et al.*, 2001). In addition to its direct effect on glucose metabolism, adiponectin decreases hepatic triglyceride content through an AMPK- and PPARα-mediated stimulation of fatty acid oxidation, leading to an increase in insulin sensitivity (Kadowaki and Yamauchi, 2005; Long and Zierath, 2006). In this sense, adiponectin has been proposed as a protective factor against the development of non-alcoholic fatty liver disease (Schäffler *et al.*, 2005).

Adiponectin and the skeletal muscle

Administration of recombinant globular adiponectin to mice produces a decrease of serum FFA, glucose and triglycerides through an increase in fatty acid oxidation (Fruebis *et al.*, 2001), together with an increase in glucose uptake and lactate production (Yamauchi *et al.*, 2002) in skeletal muscle (Fig. 7.7). One of the molecular mechanisms involved in this effect is the increase in the expression of CD36 (involved in fatty acid transport), acyl-coenzyme A oxidase (involved in fatty acid combustion) and uncoupling protein 2 (involved in energy dissipation) (Yamauchi *et al.*, 2001). The adiponectin fat-oxidizing effect is brought about also through induction of the phosphorylation of the insulin receptor, IRS-1, and Akt (Yamauchi *et al.*, 2001), increasing the expression and activation of PPARα (Yamauchi *et al.*, 2001), as well as via the activation of AMPK (Tomas *et al.*, 2002; Yamauchi *et al.*, 2002).

Interestingly, insulin resistance in lipoatrophic mice is reversed completely by the combination of physiological doses of adiponectin and leptin, but only partially by individual administration of each of the adipokines (Yamauchi *et al.*, 2001), confirming that both hormones are strongly involved in glucose homeostasis. In summary, adiponectin increases insulin sensitivity (Berg *et al.*, 2001; Yamauchi *et al.*, 2001) and protects against lipid accumulation in skeletal muscle, increasing fatty acid oxidation (Fruebis *et al.*, 2001; Yamauchi *et al.*, 2001) through AMPK, p38 MAPK and PPARα activation (Tomas *et al.*, 2002; Yamauchi *et al.*, 2002; Yoon *et al.*, 2006).

Adiponectin effect on adipose tissue

Evidence suggests a role for adiponectin as an autocrine factor in adipose tissue promoting cell proliferation and differentiation from preadipocytes to adipocytes, increasing programmed gene expression responsible for adipogenesis and elevating the lipid content and insulin responsiveness of adipocytes (Fu *et al.*, 2005). Regarding glucose metabolism, adiponectin increases glucose uptake in adipocytes without stimulating phosphorylation of the insulin receptor, IRS-1, or Akt. Adiponectin further enhances insulin-stimulated glucose uptake and reverses the inhibitory effect of TNF-α on insulin-stimulated glucose uptake (Wu *et al.*, 2003). Furthermore, adiponectin prevents the release of insulin resistance-inducing factors by adipocytes (Dietze-Schroeder *et al.*, 2005).

Resistin

Resistin has been found to be highly expressed in adipose tissue and secreted into the bloodstream in mice. However, this adipokine does not seem to be expressed at significant levels in human adipocytes (Arner, 2005; Gómez-Ambrosi and Frühbeck, 2005a). Under physiological circumstances, resistin apparently opposes insulin action in adipocytes and impairs glucose tolerance and insulin sensitivity in normal mice. Moreover, insulin-stimulated glucose uptake by adipocytes is enhanced by resistin neutralization and is reduced by resistin treatment (Steppan *et al.*, 2001; McTernan *et al.*, 2006). Different genetic and dietary models of rodent obesity exhibit increased serum concentrations of resistin (Steppan *et al.*, 2001) but, surprisingly, resistin mRNA levels in the adipose tissue of these animals are severely suppressed (Le Lay *et al.*, 2001; Way *et al.*, 2001). This apparently contradictory finding has been further confirmed, suggesting the existence of regulatory binding proteins or negative feedback mechanisms of resistin regulating its own expression (Rajala *et al.*, 2004).

Transgenic mice expressing a dominant inhibitory form of resistin show improved glucose tolerance and insulin sensitivity on either a normal or a high-fat diet (Kim *et al.*, 2004b). Moreover, specific antisense oligodeoxynucleotide directed against resistin reverses completely the hepatic insulin resistance induced by a high-fat diet (Muse *et al.*, 2004). In addition, transgenic or adenovirus-mediated chronic hyperresistinaemia leads to insulin resistance in rats (Satoh *et al.*, 2004) and mice (Rangwala *et al.*, 2004). Resistin seems to impair selectively the inhibitory action of insulin on glucose production, inducing the upregulation of all pathways involved in hepatic glucose formation, leading to an enhanced glucose output (Rajala *et al.*, 2003). In this sense, mice lacking resistin exhibit low blood glucose levels after fasting, due to reduced hepatic glucose production through the decreased expression of gluconeogenic enzymes (Banerjee *et al.*, 2004). Resistin deficiency also improves glucose tolerance and insulin sensitivity in *ob/ob* mice (Qi *et al.*, 2006). Molecular studies suggest that resistin normally acts on the liver to inhibit AMPK and this response is impaired in the resistin knockout mice, which show an increased active phosphorylated form of AMPK and reduced blood glucose concentrations (Banerjee *et al.*, 2004).

The exact role of resistin in human physiology and whether or not resistin is involved in the development of insulin resistance still needs to be clarified completely (Kusminski *et al.*, 2005; McTernan *et al.*, 2006). Several groups have described increased concentrations of resistin in obesity (Azuma *et al.*, 2003; Degawa-Yamauchi *et al.*, 2003), while others report no differences (Lee *et al.*, 2003; Silha *et al.*, 2003; Heilbronn *et al.*, 2004). Human recombinant resistin produces a small but significant decrease in glucose uptake in human cultured preadipocytes (McTernan *et al.*, 2003). However, cross-sectional epidemiological investigations in which blood resistin has been measured in individuals with T2DM (Fehmann and Heyn, 2002; Chen *et al.*, 2006) or the metabolic syndrome (Utzschneider *et al.*, 2005) have revealed that human resistin concentrations are not involved significantly in insulin resistance. Rather than a role in the development of insulin resistance, considerable evidence links resistin to inflammation (Gómez-Ambrosi and Frühbeck, 2001; Gómez-Ambrosi and Frühbeck, 2005b; Kusminski *et al.*, 2005). In this respect, it has been reported that resistin upregulates TNF-α, IL-6 and IL-12 markedly (Bokarewa *et al.*, 2005; Silswal *et al.*, 2005) and that inflammation induces resistin in primary human macrophages via a cascade involving the secretion of inflammatory cytokines (Kaser *et al.*, 2003; Lehrke *et al.*, 2004). Furthermore, it has been confirmed that resistin is associated with markers of inflammation, being predictive of coronary atherosclerosis in humans (Reilly *et al.*, 2005), and that it influences proinflammatory cytokine release from human adipocytes, potentially via the integration of nuclear factor-κB and JNK signalling pathways (Kusminski *et al.*, 2007).

TNF-α

TNF-α is a cytokine implicated in the metabolic disturbances of chronic inflammation, with biological actions including induction of insulin resistance, anorexia and weight loss (Frühbeck *et al.*, 2001). In addition, TNF-α is a potent negative regulator of adipocyte differentiation (Cawthorn *et al.*, 2007). Adipose tissue is both a source of and a target for TNF-α (Cawthorn and Sethi, 2008). It has been suggested that TNF-α is a candidate mediator of insulin resistance in obesity, as it is overexpressed in the adipose tissue of rodents and humans (Hotamisligil, 2006). TNF-α mRNA expression also exhibits a close correlation with hyperinsulinaemia, showing positive associations with fasting insulin and triglyceride concentrations. In addition, it blocks the action of insulin in adipose tissue and skeletal muscle *in vitro* and *in vivo*. In this sense, TNF-α-deficient mice exhibit decreased glucose, insulin and leptin concentrations, showing an impaired glucose clearance when challenged with a high-fat diet, as evidenced by increased circulating glucose and insulin, which does not, however, reach the concentrations of wild-type controls. Disruption of the expression of TNF-α receptors has no effect on body weight or glucose homeostasis when mice are fed a standard diet. However, absence of both receptor subtypes (p55 and p75) results in severe hyperinsulinaemia on a high-fat diet. In summary, TNF-α plays an important role in obesity-related insulin resistance, although it has been suggested recently that the link between TNF-α, obesity and insulin resistance in humans is indirect (Arner, 2005). This suggestion has been argued by a local stimulatory effect of TNF-α on adipocyte lipolysis, and thereby the release of FFA into the circulation (Arner, 2005). Recent findings suggest that TNF-α induces insulin resistance through AMPK signalling suppression in skeletal muscle (Steinberg *et al.*, 2006) and through both IRS-1 serine phosphorylation and SOCS3 induction in adipocytes (Ishizuka *et al.*, 2007).

IL-6

Interleukin-6 is an inflammatory mediator with pleiotropic effects on a variety of tissues, including stimulation of acute-phase protein synthesis and regulation of glucose and lipid metabolism (Frühbeck *et al.*, 2001; Frühbeck and Salvador, 2004). Adipose tissue secretes IL-6, with serum concentrations of IL-6 being proportional to fat mass and the degree of insulin resistance (Bastard *et al.*, 2002; Tilg and Moschen, 2008). Moreover, IL-6 blunts the ability of insulin to suppress hepatic glucose production and also reduces insulin-stimulated glucose uptake in mouse skeletal muscle. This effect has been associated with defects in insulin-stimulated IRS-1- and IRS-2-associated PI3K activation (Kim *et al.*, 2004a) and with an induction of SOCS3 expression (Senn *et al.*, 2003). Although increased concentrations of IL-6 have been detected in obese subjects, IL-6-deficient mice develop mature-onset obesity, with the obese phenotype being reversed only partly by IL-6 replacement (Wallenius *et al.*, 2002). Interestingly, acute IL-6 treatment has been reported to produce an increase in insulin-stimulated glucose disposal in humans *in vivo* and to induce fatty acid oxidation, glucose transport and GLUT4 translocation to the plasma membrane *in vitro* through activation of AMPK (Carey *et al.*, 2006). In this sense, IL-6 may be considered both an autocrine and paracrine regulator of adipocyte function, in addition to exerting broader endocrine effects, but its implication in the development of insulin resistance still has to be understood completely (Tilg and Moschen, 2008).

Other adipokines

Visfatin, identified previously as the colony-enhancing factor of pre-B-cells, is so named because it is highly secreted by the visceral fat of both mice and humans and its expression levels in serum increase during the development of obesity (Fukuhara *et al.*, 2005; Sethi and Vidal-Puig, 2005). Visfatin reportedly has insulin-like activity binding to the insulin receptor and thereby lowering blood glucose concentrations. Visfatin has been shown to stimulate glucose uptake into adipocytes and skeletal muscle cells and to suppress glucose output from hepatocytes. Moreover, it has been shown that visfatin stimulates the phosporylation of proteins downstream of the insulin receptor (Fukuhara *et al.*, 2005). However, these findings are currently controversial, with the authors having been forced to retract some of their original conclusions (Fukuhara *et al.*, 2007). Surprisingly, plasma visfatin correlates with measures of obesity but not with visceral fat mass or waist–hip ratio or variables of insulin sensitivity in humans. In addition, no differences in visfatin mRNA expression between visceral and subcutaneous adipose tissue have been observed (Berndt *et al.*, 2005). It has also been described that circulating concentrations of visfatin are increased in T2DM patients and its potential involvement in inflammation has been put forward (Moschen *et al.*, 2007). Undoubtedly, more studies clearly are needed to clarify fully the real implications of visfatin in glucose homeostasis (Arner, 2006; Chen *et al.*, 2006).

Visceral adipose tissue-derived serpin (vaspin) is a member of the serine protease inhibitor family. Vaspin is highly expressed in adipocytes of visceral adipose tissue at the same time that obesity and insulin levels peak in Otsuka Long–Evans Tokushima fatty (OLETF) rats. Administration of vaspin to obese insulin-resistant mice improves glucose tolerance and insulin sensitivity. These findings indicate that vaspin exerts an insulin-sensitizing effect in states of obesity (Hida *et al.*, 2005). Human vaspin mRNA expression in adipose tissue is not detectable in lean glucose-tolerant individuals, but can be induced by increased fat mass and decreased insulin sensitivity, which could represent a compensatory mechanism associated with obesity and T2DM (Klöting *et al.*, 2006). However, no difference in circulating concentrations of vaspin between individuals with normal glucose tolerance and T2DM have been observed (Youn *et al.*, 2008). The potential involvement of vaspin in glucose homeostasis certainly requires further investigation (Zvonic *et al.*, 2007).

RBP4 mRNA is one of the most abundant transcripts present in both rodent and human adipose tissue. Synthesis and secretion of RBP4 by adipocytes are induced by retinoic acid, showing that adipose tissue plays an important role in retinoid storage and metabolism (Frühbeck and Salvador, 2004). Serum RBP4 concentrations have been described as being elevated in insulin-resistant mice and humans with obesity and T2DM (Yang *et al.*, 2005; Cho *et al.*, 2006; Graham *et al.*, 2006). It seems that RBP4 induces hepatic expression of the gluconeogenic enzyme, PEPCK, and impairs insulin signalling in skeletal muscle. Moreover, transgenic overexpression of RBP4 or recombinant RBP4 administration in mice causes insulin resistance, while genetic deletion of the RBP4 gene enhances insulin sensitivity (Yang *et al.*, 2005). However, the true contribution in human obesity has not been clarified completely, with some studies observing normal RBP4 concentrations (Janke *et al.*, 2006; Broch *et al.*, 2007; Gómez-Ambrosi *et al.*, 2008). Further research is needed to unravel the involvement of RBP4 in the development of obesity-associated insulin resistance in humans.

Murine models of obesity show increased local formation of angiotensin II (Ang II) due to elevated secretion of its precursor, angiotensinogen (AGT), from adipocytes (Frühbeck *et al.*, 2001), with deficiency or overexpression of AGT affecting body weight regulation (Frühbeck and Gómez-Ambrosi, 2003). Given the close relationship between Ang II and insulin resistance (Katovich and Pachori, 2000), the participation of the adipose-tissue renin–angiotensin system in the development of insulin resistance is conceivable in humans, but has to be evaluated in more detail (Engeli *et al.*, 2003).

Members of the insulin-like growth factor (IGF) system are related functionally to insulin. They are expressed ubiquitously and play a role in all tissues. While insulin is a short-term regulator of glucose homeostasis, IGFs have been suggested to exert a long-term regulation of glucose homeostasis (Murphy, 2003; Clemmons, 2006a,b). Insulin and IGF-1 show cross-reactivity at the receptor level. After ligand binding-induced autophosphorylation, insulin receptor and IGF-1 receptor catalyse the phosphorylation of cellular proteins such as members of the IRS family (Saltiel and Kahn, 2001). A number of immediate second messengers have been implicated in insulin/IGF-1 action, including PI3K and PLC, among others. Adipose tissue levels of IGF-1 have been shown to be higher in both human and rodent obesity (Frystyk *et al.*, 1995), but IGF-1-induced activation of PI3K is impaired in obese mice (Le Marchand-Brustel *et al.*, 1995).

FFA

As mentioned before, the release of FFA by adipose tissue through lipolysis plays a critical role in the ability of organisms to provide energy from triglyceride stores (Frühbeck and Gómez-Ambrosi, 2005). FFA are a major fuel for peripheral organs, being an alternative source to glucose, preserving glucose for brain requirements (Arner, 2001). Most obese patients with T2DM exhibit increased serum concentrations of FFA, which induce insulin resistance in skeletal muscle, the liver, adipose tissue and the endothelium (Boden, 2003).

In the liver, FFA produce insulin resistance and stimulate endogenous glucose production, although the underlying molecular mechanisms are not yet understood clearly (Boden, 2003; Boden *et al.*, 2005). In skeletal muscle, FFA inhibit insulin-stimulated glucose uptake at the level of glucose transport and/or phosphorylation, through mechanisms that involve intramyocellular accumulation of DAG and decreased phosphorylation of IRS-1 and IRS-2, which is then followed by a reduction in both the rate of muscle glycogen synthesis and glucose oxidation (Roden *et al.*, 1996; Boden *et al.*, 2005). Although an acute elevation in FFA increases insulin secretion (Nolan *et al.*, 2006), its long-term effects are somewhat controversial. It seems that insulin resistance induced by prolonged raised levels of FFA is compensated by an increase in insulin secretion in healthy individuals, but T2DM-prone individuals fail to compensate, facilitating the emergence of T2DM (Boden, 2005). Therefore, increased circulating concentrations of FFA may contribute to or aggravate obesity-associated insulin resistance. Noteworthy, recent findings indicate that upper intestinal lipids activate an intestine–brain–liver neural axis to inhibit glucose production, thereby revealing a previously unappreciated pathway in glucose homeostasis (Wang *et al.*, 2008b).

Concluding Remarks

Glucose homeostasis is a critical physiological process that requires a complex interaction of many factors in order to provide tissues with their fuel requirements. Glycaemia results from the net integrated balance between glucose influx from the circulation and utilization by tissues. The key elements of this process traditionally have been considered to be the liver and the pancreas. However, it has been clearly established that adipose tissue is of extreme importance in the maintenance of glucose homeostasis. In this sense, adipokines, especially leptin and adiponectin, exert a relevant role in the control of whole-body glucose metabolism and insulin sensitivity. In the regulation of this process, adipose tissue, the liver, skeletal muscle, the pancreas and the brain establish a complex and dynamic crosstalk, whereby any delicate derangement in the network translates in dysfunction of the whole control system.

The study of the adipo–hepato–insular axis as an integrated system represents an interesting and fertile area of research. Further, a more exact and precise knowledge about the complex interplay between the diverse and numerous components of this axis, as well as the pathophysiological alterations that take place in obesity and T2DM, will lead to a better understanding of the causes and pathogenesis of insulin resistance.

References

Accili, D., Drago, J., Lee, E.J., Johnson, M.D., Cool, M.H., Salvatore, P., Asico, L.D., Jose, P.A., Taylor, S.I. and Westphal, H. (1996) Early neonatal death in mice homozygous for a null allele of the insulin receptor gene. *Nature Genetics* 12, 106–109.

Ahima, R.S., Qi, Y., Singhal, N.S., Jackson, M.B. and Scherer, P.E. (2006) Brain adipocytokine action and metabolic regulation. *Diabetes* 55 (Suppl. 2), S145–154.

Anderwald, C., Müller, G., Koca, G., Fürnsinn, C., Waldhäusl, W. and Roden, M. (2002) Short-term leptin-dependent inhibition of hepatic gluconeogenesis is mediated by insulin receptor substrate-2. *Molecular Endocrinology* 16, 1612–1628.

Arita, Y., Kihara, S., Ouchi, N., Takahashi, M., Maeda, K., Miyagawa, J., Hotta, K., Shimomura, I., Nakamura, T., Miyaoka, K., Kuriyama, H., Nishida, M., Yamashita, S., Okubo, K., Matsubara, K., Muraguchi, M., Ohmoto, Y., Funahashi, T. and Matsuzawa, Y. (1999) Paradoxical decrease of an adipose-specific protein, adiponectin, in obesity. *Biochemical and Biophysical Research Communications* 257, 79–83.

Arner, P. (2001) Free fatty acids – do they play a central role in type 2 diabetes? *Diabetes, Obesity and Metabolism* 3 (Suppl. 1), S11–19.

Arner, P. (2005) Resistin: yet another adipokine tells us that men are not mice. *Diabetologia* 48, 2203–2205.

Arner, P. (2006) Visfatin – a true or false trail to type 2 diabetes mellitus. *Journal of Clinical Endocrinology and Metabolism* 91, 28–30.

Azuma, K., Katsukawa, F., Oguchi, S., Murata, M., Yamazaki, H., Shimada, A. and Saruta, T. (2003) Correlation between serum resistin level and adiposity in obese individuals. *Obesity Research* 11, 997–1001.

Banerjee, R.R., Rangwala, S.M., Shapiro, J.S., Rich, A.S., Rhoades, B., Qi, Y., Wang, J., Rajala, M.W., Pocai, A., Scherer, P.E., Steppan, C.M., Ahima, R.S., Obici, S., Rossetti, L. and Lazar, M.A. (2004) Regulation of fasted blood glucose by resistin. *Science* 303, 1195–1198.

Barzilai, N., She, L., Liu, B.Q., Vuguin, P., Cohen, P., Wang, J. and Rossetti, L. (1999) Surgical removal of visceral fat reverses hepatic insulin resistance. *Diabetes* 48, 94–98.

Bastard, J.P., Maachi, M., Van Nhieu, J.T., Jardel, C., Bruckert, E., Grimaldi, A., Robert, J.J., Capeau, J. and Hainque, B. (2002) Adipose tissue IL-6 content correlates with resistance to insulin activation of glucose uptake both *in vivo* and *in vitro*. *Journal of Clinical Endocrinology and Metabolism* 87, 2084–2089.

Basu, R., Chandramouli, V., Dicke, B., Landau, B. and Rizza, R. (2005) Obesity and type 2 diabetes impair insulin-induced suppression of glycogenolysis as well as gluconeogenesis. *Diabetes* 54, 1942–1948.

Berg, A.H., Combs, T.P., Du, X., Brownlee, M. and Scherer, P.E. (2001) The adipocyte-secreted protein Acrp30 enhances hepatic insulin action. *Nature Medicine* 7, 947–953.

Berndt, J., Klöting, N., Kralisch, S., Kovacs, P., Fasshauer, M., Schön, M.R., Stumvoll, M. and Blüher, M. (2005) Plasma visfatin concentrations and fat depot-specific mRNA expression in humans. *Diabetes* 54, 2911–2916.

Boden, G. (2003) Effects of free fatty acids (FFA) on glucose metabolism: significance for insulin resistance and type 2 diabetes. *Experimental and Clinical Endocrinology and Diabetes* 111, 121–124.

Boden, G. (2005) Free fatty acids and insulin secretion in humans. *Current Diabetes Report* 5, 167–170.

Boden, G., She, P., Mozzoli, M., Cheung, P., Gumireddy, K., Reddy, P., Xiang, X., Luo, Z. and Ruderman, N. (2005) Free fatty acids produce insulin resistance and activate the proinflammatory nuclear factor-kappaB pathway in rat liver. *Diabetes* 54, 3458–3465.

Bokarewa, M., Nagaev, I., Dahlberg, L., Smith, U. and Tarkowski, A. (2005) Resistin, an adipokine with potent proinflammatory properties. *Journal of Immunology* 174, 5789–5795.

Broch, M., Vendrell, J., Ricart, W., Richart, C. and Fernández-Real, J.M. (2007) Circulating retinol-binding protein-4, insulin sensitivity, insulin secretion, and insulin disposition index in obese and non-obese subjects. *Diabetes Care* 30, 1802–1806.

Carey, A.L., Steinberg, G.R., Macaulay, S.L., Thomas, W.G., Holmes, A.G., Ramm, G., Prelovsek, O., Hohnen-Behrens, C., Watt, M.J., James, D.E., Kemp, B.E., Pedersen, B.K. and Febbraio, M.A. (2006) Interleukin-6 increases insulin-stimulated glucose disposal in humans and glucose uptake and fatty acid oxidation *in vitro* via AMP-activated protein kinase. *Diabetes* 55, 2688–2697.

Carvalheira, J.B.C., Ribeiro, E.B., Folli, F., Velloso, L.A. and Saad, M.J. (2003) Interactions between leptin and insulin signaling pathways differentially affects JAK-STAT and PI 3-kinase-mediated signaling in rat liver. *Biological Chemistry* 384, 151–159.

Cawthorn, W.P. and Sethi, J.K. (2008) TNF-alpha and adipocyte biology. *FEBS Letters* 582, 117–131.

Cawthorn, W.P., Heyd, F., Hegyi, K. and Sethi, J.K. (2007) Tumour necrosis factor-α inhibits adipogenesis via a β-catenin/TCF4(TCF7L2)-dependent pathway. *Cell Death and Differentiation* 14, 1361–1373.

Ceddia, R.B. (2005) Direct metabolic regulation in skeletal muscle and fat tissue by leptin: implications for glucose and fatty acids homeostasis. *International Journal of Obesity* 29, 1175–1183.

Ceddia, R.B., Koistinen, H.A., Zierath, J.R. and Sweeney, G. (2002) Analysis of paradoxical observations on the association between leptin and insulin resistance. *The FASEB Journal* 16, 1163–1176.

Chen, M.P., Chung, F.M., Chang, D.M., Tsai, J.C., Huang, H.F., Shin, S.J. and Lee, Y.J. (2006) Elevated plasma level of visfatin/pre-B cell colony-enhancing factor in patients with type 2 diabetes mellitus. *Journal of Clinical Endocrinology and Metabolism* 91, 295–299.

Chen, N.G., Swick, A.G. and Romsos, D.R. (1997) Leptin constrains acetylcholine-induced insulin secretion from pancreatic islets of *ob/ob* mice. *Journal of Clinical Investigation* 100, 1174–1179.

Cho, Y.M., Youn, B.S., Lee, H., Lee, N., Min, S.S., Kwak, S.H., Lee, H.K. and Park, K.S. (2006) Plasma retinol-binding protein-4 concentrations are elevated in human subjects with impaired glucose tolerance and type 2 diabetes. *Diabetes Care* 29, 2457–2461.

Clemmons, D.R. (2006a) Clinical utility of measurements of insulin-like growth factor 1. *Nature Clinical Practice Endocrinology and Metabolism* 2, 436–446.

Clemmons, D.R. (2006b) Involvement of insulin-like growth factor-I in the control of glucose homeostasis. *Current Opinion in Pharmacology* 6, 620–625.

Combs, T.P., Berg, A.H., Obici, S., Scherer, P.E. and Rossetti, L. (2001) Endogenous glucose production is inhibited by the adipose-derived protein Acrp30. *Journal of Clinical Investigation* 108, 1875–1881.

Combs, T.P., Wagner, J.A., Berger, J., Doebber, T., Wang, W.J., Zhang, B.B., Tanen, M., Berg, A.H., O'Rahilly, S., Savage, D.B., Chatterjee, K., Weiss, S., Larson, P.J., Gottesdiener, K.M., Gertz, B.J., Charron, M.J., Scherer, P.E. and Moller, D.E. (2002) Induction of adipocyte complement-related protein of 30 kilodaltons by PPARγ agonists: a potential mechanism of insulin sensitization. *Endocrinology* 143, 998–1007.

Cong, L., Chen, K., Li, J., Gao, P., Li, Q., Mi, S., Wu, X. and Zhao, A.Z. (2007) Regulation of adiponectin and leptin secretion and expression by insulin through a PI3K-PDE3B dependent mechanism in rat primary adipocytes. *Biochemical Journal* 403, 519–525.

Cota, D., Proulx, K. and Seeley, R.J. (2007) The role of CNS fuel sensing in energy and glucose regulation. *Gastroenterology* 132, 2158–2168.

Covey, S.D., Wideman, R.D., McDonald, C., Unniappan, S., Huynh, F., Asadi, A., Speck, M., Webber, T., Chua, S.C. and Kieffer, T.J. (2006) The pancreatic β cell is a key site for mediating the effects of leptin on glucose homeostasis. *Cell Metabolism* 4, 291–302.

Cryer, P.E. (2003) Glucose homeostasis and hypoglycemia. In: Larsen, P.R., Kronenberg, H.M., Melmed, S. and Polonsky, K.S. (eds) *Williams Textbook of Endocrinology.* Saunders, Philadelphia, Pennsylvania, pp. 1585–1618.

Danforth, E. Jr (2000) Failure of adipocyte differentiation causes type II diabetes mellitus? *Nature Genetics* 26, 13.

Degawa-Yamauchi, M., Bovenkerk, J.E., Juliar, B.E., Watson, W., Kerr, K., Jones, R., Zhu, Q. and Considine, R.V. (2003) Serum resistin (FIZZ3) protein is increased in obese humans. *Journal of Clinical Endocrinology and Metabolism* 88, 5452–5455.

Dietze-Schroeder, D., Sell, H., Uhlig, M., Koenen, M. and Eckel, J. (2005) Autocrine action of adiponectin on human fat cells prevents the release of insulin resistance-inducing factors. *Diabetes* 54, 2003–2011.

Drucker, D.J. (2005) Biologic actions and therapeutic potential of the proglucagon-derived peptides. *Nature Clinical Practice Endocrinology and Metabolism* 1, 22–31.

Dube, J.J., Bhatt, B.A., Dedousis, N., Bonen, A. and O'Doherty, R.M. (2007) Leptin, skeletal muscle lipids, and lipid-induced insulin resistance. *American Journal of Physiology – Regulatory Integrative and Comparative Physiology* 293, R642–650.

Duvillié, B., Cordonnier, N., Deltour, L., Dandoy-Dron, F., Itier, J.M., Monthioux, E., Jami, J., Joshi, R.L. and Bucchini, D. (1997) Phenotypic alterations in insulin-deficient mutant mice. *Proceedings of the National Academy of Sciences of the United States of America* 94, 5137–5140.

Engeli, S., Schling, P., Gorzelniak, K., Boschmann, M., Janke, J., Ailhaud, G., Teboul, M., Massiéra, F. and Sharma, A.M. (2003) The adipose-tissue renin-angiotensin-aldosterone system: role in the metabolic syndrome? *International Journal of Biochemistry and Cell Biology* 35, 807–825.

Farooqi, I.S., Jebb, S.A., Langmack, G., Lawrence, E., Cheetham, C.H., Prentice, A.M., Hughes, I.A., McCamish, M.A. and O'Rahilly, S. (1999) Effects of recombinant leptin therapy in a child with congenital leptin deficiency. *New England Journal of Medicine* 341, 879–884.

Fasshauer, M., Klein, J., Neumann, S., Eszlinger, M. and Paschke, R. (2002) Hormonal regulation of adiponectin gene expression in 3T3-L1 adipocytes. *Biochemical and Biophysical Research Communications* 290, 1084–1089.

Fehmann, H.C. and Heyn, J. (2002) Plasma resistin levels in patients with type 1 and type 2 diabetes mellitus and in healthy controls. *Hormone and Metabolic Research* 34, 671–673.

Fishman, S., Muzumdar, R.H., Atzmon, G., Ma, X., Yang, X., Einstein, F.H. and Barzilai, N. (2007) Resistance to leptin action is the major determinant of hepatic triglyceride accumulation *in vivo*. *FASEB Journal* 21, 53–60.

Fruebis, J., Tsao, T.S., Javorschi, S., Ebbets Reed, D., Erickson, M.R., Yen, F.T., Bihain, B.E. and Lodish, H.F. (2001) Proteolytic cleavage product of 30-kDa adipocyte complement-related protein increases fatty acid oxidation in muscle and causes weight loss in mice. *Proceedings of the National Academy of Sciences of the United States of America* 98, 2005–2010.

Frühbeck, G. (2006) Intracellular signalling pathways activated by leptin. *Biochemical Journal* 393, 7–20.

Frühbeck, G. and Gómez-Ambrosi, J. (2003) Control of body weight: a physiologic and transgenic perspective. *Diabetologia* 46, 143–172.

Frühbeck, G. and Gómez-Ambrosi, J. (2005) Adipose tissue. In: Caballero, B., Allen, L. and Prentice, A. (eds) *Encyclopedia of Human Nutrition 2nd Edition*, Volume 1. Elsevier Ltd., Oxford, UK, pp. 1–14.

Frühbeck, G. and Salvador, J. (2000) Relations between leptin and the regulation of glucose metabolism. *Diabetologia* 43, 3–12.

Frühbeck, G. and Salvador, J. (2004) Role of adipocytokines in metabolism and disease. *Nutrition Research* 24, 803–826.

Frühbeck, G., Aguado, M., Gómez-Ambrosi, J. and Martínez, J.A. (1998a) Lipolytic effect of *in vivo* leptin administration on adipocytes of lean and *ob/ob* mice, but not *db/db* mice. *Biochemical and Biophysical Research Communications* 250, 99–102.

Frühbeck, G., Jebb, S.A. and Prentice, A.M. (1998b) Leptin: physiology and pathophysiology. *Clinical Physiology* 18, 399–419.

Frühbeck, G., Gómez-Ambrosi, J. and Martínez, J.A. (1999) Pre- and postprandial expression of the leptin receptor splice variants OB-Ra and OB-Rb in murine peripheral tissues. *Physiological Research* 48, 189–195.

Frühbeck, G., Gómez-Ambrosi, J., Muruzábal, F.J. and Burrell, M.A. (2001) The adipocyte: a model for integration of endocrine and metabolic signaling in energy metabolism regulation. *American Journal of Physiology – Endocrinology and Metabolism* 280, E827–847.

Frystyk, J., Vestbo, E., Skjaerbaek, C., Mogensen, C.E. and Orskov, H. (1995) Free insulin-like growth factors in human obesity. *Metabolism: Clinical and Experimental* 44, 37–44.

Fu, Y., Luo, N., Klein, R.L. and Garvey, W.T. (2005) Adiponectin promotes adipocyte differentiation, insulin sensitivity, and lipid accumulation. *Journal of Lipid Research* 46, 1369–1379.

Fukuhara, A., Matsuda, M., Nishizawa, M., Segawa, K., Tanaka, M., Kishimoto, K., Matsuki, Y., Murakami, M., Ichisaka, T., Murakami, H., Watanabe, E., Takagi, T., Akiyoshi, M., Ohtsubo, T., Kihara, S., Yamashita, S., Makishima, M., Funahashi, T., Yamanaka, S., Hiramatsu, R., Matsuzawa, Y. and Shimomura, I. (2005) Visfatin: a protein secreted by visceral fat that mimics the effects of insulin. *Science* 307, 426–430.

Fukuhara, A., Matsuda, M., Nishizawa, M., Segawa, K., Tanaka, M., Kishimoto, K., Matsuki, Y., Murakami, M., Ichisaka, T., Murakami, H., Watanabe, E., Takagi, T., Akiyoshi, M., Ohtsubo, T., Kihara, S., Yamashita, S., Makishima, M., Funahashi, T., Yamanaka, S., Hiramatsu, R., Matsuzawa, Y. and Shimomura, I. (2007) Visfatin: a protein secreted by visceral fat that mimics the effects of insulin. Retraction. *Science* 318, 565.

Gallardo, N., Bonzón-Kulichenko, E., Fernández-Agulloó, T., Moltó, E., Gómez-Alonso, S., Blanco, P., Carrascosa, J.M., Ros, M. and Andrés, A. (2007) Tissue-specific effects of central leptin on the expression of genes involved in lipid metabolism in liver and white adipose tissue. *Endocrinology* 148, 5604–5610.

Garg, A. (2004) Acquired and inherited lipodystrophies. *New England Journal of Medicine* 350, 1220–1234.

Gavrila, A., Chan, J.L., Yiannakouris, N., Kontogianni, M., Miller, L.C., Orlova, C. and Mantzoros, C.S. (2003) Serum adiponectin levels are inversely associated with overall and central fat distribution but are not directly regulated by acute fasting or leptin administration in humans: cross-sectional and interventional studies. *Journal of Clinical Endocrinology and Metabolism* 88, 4823–4831.

Gómez-Ambrosi, J. and Frühbeck, G. (2001) Do resistin and resistin-like molecules also link obesity to inflammatory diseases? *Annals of Internal Medicine* 135, 306–307.

Gómez-Ambrosi, J. and Frühbeck, G. (2005a) Resistin: a promising therapeutic target for the management of type 2 diabetes mellitus? *Drug Design Reviews – Online* 2, 1–12.

Gómez-Ambrosi, J. and Frühbeck, G. (2005b) Evidence for the involvement of resistin in inflammation and cardiovascular disease. *Current Diabetes Reviews* 1, 227–234.

Gómez-Ambrosi, J., Rodríguez, A., Catalán, V., Ramírez, B., Silva, C., Rotellar, F., Gil, M.J., Salvador, J. and Frühbeck, G. (2008) Serum retinol-binding protein 4 is not increased in obesity or obesity-associated type 2 diabetes mellitus, but is reduced after relevant reductions in body fat following gastric bypass. *Clinical Endocrinology* 69, 208–215.

Graham, T.E., Yang, Q., Blüher, M., Hammarstedt, A., Ciaraldi, T.P., Henry, R.R., Wason, C.J., Oberbach, A., Jansson, P.A., Smith, U. and Kahn, B.B. (2006) Retinol-binding protein 4 and insulin resistance in lean, obese, and diabetic subjects. *New England Journal of Medicine* 354, 2552–2563.

Guilherme, A., Virbasius, J.V., Puri, V. and Czech, M.P. (2008) Adipocyte dysfunctions linking obesity to insulin resistance and type 2 diabetes. *Nature Reviews Molecular Cell Biology* 9, 367–377.

Halleux, C.M., Takahashi, M., Delporte, M.L., Detry, R., Funahashi, T., Matsuzawa, Y. and Brichard, S.M. (2001) Secretion of adiponectin and regulation of apM1 gene expression in human visceral adipose tissue. *Biochemical and Biophysical Research Communications* 288, 1102–1107.

Haque, W.A., Shimomura, I., Matsuzawa, Y. and Garg, A. (2002) Serum adiponectin and leptin levels in patients with lipodystrophies. *Journal of Clinical Endocrinology and Metabolism* 87, 2395–2398.

Heilbronn, L.K., Rood, J., Janderova, L., Albu, J.B., Kelley, D.E., Ravussin, E. and Smith, S.R. (2004) Relationship between serum resistin concentrations and insulin resistance in non-obese, obese, and obese diabetic subjects. *Journal of Clinical Endocrinology and Metabolism* 89, 1844–1848.

Hida, K., Wada, J., Eguchi, J., Zhang, H., Baba, M., Seida, A., Hashimoto, I., Okada, T., Yasuhara, A., Nakatsuka, A., Shikata, K., Hourai, S., Futami, J., Watanabe, E., Matsuki, Y., Hiramatsu, R., Akagi, S., Makino, H. and Kanwar, Y.S. (2005) Visceral adipose tissue-derived serine protease inhibitor: a unique insulin-sensitizing adipocytokine in obesity. *Proceedings of the National Academy of Sciences of the United States of America* 102, 10610–10615.

Hotamisligil, G.S. (2006) Inflammation and metabolic disorders. *Nature* 444, 860–867.

Hotta, K., Funahashi, T., Arita, Y., Takahashi, M., Matsuda, M., Okamoto, Y., Iwahashi, H., Kuriyama, H., Ouchi, N., Maeda, K., Nishida, M., Kihara, S., Sakai, N., Nakajima, T., Hasegawa, K., Muraguchi, M., Ohmoto, Y., Nakamura, T., Yamashita, S., Hanafusa, T. and Matsuzawa, Y. (2000) Plasma concentrations of a novel, adipose-specific protein, adiponectin, in type 2 diabetic patients. *Arteriosclerosis, Thrombosis, and Vascular Biology* 20, 1595–1599.

Huan, J.N., Li, J., Han, Y., Chen, K., Wu, N. and Zhao, A.Z. (2003) Adipocyte-selective reduction of the leptin receptors induced by antisense-RNA leads to increased adiposity, dyslipidemia and insulin resistance. *Journal of Biological Chemistry* 278, 45638–45650.

Hug, C., Wang, J., Ahmad, N.S., Bogan, J.S., Tsao, T.S. and Lodish, H.F. (2004) T-cadherin is a receptor for hexameric and high-molecular-weight forms of Acrp30/adiponectin. *Proceedings of the National Academy of Sciences of the United States of America* 101, 10308–10313.

Huypens, P., Moens, K., Heimberg, H., Ling, Z., Pipeleers, D. and Van de Casteele, M. (2005) Adiponectin-mediated stimulation of AMP-activated protein kinase (AMPK) in pancreatic beta cells. *Life Sciences* 77, 1273–1282.

Ishizuka, K., Usui, I., Kanatani, Y., Bukhari, A., He, J., Fujisaka, S., Yamazaki, Y., Suzuki, H., Hiratani, K., Ishiki, M., Iwata, M., Urakaze, M., Haruta, T. and Kobayashi, M. (2007) Chronic tumor necrosis factor-α treatment causes insulin resistance via insulin receptor substrate-1 serine phosphorylation and suppressor of cytokine signaling-3 induction in 3T3-L1 adipocytes. *Endocrinology* 148, 2994–3003.

Janke, J., Engeli, S., Boschmann, M., Adams, F., Bohnke, J., Luft, F.C., Sharma, A.M. and Jordan, J. (2006) Retinol-binding protein 4 in human obesity. *Diabetes* 55, 2805–2810.

Jiang, G. and Zhang, B.B. (2003) Glucagon and regulation of glucose metabolism. *American Journal of Physiology – Endocrinology and Metabolism* 284, E671–678.

Kadowaki, T. and Yamauchi, T. (2005) Adiponectin and adiponectin receptors. *Endocrine Reviews* 26, 439–451.

Kadowaki, T., Yamauchi, T., Kubota, N., Hara, K., Ueki, K. and Tobe, K. (2006) Adiponectin and adiponectin receptors in insulin resistance, diabetes, and the metabolic syndrome. *Journal of Clinical Investigation* 116, 1784–1792.

Kahn, B.B., Alquier, T., Carling, D. and Hardie, D.G. (2005) AMP-activated protein kinase: ancient energy gauge provides clues to modern understanding of metabolism. *Cell Metabolism* 1, 15–25.

Kahn, S.E., Hull, R.L. and Utzschneider, K.M. (2006) Mechanisms linking obesity to insulin resistance and type 2 diabetes. *Nature* 444, 840–846.

Kaser, S., Kaser, A., Sandhofer, A., Ebenbichler, C.F., Tilg, H. and Patsch, J.R. (2003) Resistin messenger-RNA expression is increased by proinflammatory cytokines *in vitro*. *Biochemical and Biophysical Research Communications* 309, 286–290.

Katovich, M.J. and Pachori, A. (2000) Effects of inhibition of the renin–angiotensin system on the cardiovascular actions of insulin. *Diabetes, Obesity and Metabolism* 2, 3–14.

Kazumi, T., Kawaguchi, A., Sakai, K., Hirano, T. and Yoshino, G. (2002) Young men with high-normal blood pressure have lower serum adiponectin, smaller LDL size, and higher elevated heart rate than those with optimal blood pressure. *Diabetes Care* 25, 971–976.

Kellerer, M., Koch, M., Metzinger, E., Mushack, J., Capp, E. and Haring, H.U. (1997) Leptin activates PI-3 kinase in C2C12 myotubes via janus kinase-2 (JAK-2) and insulin receptor substrate-2 (IRS-2) dependent pathways. *Diabetologia* 40, 1358–1362.

Kharroubi, I., Rasschaert, J., Eizirik, D.L. and Cnop, M. (2003) Expression of adiponectin receptors in pancreatic β cells. *Biochemical and Biophysical Research Communications* 312, 1118–1122.

Kieffer, T.J. and Habener, J.F. (2000) The adipoinsular axis: effects of leptin on pancreatic β-cells. *American Journal of Physiology – Endocrinology and Metabolism* 278, E1–E14.

Kieffer, T.J., Heller, R.S. and Habener, J.F. (1996) Leptin receptors expressed on pancreatic β-cells. *Biochemical and Biophysical Research Communications* 224, 522–527.

Kieffer, T.J., Heller, R.S., Leech, C.A., Holz, G.G. and Habener, J.F. (1997) Leptin suppression of insulin secretion by the activation of ATP-sensitive K+ channels in pancreatic β-cells. *Diabetes* 46, 1087–1093.

Kim, H.J., Higashimori, T., Park, S.Y., Choi, H., Dong, J., Kim, Y.J., Noh, H.L., Cho, Y.R., Cline, G., Kim, Y.B. and Kim, J.K. (2004a) Differential effects of interleukin-6 and -10 on skeletal muscle and liver insulin action *in vivo*. *Diabetes* 53, 1060–1067.

Kim, J.B., Sarraf, P., Wright, M., Yao, K.M., Mueller, E., Solanes, G., Lowell, B.B. and Spiegelman, B.M. (1998) Nutritional and insulin regulation of fatty acid synthetase and leptin gene expression through ADD1/SREBP1. *Journal of Clinical Investigation* 101, 1–9.

Kim, K.H., Zhao, L., Moon, Y., Kang, C. and Sul, H.S. (2004b) Dominant inhibitory adipocyte-specific secretory factor (ADSF)/resistin enhances adipogenesis and improves insulin sensitivity. *Proceedings of the National Academy of Sciences of the United States of America* 101, 6780–6785.

Klöting, N., Berndt, J., Kralisch, S., Kovacs, P., Fasshauer, M., Schön, M.R., Stumvoll, M. and Blüher, M. (2006) Vaspin gene expression in human adipose tissue: association with obesity and type 2 diabetes. *Biochemical and Biophysical Research Communications* 339, 430–436.

Koeslag, J.H., Saunders, P.T. and Terblanche, E. (2003) A reappraisal of the blood glucose homeostat which comprehensively explains the type 2 diabetes mellitus–syndrome X complex. *Journal of Physiology* 549, 333–346.

Kubota, N., Terauchi, Y., Yamauchi, T., Kubota, T., Moroi, M., Matsui, J., Eto, K., Yamashita, T., Kamon, J., Satoh, H., Yano, W., Froguel, P., Nagai, R., Kimura, S., Kadowaki, T. and Noda, T. (2002) Disruption of adiponectin causes insulin resistance and neointimal formation. *Journal of Biological Chemistry* 277, 25863–25866.

Kusminski, C.M., McTernan, P.G. and Kumar, S. (2005) Role of resistin in obesity, insulin resistance and type II diabetes. *Clinical Science* 109, 243–256.

Kusminski, C.M., da Silva, N.F., Creely, S.J., Fisher, F.M., Harte, A.L., Baker, A.R., Kumar, S. and McTernan, P.G. (2007) The *in vitro* effects of resistin on the innate immune signaling pathway in isolated human subcutaneous adipocytes. *Journal of Clinical Endocrinology and Metabolism* 92, 270–276.

Le Lay, S., Boucher, J., Rey, A., Castan-Laurell, I., Krief, S., Ferré, P., Valet, P. and Dugail, I. (2001) Decreased resistin expression in mice with different sensitivities to a high-fat diet. *Biochemical and Biophysical Research Communications* 289, 564–567.

Le Marchand-Brustel, Y., Heydrick, S.J., Jullien, D., Gautier, N. and Van Obberghen, E. (1995) Effect of insulin and insulin-like growth factor-I on glucose transport and its transporters in soleus muscle of lean and obese mice. *Metabolism: Clinical and Experimental* 44, 18–23.

Lee, J.H., Chan, J.L., Yiannakouris, N., Kontogianni, M., Estrada, E., Seip, R., Orlova, C. and Mantzoros, C.S. (2003) Circulating resistin levels are not associated with obesity or insulin resistance in humans and are not regulated by fasting or leptin administration: cross-sectional and interventional studies in normal, insulin-resistant, and diabetic subjects. *Journal of Clinical Endocrinology and Metabolism* 88, 4848–4856.

Lehrke, M., Reilly, M.P., Millington, S.C., Iqbal, N., Rader, D.J. and Lazar, M.A. (2004) An inflammatory cascade leading to hyperresistinemia in humans. *PLoS Medicine* 1, e45.

Licinio, J., Caglayan, S., Ozata, M., Yildiz, B.O., de Miranda, P.B., O'Kirwan, F., Whitby, R., Liang, L., Cohen, P., Bhasin, S., Krauss, R.M., Veldhuis, J.D., Wagner, A.J., DePaoli, A.M., McCann, S.M. and Wong, M.L. (2004) Phenotypic effects of leptin replacement on morbid obesity, diabetes mellitus, hypogonadism, and behavior in leptin-deficient adults. *Proceedings of the National Academy of Sciences of the United States of America* 101, 4531–4536.

Lin, J., Handschin, C. and Spiegelman, B.M. (2005) Metabolic control through the PGC-1 family of transcription coactivators. *Cell Metabolism* 1, 361–370.

Long, Y.C. and Zierath, J.R. (2006) AMP-activated protein kinase signaling in metabolic regulation. *Journal of Clinical Investigation* 116, 1776–1783.

McTernan, P.G., Fisher, F.M., Valsamakis, G., Chetty, R., Harte, A., McTernan, C.L., Clark, P.M., Smith, S.A., Barnett, A.H. and Kumar, S. (2003) Resistin and type 2 diabetes: regulation of resistin expression by insulin and rosiglitazone and the effects of recombinant resistin on lipid and glucose metabolism in human differentiated adipocytes. *Journal of Clinical Endocrinology and Metabolism* 88, 6098–6106.

McTernan, P.G., Kusminski, C.M. and Kumar, S. (2006) Resistin. *Current Opinion in Lipidology* 17, 170–175.

Maeda, N., Shimomura, I., Kishida, K., Nishizawa, H., Matsuda, M., Nagaretani, H., Furuyama, N., Kondo, H., Takahashi, M., Arita, Y., Komuro, R., Ouchi, N., Kihara, S., Tochino, Y., Okutomi, K., Horie, M., Takeda, S., Aoyama, T., Funahashi, T. and Matsuzawa, Y. (2002) Diet-induced insulin resistance in mice lacking adiponectin/ACRP30. *Nature Medicine* 8, 731–737.

Mallamaci, F., Zoccali, C., Cuzzola, F., Tripepi, G., Cutrupi, S., Parlongo, S., Tanaka, S., Ouchi, N., Kihara, S., Funahashi, T. and Matsuzawa, Y. (2002) Adiponectin in essential hypertension. *Journal of Nephrology* 15, 507–511.

Matsubara, M., Maruoka, S. and Katayose, S. (2002a) Inverse relationship between plasma adiponectin and leptin concentrations in normal-weight and obese women. *European Journal of Endocrinology* 147, 173–180.

Matsubara, M., Maruoka, S. and Katayose, S. (2002b) Decreased plasma adiponectin concentrations in women with dyslipidemia. *Journal of Clinical Endocrinology and Metabolism* 87, 2764–2769.

Matsuzawa, Y., Funahashi, T., Kihara, S. and Shimomura, I. (2004) Adiponectin and metabolic syndrome. *Arteriosclerosis, Thrombosis, and Vascular Biology* 24, 29–33.

Minokoshi, Y., Kim, Y.B., Peroni, O.D., Fryer, L.G., Muller, C., Carling, D. and Kahn, B.B. (2002) Leptin stimulates fatty-acid oxidation by activating AMP-activated protein kinase. *Nature* 415, 339–343.

Morioka, T., Asilmaz, E., Hu, J., Dishinger, J.F., Kurpad, A.J., Elias, C.F., Li, H., Elmquist, J.K., Kennedy, R.T. and Kulkarni, R.N. (2007) Disruption of leptin receptor expression in the pancreas directly affects beta cell growth and function in mice. *Journal of Clinical Investigation* 117, 2860–2868.

Moschen, A.R., Kaser, A., Enrich, B., Mosheimer, B., Theurl, M., Niederegger, H. and Tilg, H. (2007) Visfatin, an adipocytokine with proinflammatory and immunomodulating properties. *Journal of Immunology* 178, 1748–1758.

Mueller, W.M., Gregoire, F.M., Stanhope, K.L., Mobbs, C.V., Mizuno, T.M., Warden, C.H., Stern, J.S. and Havel, P.J. (1998) Evidence that glucose metabolism regulates leptin secretion from cultured rat adipocytes. *Endocrinology* 139, 551–558.

Muoio, D.M. and Newgard, C.B. (2008) Molecular and metabolic mechanisms of insulin resistance and β-cell failure in type 2 diabetes. *Nature Reviews Molecular Cell Biology* 9, 193–205.

Murakami, T., Otani, S., Honjoh, T., Doi, T. and Shima, K. (2001) Influence of the presence of OB-Re on leptin radioimmunoassay. *Journal of Endocrinology* 168, 79–86.

Murphy, L.J. (2003) The role of the insulin-like growth factors and their binding proteins in glucose homeostasis. *Experimental Diabetes Research* 4, 213–224.

Muse, E.D., Obici, S., Bhanot, S., Monia, B.P., McKay, R.A., Rajala, M.W., Scherer, P.E. and Rossetti, L. (2004) Role of resistin in diet-induced hepatic insulin resistance. *Journal of Clinical Investigation* 114, 232–239.

Muzumdar, R., Ma, X., Yang, X., Atzmon, G., Bernstein, J., Karkanias, G. and Barzilai, N. (2003) Physiologic effect of leptin on insulin secretion is mediated mainly through central mechanisms. *The FASEB Journal* 17, 1130–1132.

Myers, M.G., Cowley, M.A. and Münzberg, H. (2008) Mechanisms of leptin action and leptin resistance. *Annual Review of Physiology* 70, 537–556.

Nemecz, M., Preininger, K., Englisch, R., Fürnsinn, C., Schneider, B., Waldhäusl, W. and Roden, M. (1999) Acute effect of leptin on hepatic glycogenolysis and gluconeogenesis in perfused rat liver. *Hepatology* 29, 166–172.

Nolan, C.J., Madiraju, M.S., Delghingaro-Augusto, V., Peyot, M.L. and Prentki, M. (2006) Fatty acid signaling in the β-cell and insulin secretion. *Diabetes* 55 (Suppl. 2), S16–23.

Ookuma, M., Ookuma, K. and York, D.A. (1998) Effects of leptin on insulin secretion from isolated rat pancreatic islets. *Diabetes* 47, 219–223.

Oral, E.A., Simha, V., Ruiz, E., Andewelt, A., Premkumar, A., Snell, P., Wagner, A.J., DePaoli, A.M., Reitman, M.L., Taylor, S.I., Gorden, P. and Garg, A. (2002) Leptin-replacement therapy for lipodystrophy. *New England Journal of Medicine* 346, 570–578.

Orci, L., Cook, W.S., Ravazzola, M., Wang, M.Y., Park, B.H., Montesano, R. and Unger, R.H. (2004) Rapid transformation of white adipocytes into fat-oxidizing machines. *Proceedings of the National Academy of Sciences of the United States of America* 101, 2058–2063.

Ouchi, N., Kihara, S., Arita, Y., Maeda, K., Kuriyama, H., Okamoto, Y., Hotta, K., Nishida, M., Takahashi, M., Nakamura, T., Yamashita, S., Funahashi, T. and Matsuzawa, Y. (1999) Novel modulator for endothelial adhesion molecules: adipocyte-derived plasma protein adiponectin. *Circulation* 100, 2473–2476.

Pérez, C., Fernández-Galaz, C., Fernández-Agullo, T., Arribas, C., Andrés, A., Ros, M. and Carrascosa, J.M. (2004) Leptin impairs insulin signaling in rat adipocytes. *Diabetes* 53, 347–353.

Petersen, K.F., Oral, E.A., Dufour, S., Befroy, D., Ariyan, C., Yu, C., Cline, G.W., DePaoli, A.M., Taylor, S.I., Gorden, P. and Shulman, G.I. (2002) Leptin reverses insulin resistance and hepatic steatosis in patients with severe lipodystrophy. *Journal of Clinical Investigation* 109, 1345–1350.

Qi, Y., Nie, Z., Lee, Y.-S., Singhal, N.S., Scherer, P.E., Lazar, M.A. and Ahima, R.S. (2006) Loss of resistin improves glucose homeostasis in leptin deficiency. *Diabetes* 55, 3083–3090.

Rajala, M.W., Obici, S., Scherer, P.E. and Rossetti, L. (2003) Adipose-derived resistin and gut-derived resistin-like molecule-β selectively impair insulin action on glucose production. *Journal of Clinical Investigation* 111, 225–230.

Rajala, M.W., Qi, Y., Patel, H.R., Takahashi, N., Banerjee, R., Pajvani, U.B., Sinha, M.K., Gingerich, R.L., Scherer, P.E. and Ahima, R.S. (2004) Regulation of resistin expression and circulating levels in obesity, diabetes, and fasting. *Diabetes* 53, 1671–1679.

Rakatzi, I., Mueller, H., Ritzeler, O., Tennagels, N. and Eckel, J. (2004) Adiponectin counteracts cytokine- and fatty acid-induced apoptosis in the pancreatic beta-cell line INS-1. *Diabetologia* 47, 249–258.

Rangwala, S.M., Rich, A.S., Rhoades, B., Shapiro, J.S., Obici, S., Rossetti, L. and Lazar, M.A. (2004) Abnormal glucose homeostasis due to chronic hyperresistinemia. *Diabetes* 53, 1937–1941.

Reilly, M.P., Lehrke, M., Wolfe, M.L., Rohatgi, A., Lazar, M.A. and Rader, D.J. (2005) Resistin is an inflammatory marker of atherosclerosis in humans. *Circulation* 111, 932–939.

Roden, M., Price, T.B., Perseghin, G., Petersen, K.F., Rothman, D.L., Cline, G.W. and Shulman, G.I. (1996) Mechanism of free fatty acid-induced insulin resistance in humans. *Journal of Clinical Investigation* 97, 2859–2865.

Rosen, E.D. and Spiegelman, B.M. (2006) Adipocytes as regulators of energy balance and glucose homeostasis. *Nature* 444, 847–853.

Rother, E., Könner, A.C. and Brüning, J. (2008) Neurocircuits integrating hormone and nutrient signaling in control of glucose metabolism. *American Journal of Physiology – Endocrinology and Metabolism* 294, E810–816.

Saltiel, A.R. and Kahn, C.R. (2001) Insulin signalling and the regulation of glucose and lipid metabolism. *Nature* 414, 799–806.

Satoh, H., Nguyen, M.T., Miles, P.D., Imamura, T., Usui, I. and Olefsky, J.M. (2004) Adenovirus-mediated chronic 'hyper-resistinemia' leads to *in vivo* insulin resistance in normal rats. *Journal of Clinical Investigation* 114, 224–231.

Schäffler, A., Schölmerich, J. and Büchler, C. (2005) Mechanisms of disease: adipocytokines and visceral adipose tissue – emerging role in non-alcoholic fatty liver disease. *Nature Clinical Practice Gastroenterology and Hepatology* 2, 273–280.

Segal, K.R., Landt, M. and Klein, S. (1996) Relationship between insulin sensitivity and plasma leptin concentration in lean and obese men. *Diabetes* 45, 988–991.

Sell, H., Dietze-Schroeder, D. and Eckel, J. (2006) The adipocyte–myocyte axis in insulin resistance. *Trends in Endocrinology and Metabolism* 17, 416–422.

Senn, J.J., Klover, P.J., Nowak, I.A., Zimmers, T.A., Koniaris, L.G., Furlanetto, R.W. and Mooney, R.A. (2003) Suppressor of cytokine signaling-3 (SOCS-3), a potential mediator of interleukin-6-dependent insulin resistance in hepatocytes. *Journal of Biological Chemistry* 278, 13740–13746.

Sethi, J.K. and Vidal-Puig, A. (2005) Visfatin: the missing link between intra-abdominal obesity and diabetes? *Trends in Molecular Medicine* 11, 344–347.

Seufert, J. (2004) Leptin effects on pancreatic β-cell gene expression and function. *Diabetes* 53 (Suppl. 1), S152–158.

Shimomura, I., Hammer, R.E., Ikemoto, S., Brown, M.S. and Goldstein, J.L. (1999) Leptin reverses insulin resistance and diabetes mellitus in mice with congenital lipodystrophy. *Nature* 401, 73–76.

Siegrist-Kaiser, C.A., Pauli, V., Juge-Aubry, C.E., Boss, O., Pernin, A., Chin, W.W., Cusin, I., Rohner-Jeanrenaud, F., Burger, A.G., Zapf, J. and Meier, C.A. (1997) Direct effects of leptin on brown and white adipose tissue. *Journal of Clinical Investigation* 100, 2858–2864.

Silha, J.V., Krsek, M., Skrha, J.V., Sucharda, P., Nyomba, B.L. and Murphy, L.J. (2003) Plasma resistin, adiponectin and leptin levels in lean and obese subjects: correlations with insulin resistance. *European Journal of Endocrinology* 149, 331–335.

Silswal, N., Singh, A.K., Aruna, B., Mukhopadhyay, S., Ghosh, S. and Ehtesham, N.Z. (2005) Human resistin stimulates the pro-inflammatory cytokines TNF-α and IL-12 in macrophages by NF-κB-dependent pathway. *Biochemical and Biophysical Research Communications* 334, 1092–1101.

Spranger, J., Kroke, A., Mohlig, M., Bergmann, M.M., Ristow, M., Boeing, H. and Pfeiffer, A.F. (2003) Adiponectin and protection against type 2 diabetes mellitus. *Lancet* 361, 226–228.

Steinberg, G.R., Dyck, D.J., Calles-Escandon, J., Tandon, N.N., Luiken, J.J., Glatz, J.F. and Bonen, A. (2002a) Chronic leptin administration decreases fatty acid uptake and fatty acid transporters in rat skeletal muscle. *Journal of Biological Chemistry* 277, 8854–8860.

Steinberg, G.R., Parolin, M.L., Heigenhauser, G.J.F. and Dyck, D.J. (2002b) Leptin increases FA oxidation in lean but not obese human skeletal muscle: evidence of peripheral leptin resistance. *American Journal of Physiology – Endocrinology and Metabolism* 283, E187–192.

Steinberg, G.R., Michell, B.J., van Denderen, B.J., Watt, M.J., Carey, A.L., Fam, B.C., Andrikopoulos, S., Proietto, J., Gorgun, C.Z., Carling, D., Hotamisligil, G.S., Febbraio, M.A., Kay, T.W. and Kemp, B.E. (2006) Tumor necrosis factor α-induced skeletal muscle insulin resistance involves suppression of AMP-kinase signaling. *Cell Metabolism* 4, 465–474.

Steppan, C.M., Bailey, S.T., Bhat, S., Brown, E.J., Banerjee, R.R., Wright, C.M., Patel, H.R., Ahima, R.S. and Lazar, M.A. (2001) The hormone resistin links obesity to diabetes. *Nature* 409, 307–312.

Stumvoll, M., Goldstein, B.J. and van Haeften, T.W. (2005) Type 2 diabetes: principles of pathogenesis and therapy. *Lancet* 365, 1333–1346.

Szanto, I. and Kahn, C.R. (2000) Selective interaction between leptin and insulin signaling pathways in a hepatic cell line. *Proceedings of the National Academy of Sciences of the United States of America* 97, 2355–2360.

Tartaglia, L.A., Dembski, M., Weng, X., Deng, N., Culpepper, J., Devos, R., Richards, G.J., Campfield, L.A., Clark, F.T., Deeds, J., Muir, C., Sanker, S., Moriarty, A., Moore, K.J., Smutko, J.S., Mays, G.G., Woolf, E.A., Monroe, C.A. and Tepper, R.I. (1995) Identification and expression cloning of a leptin receptor, OB-R. *Cell* 83, 1263–1271.

Tilg, H. and Moschen, A.R. (2008) Inflammatory mechanisms in the regulation of insulin resistance. *Molecular Medicine* 14, 222–231.

Tomas, E., Tsao, T.S., Saha, A.K., Murrey, H.E., Zhang Cc, C., Itani, S.I., Lodish, H.F. and Ruderman, N.B. (2002) Enhanced muscle fat oxidation and glucose transport by ACRP30 globular domain: acetyl-CoA carboxylase inhibition and AMP-activated protein kinase activation. *Proceedings of the National Academy of Sciences of the United States of America* 99, 16309–16313.

Trujillo, M.E. and Scherer, P.E. (2005) Adiponectin – journey from an adipocyte secretory protein to biomarker of the metabolic syndrome. *Journal of Internal Medicine* 257, 167–175.

Trujillo, M.E. and Scherer, P.E. (2006) Adipose tissue-derived factors: impact on health and disease. *Endocrine Reviews* 27, 762–778.

Unger, R.H. (2003) The physiology of cellular liporegulation. *Annual Review of Physiology* 65, 333–347.

Utzschneider, K.M., Carr, D.B., Tong, J., Wallace, T.M., Hull, R.L., Zraika, S., Xiao, Q., Mistry, J.S., Retzlaff, B.M., Knopp, R.H. and Kahn, S.E. (2005) Resistin is not associated with insulin sensitivity or the metabolic syndrome in humans. *Diabetologia* 48, 2330–2333.

Walder, K., Filippis, A., Clark, S., Zimmet, P. and Collier, G.R. (1997) Leptin inhibits insulin binding in isolated rat adipocytes. *Journal of Endocrinology* 155, R5–7.

Wallenius, V., Wallenius, K., Ahren, B., Rudling, M., Carlsten, H., Dickson, S.L., Ohlsson, C. and Jansson, J.O. (2002) Interleukin-6-deficient mice develop mature-onset obesity. *Nature Medicine* 8, 75–79.

Wang, A.Y., Hickman, I.J., Richards, A.A., Whitehead, J.P., Prins, J.B. and Macdonald, G.A. (2005a) High molecular weight adiponectin correlates with insulin sensitivity in patients with hepatitis C genotype 3, but not genotype 1 infection. *American Journal of Gastroenterology* 100, 2717–2723.

Wang, J.L., Chinookoswong, N., Scully, S., Qi, M. and Shi, Z.Q. (1999a) Differential effects of leptin in regulation of tissue glucose utilization *in vivo*. *Endocrinology* 140, 2117–2124.

Wang, M.Y., Lee, Y. and Unger, R.H. (1999b) Novel form of lipolysis induced by leptin. *Journal of Biological Chemistry* 274, 17541–17544.

Wang, M.Y., Orci, L., Ravazzola, M. and Unger, R.H. (2005b) Fat storage in adipocytes requires inactivation of leptin's paracrine activity: implications for treatment of human obesity. *Proceedings of the National Academy of Sciences of the United States of America* 102, 18011–18016.

Wang, M.Y., Grayburn, P., Chen, S., Ravazzola, M., Orci, L. and Unger, R.H. (2008a) Adipogenic capacity and the susceptibility to type 2 diabetes and metabolic syndrome. *Proceedings of the National Academy of Sciences of the United States of America* 105, 6139–6144.

Wang, P.Y., Caspi, L., Lam, C.K., Chari, M., Li, X., Light, P.E., Gutierrez-Juarez, R., Ang, M., Schwartz, G.J. and Lam, T.K. (2008b) Upper intestinal lipids trigger a gut–brain–liver axis to regulate glucose production. *Nature* 452, 1012–1016.

Way, J.M., Görgün, C.Z., Tong, Q., Uysal, K.T., Brown, K.K., Harrington, W.W., Oliver, W.R. Jr, Willson, T.M., Kliewer, S.A. and Hotamisligil, G.S. (2001) Adipose tissue resistin expression is severely suppressed in obesity and stimulated by peroxisome proliferator-activated receptor γ agonists. *Journal of Biological Chemistry* 276, 25651–25653.

Weyer, C., Funahashi, T., Tanaka, S., Hotta, K., Matsuzawa, Y., Pratley, R.E. and Tataranni, P.A. (2001) Hypoadiponectinemia in obesity and type 2 diabetes: close association with insulin resistance and hyperinsulinemia. *Journal of Clinical Endocrinology and Metabolism* 86, 1930–1935.

Winzell, M.S., Nogueiras, R., Dieguez, C. and Ahrén, B. (2004) Dual action of adiponectin on insulin secretion in insulin-resistant mice. *Biochemical and Biophysical Research Communications* 321, 154–160.

Wu, X., Motoshima, H., Mahadev, K., Stalker, T.J., Scalia, R. and Goldstein, B.J. (2003) Involvement of AMP-activated protein kinase in glucose uptake stimulated by the globular domain of adiponectin in primary rat adipocytes. *Diabetes* 52, 1355–1363.

Yamauchi, T., Kamon, J., Waki, H., Terauchi, Y., Kubota, N., Hara, K., Mori, Y., Ide, T., Murakami, K., Tsuboyama Kasaoka, N., Ezaki, O., Akanuma, Y., Gavrilova, O., Vinson, C., Reitman, M.L., Kagechika, H., Shudo, K., Yoda, M., Nakano, Y., Tobe, K., Nagai, R., Kimura, S., Tomita, M., Froguel, P. and Kadowaki, T. (2001) The fat-derived hormone adiponectin reverses insulin resistance associated with both lipoatrophy and obesity. *Nature Medicine* 7, 941–946.

Yamauchi, T., Kamon, J., Minokoshi, Y., Ito, Y., Waki, H., Uchida, S., Yamashita, S., Noda, M., Kita, S., Ueki, K., Eto, K., Akanuma, Y., Froguel, P., Foufelle, F., Ferre, P., Carling, D., Kimura, S., Nagai, R., Kahn, B.B. and Kadowaki, T. (2002) Adiponectin stimulates glucose utilization and fatty-acid oxidation by activating AMP-activated protein kinase. *Nature Medicine* 8, 1288–1295.

Yamauchi, T., Kamon, J., Ito, Y., Tsuchida, A., Yokomizo, T., Kita, S., Sugiyama, T., Miyagishi, M., Hara, K., Tsunoda, M., Murakami, K., Ohteki, T., Uchida, S., Takekawa, S., Waki, H., Tsuno, N.H., Shibata, Y., Terauchi, Y., Froguel, P., Tobe, K., Koyasu, S., Taira, K., Kitamura, T., Shimizu, T., Nagai, R. and Kadowaki, T. (2003) Cloning of adiponectin receptors that mediate antidiabetic metabolic effects. *Nature* 423, 762–769.

Yamauchi, T., Nio, Y., Maki, T., Kobayashi, M., Takazawa, T., Iwabu, M., Okada-Iwabu, M., Kawamoto, S., Kubota, N., Kubota, T., Ito, Y., Kamon, J., Tsuchida, A., Kumagai, K., Kozono, H., Hada, Y., Ogata, H., Tokuyama, K., Tsunoda, M., Ide, T., Murakami, K., Awazawa, M., Takamoto, I., Froguel, P., Hara, K., Tobe, K., Nagai, R., Ueki, K. and Kadowaki, T. (2007) Targeted disruption of AdipoR1 and AdipoR2 causes abrogation of adiponectin binding and metabolic actions. *Nature Medicine* 13, 332–339.

Yang, Q., Graham, T.E., Mody, N., Preitner, F., Peroni, O.D., Zabolotny, J.M., Kotani, K., Quadro, L. and Kahn, B.B. (2005) Serum retinol binding protein 4 contributes to insulin resistance in obesity and type 2 diabetes. *Nature* 436, 356–362.

Yang, W.S., Lee, W.J., Funahashi, T., Tanaka, S., Matsuzawa, Y., Chao, C.L., Chen, C.L., Tai, T.Y. and Chuang, L.M. (2001) Weight reduction increases plasma levels of an adipose-derived anti-inflammatory protein, adiponectin. *Journal of Clinical Endocrinology and Metabolism* 86, 3815–3819.

Yaspelkis, B.B., 3rd, Singh, M.K., Krisan, A.D. and Collins, D.E. (2004) Chronic leptin administration enhances insulin-stimulated glucose disposal in skeletal muscle of high-fat fed rodents. *Life Sciences* 74, 1801–1816.

Yoon, M.J., Lee, G.Y., Chung, J.J., Ahn, Y.H., Hong, S.H. and Kim, J.B. (2006) Adiponectin increases fatty acid oxidation in skeletal muscle cells by sequential activation of AMP-activated protein kinase, p38 mitogen-activated protein kinase, and peroxisome proliferator-activated receptor alpha. *Diabetes* 55, 2562–2570.

Youn, B.S., Klöting, N., Kratzsch, J., Lee, N., Park, J.W., Song, E.S., Ruschke, K., Oberbach, A., Fasshauer, M., Stumvoll, M. and Blüher, M. (2008) Serum vaspin concentrations in human obesity and type 2 diabetes. *Diabetes* 57, 372–377.

Zhang, Y., Proenca, R., Maffei, M., Barone, M., Leopold, L. and Friedman, J.M. (1994) Positional cloning of the mouse obese gene and its human homologue. *Nature* 372, 425–432.

8 Adipokines in the Immune–Stress Response

Rana Madani, Nicola C. Ogston and Vidya Mohamed-Ali

Adipokines and Metabolism Research Group, Centre for Clinical Pharmacology, University College, London, UK

Obesity and the Immune System

Immunity refers to the defence mechanisms that come into play when an organism is confronted by any harmful stimuli. It is pivotal in the discrimination between infectious non-self and non-infectious self, with the decision to respond to a particular ligand being determined largely by the genome-encoded innate immune receptors. Inflammatory responses are part of innate immunity and cells that trigger this response include macrophages, polymorphonuclear leukocytes and mast cells that bind ligands through their immune receptors. These receptors may be expressed on the cell surface, in intracellular compartments or secreted into the systemic circulation. Their functions include opsonization, activation of complement and coagulation cascades, phagocytosis, activation of proinflammatory pathways and induction of apoptosis. The innate immune response is acute, self-limiting and aimed principally at restoring the balance disturbed by an acute stressor. However, in response to chronic stressors, the system's resources are overloaded and it breaks down, with maladaptive consequences.

Innate and cell-mediated immune responses are altered markedly by an organism's nutritional status. In mice and humans, nutritional factors, specifically during states of excess intake, function as chronic activators of the immune response and several lines of evidence suggest that obesity is an inflammatory condition (Chandra and Chandra, 1986). It is associated with certain non-specific markers of activation of the immune system, such as elevated body temperature and white blood cell count (Pratley *et al.*, 1995). Also, markers of low-grade inflammation, several of which are also adipose tissue-derived factors or adipokines, predict those at high risk for obesity-associated pathologies, such as type 2 diabetes (T2DM), coronary heart disease (CHD) and atherosclerosis (Ridker *et al.*, 1998, 2000; Festa *et al.*, 2000). Excess adipose tissue, especially in visceral depots, is associated with a cluster of metabolic disturbances such as insulin resistance, hyperinsulinaemia, glucose intolerance, hypertriglyceridaemia, as well as low high-density lipoprotein (HDL) cholesterol levels (Despres, 1998). The immune response frequently is accompanied by changes in glucose and lipid metabolism and the ability of the adipokines to mediate these is relevant. In this chapter, we outline the evidence for the contribution of the adipose tissue to both local and systemic elevations in concentrations of many of these adipokines and their role in the immune–stress response.

Cellular Composition of White Adipose Tissue

White adipose tissue (WAT) is composed of many cell types, with about 50% being adipocytes, a further 10% being $CD14^+CD31^+$ macrophages and the remaining 40% comprising preadipocytes, endothelial and epithelial cells. In obesity there is macrophage infiltration into WAT and the number of these macrophages is correlated directly to adipocyte size (Cousin *et al.*, 1999). Evidence shows that WAT macrophages are bone marrow-derived, rather than resident cells, and the increase of the cells into the tissue is, therefore, from circulating monocytes (Weisberg *et al.*, 2003). That these macrophages are also activated is supported by data showing they are large, multinucleated and secrete cytokines (Xu *et al.*, 2003). However, there is also ample evidence that preadipocytes themselves have many macrophage-like qualities, along with their ability to differentiate into adipocytes. For example, in response to lipopolysaccharide (LPS), levels of interleukin-6 (IL-6), tumour necrosis factor-alpha (TNF-α), IL-1β, IL-8 and monocyte chemoattractant protein-1 (MCP-1) mRNA are elevated primarily in the preadipocytes, as opposed to macrophages. Concomittantly, this treatment decreases adipogenic gene expression, PPARγ activity and insulin responsiveness in human adipocytes (Chung *et al.*, 2006). Thus, triggering the innate immune response in adipose tissue appears to favour both macrophage infiltration into the tissue and maintenance of a preadipocyte phenotype.

Lymphocytes, although not present in WAT, are often in close proximity to adipose tissue, particularly in lymph nodes which are surrounded by perivascular adipose tissue. The omentum and bone marrow are also sites of close association between adipose and lymphoid tissue. The lymphoid tissue there can take up lipolytic products released by adjacent adipocytes and, conversely, adipocytes in close proximity to lymph nodes (perinodal) are smaller than those more distal, with consequences for adipokine secretion (Pond and Mattacks, 1998).

Signals from WAT

It is rapidly becoming apparent that adipose tissue is a significant source of several innate immune receptors. These may be expressed within the cell and function mainly in an autocrine or paracrine fashion, or be released in significant quantities extracellularly with more endocrine effects. These adipokines include the cytokines (leptin, IL-6 and TNFα), complement-like molecules (adiponectin, acylation stimulating protein (ASP) and adipsin), acute-phase proteins (plasminogen activator inhibitor-1 (PAI-1), serum amyloid A (SAA), C-reactive protein (CRP)), chemokines (MCP-1 and regulated-upon activation normal T-cell expressed and secreted (RANTES)) and the toll-like receptors (TLR2 and 4). These will be discussed more extensively below.

Cytokines and cytokine-like molecules

Cytokines are a diverse group of proteins that are produced by a wide variety of cell types (Balkwill, 1993). They are mainly autocrine and/or paracrine factors that control many aspects of both innate and adaptive immune responses, as well as a variety of other processes such as tissue remodelling. They are pleiotropic, that is they have multiple biological effects, acting on many different cell types and there is considerable redundancy among cytokines, with many cytokines sharing similar functions (McDermott, 2001; Lago *et al.*, 2007a,b).

Leptin

Leptin is a 16-kDa peptide which has a cytokine-like structure (Zhang *et al.*, 1994; Gaucher *et al.*, 2003; Lago *et al.*, 2008). Its receptor, consisting of a large, single, membrane-spanning protein, is a member of the gp130 class I cytokine receptor family (Baumann *et al.*, 1996). Consequently, leptin is occasionally grouped with the more traditional cytokines, such as IL-6 and TNF-α.

Leptin is encoded by the *ob* gene and is expressed predominantly in adipose tissue, in correlation with the amount of fat present in adipocytes, with the larger, more lipid engorged cells secreting more leptin (Friedman and Halass, 1998). It has been shown to have a role in a range of processes, including the regulation of appetite and energy expenditure, glucose homeostasis, bone formation, regulation of puberty and reproduction, immunity and inflammation (Rajala and Scherer, 2003; O'Rahilly *et al.*, 2003; Lago *et al.*, 2007a,b, 2008). Systemic levels of leptin in various conditions are shown in Table 8.1 for comparison.

Table 8.1. Comparison of cytokine levels in obesity and in response to acute inflammatory stimuli.

Factor	Lean	Obese	Immune reaction/ inflammation
Leptin (ng/ml)	9.2 (5.2)	26.9 (3.9)	36.8 (11.2)
Adiponectin (μg/ml)	9.2 (3.0)	5.6 (2.5)	Not known
IL-6 (pg/ml)	0.45 (0.99)	2.9 (7.7)	3321 (1821)
TNFα (pg/ml)	0.4 (0.12)	2.0 (0.4)	422.4 (214.5)
Resistin (ng/ml)	21.5 (3.2)	28.8 (5.8)	Not known
Visfatin (pM)	6.9 (1.4) (mouse)	↑	Not known
MCP-1 (pg/ml)	119.8 (28.3)	183.3 (112.1)	Not known
RANTES (ng/ml)	7.96 (5.56–20.18)	16.99 (11.37–28.39)	Not known
CRP (μg/ml)	0.7 (0.3–2.0)	8.9 (5.1–16.6)	240.8 (86.6)

Note: Data are shown as mean (SD) or median (IQR).

Leptin is involved in regulating many inflammatory and immune processes, with leptin levels being modulated by inflammatory stimuli in both animals and humans (Fantuzzi *et al*., 2005; Bernotiene *et al*., 2006; Lago *et al*., 2008). Elevated leptin concentrations have been observed in sepsis patients compared to healthy subjects (Torpy *et al*., 1998; Arnalich *et al*., 1999), while being reduced significantly in patients with tuberculosis (van Crevel *et al*., 2002). Inflammatory stimuli have been found to induce leptin expression acutely and increase serum levels in animals, but not always in humans (Fantuzzi and Faggioni, 2000). Starvation and malnutrition are associated with lowered leptin levels and impaired immune function. This immunosuppression can be reversed with the administration of exogenous leptin (Lord *et al*., 1998).

The leptin receptor is expressed in T-cells, B-cells, macrophages and haematopoietic cells, with the ligand having direct effects on these cells (Bernotiene *et al*., 2006). Leptin protects T-cells from apoptosis and regulates their proliferation and activation, in part by influencing the production of other cytokines by T-cells (Fantuzzi, 2005). Leptin-deficient (*ob/ob)* mice have been shown to have impaired T-cell immunity. Furthermore, cytokine production from T-cells is suppressed in leptin-deficient children, but is restored when leptin is administered subcutaneously (Farooqi *et al*., 2002). Since its cloning in 1994, leptin's role in regulating immune and inflammatory response has become evident. Actually, the increase of leptin production that occurs during infection and inflammation strongly suggests that leptin is a part of the loop governing the inflammatory–immune response and the host defence mechanisms (Lago *et al*., 2007a,b). In this context, leptin stimulates the production of proinflammatory cytokines from cultured monocytes and enhances the production of Th1-type cytokines from stimulated lymphocytes. Several studies have implicated leptin in the pathogenesis of autoimmune inflammatory conditions such as T1DM, rheumatoid arthritis and chronic bowel disease. Some of these immune functions of leptin are summarized in Table 8.2.

Recently, leptin has been shown to activate macrophages directly and induce the formation of adipose differentiation-related protein-enriched lipid bodies (Maya-Monteiro *et al*., 2008). Newly formed lipid bodies were sites of 5-lipoxygenase localization and correlated with an enhanced capacity of leukotriene B(4) production. It was further demonstrated that leptin-induced macrophage activation was dependent on phosphatidylinositol 3-kinase (PI3K) activity. Leptin induces phosphorylation of p70(S6K) and 4EBP1, key downstream signalling intermediates of the mammalian target of the rapamycin (mTOR) pathway in a rapamycin-sensitive mechanism (Maya-Monteiro *et al*., 2008). The mTOR inhibitor, rapamycin, inhibited leptin-induced lipid body formation, both *in vivo* and *in vitro*. In addition, rapamycin inhibited leptin-induced adipose differentiation-related protein accumulation in macrophages and lipid body-dependent leukotriene synthesis, providing evidence for a key role for mTOR in lipid body biogenesis and function. These findings establish PI3K/mTOR as an important signalling pathway for leptin-induced cytoplasmic lipid body biogenesis and adipose differentiation-related protein accumulation. Furthermore, a previously unrecognized link between intracellular (mTOR) and systemic (leptin) nutrient sensors in macrophage lipid metabolism has been highlighted. Leptin-induced

Table 8.2. Immunomodulatory function of adipokines.

Adipokine	Immune functions
Leptin	Protects T-cells from corticosteroid-induced apoptosis Increases phagocyte activity Activates monocytes
Adiponectin	Reduces activity and production of TNF-α Inhibits production of IL-6 Induces IL-10 and IL-1 receptor antagonists
IL-6	Stimulates synthesis of acute phase proteins by liver Inhibits secretion of TNF-α and IL-1 Inhibits secretion of visfatin Stimulates monocyte to macrophage differentiation
TNFα	Antiviral activity Stimulates leptin and IL-6 secretion Stimulates expression of haptoglobin (an acute phase protein)
Resistin	Upregulates expression of MCP-1, vascular cell adhesion molecule 1 and ICAM-1 in endothelial cells
Visfatin	Inhibits neutrophil apoptosis Promotes cell proliferation
MCP-1	Attracts monocytes and macrophages
RANTES	Recruits, activates and stimulates T-cells and monocytes

increased formation of cytoplasmic lipid bodies and enhanced inflammatory mediator production in macrophages may have implications for obesity-related cardiovascular diseases.

TNF-α

TNF-α is a 17-kDa proinflammatory cytokine secreted by adipose tissue (Hotamisligil *et al.*, 1993), as well as by activated monocytes and macrophages. It has a broad range of biological and immunological effects including immune modulation, growth regulation and antiviral activity (Rosenblum and Donato, 1989).

TNF-α has been demonstrated to regulate adipocyte metabolism at numerous sites, including transcriptional regulation, glucose and fatty acid metabolism and hormone receptor signalling (Sethi and Hotamisligil, 1999). Hotamisligil *et al.* (1995) reported that TNF-α regulated cell size in mature adipocytes and it was suggested that as the cells got bigger they produced more TNF-α, which subsequently initiated changes to limit adipocyte size or to induce apoptosis. TNF-α has been shown to induce apoptosis in preadipocytes and mature adipocytes (Prins *et al.*, 1997) and can block differentiation in preadipocytes and induce de-differentiation in both preadipocytes and mature adipocytes (Petruschke and Hauner, 1993). The synthesis of a number of other adipokines is also regulated by TNF-α; it has been shown to stimulate leptin and IL-6 production, while having an inhibitory effect on adiponectin (Wang *et al.*, 2005). At the early stages of atherosclerosis, TNF-α activates endothelial cells. This results in the synthesis of various adhesion molecules such as vascular cell adhesion molecule-1 (VCAM-1), endothelial-leukocyte adhesion molecule-1 (E-selectin) and intracellular adhesion molecule-1 (ICAM-1), which regulate the inflammatory reactions in different cell types and increase the adherence of monocytes (Ross, 1993; Springer, 1994; Vora *et al.*, 1997).

Two structurally distinct forms of the TNF receptor have been identified, namely TNFR-I and TNFR-II (Hotamisligil, 1999). In humans, these are known as gp60 or TNFR60 and gp80 or TNFR80, respectively (Coppack, 2001). Both receptors are expressed by adipose tissue, as well as by numerous other tissues, and both exist in a soluble form, with TNFR-I being released from the tissue (Mohamed-Ali *et al.*, 1999; Sethi and Hotamisligil, 1999). It is thought that TNFR-I mediates the majority of the actions of TNF-α, but there is still some debate as to how the different receptors relate to the different actions of TNF-α. The soluble forms of the receptor act as antagonists, inhibiting the ligand-binding cell surface receptor (Mohamed-Ali *et al.*, 1999).

In obese mice, large amounts of TNF-α are produced by adipose tissue and are released into the circulation (Hotamisligil *et al.*, 1993). In humans, however, TNF-α secreted by adipose tissue is not always detected in the circulation but instead appears to act locally within the adipose tissue itself (Fain *et al.*, 2004).

A rise in TNF-α levels, as observed in both murine and human obesity (Hotamisligil and Spiegelman, 1994; Kern *et al.*, 1995), is associated with insulin resistance and has been shown to interfere with the insulin signalling pathway (Hotamisligil, 1999). In obese mice, insulin resistance can be counteracted by neutralizing TNF-α with antibodies (Hotamisligil *et al.*, 1993). However, a similar effect has not been observed in obese human subjects (Ofei *et al.*, 1996).

TNF-α stimulates adipose tissue lipolysis *in vitro*, resulting in the release of non-esterified fatty acids (NEFA) into the media. TNF-α administration increases serum triglyceride levels acutely by stimulating hepatic *de novo* fatty acid and triglyceride synthesis (Sethi and Hotamisligil, 1999). Thus, it is clear that TNF-induced stimulation of hepatic lipid synthesis and secretion contributes to TNF-induced hyperlipidaemia (Feingold *et al.*, 1989).

IL-6

IL-6 is a 26-kDa cytokine that is produced by numerous immune cell types such as monocytes, lymphocytes, macrophages and mast cells during inflammation, mainly in response to induction by TNF-α and IL-1β (Papanicolaou *et al.*, 1998). Non-immune cells and tissues, in particular adipose tissue and skeletal muscle, also release significant quantities of this cytokine (Mohamed-Ali *et al.*, 1997; Keller *et al.*, 2001).

IL-6 is a key regulator of the hepatic acute-phase response by controlling the expression of various hepatic inflammatory markers, such as C-reactive protein

(CRP) and fibrinogen. It also stimulates the differentiation of monocytes to macrophages, rather than antigen-presenting dendritic cells (Chomarat *et al*., 2000). IL-6 is associated with the suppression of thyroid function, stimulates vasopressin secretion and affects the basal metabolic rate by stimulating thermogenesis (Papanicolaou *et al*., 1998). It activates the hypothalamic–pituitary–adrenal axis to help control inflammation (Papanicolaou *et al*., 1998). It is also thought that IL-6 may promote atherosclerosis in obese individuals by increasing thrombocyte aggregation and expression of adhesion molecules by endothelial cells (Woods *et al*., 2000).

IL-6 exerts its broad range of actions through the IL-6 receptor. The receptor consists of two membrane-bound glycoproteins, an 80-kDa receptor subunit (gp80, IL-6R) and a 130-kDa transmembrane signal-transducing element (gp130) (Jones *et al*., 2001). Most, if not all, cells express the gp130 element, whereas expression of the gp80 subunit is restricted to hepatocytes and some leukocytes (including monocytes, neutrophils, B-cells and T-cells) (Taga, 1992). A soluble form of IL-6R (sIL-6R) also exists and is abundant in the circulation (Mohamed-Ali *et al*., 1999). Unlike many other soluble cytokine receptors, the sIL-6R is a receptor agonist and can bind IL-6 alone and in the IL-6–sIL-6R complexed form activates gp130, the signal transducing component, on cells which do not express membrane-bound IL-6R, thus initiating signalling in cells that otherwise would be unresponsive to IL-6 (Mackiewicz *et al*., 1992). The gp130 receptor is shared by many cytokines and growth factors for signal transduction, including IL-11, oncostatin-M, leukaemia inhibitory factor (LIF), ciliary neurotrophic factor (CNTF), cardiotrophin-1 and leptin (Papanicolaou *et al*., 1998). Consequently, these factors possess overlapping activities and may explain why the phenotypic characteristics of mice lacking IL-6, IL-11, LIF or CNTF are less severe than the apparent pleiotropic properties of these factors would otherwise suggest (Jones *et al*., 2001).

Significant amounts of IL-6 have been shown to be released from adipose tissue, predominantly from visceral fat (by preadipocytes, adipocytes and macrophages), with a significant contribution to systemic IL-6 in obesity being adipose tissue-derived (Mohamed-Ali *et al*., 1997; Fried *et al*., 1998). Obesity in both humans and animals is associated with a chronic low-level rise in plasma IL-6 concentrations, while weight loss leads to a significant decrease in IL-6 levels (Bastard *et al*., 2000).

Changes in hepatic triglyceride metabolism and insulin sensitivity can be mediated by IL-6 (Nonogaki *et al*., 1995). IL-6 has been shown to inhibit adipose tissue lipoprotein lipase (LPL) activity and production and increase lipolysis, and this has been implicated in the fat wasting that occurs during cancer cachexia (Greenberg *et al*., 1992; Strassmann *et al*., 1992; Hardardottir *et al*., 1994; van Hall *et al*., 2003). Intravenous administration of IL-6 in rats increased serum triglyceride levels acutely in a dose-dependent manner. IL-6 treatment increased hepatic triglyceride secretion without decreasing the clearance of triglyceride-rich lipoproteins, indicating that the hypertriglyceridaemia was due to increased secretion by the liver (Feingold *et al*., 1991). In mice, it has been shown that chronic administration of IL-6, at levels equivalent to those observed in obese individuals, leads to insulin resistance and hyperlipidaemia. Furthermore, injection of IL-6-neutralizing antibodies into two murine models of obesity has been shown to improve insulin sensitivity, predominantly affecting the liver (Klover *et al.*, 2005). These results suggest that IL-6 contributes to the development of the insulin resistance associated with obesity.

Conversely, administration of recombinant IL-6 also decreased serum cholesterol in cancer patients (Weber *et al.*, 1993; van Gameren *et al.*, 1994). Treatment of rheumatoid arthritis with humanized anti-IL-6 receptor antibody over 3 months increased total cholesterol, triglycerides and, surprisingly, HDL-cholesterol (Nishimoto *et al.*, 2004). Furthermore, IL-6-deficient mice also showed elevated serum triglycerides and very low-density lipoprotein (VLDL), along with increased food intake and lower energy expenditure (Wallenius *et al.*, 2002).

Altogether, while IL-6 is a potent regulator of lipid metabolism and insulin resistance, depending on concentrations achieved, duration of elevation and whether this is in combination with other underlying pathologies, the effects are considerably varied, often opposing. These results may be explained by the fact that, while in the initial phases of the immune response IL-6 is proinflammatory, in the later stages it is anti-inflammatory, inhibiting the release of TNFα and stimulating IL-10 and IL-1 receptor antagonist (IL-1RA). Also, levels and duration of elevated IL-6 are significantly different under these various conditions and it is conceivable that these mediate opposing effects, either in the same cells or by recruiting other cells and tissues (Fig. 8.1 and Table 8.3).

Resistin

Resistin is a polypeptide that, in rodents, is derived primarily from adipocytes, with its expression being proportional to adipocyte differentiation and the amount of adipose tissue (Steppan *et al.*, 2001). In humans, however, resistin comes principally from peripheral blood monocytes, with very little being produced by adipocytes (Janke *et al.*, 2002; Fain *et al.*, 2003; Patel *et al.*, 2003).

In mice, obesity is associated with a rise in circulating resistin levels (Steppan *et al.*, 2001). It has been proposed that this adipokine induces insulin resistance in rodents. Administration of exogenous resistin into wild-type mice impaired glucose tolerance and insulin sensitivity, whereas administration of resistin-neutralizing antibodies reduced hyperglycaemia in diet-induced insulin-resistant mice (Steppan *et al.*, 2001). In resistin-null mice, it was demonstrated that the effects on insulin action were due to stimulation of glucose production in the liver (Banerjee *et al.*, 2004). Thiazolidinediones, hypoglycaemic agents, suppress resistin production in 3T3-L1 adipocytes, again suggesting that resistin could be a link between obesity and insulin resistance (Steppan *et al.*, 2001).

The physiological role of resistin in humans is still unclear. A possible role in inflammation has been suggested following the discovery that resistin levels correlate with IL-6 concentrations (Savage *et al.*, 2001; Steppan and Lazar, 2004). Concentrations of resistin are increased in atherosclerosis, suggesting a role in the pathogenesis of atherosclerosis in humans (Reilly *et al.*, 2005). However, its

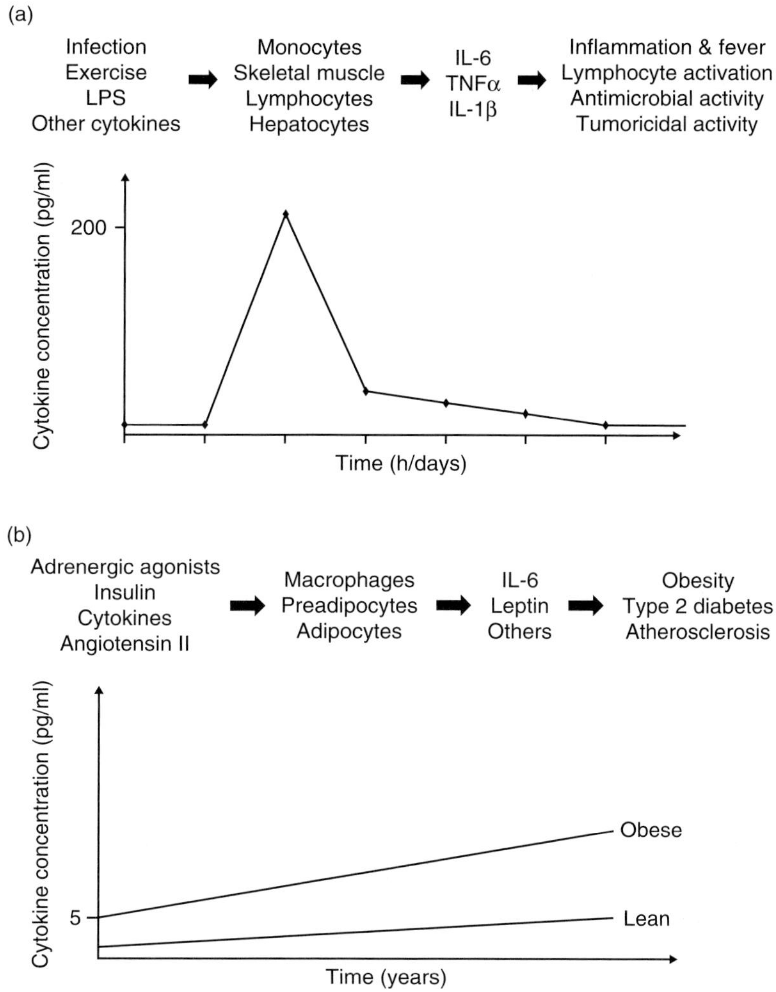

Fig. 8.1. Differences in acute versus chronic cytokine production. (a) Acute cytokine release. (b) Chronic cytokine release. LPS, lipopolysaccharide; IL-6, interleukin-6; TNF-α, tumour necrosis factor-α; IL-1β, interleukin-1β.

role in human obesity and insulin resistance is unclear (Vidal-Puig and O'Rahilly, 2001; Savage *et al.*, 2001; McTernan *et al.*, 2002; Rea and Donnelly, 2004).

Visfatin

Visfatin is a 52-kDa protein produced predominantly by visceral adipose tissue (Fukuhara *et al.*, 2005). It is identical to pre-B-cell colony-enhancing factor, a cytokine expressed by lymphocytes. Plasma concentrations correlate with the amount of visceral fat and visfatin levels increase in obesity in mice, as well as in humans (Fukuhara *et al.*, 2005). This adipokine is thought to have both paracrine

Table 8.3. Effect of weight change and diet on systemic adipokine levels.

Factor	Weight loss	Weight gain	EPA/DHA	Refs
Leptin	↓	↑	↔ (mice)	Friedman and Halaas, 1998; Wolfe *et al.*, 2004; Hauner, 2005; Flachs *et al.*, 2006.
Adiponectin	↑	↓	↑	Yang *et al.*, 2001; Weyer *et al.*, 2001; Flachs *et al.*, 2006
IL-6	↓	↑	↓ (not all)	Hauner, 2005; Panagiotakos *et al.*, 2005; Calder, 2006; Salas-Salvadó *et al.*, 2006;
TNFα	↓	↑	↓ (not all)	Hauner, 2005; Panagiotakos *et al.*, 2005; Calder, 2006; Salas-Salvadó *et al.*, 2006;
Resistin	↔ (minimal weight loss, calorie restriction)	↔↑↓	Not known	Steppan *et al.*, 2001; Wolfe *et al.*, 2004; Fantuzzi, 2005
Visfatin	Not known	↑	Not known	Fukuhara *et al.*, 2005, 2007
MCP-1	↓	↑	↑	Christiansen *et al.*, 2005; Hagiwara *et al.*, 2006

and endocrine actions (Sethi and Vidal-Puig, 2005), and in mice, it apparently exerts insulin-mimetic effects on various tissues, including the liver, muscle and fat (Fukuhara *et al.*, 2005). Visfatin gene expression in 3T3-L1 adipocytes is suppressed significantly by IL-6 (Kralisch *et al.*, 2005). It is noteworthy that some of these findings have not been replicated subsequently, with the authors having been forced to retract some of their original conclusions (Fukuhara *et al.*, 2007). Thus, the exact physiological function of visfatin remains controversial as regards some of the effects in animal models, and even much more so in humans.

Other cytokines

Several other cytokines have also been shown to be expressed or secreted by adipose tissue. Some of the most prominent and emerging adipokines are discussed briefly below.

Studies have shown IL-8 to be produced and released from isolated human adipocytes and whole adipose tissue cultures (Bruun *et al.*, 2001). As in other immune cell types, the release is stimulated by IL-1β and TNF-α (Matsushima *et al.*, 1989; Bruun *et al.*, 2001). Circulating levels of IL-8 are higher in abdominally obese subjects compared to lean individuals (Straczkowski *et al.*, 2002) and in patients with T1DM and T2DM (Zozulińska *et al.*, 1999; Erbagci *et al.*, 2001). Visceral fat explants secrete greater amounts of IL-8 compared to subcutaneous fat (Bruun *et al.*, 2004). The circulating levels of this cytokine correlate with measures of adiposity and insulin sensitivity (Bruun *et al.*, 2003), as well as being associated with that of atherosclerosis (Moreau *et al.*, 1999) and CHD (Romuk *et al.*,

2002). IL-8 is released from the macrophages in atherosclerotic lesions after stimulation with oxidized low-density lipoprotein (Liu *et al.*, 1997) and may contribute to atherosclerosis through leukocyte recruitment and increasing the release of matrix-degrading metalloproteinases by decreasing specific tissue inhibitors of metalloproteinases, resulting in instability of an advanced atherosclerotic plaque (Moreau *et al.*, 1999). The role of IL-8 from adipose tissue is not clear but, as with MCP-1, it has been suggested that it may be involved in the recruitment of monocytes to adipose tissue, where they are transformed into macrophages (Fain *et al.*, 2005).

IL-10 is secreted by both human adipocytes and the stromovascular cells and is elevated in obese individuals (Esposito *et al.*, 2003; Fain *et al.*, 2004). IL-18 also might be produced by adipose tissue, as its circulating levels were raised in obese subjects and decreased with weight loss (Esposito *et al.*, 2002).

Vaspin was discovered by Hida *et al.* (2005) as a serpin (serine protease inhibitor) that was produced by visceral adipose tissue. Interestingly, administration of vaspin to obese mice improved glucose tolerance and insulin sensitivity, and reversed altered expression of genes that might promote insulin resistance. However, the exact regulation and relevance of vaspin in obesity, as well as its potential role in the immune response and inflammatory complications of excess visceral adiposity, need to be clarified.

Omentin is a protein of 40 kDa, secreted by omental adipose tissue and highly abundant in human plasma, which previously had been identified as intelectin, a new type of Ca^{2+}-dependent lectin with affinity to galactofuranosyl residues, which are constituents of pathogens and dominant inmunogens (Lago *et al.*, 2007a,b). Thus, an important role in the innate immune response to parasite infection consisting of the specific recognition of pathogens and bacterial components was attributed to omentin/intelectin. Several studies have shown that omentin gene expression is altered by inflammatory states and obesity. The range of the biological actions of omentin seems to be similar to those of adiponectin, but still needs to be studied in more detail.

Apelin is a bioactive peptide and originally was identified as the endogenous ligand of the orphan G protein-coupled receptor, APJ (Lago *et al.*, 2007a,b). TNF-α increases both apelin production in adipose tissue and circulating apelin concentrations when administered to mice. In mice with diet-induced obesity, macrophage counts and the levels of proinflammatory factors such as TNF-α increase progressively in adipose tissue. In this context, apelin overproduction in obese states might be viewed as an adaptive response attempting to forestall the onset of obesity-related disorders, such as mild chronic inflammation.

Hepcidin was identified originally as a urinary antimicrobial peptide synthesized in the liver, being recognized only later as an adipokine (Lago *et al.*, 2007a,b). It has been described as a key regulator of iron homeostasis. None the less, hepatic hepcidin production depends not only on iron homeostasis, but also on hypoxia and inflammatory stimuli. Hepcidin concentrations increase in disorders involving generalized inflammation, which results in hypoferraemia due to a combination of decreased duodenal iron absorption and elevated sequestration of iron by macrophages. The induction of hepcidin in cultured cell lines and in a murine model by acute inflammatory stimuli has been shown to be mediated mainly by IL-6 via a STAT3 mechanism. The resulting decrease in plasma iron levels eventually limits iron availability to erythropoiesis and contributes to the anaemia associated with infection and inflammation. On the other hand, decrease in extracellular iron concentrations due to hepcidin probably limits iron availability to invading microorganisms, thereby contributing to host defence.

Chemerin is a novel and promising adipokine whose plasma levels in humans have been found to be associated significantly with body mass index (BMI), circulating triglycerides and blood pressure (Bozaoglu *et al.*, 2007). It has been demonstrated that chemerin or chemerin receptor knockdown impairs differentiation of 3T3-L1 cells into adipocytes, reduces the expression of adipocyte genes involved in glucose and lipid homeostasis, including adiponectin and leptin, and alters metabolic functions in mature adipocytes (Goralski *et al.*, 2007).

Complement and complement-like proteins

Adiponectin

Adiponectin is a 30-kDa polypeptide that was identified by different groups in 1995, with each group naming it differently; adipose most abundant gene transcript 1 (apM1), adipocyte complement-related protein of 30 kDa (Acrp30), adipoQ and gelatin-binding protein of 28 kDa (GBP28) (Scherer *et al.*, 1995; Hu *et al.*, 1996; Maeda *et al.*, 1996; Nakano *et al.*, 1996).

Adiponectin is structurally similar to complement 1q with an amino-terminal collagen-like, variable domain and a carboxy-terminal head domain (Shapiro and Scherer, 1998; Yokota *et al.*, 2000). Several isoforms of the molecule have been reported in the circulation, with the trimers forming the low molecular weight (LMW) form, with the other two oligmeric forms, 12- to 18-mer, being designated middle molecular weight (MMW) and high molecular weight (HMW) polymers (Pajvani *et al.*, 2003; Waki *et al.*, 2003). The higher molecular weight forms of adiponectin are the predominant species found in the circulation (Chandran *et al.*, 2003; Tilg and Moschen, 2008).

There are two adiponectin receptors, AdipoR1 and AdipoR2, containing seven transmembrane domains through which the biological effects of adiponectin are mediated. The receptors are different from G protein-coupled receptors as the N-terminus is internal and the C-terminus external. AdipoR1 is expressed primarily in the muscle and AdipoR2 in the liver (Yamauchi *et al.*, 2003). Therefore, as well as circulating levels of adiponectin and its isoforms, its receptor subtypes may play a role in the tissue-specific effects exerted.

Adiponectin circulates at extremely high levels (3–30 μg/ml: Chandran *et al.*, 2003; Fantuzzi, 2005) and is higher in females, suggesting a possible regulation by sex hormones (Nishizawa *et al.*, 2002; Combs *et al.*, 2003; Adamczak *et al.*, 2005; Xu *et al.*, 2005). This adipokine is known to have anti-inflammatory effects (Lyon *et al.*, 2003; Steffens and Mach, 2008; Zhu *et al.*, 2008) such as reducing the activity and production of TNF-α (Masaki *et al.*, 2004), inhibiting production of IL-6 and induction of the anti-inflammatory cytokines, IL-10 and IL-1 receptor antagonists (Kumada *et al.*, 2004; Wolf *et al.*, 2004; Wulster-Radcliffe *et al.*,

2004). Adiponectin also inhibited TNF-α-induced THP-1 adhesion dose-dependently and expression of VCAM-1, E-selectin and ICAM-1 on human aortic endothelial cells (Ouchi *et al.*, 1999). As the size of fat depots increase, it has been suggested that the hypertrophied fat depots become hypoxic (Kabon *et al.*, 2004; Trayhurn *et al.*, 2004; Fleischmann *et al.*, 2005). Adipocytes have been found to express hypoxia-inducible factor (HIF)-1α, a key transcription factor involved in cellular response to hypoxia (Lolmede *et al.*, 2003; Trayhurn and Wood, 2004). Also, hypoxia has been shown to increase the expression of some angiogenic factors like vascular endothelial growth factor (VEGF) and leptin via the same pathway (Lolmede *et al.*, 2003). Both reactive oxygen species (ROS) and HIF-1α trigger the production of adiponectin and PAI-1 (Chen *et al.*, 2006). However, other data show that various proinflammatory cytokines like TNF-α, IL-6 and ROS increase PAI-1 production and decrease that of adiponcetin (Furukawa *et al.*, 2004; Soares *et al.*, 2005; de Taeye *et al.*, 2005). In overweight and obese women, PAI-1 activity was related inversely to serum adiponectin, independent of visceral adipose tissue (Mertens *et al.*, 2005).

In skeletal muscle, adiponectin increases fatty acid oxidation by elevating the expression of acyl-coenzyme A oxidase, increases the expression of uncoupling protein 2 (UCP-2), which helps dissipate energy, and increases CD36 that is involved in fatty acid transport (Yamauchi *et al.*, 2001). All these help reduce tissue triglycerides and prevent insulin resistance (Shulman, 2000). It has also been shown to activate PPARα, hence increasing fatty acid oxidation and energy consumption and therefore reducing triglyceride in the liver and in muscle (Yamauchi *et al.*, 2001). It further activates AMP kinase, therefore stimulating phosphorylation of acetyl coenzyme-A carboxylase, fatty acid oxidation and glucose uptake (Yamauchi *et al.*, 2002).

Contrary to other adipokines, adiponectin expression and circulatory levels are related inversely to BMI and insulin resistance. Therefore, reduced plasma levels are found in conditions associated with insulin resistance, such as T2DM, cardiovascular disease and hypertension (Adamczak *et al.*, 2003; Kumada *et al.*, 2003; Ouchi *et al.*, 2003; Pischon *et al.*, 2004). Adiponectin prevents the progression of atherosclerosis and low adiponectin levels have been shown to be associated independently with the metabolic syndrome while being elevated by insulin-sensitizing drugs and weight loss (Weyer *et al.*, 2001; Yang *et al.*, 2001; Chandran *et al.*, 2003; Diez and Iglesias, 2003; Matsushita *et al.*, 2006).

Acylation stimulation protein (ASP) and adipsin

ASP is a small 9-kDa protein, identical to a fragment of the third component of the complement system, C3adesArg, involved in lipid and glucose metabolism. It is generated through cleavage of the terminal Arg of C3a through the combined action of adipsin and cofactor B (Cianflone *et al.*, 1989; Baldo *et al.*, 1993). The activity of ASP relative to C3a, in its immunological role, is reduced profoundly (Zwirner *et al.*, 1998). In normal human plasma, only the ASP form of the protein is present (Saleh *et al.*, 1998).

ASP has two main effects mediated by binding to the cell surface (Murray *et al.*, 1997) and triggering a signalling pathway that results in activation of protein kinase C (Baldo *et al.*, 1995). First, it increases the translocation of glucose transporters (GLUT1, GLUT3 and GLUT4) from intracellular sites to the plasma membrane (Germinario *et al.*, 1993; Maslowska *et al.*, 1997) and hence increases glucose transport into the adipocytes, as well as preadipocytes, smooth muscles and fibroblasts. Secondly, ASP increases LPL and diacylglycerol acyl-transferase activity, resulting in fatty acid uptake, triglyceride synthesis and lipolysis and release of NEFA from adipocytes (Cianflone *et al.*, 2003). ASP is reported as being more potent than insulin in stimulating the esterification of fatty acids into intracellular triglyceride in human fibroblasts and adipocytes (Cianflone *et al.*, 1989).

Adipsin, or complement factor D, is an adipose-specific serine protease enzyme secreted by adipose tissue and is needed for ASP production. It is one of the proteins involved in the alternate complement pathway and homologous to human plasma factor D (Choy *et al.*, 1992). Murine 3T3-L1 adipocytes produce and secrete adipsin (White *et al.*, 1992). Serum adipsin and its mRNA expression in adipose tissue decrease in obesity (Flier *et al.*, 1987; Rosen *et al.*, 1989). However, ASP is increased in obesity as well as in insulin resistance, dyslipidaemia and cardiovascular disease.

Acute-phase proteins

During the early stages of infection, the acute-phase response is initiated by various cytokines such as IL-6, TNFα and IL-1β, leading to the mainly hepatic and, to a lesser extent, adipose tissue release of CRP, PAI-1 and SAA. CRP and SAA are members of the pentraxin family that can function as opsonins and also bind to C1q, and thus activate the classical complement pathway.

CRP

While serum levels of CRP above 10 mg/l indicate acute inflammation, in obese patients and those with T2DM sub-inflammatory elevations of this molecule have been reported in the systemic circulation (see Table 8.1; Ridker *et al.*, 1998, 2000; Yudkin *et al.*, 1999). Furthermore, these smaller circulating elevations in concentration appear to predict future cardiovascular events (Danesh *et al.*, 2000; De Ferranti *et al.*, 2002). In a 3-year follow-up study of healthy women, those with CRP concentrations in the highest quartile had a 4.4-fold increased risk of having a cardiovascular event when compared with those with concentrations in the lowest quartile (Ridker *et al.*, 2000). CRP is also raised in obesity, correlating with BMI in both the elderly and young and is reduced after weight loss (Rohde *et al.*, 1999; Barinas-Mitchell *et al.*, 2001; Dietrich and Jialal, 2005).

PAI-1

PAI-1 is a member of the serine protease inhibitor family and is the primary inhibitor of plasminogen activation and, hence, inhibits fibrinolysis and promotes thrombosis (Yamamoto and Saito, 1998). Predictably, increased PAI-1 can lead to coronary artery disease (Kohler and Grant, 2000). PAI-1 was first shown to be expressed in rodent epididymal fat pads (Sawdey and Loskutoff, 1991). Subsequently, it was found in human adipose tissue as well, with higher levels in visceral fat compared to subcutaneous fat (Alessi *et al.*, 1997; Eriksson *et al.*, 1998; Gottschling-Zeller *et al.*, 2000). Studies have shown that PAI-1 expression in cultured adipocytes is upregulated by insulin (Samad *et al.*, 1997), TNF-α (Samad *et al.*, 1996), TGF-β (Samad *et al.*, 1997) and glucocorticoids (Konkle *et al.*, 1992). Serum PAI-1 levels are elevated in both obesity and insulin resistance and correlate with features of the metabolic syndrome (Mertens and Van Gaal, 2002; Juhan-Vague *et al.*, 2003). Serum PAI-1 concentrations decrease after weight loss and treatment with insulin-sensitizing drugs such as metformin and thiazoladinediones (Mertens *et al.*, 2005).

Serum amyloid A

There are two types of SAA, namely SAA1 and SAA2. It increases in response to infection, inflammation, injury and stress (Malle and De Beer, 1996). SAA is associated with cardiovascular disease (Johnson *et al.*, 2004). It has been shown that SAA is expressed in human adipocytes (Poitou *et al.*, 2005; Sjoholm *et al.*, 2005). Serum SAA (Poitou *et al.*, 2005) and mRNA expression in visceral fat (Gómez-Ambrosi *et al.*, 2006) is higher in obese compared to normal subjects. Studies have also shown that circulating SAA levels correlate with body fat (Gómez-Ambrosi *et al.*, 2006) and both serum and subcutaneous WAT expression of SAA has been found to decrease after weight loss (Sjöholm *et al.*, 2005).

Chemokines

Chemokines are secreted LMW proteins with crucial roles in physiological and pathophysiological processes (Baggiolini *et al.*, 1997; Gerard and Rollins, 2001), but whose eponymous function is represented by induction of leukocyte migration (Mantovani, 1999). Over 50 distinct molecules are known in humans (Bacon and Harrison, 2000) and classified into four subclasses according to the position of their cysteines; CXC, C, CX3C and CC (Laing and Secombes, 2004). Four chemokines have been identified in adipose tissue and two belong to CC chemokines; MCP-1 (or CCL2) and RANTES (or CCL5). The other two belong to the CXC subclass; interferon-γ (IFNγ) inducible protein 10 (IP-10, or CXCL10) and IL-8 (or CXCL8) (Laing and Secombes, 2004).

MCP-1 and RANTES

The infusion of MCP-1 into mice increases circulating monocytes, as well as accumulation of monocytes in collateral arteries and neointimal formation (Takahashi *et al.*, 2003; van Royen *et al.*, 2003). MCP-1 also plays an important role in the development of atherosclerosis as its expression is increased in atherosclerotic lesions (Ylä-Herttuala *et al.*, 1991; Takeya *et al.*, 1993). Atheroma formation in hypercholesterolaemic mice is reduced by inhibition of MCP-1 expression or that of its receptor (Boring *et al.*, 1998; Gu *et al.*, 1998). In apolipoprotein E-knockout mice that are prone to severe atheroma formation this process was decreased when MCP-1 release was blocked through transfection of an N-terminal deletion mutant in the MCP-1 gene (Inoue *et al.*, 2002).

MCP-1 is produced and released by the stromovascular fraction of WAT, preadipocytes and isolated mature adipocytes (Gerhardt *et al.*, 2001; Xu *et al.*, 2003). The release of MCP-1 from preadipocytes is triggered by levels of TNF-α secreted by adipocytes in obesity (Xu *et al.*, 2003). While it is not clear whether the MCP-1 from endothelial cells or that from adipocytes attracts the macrophages into adipose tissue, its expression precedes the appearance of macrophage markers (Xu *et al.*, 2003).

Circulating MCP-1 levels are elevated in genetic (*ob/ob* mice) and diet-induced obese (DIO) mice and reduced after weight loss (Table 8.3) (Sartipy and Loskutoff, 2003; Takahashi *et al.*, 2003). It has been suggested that overnutrition causes a metabolic overload with increased demands on the endoplasmic reticulum and on the mitochondria, resulting in release of proinflammatory mediators via excess production of reactive oxygen species (Wellen and Hotamisligil, 2005). However, it has also been shown that treating adipocytes with MCP-1 causes a decrease in lipid accumulation, as well as stimulation of leptin secretion by post-transcriptional mechanisms (Sartipy and Loskutoff, 2003). In obese children and adolescents, no significant correlation between circulating MCP-1 and BMI was found. Moreover, it has been observed that MCP-1 levels, both circulating and mRNA content of subcutaneous adipose tissue, are elevated in human obesity and are greater in visceral compared to that in subcutaneous tissue (Bruun *et al.*, 2003; Christiansen *et al.*, 2005).

MCP-1 is also elevated in T2DM patients and, in these patients, it is associated with an increase in cardiovascular disease (Nomura *et al.*, 2000; Piemonti *et al.*, 2003). In adipocyte cell lines, MCP-1 has been found to decrease insulin-stimulated glucose uptake and insulin-induced insulin receptor tyrosine phosphorylation, suggesting an important role in insulin resistance of adipose tissue (Gerhardt *et al.*, 2001; Sartipy and Loskutoff, 2003).

RANTES recruits, activates and co-stimulates T-cells and monocytes and so plays a role in immunoregulatory and inflammatory processes (Luster, 1998; Gerard and Rollins, 2001; Economou *et al.*, 2004). However, the true contribution of this chemokine in obesity needs to be fully disentangled. In one study looking at prepubertal children, the obese group had higher circulating RANTES levels.

IP-10 (interferon-γ inducible protein 10)

IP-10 is expressed by several cell types, including neutrophils, monocytes, endothelial cells, fibroblasts and keratinocytes (Luster and Ravetch, 1987; Dufour *et al.*, 2002; Villagomez *et al.*, 2004). Its expression is regulated by IFN-γ and it has chemoattractant properties for activated T-cells, monocytes, natural killer cells, dendritic cells and eosinophils (Taub *et al.*, 1993; Robertson, 2002). Mature human adipocytes also express and secrete IP-10 in response to IFN-γ (Herder *et al.*, 2006, 2007). Although some studies have reported positive correlations between serum IP-10 concentrations with BMI and other parameters of obesity, this has not been consistent (Herder *et al.*, 2005, 2006; Rothenbacher *et al.*, 2006). Its presence in human atherosclerotic plaques, but not in normal vessels, suggests a role in the development of atherosclerosis (Mach *et al.*, 1999).

Toll-like receptors

The Toll-like receptors (TLRs) are essential innate immune receptors that alert the immune system to the presence of invading microbes (Zhang and Schluesener, 2006). To date, ten TLRs have been described in human and mice tissues. The first TLR shown to be expressed in adipose tissue was the human homologue of Drosophila Toll, now designated TLR4, closely followed by TLR2 (Lin *et al.*, 2000; Bès-Houtmann *et al.*, 2007). Studies have shown that TLR4 plays an essential role in LPS responsiveness (Lin *et al.*, 2000), although not on its own. TLR2 recognizes a large number of ligands, which includes LPS, peptidoglycans, zymosan and glycosylphosphatidylinositol lipid. TLR2 forms heterodimers with either TLR1 or TLR6, to recognize pathogen-associated molecular patterns, and this increases its repertoire of ligand specificities (Ozinsky *et al.*, 2000). TLR4, on the other hand, appears to function mostly as homodimers (Lorenz, 2006). On recognition of their ligands, TLRs are capable of inducing the expression of a variety of host defence genes. These include the inflammatory cytokines and chemokines in the cytosol, the major histocompatibility complex molecules on the cell surface and other effectors necessary to arm the host cell against any non-self ligands. These molecules can both attract naive T-cells through the secretions of chemokines and, furthermore, activate them to respond to specific antigens. Mutations in some of the TLRs, especially in TLR4 and TLR2, have been associated with increased susceptibility to infectious diseases (Lorenz, 2006).

Diet and Adipokines

In humans, as well as in experimental animals, a high content of dietary fat promotes WAT accumulation and obesity. Adipokines respond to dietary modulation (Table 8.3) with energy-restricted low-fat and very low-carbohydrate diets reducing inflammatory markers such as TNF-α, IL-6, CRP and sICAM (soluble-ICAM-1) (Sharman and Volek, 2004). Increased dietary intake of saturated fat or cholesterol results in higher serum CRP and IL-6 concentrations, while a decrease in circulating levels is observed after dietary restriction of these fats (Han *et al.*, 2002; Baer *et al.*, 2004; Mozaffarian *et al.*, 2004). A very low caloric diet decreased both SAA mRNA and circulating levels (Viguerie *et al.*, 2005). Adipsin mRNA in diet-induced obese rats was reduced after weaning the animal off a high-fat diet (Dugail *et al.*, 1989).

Previous studies by Storlien *et al.* (1987) have demonstrated that the substitution of saturated fatty acids by fish oil rich in *n*-3 long-chain polyunsaturated fatty acids (*n*-3 LC-PUFA, i.e. eicosapentaenoic, EPA, and docosahexaenoic acid, DHA), or by *n*-6 PUFA contained in vegetable oils, prevented liver and muscle insulin resistance induced by the diet. *n*-3 LC-PUFA supplementation alleviates symptoms in patients with chronic inflammatory conditions such as rheumatoid arthritis (Geusens *et al.*, 1994), asthma (Broughton *et al.*, 1997), Crohn's disease (Belluzzi *et al.*, 1996) and psoriasis (Mayser *et al.*, 1998). In patients with CHD, *n*-3 PUFA decrease the overall mortality due to myocardial infarction and sudden death (Bucher *et al.*, 2002). The reduction of saturated fat counterbalanced by an increase in *n*-3 PUFA is antithrombotic, antifibrinolytic, hypotriacylglycerolaemic and anti-inflammatory (Saynor and Gillott, 1992; Eritsland *et al.*, 1994; Vognild *et al.*, 1998). EPA/DHA supplementation increases plasma adiponectin levels, independent of food intake, reflecting the stimulation of the expression of adiponectin in adipocytes, and its release from epididymal, but less from subcutaneous fat. Expression of leptin and its release from adipose tissue explants *ex vivo*, while being exquisitely sensitive to caloric restriction, was not affected by EPA/DHA (Table 8.3). Thus, EPA and DHA intake leads to induction of adiponectin, in a manner largely independent of food intake or adiposity, and this may partially explain their antidiabetic effects (Flachs *et al.*, 2006).

Conclusions

The increasing body of knowledge provides clear evidence to consider obesity an inflammatory condition leading to chronic activation of the innate immune system, which ultimately causes progressive impairment of glucose tolerance. In this chapter, we have outlined the several expressed and secreted factors from adipose tissue that are classically part of the innate immune response. WAT also encompasses the cellular components of the innate immune system. The expressed and secretory repertoire of adipose tissue includes many immune modulators. However, what does seem to set apart, to some extent, the secretion from the adipose tissue as opposed to their release from the liver, circulating macrophages or other cells outside the adipose tissue is that many adipokines are, at least in part, expressed and secreted constitutively. While in the lean state this constitutive release may not be of physiological or pathological significance, the enlargement of the adipose organ, as seen in obesity, often leads to a significant contribution of these adipokines to the systemic circulation. In addition, in obesity, the adipose tissue also lies in greater proximity to visceral organs and skeletal muscle, as well as the blood vessels. Thus, local concentrations of these mainly autocrine/paracrine factors would far exceed the reported systemic concentrations and, therefore, may be able to induce several components of the immune cascade. Experimental studies in animals and evidence from prospective and longitudinal studies in humans are consistent with an aetiologic role for the subclinical inflammation in obesity-induced insulin resistance. However, the exact chain of molecular events linking overnutrition and the role of adipose tissue in activation of the immune system and impairment of insulin sensitivity remains incompletely understood.

The interdependence of the various adipokines has to be recognized, with both *in vitro* and *in vivo* data pointing to adipokine activity and function comprising a complex regulatory network. Their expression and secretion may follow specific timelines dependent on the existing immune, hormonal and metabolic milieu. Despite the complexities involved in adipokine function, the significant progress made in our understanding of the role these proteins play in human obesity indicates their potential value as therapeutic agents. Therefore, treating the underlying inflammation of obesity, by dietary and/or pharmacological means, may constitute a novel approach in the prevention and treatment of its associated pathologies.

References

Adamczak, M., Wiecek, A., Funahashi, T., Chudek, J., Kokot, F. and Matsuzawa, Y. (2003) Decreased plasma adiponectin concentration in patients with essential hypertension. *American Journal of Hypertension* 16, 72–75.

Adamczak, M., Rzepka, E., Chudek, J. and Wiecek A. (2005) Ageing and plasma adiponectin concentrations in apparently healthy males and females. *Clinical Endocrinology* 62, 114–118.

Alessi, M.C., Pirelli, F., Morange, P., Henry, M., Nalbone, G. and Juhan-Vague, I. (1997) Production of plasminogen activator inhibitor-1 by adipose tissue, possible link between visceral fat accumulation and vascular disease. *Diabetes* 46, 860–867.

Arnalich, F., Lopez, J., Codoceo, R., Jimenez, M., Madero, R. and Montiel, C. (1999) Relationship of plasma leptin to plasma cytokines and human survival in sepsis and septic shock. *Journal of Infectious Diseases* 180, 908–911.

Bacon, K.B. and Harrison, J.K. (2000) Chemokines and their receptors in neurobiology, perspectives in physiology and homeostasis. *Journal of Neuroimmunology* 104, 92–97.

Baer, D.J., Judd, J.T., Clevidence, B.A. and Tracy, R.P. (2004) Dietary fatty acids affect plasma markers of inflammation in healthy men fed controlled diets, a randomized crossover study. *American Journal of Clinical Nutrition* 79, 969–973.

Baggiolini, M., Dewald, B. and Moser, B. (1997) Human chemokines, an update. *Annual Review of Immunology* 15, 675–705.

Baldo, A., Sniderman, A.D., St-Luce, S., Avramoglu, R.K., Maslowska, M., Hoang, B., Monge, J.C., Bell, A., Mulay, S. and Cianflone, K. (1993) The adipsin–acylation stimulating protein system and regulation of intracellular triglyceride synthesis. *Journal of Clinical Investigation* 92, 1543–1547.

Baldo, A., Sniderman, A.D., Yazruel, Z. and Cianflone, K. (1995) *Journal of Lipid Research* 36, 1415–1426.

Balkwill, F. (1993) Cytokines in health and disease. *Immunology Today* 14, 149–150.

Banerjee, R.R., Rangwala, S.M., Shapiro, J.S., Rich, A.S., Rhoades, B., Qi, Y., Wang, J., Rajala, M.W., Pocai, A., Scherer, P.E., Steppan, C.M., Ahima, R.S., Obici, S., Rossetti, L. and Lazar, M.A. (2004) Regulation of fasted blood glucose by resistin. *Science* 303, 1195–1198.

Barinas-Mitchell, E., Cushman, M., Meilahn, E.N., Tracy, R.P. and Kuller, L.H. (2001) Serum levels of C-reactive protein are associated with obesity, weight gain, and hormone replacement therapy in healthy postmenopausal women. *American Journal of Epidemiology* 153, 1094–1101.

Bastard, J.P., Jardel, C., Bruckert, E., Blondy, P., Capeau, J., Laville, M., Vidal, H. and Hainque, B. (2000) Elevated levels of interleukin-6 are reduced in serum and subcutaneous adipose tissue of obese women after weight loss. *Journal of Clinical Endocrinology and Metabolism* 85, 3338–3342.

Baumann, H., Morella, K.K., White, D.W., Demdski, M., Bailon, P.S., Kim, H., Lai, C.F. and Tartaglia, L.A. (1996) The full-length leptin receptor has signalling capabilities of interleukin 6-type cytokine receptors. *Proceedings of the National Academy of Sciences of the United States of America* 93, 8374–8378.

Belluzzi, A., Brignola, C., Campieri, M., Pera, A., Boschi, S. and Miglioli, M. (1996) Effect of an enteric-coated fish-oil preparation on relapses in Crohn's disease. *New England Journal of Medicine* 334, 1557–1560.

Bernotiene, E., Palmer, G. and Gabay, C. (2006) The role of leptin in innate and adaptive immune responses. *Arthritis, Research and Therapy* 8, 217–226.

Bès-Houtmann, S., Roche, R., Hoareau, L., Gonthier, M., Festy, F., Caillens, H., Gasque, P., Lefebvre d'Hellencourt, C. and Cesari, M. (2007) Presence of functional TLR2 and TLR4 on human adipocytes. *Histochemistry and Cell Biology* 127, 131–137.

Boring, M., Gosling, J., Cleary, M. and Charo, I.F. (1998) Decreased lesion formation in $CCR^{2-/-}$ mice reveals a role for chemokines in the initiation of atherosclerosis. *Nature* 394, 894–897.

Bozaoglu, K., Bolton, K., McMillan, J., Zimmet, P., Jowett, J., Collier, G., Walder, K. and Segal, D. (2007) Chemerin is a novel adipokine associated with obesity and metabolic syndrome. *Endocrinology* 148, 4687–4694.

Broughton, K.S., Johnson, C.S., Pace, B.K., Liebman, M. and Kleppinger, K.M. (1997) Reduced asthma symptoms with *n*-3 fatty acid ingestion are related to 5-series leukotriene production. *American Journal of Clinical Nutrition* 65, 1011–1017.

Bruun, J.M., Pedersen, S.B. and Richelsen, B. (2001) Regulation of interleukin 8 production and gene expression in human adipose tissue *in vitro*. *Journal of Clinical Endocrinology and Metabolism* 86, 1267–1273.

Bruun, J.M., Verdich, C., Toubro, S., Astrup, A. and Richelsen, B. (2003) Association between measures of insulin sensitivity and circulating levels of interleukin-8, interleukin-6 and tumor necrosis factor-alpha. Effect of weight loss in obese men. *European Journal of Endocrinology* 148, 535–542.

Bruun, J.M., Lihn, A.S., Madan, A.K., Pedersen, S.B., Schiott, K.M., Fain, J.N. and Richelsen, B. (2004) Higher production of IL-8 in visceral vs. subcutaneous adipose tissue. Implication of non-adipose cells in adipose tissue. *American Journal of Physiology Endocrinology and Metabolism* 286, E8–E13.

Bucher, H.C., Hengstler, P., Schindler, C. and Meier, G. (2002) *N*-3 Polyunsaturated fatty acids in coronary heart disease: a meta-analysis of randomized controlled trials. *The American Journal of Medicine* 112, 298–304.

Calder, P.C. (2006) *n*-3 Polyunsaturated fatty acids, inflammation, and inflammatory diseases. *The American Journal of Clinical Nutrition* 83, 1505S–1519S.

Chandra, S. and Chandra, R.K. (1986) Nutrition, immune response, and outcome. *Progress in Food and Nutrition Science* 986, 1–65.

Chandran, M., Phillips, S.A., Ciaraldi, T. and Henry, R.R. (2003) Adiponectin, more than just another fat cell hormone? *Diabetes Care* 26, 2442–2450.

Chen, B., Lam, K.S., Wang, Y., Wu, D., Lam, M.C., Shen, J., Wong, L., Hoo, R.L., Zhang, J. and Xu, A. (2006) Hypoxia dysregulates the production of adiponectin and plasminogen activator inhibitor-1 independent of reactive oxygen species in adipocytes. *Biochemical and Biophysical Research Communications* 341, 549–556.

Chomarat, P., Banchereau, J., Davoust, J. and Palucka, A.K. (2000) IL-6 switches the differentiation of monocytes from dendritic cells to macrophages. *Nature Immunology* 1, 510–514.

Choy, L.N., Rosen, B.S. and Spiegelman, B.M. (1992) Adipsin and an endogenous pathway of complement from adipose cells. *Journal of Biological Chemistry* 267, 12736–12741.

Christiansen, T., Richelsen, B. and Bruun, J.M. (2005) Monocyte chemoattractant protein-1 is produced in isolated adipocytes, associated with adiposity and reduced after weight loss in morbid obese subjects. *International Journal of Obesity and Related Metabolic Disorders* 29, 146–150.

Chung, S., Lapoint, K., Martinez, K., Kennedy, A., Boysen Sandberg, M. and McIntosh, M.K. (2006) Preadipocytes mediate lipopolysaccharide-induced inflammation and insulin resistance in primary cultures of newly differentiated human adipocytes. *Endocrinology* 147, 5340–5351.

Cianflone, K., Sniderman, A.D., Walsh, M.J., Vu, H., Gagnon, J. and Rodriguez, M.A. (1989) Purification and characterization of acylation stimulating protein. *Journal of Biological Chemistry* 264, 426–430.

Cianflone, K., Xia, Z. and Chen, L.Y. (2003) Critical review of acylation-stimulating protein physiology in humans and rodents. *Biochimica Biophysica Acta* 1609, 127–143.

Combs, T.P., Berg, A.H., Rajala, M.W., Klebanov, S., Iyengar, P., Jimenez-Chillaron, J.C., Patti, M.E., Klein, S.L., Weinstein, R.S. and Scherer, P.E. (2003) Sexual differentiation, pregnancy, calorie restriction, and aging affect the adipocyte-specific secretory protein adiponectin. *Diabetes* 52, 268–276.

Coppack, S.W. (2001) Pro-inflammatory cytokines and adipose tissue. *Proceedings of the Nutrition Society* 60, 349–356.

Cousin, B., Munoz, O., Andre, M., Fontanilles, A.M., Dani, C., Cousin, J.L., Laharrague, P., Casteilla, L. and Pénicaud, L. (1999) A role for preadipocytes as macrophage-like cells. *FASEB Journal* 13, 305–312.

Crevel, R. van, Karyadi, E., Netea, M.G., Verhoef, H., Nelwan, R.H., West, C.E. and van der Meer, J.W. (2002) Decreased plasma leptin concentrations in tuberculosis patients are associated with wasting and inflammation. *Journal of Clinical Endocrinology and Metabolism* 87, 758–763.

Danesh, J., Whincup, P., Walker, M., Lennon, L., Thomson, A., Appleby, P., Gallimore, J.R. and Pepys, M.B. (2000) Low grade inflammation and coronary heart disease, prospective study and updated meta analyses. *British Medical Journal* 321, 199–204.

De Ferranti, S. and Rifai, N. (2002) C-reactive protein and cardiovascular disease, a review of risk prediction and interventions. *Clinica Chimica Acta* 317, 1–15.

Despres, J.P. (1998) The insulin resistance-dyslipidemic syndrome of visceral obesity, effect on patients' risk. *Obesity Research* 6, 8S–17S.

De Taeye, B., Smith, L.H. and Vaughan, D.E. (2005) Plasminogen activator inhibitor-1, a common denominator in obesity, diabetes and cardiovascular disease. *Current Opinion in Pharmacology* 5, 149–154.

Dietrich, M. and Jialal, I. (2005) The effect of weight loss on a stable biomarker of inflammation, C-reactive protein. *Nutrition Reviews* 63, 22–28.

Diez, J.J. and Iglesias, P. (2003) The role of the novel adipocyte-derived hormone adiponectin in human disease. *European Journal of Endocrinology* 148, 293–300.

Dufour, J.H., Dziejman, M., Liu, M.T., Leung, J.H., Lane, T.E. and Luster, A.D. (2002) IFN-γ-inducible protein 10 (IP-10; CXCL10)-deficient mice reveal a role for IP-10 in effector T-cell generation and trafficking. *Journal of Immunology* 168, 3195–3204.

Dugail, I., Le Liepvre, X., Quignard-Boulange, A., Pairaultt, J. and Lavau, M. (1989) Adipsin mRNA amounts are not decreased in the genetically obese Zucker rat. *Biochemical Journal* 257, 917–919.

Economou, E.V., Malamitsi-Puchner, A.V., Pitsavos, C.P., Kouskoun, E.E., Magaziotou-Elefsinioti, I., Damianaki-Uranou, D., Stefanadi, C.I. and Creatsas, G. (2004) Negative association between circulating total homocysteine and proinflammatory chemokines MCP-1 and RANTES in prepubertal lean, but not in obese, children. *Journal of Cardiovascular Pharmacology* 44, 310–315.

Erbagci, A.B., Tarakcioglu, M., Coskun, Y., Sivasli, E. and Sibel, N.E. (2001) Mediators of inflammation in children with type I diabetes mellitus, cytokines in type I diabetic children. *Clinical Biochemistry* 34, 645–650.

Eriksson, P., Reynisdottir, S., Lönnqvist, F., Stemme, V., Hamsten, A. and Arner, P. (1998) Adipose tissue secretion of plasminogen activator inhibitor-1 in non-obese and obese individuals. *Diabetologia* 41, 65–71.

Eritsland, J., Seljeflot, I., Abdelnoor, M., Arnesen, H. and Torjesen, P.A. (1994) Long-term effects of *n*-3 fatty acids on serum lipids and glycaemic control. *Scandinavian Journal of Clinical and Laboratory Investigation* 54, 273–280.

Esposito, K., Pontillo, A., Ciotola, M., Di Palo, C., Grella, E., Nicoletti, G. and Giugliano, D. (2002) Weight loss reduces interleukin-18 levels in obese women. *Journal of Clinical Endocrinology and Metabolism* 87, 3864–3866.

Esposito, K., Pontillo, A., Giugliano, F., Giugliano, G., Marfella, R., Nicoletti, G. and Giugliano, D. (2003) Association of low interleukin-10 levels with metabolic syndrome in obese women. *Journal of Clinical Endocrinology and Metabolism* 88, 1055–1058.

Fain, J.N., Cheema, P.S., Bahouth, S.W. and Lloyd Hiler, M. (2003) Resistin release by human adipose tissue explants in primary culture. *Biochemical and Biophysical Research Communications* 300, 674–678.

Fain, J.N., Madan, A.K., Hiler, M.L., Cheema, P. and Bahouth, S.W. (2004) Comparison of the release of adipokines by adipose tissue, adipose tissue matrix, and adipocytes from visceral and subcutaneous abdominal adipose tissues of obese humans. *Endocrinology* 145, 2273–2282.

Fain, J.N., Bahouth, S.W. and Madan, A.K. (2005) Involvement of multiple signaling pathways in the post-bariatric induction of IL-6 and IL-8 mRNA and release in human visceral adipose tissue. *Biochemical Pharmacology* 69, 1315–1324.

Fantuzzi, G. (2005) Adipose tissue, adipokines, and inflammation. *Journal of Allergy and Clinical Immunology* 115, 911–919.

Fantuzzi, G. and Faggioni, R. (2000) Leptin in the regulation of immunity, inflammation and hematopoiesis. *Journal of Leukocyte Biology* 68, 437–446.

Fantuzzi, G., Sennello, J.A., Batra, A., Fedke, I., Lehr, H.A., Zeitz, M. and Siegmund, B. (2005) Defining the role of T-cell-derived leptin in the modulation of hepatic or intestinal inflammation in mice. *Immunology* 142, 31–38.

Farooqi, I.S., Matarese, G., Lord, G.M., Keogh, J.M., Lawrence, E., Agwu, C., Sanna, V., Jebb, S.A., Perna, F., Fontana, S., Lechler, R.I., DePaoli, A.M. and O'Rahilly, S. (2002) Beneficial effects of leptin on obesity, T-cell hyporesponsiveness, and neuroendocrine/metabolic dysfunction of human congenital leptin deficiency. *Journal of Clinical Investigation* 110, 1093–1103.

Feingold, K.R., Serio, M.K., Adi, S., Moser, A.H. and Grunfeld, C. (1989) Tumor necrosis factor stimulates hepatic lipid synthesis and secretion. *Endocrinology* 124, 2336–2342.

Feingold, K.R., Soued, M., Adi, S., Staprans, I., Neese, R., Shigenaga, J., Doerrler, W., Moser, A., Dinarello, C.A. and Grunfeld, C. (1991) Effect of interleukin-1 on lipid metabolism in the rat. Similarities to and differences from tumor necrosis factor. *Arteriosclerosis and Thrombosis* 11, 495–500.

Festa, A., D'Agostino, R. Jr, Howard, G., Mykkanen, L., Tracy, R.P. and Haffner, S.M. (2000) Chronic subclinical inflammation as part of the insulin resistance syndrome, the Insulin Resistance Atherosclerosis Study (IRAS). *Circulation* 102, 42–47.

Flachs, P., Mohamed-Ali, V., Horakova, O., Rossmeisl, M., Hosseinzadeh-Attar, M.J., Hensler, M., Ruzickova, J. and Kopecky, J. (2006) Polyunsaturated fatty acids of marine origin induce adiponectin in mice fed a high-fat diet. *Diabetologia* 49, 394–397.

Fleischmann, E., Kurz, A., Niedermayr, M., Schebesta, K., Kimberger, O., Sessler, D.I., Kabon, B. and Prager, G. (2005) Tissue oxygenation in obese and non-obese patients during laparoscopy. *Obesity Surgery* 15, 813–819.

Flier, J.S., Cook, K.S., Usher, P. and Spiegelman, B.M. (1987) Severely impaired adipsin expression in genetic and acquired obesity. *Science* 237, 405–408.

Fried, S.K., Bunkin, D.A. and Greenberg, A.S. (1998) Omental and subcutaneous adipose tissues of obese subjects release interleukin-6, depot difference and regulation by glucocorticoid. *Journal of Clinical Endocrinology and Metabolism* 83, 847–850.

Friedman, J.M. and Halaass, J.L. (1998) Leptin and the regulation of body weight in mammals. *Nature* 395, 763–770.

Fukuhara, A., Matsuda, M., Nishizawa, M., Segawa, K., Tanaka, M., Kishimoto, K., Matsuki, Y., Murakami, M., Ichisaka, T., Murakami, H., Watanabe, E., Takagi, T., Akiyoshi, M., Ohtsubo, T., Kihara, S., Yamashita, S., Makishima, M., Funahashi, T., Yamanaka, S., Hiramatsu, R., Matsuzawa, Y. and Shimomura, I. (2005) Visfatin, a protein secreted by visceral fat that mimics the effects of insulin. *Science* 307, 426–430.

Fukuhara, A., Matsuda, M., Nishizawa, M., Segawa, K., Tanaka, M., Kishimoto, K., Matsuki, Y., Murakami, M., Ichisaka, T., Murakami, H., Watanabe, E., Takagi, T., Akiyoshi, M., Ohtsubo, T., Kihara, S., Yamashita, S., Makishima, M., Funahashi, T., Yamanaka, S., Hiramatsu, R., Matsuzawa, Y. and Shimomura, I. (2007) Visfatin, a protein secreted by visceral fat that mimics the effects of insulin. Retraction. *Science* 318, 565.

Furukawa, S., Fujita, T., Shimabukuro, M., Iwaki, M., Yamada, Y., Nakajima, Y., Nakayama, O., Makishima, M., Matsuda, M. and Shimomura, I. (2004) Increased oxidative stress in obesity and its impact on metabolic syndrome. *Journal of Clinical Investigation* 114, 1752–1761.

Gameren, M.M. van, Willemse, P.H., Mulder, N.H., Limburg, P.C., Groen, H.J., Vellenga, E. and de Vries, E.G. (1994) Effects of recombinant human interleukin-6 in cancer patients, a phase I-II study. *Blood* 84, 1434–1441.

Gaucher, E.A., Miyamoto, M.M. and Benner, S.A. (2003) Evolutionary, structural and biochemical evidence for a new interaction site of the leptin obesity protein. *Genetics* 163, 1549–1553.

Gerard, C. and Rollins, B.J. (2001) Chemokines and disease. *Nature Immunology* 2, 108–115.

Gerhardt, C.C., Romero, I.A., Cancello, R., Camoin, L. and Strosberg, A.D. (2001) Chemokines control fat accumulation and leptin secretion by cultured human adipocytes. *Molecular and Cellular Endocrinology* 175, 81–92.

Germinario, R., Sniderman, A.D., Manuel, S., Pratt, S., Baldo, A. and Cianflone, K. (1993) Coordinate regulation of triacylglycerol synthesis and glucose transport by acylation-stimulating protein. *Metabolism* 40, 574–580.

Geusens, P., Wouters, C., Nijs, J., Jiang, Y. and Dequeker, J. (1994) Long-term effect of omega-3 fatty acid supplementation in active rheumatoid arthritis. A 12 month, double blind, controlled study. *Arthritis and Rheumatism* 37, 824–829.

Gómez-Ambrosi, J., Salvador, J., Rotellar, F., Silva, C., Catalán, V., Rodríguez, A., Gil, M.J. and Frühbeck, G. (2006) Increased serum amyloid A concentrations in morbid obesity decrease after gastric bypass. *Obesity Surgery* 16, 262–269.

Goralski, K.B., McCarthy, T.C., Hanniman, E.A., Zabel, B.A., Butcher, E.C., Parlee, S.D., Muruganandan, S. and Sinal, C.J. (2007) Chemerin, a novel adipokine that regulates adipogenesis and adipocyte metabolism. *Journal of Biological Chemistry* 282, 28175–28188.

Gottschling-Zeller, H., Birgel, M., Röhrig, K. and Hauner, H. (2000) Effect of tumor necrosis factor alpha and transforming growth factor beta 1 on plasminogen activator inhibitor-1 secretion from subcutaneous and omental human fat cells in suspension culture. *Metabolism* 49, 666–671.

Greenberg, A.S., Nordan, R.P., McIntosh, J., Calvo, J.P., Scow, R.O. and Jablons, D. (1992) Interleukin-6 reduces lipoprotein lipase activity in adipose tissue of mice *in vivo* and in 3T3-L1 adipocytes, a possible role for interleukin 6 in cancer cachexia. *Cancer Research* 52, 4113–4116.

Gu, L., Okada, Y., Clinton, S.K., Gerard, C., Sukhova, G.K., Libby, P. and Rollins, B.J. (1998) Absence of monocyte chemoattractant protein-1 reduces atherosclerosis in low density lipoprotein receptor-deficient mice. *Molecular Cell* 2, 275–281.

Hagiwara, S., Makita, Y., Gu, L., Tanimoto, M., Zhang, M., Nakamura, S., Kaneko, S., Itoh, T., Gohda, T., Horikoshi, S. and Tomino, Y. (2006) Eicosapentaenoic acid ameliorates diabetic nephropathy of type 2 diabetic KKAy/Ta mice: involvement of MCP-1 suppression and decreased ERK1/2 and p38 phosphorylation. *Nephrology, Dialysis, Transplantation* 21, 605–615.

Hall, G. van, Steensberg, A., Sacchetti, M., Fischer, C., Keller, C., Schjerling, P., Hiscock, N., Moller, K., Saltin, B., Febbraio, M.A. and Pedersen, B.K. (2003) Interleukin-6 stimulates lipolysis and fat oxidation in humans. *Journal of Clinical Endocrinology and Metabolism* 88, 3005–3010.

Han, S.N., Leka, L.S., Lichtenstein, A.H., Ausman, L.M., Schaefer, E.J. and Meydani, S.N. (2002) Effect of hydrogenated and saturated, relative to polyunsaturated, fat on immune and inflammatory responses of adults with moderate hypercholesterolemia. *Journal of Lipid Research* 43, 445–452.

Hardardottir, I., Grunfeld, C. and Feingold, K.R. (1994) Effects of endotoxin and cytokines on lipid metabolism. *Lipidology* 5, 207–215.

Hauner, H. (2005) Secretory factors from human adipose tissue and their functional role. *Proceedings of the Nutrition Society* 64, 163–169.

Herder, C., Haastert, B., Müller-Scholze, S., Koenig, W., Thorand, B., Holle, R., Wichmann, H.E., Scherbaum, W.A., Martin, S. and Kolb, H. (2005) Association of systemic chemokine concentrations with impaired glucose tolerance and type 2 diabetes. Results from the Cooperative Health Research in the Region of Augsburg Survey S4 (KORA S4). *Diabetes* 54 (Suppl. 2), S11–S17.

Herder, C., Baumert, J., Thorand, B., Koenig, W., de Jager, W., Meisinger, C., Illig, T., Martin, S. and Kolb, H. (2006) Chemokines as risk factors for type 2 diabetes – results from the MONICA/KORA Augsburg Study, 1984–2002. *Diabetologia* 49, 921–929.

Herder, C., Hauner, H., Kempf, K., Kolb, H. and Skurk, T. (2007) Constitutive and regulated expression and secretion of interferon-c-inducible protein 10 (IP-10/CXCL10) in human adipocytes. *International Journal of Obesity* 31, 403–410.

Hida, K., Wada, J., Eguchi, J., Zhang, H., Baba, M., Seida, A., Hashimoto, I., Okada, T., Yasuhara, A., Nakatsuka, A., Shikata, K., Hourai, S., Futami, J., Watanabe, E., Matsuki, Y., Hiramatsu, R., Akagi, S., Makino, H. and Kanwar, Y.S. (2005) Visceral adipose tissue-derived serine protease inhibitor, a unique insulin-sensitizing adipocytokine in obesity. *Proceedings of the National Academy of Sciences of the United States of America* 102, 10610–10615.

Hotamisligil, G.S. (1999) The role of TNFα and TNF receptors in obesity and insulin resistance. *Journal of Internal Medicine* 245, 621–625.

Hotamisligil, G.S. and Spiegelman, B.M. (1994) Tumor necrosis factor α, a key component of the obesity-diabetes link. *Diabetes* 43, 1271–1278.

Hotamisligil, G.S., Shargill, N.S. and Spiegelman, B.M. (1993) Adipose expression of tumor necrosis factor-alpha – direct role in obesity-linked insulin resistance. *Science* 259, 87–91.

Hotamisligil, G.S., Arner, P., Caro, J.F., Atkinson, R.L. and Spiegelman, B.M. (1995) Increased adipose tissue expression of tumor necrosis factor-α in human obesity and insulin resistance. *Journal of Clinical Investigation* 95, 2409–2415.

Hu, E., Liang, P. and Spiegelman, B.M. (1996) AdipoQ is a novel adipose-specific gene dysregulated in obesity. *Journal of Biological Chemistry* 271, 10697–10703.

Inoue, S., Egashira, K., Ni, W., Kitamoto, S., Usui, M., Otani, K., Ishibashi, M., Hiasa, K., Nishida, K. and Takeshita, A. (2002) Anti-monocyte chemoattractant protein-1 gene therapy limits progression and destabilization of established atherosclerosis in apolipoprotein E-knockout mice. *Circulation* 106, 2700–2706.

Janke, J., Engeli, S., Gorzelniak, K., Luft, F.C. and Sharma, A.M. (2002) Resistin gene expression in human adipocytes is not related to insulin resistance. *Obesity Research* 10, 1–5.

Johnson, B.D., Kip, K.E., Marroquin, O.C., Ridker, P.M., Kelsey, S.F., Shaw, L.J., Pepine, C.J., Sharaf, B., Bairey Merz, C.N., Sopko, G., Olson, M.B., Reis, S.E., National Heart, Lung, and Blood Institute. (2004) Serum amyloid A as a predictor of coronary artery disease and cardiovascular outcome in women. The National Heart, Lung, and Blood Institute-Sponsored Women's Ischemia Syndrome Evaluation (WISE). *Circulation* 109, 726–732.

Jones, S.A., Horiuchi, S., Topley, N., Yamamoto, N. and Fuller, G.M. (2001) The soluble interleukin 6 receptor, mechanisms of production and implications in disease. *FASEB Journal* 15, 43–58.

Juhan-Vague, I., Alessi, M.C., Mavri, A. and Morange, P.E. (2003) Plasminogen activator inhibitor-1, inflammation, obesity, insulin resistance and vascular risk. *Journal of Thrombosis and Haemostasis* 1, 1575–1579.

Kabon, B., Nagele, A., Reddy, D., Eagon, C., Fleshman, J.W., Sessler, D.I. and Kurz, A. (2004) Obesity decreases perioperative tissue oxygenation. *Anesthesiology* 100, 274–280.

Keller, C., Steensberg, A., Pilegaard, H., Osada, T., Saltin, B., Pedersen, B.K. and Neufer, P.D. (2001) Transcriptional activation of the IL-6 gene in human contracting skeletal muscle, influence of muscle glycogen content. *FASEB Journal* 15, 2748–2750.

Kern, P.A., Saghizadech, M., Ong, J.M., Bosch, R.J., Deem, R. and Simsolo, R.B. (1995) The expression of tumor necrosis factor in human adipose tissue. Regulation by obesity, weight loss and relationship to lipoprotein lipase. *Journal of Clinical Investigation* 95, 2111–2119.

Klover, P.J., Clementi, A.H. and Mooney, R.A. (2005) Interleukin-6 depletion selectively improves hepatic insulin action in obesity. *Endocrinology* 146, 3417–3427.

Kohler, H.P. and Grant, P.J. (2000) Plasminogen-activator inhibitor type 1 and coronary artery disease. *New England Journal of Medicine* 342, 1792–801.

Konkle, B.A., Schuster, S.J., Kelly, M.D., Harjes, K., Hassett, D.E., Bohrer, M. and Tavassoli, M. (1992) Plasminogen activator inhibitor-1 messenger RNA expression is induced in rat hepatocytes *in vivo* by dexamethasone. *Blood* 79, 2636–2642.

Kralisch, S., Klein, J., Lossner, U., Blüher, M., Paschke, R., Stumvoll, M. and Fasshauer, M. (2005) Interleukin-6 is a negative regulator of visfatin gene expression in 3T3-L1 adipocytes. *American Journal of Physiology – Endocrinology and Metabolism* 289, 586–590.

Kumada, M., Kihara, S., Sumitsuji, S., Kawamoto, T., Matsumoto, S., Ouchi, N., Arita, Y., Okamoto, Y., Shimomura, I., Hiraoka, H., Nakamura, T., Funahashi, T., Matsuzawa, Y.; Osaka CAD Study Group. Coronary artery disease. (2003) Association of hypoadiponectinemia with coronary artery disease in men. *Arteriosclerosis, Thrombosis, and Vascular Biology* 23, 85–89.

Kumada, M., Kihara, S., Ouchi, N., Kobayashi, H., Okamoto, Y., Ohashi, K., Maeda, K., Nagaretani, H., Kishida, K., Maeda, N., Nagasawa, A., Funahashi, T. and Matsuzawa, Y.

(2004) Adiponectin specifically increased tissue inhibitor of metalloprotenase-1 through interleukin-10 expression in human macrophages. *Circulation* 109, 2046–2049.

Lago, F., Diéguez, C., Gómez-Reino, J. and Gualillo, O. (2007a) The emerging role of adipokines as mediators of inflammation and immune responses. *Cytokine and Growth Factor Reviews* 18, 313–325.

Lago, F., Diéguez, C., Gómez-Reino, J. and Gualillo, O. (2007b) Adipokines as emerging mediators of immune response and inflammation. *Nature Clinical Practice Rheumatology* 3, 716–724.

Lago, R., Gómez, R., Lago, F., Gómez-Reino, J. and Gualillo, O. (2008) Leptin beyond body weight regulation – current concepts concerning its role in immune function and inflammation. *Cell Immunology* 252, 139–145.

Laing, K.J. and Secombes, C.J. (2004) Chemokines. *Developmental and Comparative Immunology* 28, 443–460.

Lin, Y., Lee, H., Berg, A.H., Lisanti, M.P., Shapiro, L. and Scherer, P.E. (2000) The lipopolysaccharide-activated Toll-like receptor (TLR)-4 induces synthesis of the closely related receptor TLR-2 in adipocytes. *Journal of Biological Chemistry* 275, 24255–24263.

Liu, Y., Hulten, L.M. and Wiklund, O. (1997) Macrophages isolated from human atherosclerotic plaques produce IL-8, and oxysterols may have a regulatory function for IL-8 production. *Arteriosclerosis Thrombosis and Vascular Biology* 17, 317–323.

Lolmede, K., Durand de Saint Front, V., Galitzky, J., Lafontan, M. and Bouloumie, A. (2003) Effects of hypoxia on the expression of proangiogenic factors in differentiated 3T3-F442A adipocytes. *International Journal of Obesity* 27, 1187–1195.

Lord, G.M., Matarese, G., Howard, J.K., Baker, R.J., Bloom, S.R. and Lechler, R.I. (1998) Leptin modulates the T-cell immune response and reverses starvation-induced-immunosuppression. *Nature* 394, 897–901.

Lorenz, E. (2006) TLR2 and TLR4 expression during bacterial infections. *Current Pharmaceutical Design* 12, 4185–4193.

Luster, A.D. (1998) Chemokines – chemotactic cytokines that mediate inflammation. *New England Journal of Medicine* 338, 436–445.

Luster, A.D. and Ravetch, J.V. (1987) Biochemical characterization of a γ interferon-inducible cytokine (IP-10). *Journal of Experimental Medicine* 166, 1084–1097.

Lyon, C.L., Law, R.E. and Hsueh, W.A. (2003) Minireview: adiposity, inflammation and atherogenesis. *Endocrinology* 144, 2195–2200.

McDermott, M.F. (2001) TNF and TNFR biology in health and disease. *Cellular and Molecular Biology* 47, 619–635.

Mach, F., Sauty, A., Iarossi, A.S., Sukhova, G.K., Neote, K., Libby, P. and Luster, A.D. (1999) Differential expression of three T lymphocyte-activating CXC chemokines by human atheroma-associated cells. *Journal of Clinical Investigation* 104, 1041–1050.

Mackiewicz, A., Schooltink, H., Heinrich, P.C. and Rose-John, S. (1992) Complex of soluble human IL-6-receptor/IL-6 upregulates expression of acute-phase proteins. *Journal of Immunology* 149, 2021–2027.

McTernan, C.L., McTernan, P.G., Harte, A.L., Levick, P.L., Barnett, A.H. and Kumar, S. (2002) Resistin, central obesity, and type 2 diabetes. *Lancet* 359, 46–47.

Maeda, K., Okubo, K., Shimomura, I., Funahashi, T., Matsuzawa, Y. and Matsubara, K. (1996) cDNA cloning and expression of a novel adipose specific collagen-like factor, apM1 (AdiPose Most abundant Gene transcript 1). *Biochemical and Biophysical Research Communications* 221, 286–289.

Malle, E. and De Beer, F.C. (1996) Human serum amyloid A (SAA) protein, a prominent acute-phase reactant for clinical practice. *European Journal of Clinical Investigation* 26, 427–435.

Mantovani, A. (1999) The chemokine system, redundancy for robust outputs. *Immunology Today* 20, 254–257.

Masaki, T., Chiba, S., Tatsukawa, H., Yasuda, T., Noguchi, H., Seike, M. and Yoshimatsu, H. (2004) Adiponectin protects LPS-induced liver injury through modulation of TNF-alpha in KK-Ay obese mice. *Hepatology* 40, 177–184.

Maslowska, M., Sniderman, A.D., Germinario, R. and Cianflone, K. (1997) ASP stimulates glucose transport in cultured human adipocytes. *International Journal of Obesity* 21, 261–266.

Matsushima, K. and Oppenheim, J.J. (1989) Interleukin 8 and MCAF, novel inflammatory cytokines inducible by IL 1 and TNF. *Cytokine* 1, 2–13.

Matsushita, K., Yatsuya, H., Tamakoshi, K., Wada, K., Otsuka, R., Takefuji, S., Sugiura, K., Kondo, T., Murohara, T. and Toyoshima, H. (2006) Comparison of circulating adiponectin and proinflammatory markers regarding their association with metabolic syndrome in Japanese men. *Arteriosclerosis, Thrombosis, and Vascular Biology* 26, 871–876.

Maya-Monteiro, C.M., Almeida, P.E., D'Avila, H., Martins, A.S., Rezende, A.P., Castro-Faria-Neto, H. and Bozza, P.T. (2008) Leptin induces macrophage lipid body formation by a phosphatidylinositol 3-kinase- and mammalian target of rapamycin-dependent mechanism. *Journal of Biological Chemistry* 283, 2203–2210.

Mayser, P., Mrowietz, U., Arenberger, P., Bartak, P., Buchvald, J., Christophers, E., Jablonska, S., Salmhofer, W., Schill, W.B., Kramer, H.J., Schlotzer, E., Mayer, K., Seeger, W. and Grimminger, F. (1998) Omega-3 fatty acid-based lipid infusion in patients with chronic plaque psoriasis, results of a double-blind, randomized, placebo-controlled, multicenter trial. *Journal of the American Academy of Dermatology* 38, 539–547.

Mertens, I. and Van Gaal, L.F. (2002) Obesity, haemostasis and the fibrinolytic system. *Obesity Reviews* 3, 85–101.

Mertens, I., Ballaux, D., Funahashi, T., Matsuzawa, Y., Van der Planken, M., Verrijken, A., Ruige, J.B. and VanGaal, L.F. (2005) Inverse relationship between plasminogen activator inhibitor-1 activity and adiponectin in overweight and obese women. Interrelationship with visceral adipose tissue, insulin resistance, HDL-cholesterol and inflammation. *Thrombosis and Haemostasis* 94, 1190–1195.

Mohamed-Ali, V., Goodrick, S., Rawesh, A., Katz, D.R., Miles, J.M., Klein, S. and Coppack, S.W. (1997) Subcutaneous adipose tissue releases interleukin-6 but not tumor necrosis factor-alpha, *in vivo*. *Journal of Clinical Endocrinology and Metabolism* 82, 4196–4200.

Mohamed-Ali, V., Goodrick, S., Bulmer, K., Holly, J.M.P., Yudkin, J.S. and Coppack, S.W. (1999) Production of soluble tumor necrosis factor receptors by human subcutaneous adipose tissue *in vivo*. *American Journal of Physiology* 277, 971–975.

Moreau, M., Brocheriou, I., Petit, L., Ninio, E., Chapman, M.J. and Rouis, M. (1999) Interleukin-8 mediates downregulation of tissue inhibitor of metalloproteinase-1 expression in cholesterol-loaded human macrophages, relevance to stability of atherosclerotic plaque. *Circulation* 99, 420–426.

Mozaffarian, D., Pischon, T., Hankinson, S.E., Rifai, N., Joshipura, K., Willett, W.C. and Rimm, E.B. (2004) Dietary intake of trans fatty acids and systemic inflammation in women. *American Journal of Clinical Nutrition* 79, 606–612.

Murray, I., Parker, R.A., Kirchgessner, T.G., Tran, J., Zhang, Z.J., Westerlund, J. and Ciaflone, K. (1997) Functional bioactive recombinant acylation stimulating protein is distinct from C3a anaphylatoxin. *Journal of Lipid Research* 38, 2492–2501.

Nakano, Y., Tobe, T., Choi-Miura, N.H., Mazda, T. and Tomita, M. (1996) Isolation and characterization of GBP28, a novel gelatin-binding protein purified from human plasma. *Journal of Biochemistry* 120, 803–812.

Nishimoto, N., Yoshizaki, K., Miyasaka, N., Yamamoto, K., Kawai, S., Takeuchi, T., Hashimoto, J., Azuma, J. and Kishimoto, T. (2004) Treatment of rheumatoid arthritis with humanized anti-interleukin-6 receptor antibody, a multicenter, double-blind, placebo-controlled trial. *Arthritis and Rheumatism* 50, 1761–1769.

Nishizawa, H., Shimomura, I., Kishida, K., Maeda, N., Kuriyama, H., Nagaretani, H., Matsuda, M., Kondo, H., Furuyama, N., Kihara, S., Nakamura, T., Tochino, Y., Funahashi, T. and Matsuzawa, Y. (2002) Androgens decrease plasma adiponectin, an insulin-sensitizing adipocyte-derived protein. *Diabetes* 51, 2734–2741.

Nomura, S., Shouzu, A., Omoto, S., Nishikawa, M. and Fukuhara, S. (2000) Significance of chemokines and activated platelets in patients with diabetes. *Clinical and Experimental Immunology* 121, 437–443.

Nonogaki, K., Fuller, G.M., Fuentes, N.L., Moser, A.H., Staprans, I., Grunfeld, C. and Feingold, K.R. (1995) Interleukin-6 stimulates hepatic triglyceride secretion in rats. *Endocrinology* 136, 2143–2149.

Ofei, F., Hurel, S., Newkirk, J., Sopwith, M. and Taylor, R. (1996) Effects of an engineered human anti-TNF-alpha antibody (CDP571) on insulin sensitivity and glycemic control in patients with NIDDM. *Diabetes* 45, 881–885.

O'Rahilly, S., Farooqi, S.I., Yeo, G.S.H. and Challis, B.G. (2003) Minireview: human obesity – lessons from monogenic disorders. *Endocrinology* 144, 3757–3764.

Ouchi, N., Kihara, S., Arita, Y., Maeda, K., Kuriyama, H., Okamoto, Y., Hotta, K., Nishida, M., Takahashi, M., Nakamura, T., Yamashita, S., Funahashi, T. and Matsuzawa, Y. (1999) Novel modulator for endothelial adhesion molecules, adipocyte-derived plasma protein adiponectin. *Circulation* 100, 2473–2476.

Ouchi, N., Ohishi, M., Kihara, S., Funahashi, T., Nakamura, T., Nagaretani, H., Kumada, M., Ohashi, K., Okamoto, Y., Nishizawa, H., Kishida, K., Maeda, N., Nagasawa, A., Kobayashi, H., Hiraoka, H., Komai, N., Kaibe, M., Rakugi, H., Ogihara, T. and Matsuzawa, Y. (2003) Association of hypoadiponectinemia with impaired vasoreactivity. *Hypertension* 42, 231–234.

Ozinsky, A., Underhill, D.M., Fontenot, J.D., Hajjar, A.M., Smith, K.D., Wilson, C.B., Schroeder, L. and Aderem, A. (2000) The repertoire for pattern recognition of pathogens by the innate immune system is defined by cooperation between toll-like receptors. *Proceedings of the National Academy of Sciences of the United States of America* 97, 13766–13771.

Pajvani, U.B., Du, X., Combs, T.P., Berg, A.H., Rajala, M.W., Schulthess, T., Engel, J., Brownlee, M. and Scherer, P.E. (2003) Structure-function studies of the adipocyte-secreted hormone Acrp30/adiponectin. Implications for metabolic regulation and bioactivity. *Journal of Biological Chemistry* 278, 9073–9085.

Panagiotakos, D.B., Pitsavos, C., Yannakoulia, M., Chrysohoou, C. and Stefanadis, C. (2005) The implication of obesity and central fat on markers of chronic inflammation: the ATTICA study. *Atherosclerosis* 183, 308–315.

Papanicolaou, D.A., Wilder, R.L., Manolagas, S.C. and Chrousos, G.P. (1998) The pathophysiologic roles of interleukin-6 in human disease. *Annals of Internal Medicine* 128, 127–137.

Patel, L., Buckels, A.C., Kinghorn, I.J., Murdock, P.R., Holbrook, J.D., Plumpton, C., Macphee, C.H. and Smith, S.A. (2003) Resistin is expressed in human macrophages and directly regulated by PPAR γ activators. *Biochemical and Biophysical Research Communications* 300, 472–476.

Petruschke, T. and Hauner, H. (1993) Tumor necrosis factor-alpha prevents the differentiation of human adipocyte precursor cells and causes delipidation of newly developed fat cells. *Journal of Clinical Endocrinology and Metabolism* 76, 742–747.

Piemonti, L., Calori, G., Mercalli, A., Lattuada, G., Monti, P., Garancini, M.P., Constantino, F., Ruotolo, G., Luzi, L. and Perseghin, G. (2003) Fasting plasma leptin, tumor necrosis factor-a receptor 2, and monocyte chemoattracting protein 1 concentration in a population of glucose-tolerant and glucose-intolerant women. *Diabetes Care* 26, 2883–2889.

Pischon, T., Girman, C.J., Hotamisligil, G.S., Rifai, N., Hu, F.B. and Rimm, E.B. (2004) Plasma adiponectin levels and risk of myocardial infarction in men. *Journal of the American Medical Association* 291, 1730–1737.

Poitou, C., Viguerie, N., Cancello, R., De Matteis, R., Cinti, S., Stich, V., Coussieu, C., Gauthier, E., Courtine, M., Zucker, J.D., Barsh, G.S., Saris, W., Bruneval, P., Basdevant, A., Langin, D. and Clément, K. (2005) Serum amyloid A, production by human white adipocyte and regulation by obesity and nutrition. *Diabetologia* 48, 519–528.

Pond, C.M. and Mattacks, C.A. (1998) *In vivo* evidence for the involvement of the adipose tissue surrounding lymph nodes in immune responses. *Immunology Letters* 63, 159–167.

Pratley, R.E., Wilson, C. and Bogardus, C. (1995) Relation of the white blood cell count to obesity and insulin resistance, effect of race and gender. *Obesity Research* 3, 563–571.

Prins, J.B., Niesler, C.U., Winterford, C.M., Bright, N.A., Siddle, K., O'Rahilly, S., Walker, N.I. and Cameron, D.P. (1997) Tumor necrosis factor-α induces apoptosis of human adipose cells. *Diabetes* 46, 1939–1944.

Rajala, M.W. and Scherer, P.E. (2003) Minireview: the adipocyte – at the crossroads of energy homeostasis, inflammation, and atherosclerosis. *Endocrinology* 144, 3765–3773.

Rea, R. and Donnelly, R. (2004) Resistin, an adipocyte-derived hormone. Has it a role in diabetes and obesity? *Diabetes, Obesity and Metabolism* 6, 163–170.

Reilly, M.P., Lehrke, M., Wolfe, M.L., Rohatgi, A., Lazar, M.A. and Rader, D.J. (2005) Resistin is an inflammatory marker of atherosclerosis in humans. *Circulation* 111, 932–939.

Ridker, P.M., Buring, J.E., Shih, J., Matias, M. and Hennekens, C.H. (1998) Prospective study of C-reactive protein and the risk of future cardiovascular events among apparently healthy women. *Circulation* 98, 731–733.

Ridker, P.M., Hennekens, C.H., Buring, J.E. and Rifai, N. (2000) C-reactive protein and other markers of inflammation in the prediction of cardiovascular disease in women. *New England Journal of Medicine* 342, 836–843.

Robertson, M.J. (2002) Role of chemokines in the biology of natural killer cells. *Journal of Leukocyte Biology* 71, 173–184.

Rohde, L.E., Hennekens, C.H. and Ridker, P.M. (1999) Survey of C-reactive protein and cardiovascular risk factors in apparently healthy men. *American Journal of Cardiology* 84, 1018–1022.

Romuk, E., Skrzep-Poloczek, B., Wojciechowska, C., Tomasik, A., Birkner, E., Wodniecki, J., Gabrylewicz, B., Ochala, A. and Tendera, M. (2002) Selectin-P and interleukin-8 plasma levels in coronary heart disease patients. *European Journal of Clinical Investigation* 32, 657–661.

Rosen, R.S., Cook, K.S., Yaglom, I., Groves, I., Volanakis, I.K., Damm, D., White, T. and Spiegelmann, B.M. (1989) Adipsin and complement factor D activity, an immune-related defect in obesity. *Science* 244, 1483–1487.

Rosenblum, M.G. and Donato, N.J. (1989) Tumor necrosis factor alpha, a multifaceted peptide hormone. *Critical Reviews in Immunology* 9, 21–44.

Ross, R. (1993) The pathogenesis of atherosclerosis, a perspective for the 1990s. *Nature* 362, 801–809.

Rothenbacher, D., Müller-Scholze, S., Herder, C., Koenig, W. and Kolb, H. (2006) Differential expression of chemokines, risk of stable coronary heart disease and correlation with established cardiovascular risk markers. *Arteriosclerosis, Thrombosis, and Vascular Biology* 26, 194–199.

Royen, N. van, Hoefer, I., Buschmann, I., Kostin, S., Voskuil, M., Bode, C., Schaper, W. and Piek, J.J. (2003) Effects of local MCP-1 protein therapy on the development of the collateral circulation and atherosclerosis in Watanabe hyperlipidemic rabbits. *Cardiovascular Research* 57, 178–185.

Salas-Salvadó, J., Bulló, M., García-Lorda, P., Figueredo, R., Del Castillo, D., Bonada, A. and Balanzà, R. (2006) Subcutaneous adipose tissue cytokine production for the restoration of systemic inflammation markers during weight loss. *International Journal of Obesity* 30, 1714–1720.

Saleh, J., Summers, L.K., Cianflone, K., Fielding, B.A., Sniderman, A.D. and Frayn, K.N. (1998) Coordinated release of acylation stimulating protein (ASP) and triacylglycerol clearance by human adipose tissue *in vivo* in the postprandial period. *Journal of Lipid Research* 39, 884–891.

Samad, F., Yamamoto, K. and Loskutoff, D.J. (1996) Distribution and regulation of plasminogen activator inhibitor-1 in murine adipose tissue *in vivo*, induction by tumor necrosis factor-a and lipopolysaccharide. *Journal of Clinical Investigation* 97, 37–46.

Samad, F., Yamamoto, K., Pandey. M. and Loskutoff, D.J. (1997) Elevated expression of transforming growth factor-b in adipose tissue from obese mice. *Molecular Medicine* 3, 36–47.

Sartipy, P. and Loskutoff, D.J. (2003) Monocyte chemoattractant protein 1 in obesity and insulin resistance. *Proceedings of the National Academy of Sciences of the United States of America* 100, 7265–7270.

Savage, D.B., Sewter, C.P., Klenk, E.S., Segal, D.G., Vidal-Puig, A., Considine, R.V. and O'Rahilly, S. (2001) Resistin/Fizz3 expression in relation to obesity and peroxisome proliferators-activated receptor-action in humans. *Diabetes* 50, 2199–2202.

Sawdey, M.S. and Loskutoff, D.J. (1991) Regulation of murine type 1 plasminogen activator inhibitor gene expression *in vivo*, tissue specificity and induction by lipopolysaccharide, tumor necrosis factor-alpha, and transforming growth factor-beta. *Journal of Clinical Investigation* 88, 1346–1353.

Saynor, R. and Gillott, T. (1992) Changes in blood lipids and fibrinogen with a note on safety in a long-term study of the effects of *n*-3 fatty acids in subjects receiving fish oil supplements and followed for seven years. *Lipids* 27, 533–538.

Scherer, P.E., Williams, S., Fogliano, M., Baldini, G. and Lodish, H.F. (1995) A novel serum protein similar to C1q, produced exclusively in adipocytes. *Journal of Biological Chemistry* 270, 26746–26749.

Sethi, J.K. and Hotamisligil, G.S. (1999) The role of TNF alpha in adipocytes metabolism. *Seminars in Cell Development and Biology* 10, 19–29.

Sethi, J.K. and Vidal-Puig, A. (2005) Visfatin, the missing link between intra-abdominal obesity and diabetes? *Trends in Molecular Medicine* 11, 344–347.

Shapiro, L. and Scherer, P.E. (1998) The crystal structure of a complement-1q family protein suggests an evolutionary link to tumor necrosis factor. *Current Biology* 8, 335–338.

Sharman, M.J. and Volek, J.S. (2004) Weight loss leads to reductions in inflammatory biomarkers after a very-low-carbohydrate diet and a low-fat diet in overweight men. *Clinical Science* 107, 365–369.

Shulman, G.I. (2000) Cellular mechanisms of insulin resistance. *Journal of Clinical Investigation* 106, 171–176.

Sjöholm, K., Palming, J., Olofsson, L.E., Gummesson, A., Svensson, P.A., Lystig, T.C., Jennische, E., Brandberg, J., Torgerson, J.S., Carlsson, B. and Carlsson, L.M. (2005) A microarray search for genes predominantly expressed in human omental adipocytes: adipose tissue as a major production site of serum amyloid A. *Journal of Clinical Endocrinology and Metabolism* 90, 2233–2239.

Soares, A.F., Guichardant, M., Cozzone, D., Bernoud-Hubac, N., Bouzaïdi-Tiali, N., Lagarde, M. and Géloën, A. (2005) Effects of oxidative stress on adiponectin secretion and lactate production in 3T3-L1 adipocytes. *Free Radical Biology and Medicine* 38, 882–889.

Springer, T.A. (1994) Traffic signals for lymphocyte recirculation and leukocyte emigration, the multistep paradigm. *Cell* 76, 301–314.

Steffens, S. and Mach, F. (2008) Adiponectin and adaptive immunity, linking the bridge from obesity to atherogenesis. *Circulation Research* 102, 140–142.

Steppan, C.M. and Lazar, M.A. (2004) The current biology of resistin. *Journal of Internal Medicine* 255, 439–447.

Steppan, C.M., Bailey, S.T., Bhat, S., Brown, E.J., Banerjee, R.R., Wright, C.M., Patel, H.R., Ahima, R.S. and Lazar, M.A. (2001) The hormone resistin links obesity to diabetes. *Nature* 409, 307–312.

Storlien, L.H., Kraegen, E.W., Chisholm, D.J., Ford, G.L., Bruce, D.G and Pascoe, W.S. (1987) Fish oil prevents insulin resistance induced by high-fat feeding in rats. *Science* 237, 885–888.

Straczkowski, M., Dzienis-Straczkowska, S., Stepien, A., Kowalska, I., Szelachowska, M. and Kinalska, I. (2002) Plasma interleukin-8 concentrations are increased in obese subjects and related to fat mass and tumor necrosis factor-α system. *Journal of Clinical Endocrinology and Metabolism* 87, 4602–4606.

Strassmann, G., Fong, M., Kenney, J.S. and Jacob, C.O. (1992) Evidence for the involvement of interleukin 6 in experimental cancer cachexia. *Journal of Clinical Investigation* 89, 1681–1684.

Taga, T. (1992) IL6 signalling through IL6 receptor and receptor-associated signal transducer, gp130. *Research in Immunology* 143, 737–739.

Takahashi, K., Mizuarai, S., Araki, H., Mashiko, S., Ishihara, A., Kanatani, A., Itadani, H. and Kotani, H. (2003) Adiposity elevates plasma MCP-1 levels leading to the increased CD11b-positive monocytes in mice. *Journal of Biological Chemistry* 278, 46654–46660.

Takeya, M., Yoshimura, T., Leonard, E.J. and Takahashi, K. (1993) Detection of monocyte chemoattractant protein-1 in human atherosclerotic lesions by an anti-monocyte chemoattractant protein-1 monoclonal antibody. *Human Pathology* 24, 534–539.

Taub, D.D., Lloyd, A.R., Conlon, K., Wang, J.M., Ortaldo, J.R., Harada, A., Matsushima, K., Kelvin, D.J. and Oppenheim, J.J. (1993) Recombinant human interferon-inducible protein 10 is a chemoattractant for human monocytes and T lymphocytes and promotes T-cell adhesion to endothelial cells. *Journal of Experimental Medicine* 177, 1809–1814.

Tilg, H. and Moschen, A.R. (2008) Role of adiponectin and PBEF/visfatin as regulators of inflammation, involvement in obesity-associated diseases. *Clinical Science* 114, 275–288.

Torpy, D.J., Bornstein, S.R. and Chrousos, G.P. (1998) Leptin and interleukin-6 in sepsis. *Hormone and Metabolic Research* 30, 726–729.

Trayhurn, P. and Wood, I.S. (2004) Adipokines, inflammation and the pleiotropic role of white adipose tissue. *British Journal of Nutrition* 92, 347–355.

Vidal-Puig, A. and O'Rahilly, S. (2001) Resistin, a new link between obesity and insulin resistance? *Clinical Endocrinology* 55, 437–438.

Viguerie, N., Poitou, C., Cancello, R., Stich, V., Clément, K. and Langin, D. (2005) Transcriptomics applied to obesity and caloric restriction. *Biochimie* 87, 117–123.

Villagomez, M.T., Bae, S.J., Ogawa, I., Takenaka, M. and Katayama, I. (2004) Tumour necrosis factor-alpha but not interferon-gamma is the main inducer of inducible protein-10 in skin fibroblasts from patients with atopic dermatitis. *British Journal of Dermatology* 150, 910–916.

Vognild, E., Elvevoll, E.O., Brox, J., Olsen, R.L., Barstad, H., Aursand, M. and Osterud, B. (1998) Effects of dietary marine oils and olive oil on fatty acid composition, platelet membrane fluidity, platelet responses, and serum lipids in healthy humans. *Lipids* 33, 427–436.

Vora, D.K., Fang, Z.T., Liva, S.M., Tyner, T.R., Parhami, F., Watson, A.D., Drake, T.A., Territo, M.C. and Berliner, J.A. (1997) Induction of P-selectin by oxidized lipoproteins, separate effects on synthesis and surface expression. *Circulation Research* 80, 810–818.

Waki, H., Yamauchi, T., Kamon, J., Ito, Y., Uchida, S., Kita, S., Hara, K., Hada, Y., Vasseur, F., Froguel, P., Kimura, S., Nagai, R. and Kadowaki, T. (2003) Impaired multimerization of human adiponectin mutants associated with diabetes, molecular structure and multimer formation of adiponectin. *Journal of Biological Chemistry* 278, 40352–40363.

Wallenius, V., Wallenius, K., Ahren, B., Rudling, M., Carlsten, H., Dickson, S.L., Ohlsson, C. and Jansson, J.O. (2002) Interleukin-6-deficient mice develop mature-onset obesity. *Nature Medicine* 8, 75–79.

Wang, B., Jenkins, J.R. and Trayhurn, P. (2005) Expression and secretion of inflammation-related adipokines by human adipocytes differentiated in culture, integrated response to TNF-α. *American Journal of Physiology – Endocrinology and Metabolism* 288, E731–E740.

Weber, J., Yang, J.C., Topalian, S.L., Parkinson, D.R., Schwartzentruber, D.S., Ettinghausen, S.E., Gunn, H., Mixon, A., Kim, H. and Cole, D. (1993) Phase I trial of subcutaneous interleukin-6 in patients with advanced malignancies. *Journal of Clinical Oncology* 11, 499–506.

Weisberg, S.P., McCann, D., Desai, M., Rosenbaum, M., Leibel, R.L. and Ferrante, A.W. Jr (2003) Obesity is associated with macrophage accumulation in adipose tissue. *Journal of Clinical Investigation* 112, 1796–1808.

Wellen, K.E. and Hotamisligil, G.S. (2005) Inflammation, stress, and diabetes. *Journal of Clinical Investigation* 115, 1111–1119.

Weyer, C., Funahashi, T., Tanaka, S., Hotta, K., Matsuzawa, Y., Pratley, R.E. and Tataranni, P.A. (2001) Hypoadiponectinemia in obesity and type 2 diabetes, association with insulin resistance and hyperinsulinemia. *Journal of Clinical Endocrinology and Metabolism* 86, 1930–1935.

White, R.T., Damm, D., Hancock, N., Rosen, S., Lowwell, B.B., Usher, P., Flier, J.S. and Spiegelman, B.M. (1992) Human adipsin is identical to complement factor D and is expressed at high levels in adipose tissue. *Journal of Biological Chemistry* 267, 9210–9213.

Wolf, A.M., Wolf, D., Rumpold, H., Enrich, B. and Tilg, H. (2004) Adiponectin induces the anti-inflammatory cytokines IL-10 and IL-1RA in human leukocytes. *Biochemical and Biophysical Research Communications* 323, 630–635.

Wolfe, B.E., Jimerson, D.C., Orlova, C. and Mantzoros, C.S. (2004) Effect of dieting on plasma leptin, soluble leptin receptor, adiponectin and resistin levels in healthy volunteers. *Clinical Endocrinology* 61, 332–338.

Woods, A., Brull, D.J., Humphries, S.E. and Montgomery, H.E. (2000) Genetics of inflammation and risk of coronary artery disease, the central role of interleukin-6. *European Heart Journal* 21, 1574–1583.

Wulster-Radcliffe, M.C., Ajuwon, K.M., Wang, J., Christian, J.A. and Spurlock, M.E. (2004) Adiponectin differentially regulate cytokines in porcine macrophages. *Biochemical and Biophysical Research Communications* 316, 924–929.

Xu, A., Chan, K.W., Hoo, R.L., Wang, Y., Tan, K.C., Zhang, J., Chen, B., Lam, M.C., Tse, C., Cooper, G.J. and Lam, K.S. (2005) Testosterone selectively reduces the high molecular weight form of adiponectin by inhibiting its secretion from adipocytes. *Journal of Biological Chemistry* 280, 18073–18080.

Xu, H., Barnes, G.T., Yang, Q., Tan, G., Yang, D., Chou, C.J., Sole, J., Nichols, A., Ross, J.S., Tartiglia, L.A. and Chen, H. (2003) Chronic inflammation in fat plays a crucial role in the development of obesity-related insulin resistance. *Journal of Clinical Investigation* 112, 1821–1830.

Yamamoto, K. and Saito, H. (1998) A pathological role of increased expression of plasminogen activator inhibitor-1 in human or animal disorders. *International Journal of Hematology* 68, 371–385.

Yamauchi, T., Kamon, J., Waki, H., Terauchi, Y., Kubota, N., Hara, K., Mori, Y., Ide, T., Murakami, K., Tsuboyama-Kasaoka, N., Ezaki, O., Akanuma, Y., Gavrilova, O., Vinson, C., Reitman, M.L., Kagechika, H., Shudo, K., Yoda, M., Nakano, Y., Tobe, K., Nagai, R., Kimura, S., Tomita, M., Froguel, P. and Kadowaki, T. (2001) The fat-derived hormone adiponectin reverses insulin resistance associated with both lipoatrophy and obesity. *Nature Medicine* 7, 941–946.

Yamauchi, T., Kamon, J., Minokoshi, Y., Ito, Y., Waki, H., Uchida, S., Yamashita, S., Noda, M., Kita, S., Ueki, K., Eto, K., Akanuma, Y., Froguel, P., Foufelle, F., Ferre, P., Carling, D., Kimura, S., Nagai, R., Kahn, B.B. and Kadowaki, T. (2002) Adiponectin stimulates glucose utilization and fatty-acid oxidation by activating AMP-activated protein kinase. *Nature Medicine* 8, 1288–1295.

Yamauchi, T., Kamon, J., Ito, Y., Tsuchida, A., Yokomizo, T., Kita, S., Sugiyama, T., Miyagishi, M., Hara, K., Tsunoda, M., Murakami, K., Ohteki, T., Uchida, S., Takekawa, S., Waki, H., Tsuno, N.H., Shibata, Y., Terauchi, Y., Froguel, P., Tobe, K., Koyasu, S., Taira, K., Kitamura, T., Shimizu, T., Nagai, R. and Kadowaki, T. (2003) Cloning of adiponectin receptors that mediate antidiabetic metabolic effects. *Nature* 423, 762–769.

Yang, W.S., Lee, W.J., Funahashi, T., Tanaka, S., Matsuzawa, Y., Chao, C.L., Chen, C.L., Tai, T.Y. and Chuang, L.M. (2001) Weight reduction increases plasma levels of an adipose-derived anti-inflammatory protein, adiponectin. *Journal of Clinical Endocrinology and Metabolism* 86, 3815–3819.

Ylä-Herttuala, S., Lipton, B.A., Rosenfeld, M.E., Särkioja, T., Yoshimura, T., Leonard, E.J., Witztum, J.L. and Steinberg, D. (1991) Expression of monocyte chemoattractant protein 1 in macrophage-rich areas of human and rabbit atherosclerotic lesions. *Proceedings of the National Academy of Sciences of the United States of America* 88, 5252–5256.

Yokota, T., Oritani, K., Takahashi, I., Ishikawa, J., Matsuyama, A., Ouchi, N., Kihara, S., Funahashi, T., Tenner, A.J., Tomiyama, Y. and Matsuzawa, Y. (2000) Adiponectin, a new member of the family of soluble defense collagens, negatively regulates the growth of myelomonocytic progenitors and the functions of macrophages. *Blood* 96, 1723–1732.

Yudkin, J.S., Stehouwer, C.D., Emeis, J.J. and Coppack, S.W. (1999) C-reactive protein in healthy subjects, associations with obesity, insulin resistance, and endothelial dysfunction, a potential role for cytokines originating from adipose tissue? *Arteriosclerosis, Thrombosis, and Vascular Biology* 19, 972–978.

Zhang, Y., Proenca, R., Maffei, M., Barone, M., Leopold, L. and Friedman, J.M. (1994) Positional cloning of the mouse obese gene and its human homologue. *Nature* 372, 425–432.

Zhang, Z. and Schluesener, H.J. (2006) Mammalian toll-like receptors, from endogenous ligands to tissue regeneration. *Cellular and Molecular Life Sciences* 63, 2901–2907.

Zhu, W., Cheng, K.K., Vanhoutte, P.M., Lam, K.S. and Xu, A. (2008) Vascular effects of adiponectin, molecular mechanisms and potential therapeutic intervention. *Clinical Science* 114, 361–374.

Zozulińska, D., Majchrzak, A., Sobieska, M., Wiktorowicz, K. and Wierusz-Wysocka, B. (1999) Serum interleukin-8 level is increased in diabetic patients. *Diabetologia* 42, 117–118.

Zwirner, J., Werfel, T., Wilken, H.C., Theile, E. and Gotze, O. (1998) Anaphylatoxin C3a but not C3a(desArg) is a chemotaxin for the mouse macrophage cell line J774. *European Journal of Immunology* 28, 1570–1577.

9 Peptides Involved in Vascular Homeostasis

Amaia Rodríguez[1] and Gema Frühbeck[1,2]

[1]Metabolic Research Laboratory, Clínica Universitaria de Navarra, University of Navarra, Spain; [2]Department of Endocrinology, Clínica Universitaria de Navarra, University of Navarra, Spain

Obesity and Cardiovascular Disease

Obesity has been classified as a major modifiable risk factor for cardiovascular diseases (CVDs) by the American Heart Association, as well as by the American College of Cardiology guidelines for secondary prevention of coronary artery disease (CAD) (Eckel and Krauss, 1998; Smith *et al.*, 2001). Obesity is related to the development of several different comorbidities such as hypertension, type 2 diabetes mellitus (T2DM) and dyslipidaemia, all well-documented risk factors for CVD, which cluster together as the metabolic syndrome (Eckel *et al.*, 2005). In this regard, regional fat distribution is particularly relevant to the development of the metabolic syndrome and its accompanying cardiovascular complications (Rodríguez *et al.*, 2007a). Upper-body obesity (i.e. visceral or 'android' obesity), as determined by an increased waist circumference and waist–hip ratio or elevated visceral fat area by image analysis at the lumbosacral level, is associated with an increased incidence of metabolic disturbances, elevated risk of CVD and premature death (Yusuf *et al.*, 2005; Kuk *et al.*, 2006).

Weight gain is accompanied by progressive physiological changes in cardiovascular function that can lead to heart failure (HF) (Kopelman, 2000). The increased lean and fat mass as well as body surface area characteristic of obesity determine an elevation in total blood volume, which, in turn, contributes to an increase in left ventricular (LV) preload and in resting cardiac output. The augmented demand for cardiac output is achieved by an increase in stroke volume, while the heart rate (HR) remains comparatively unchanged. The obesity-related increase in stroke volume results from an increase in LV diastolic filling. The elevated circulatory preload and afterload lead to LV dilatation (Fig. 9.1). An increased cardiac output is a common finding in moderate obesity, whereas not all obese individuals are hypertensive. In patients with raised systemic vascular resistance, the combination of obesity and hypertension results in a disproportionate increase in LV wall dimensions to the chamber radius, which leads to LV

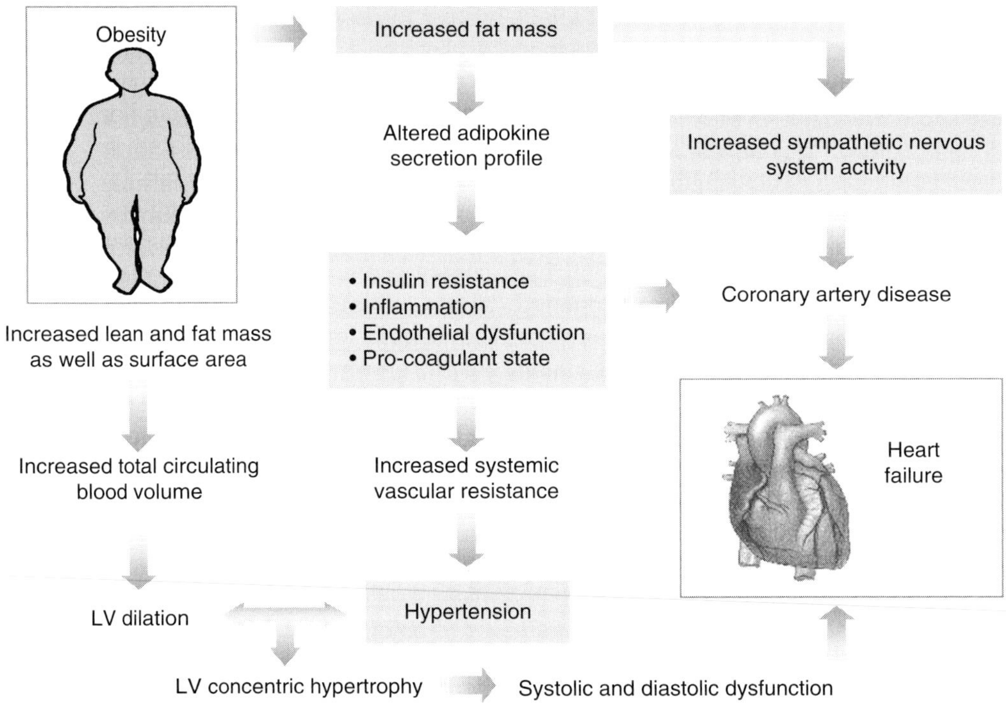

Fig. 9.1. Schematic diagram of obesity-associated cardiovascular alterations leading to heart failure.

concentric hypertrophy. In addition to increased blood pressure (BP) values, obese subjects exhibit an elevation of circulating concentrations of cardiovascular risk factors, which alters vascular function, adding further to the pressure load of the heart (Frühbeck, 2004). In spite of the increased cardiac output, obese individuals exhibit a decreased myocardial contractility proportional to excess body weight. LV hypertrophy, together with reduced ventricular compliance, results in diastolic dysfunction; a combination of systolic and diastolic dysfunction progresses to clinically significant risk of HF.

Adipokines and Cardiovascular Function

Adipose tissue acts as a metabolic active endocrine organ, secreting a large number of hormones, growth factors, enzymes, cytokines, complement factors and matrix proteins, collectively termed 'adipokines' (Frühbeck, 2004; Gualillo *et al.*, 2007). The physiological and pathophysiological relevance of adipokines in the homeostasis of the cardiovascular system resides in their effects on BP, fibrinolysis, angiogenesis, coagulation, vascular remodelling, insulin sensitivity and immunity, among others (Frühbeck, 2004; Wisse, 2004; Berg and Scherer, 2005;

Klein *et al.*, 2006; Sharma, 2006). In this respect, adipokines participate either directly or indirectly in the regulation of several processes that contribute to the development of inflammation, atherogenesis, hypertension and insulin resistance, as summarized in Table 9.1.

Table 9.1. Main adipokines implicated in cardiovascular homeostasis.

Adipokine	Cardiovascular effect	Reference
Adiponectin	Hormone with insulin-sensitizing, anti-inflammatory and anti-atherogenic properties	(Bodary and Eitzman, 2006)
Adipsin	Protein involved in the complement cascade	(Cianflone *et al.*, 2003)
Angiotensin II	Vasoconstrictor peptide that increases BP values and also participates in vascular remodelling	(Karlsson *et al.*, 1998)
Apelin	Vasoactive peptide that participates in the control of BP and stimulates cardiac contractility potently	(Tatemoto *et al.*, 1998)
ASP	Adipokine produced by the complement pathways that regulate whole-body glucose and lipid metabolism	(Cianflone *et al.*, 1989)
Cardiotrophin-1	Cytokine involved in the hypertrophy of cardiomyocytes	(Natal *et al.*, 2008)
Chemerin	Chemoattractant protein involved in adaptive and innate immunity	(Goralski *et al.*, 2007)
CRP	Acute-phase reactant involved in inflammatory processes	(Ouchi *et al.*, 2003a)
Ghrelin	Orexigenic hormone that exerts a depressor effect on BP and also exhibits cardioprotective properties	(Lin *et al.*, 2004)
IL-6	Proinflammatory cytokine implicated in inflammation and the acute-phase response	(Mohamed-Ali *et al.*, 1997)
Leptin	Anorexigenic hormone that participates in the inflammatory responses and contributes to the regulation of BP and other cardiovascular functions	(Frühbeck, 2004)
Osteopontin	Proinflammatory factor involved in vascular and myocardial remodelling	(Gómez-Ambrosi *et al.*, 2007)
PAI-1	Potent inhibitor of fibrinolysis that is implicated in atherosclerotic plaque formation	(De Taeye *et al.*, 2005)
RBP4	Protein apparently involved in the development of insulin resistance	(Quadro *et al.*, 1999)
Resistin	Hormone involved in insulin resistance also participating in the proinflammatory response	(Lehrke *et al.*, 2004)
SAA	Acute-phase reactant produced in response to injury, infection or inflammation	(Gómez-Ambrosi *et al.*, 2006)
TNF-α	Proinflammatory cytokine involved in systemic inflammation and the development of insulin resistance in obesity	(Moller, 2000)
Visfatin	Adipokine with apparent insulin-mimetic properties	(Fukuhara *et al.*, 2005)

Note: ASP, acylation-stimulating protein; BP, blood pressure; CRP, C-reactive protein; IL, interleukin; PAI-1, plasminogen activator inhibitor-1; RBP4, retinol-binding protein 4; SAA, serum amyloid A; TNF-α, tumour necrosis factor-α.

Inflammation and atherogenesis

Growing evidence highlights the relevant link between excess adiposity, inflammation and obesity-associated CVD. Adipose tissue constitutes an important source of circulating mediators of inflammation that participate in the mechanisms underlying vascular injury and atheromateous changes (Fig. 9.2). In addition to adipocytes, adipose tissue contains fibroblasts, preadipocytes, vascular constituents and, most importantly, macrophages. The resident macrophage population in adipose tissue ranges from 10% in lean humans to nearly 40% in obese subjects (Weisberg *et al.*, 2003). Macrophages are known to be crucial contributors to inflammation. However, adipocytes have also been recognized as key players in the chronic low-grade inflammation observed in obesity. In response to infectious and inflammatory signals, adipocytes synthesize and secrete several acute-phase reactants and mediators of inflammation, including

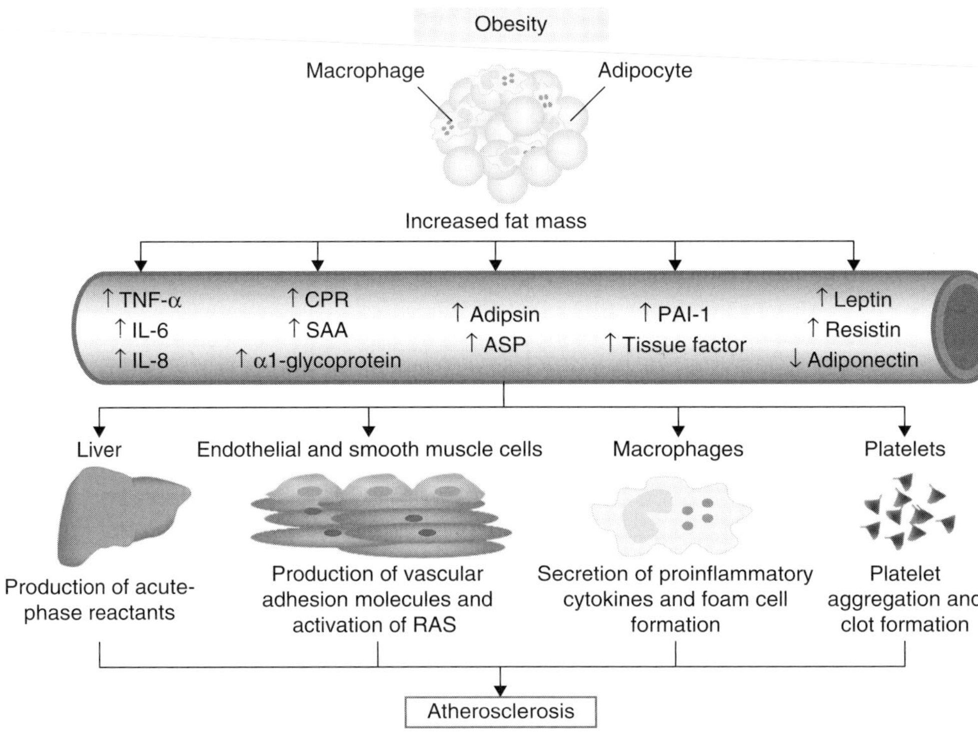

Fig. 9.2. Role of adipokines in the pathogenesis of atherosclerosis. Adipocytes and adipose tissue-embedded macrophages secrete proinflammatory cytokines, acute-phase reactants, complement factors, prothrombotic molecules and hormones implicated in the regulation of inflammation. The decrease of adiponectin secretion together with the excessive synthesis of the other prothrombotic, proinflammatory factors have been found to be associated with inflammation and vascular injury that leads to atherosclerotic plaque formation. RAS, renin–angiotensin system.

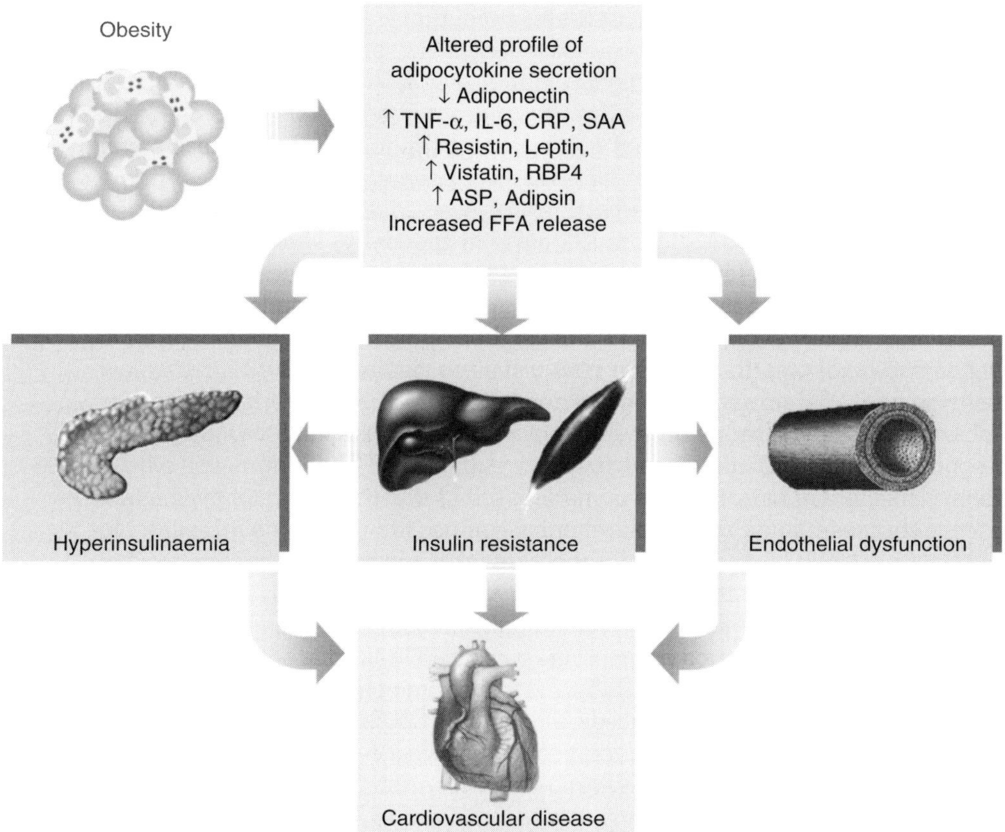

Fig. 9.3. Links between obesity-associated insulin resistance and cardiovascular disease. Excess free fatty acid (FFA) release in obesity overloads muscle, the liver and pancreatic β-cells. This ectopic lipid accumulation contributes to the development of insulin resistance, atherogenic dyslipidaemia and hyperinsulinaemia.

tumour necrosis factor-α (TNF-α), plasminogen activator inhibitor-1 (PAI-1), interleukin (IL)-1β, IL-6, IL-8, IL-10 and IL-15, leukaemia inhibitory factor (LIF), serum amyloid A (SAA), complement factors B, D, C3 and prostaglandin E2, tissue factor and other inflammatory modulators such as adiponectin, leptin and resistin. These adipokines not only exert autocrine and paracrine effects, but are also secreted to the bloodstream, contributing to systemic inflammation that favours the acceleration of CVD development (Fig. 9.3).

Tumour necrosis factor-α

TNF-α is a proinflammatory cytokine that has been implicated in the pathogenesis of insulin resistance and obesity in both mice and humans (Hotamisligil *et al.*, 1995; Moller, 2000). Adipose tissue constitutes the main source of circulating

TNF-α since it is secreted primarily by the fat-embedded macrophages and, to a lesser extent, by adipocytes, highlighting the relevance of paracrine effects (Weisberg *et al.*, 2003). One of the mechanisms whereby TNF-α promotes insulin resistance constitutes the impairment of insulin signalling in adipocytes and skeletal muscle by interference with the insulin signalling cascade at early steps and, hence, impairment of insulin-stimulated glucose transport (Hotamisligil *et al.*, 1994; Hernández *et al.*, 2004). A second mechanism used by TNF-α to contribute to insulin resistance is through elevations in circulating free fatty acids (FFAs) caused by the stimulation of lipolysis and hepatic lipogenesis (Moller, 2000).

TNF-α is a well-known biomarker of systemic inflammation. Obesity and insulin resistance are correlated with increased circulating TNF-α concentrations (Hotamisligil *et al.*, 1995). Weight loss in obese subjects is accompanied by an improvement in insulin sensitivity and is also associated with a decrease in adipose tissue TNF-α mRNA expression. Moreover, circulating TNF-α has been shown to stimulate hepatic C-reactive protein (CRP) production, which, in turn, exerts an impact on the vasculature. TNF-α also exhibits a direct vascular effect through stimulation of the production of vascular adhesion molecules and cytokines in the endothelium and vascular wall, resulting in vascular inflammation, monocyte adhesion to the vessel wall and foam cell accumulation. The sustained expression of proinflammatory cytokines in both preclinical and clinical HF has prompted the study of their effects on LV function, remodelling and cardiomyopathy. The detrimental actions of TNF-α on LV dysfunction have been described as taking place within minutes, as well as after hours or days (Oral *et al.*, 1997). In this respect, elevated local TNF-α levels in the infarcted myocardium contribute to chronic LV dysfunction and acute myocardial rupture by inducing a marked local inflammatory response, matrix and collagen degradation, increased matrix metalloproteinase activity and apoptosis (Sun *et al.*, 2004).

Interleukin-6

Within adipose tissue, both adipocytes and macrophages secrete IL-6 and studies measuring arteriovenous increases of IL-6 levels have shown that adipose tissue accounts for approximately 30% of circulating IL-6 concentrations in humans (Mohamed-Ali *et al.*, 1997; Weisberg *et al.*, 2003). The production of IL-6 increases with increasing adiposity, with circulating IL-6 concentrations being highly correlated with the percentage of body fat. The proinflammatory role of IL-6 is based on the induction of the acute-phase reactant CRP in the liver, contributing to the chronic inflammatory state linked to obesity (Wisse, 2004). Although increased CRP production is the most recognized marker of IL-6, there are other IL-6-dependent factors that may contribute to the cardiovascular risk. IL-6 contributes to the risk of clot formation, enhancing the hepatic production of fibrinogen, another acute-phase reactant, as well as increasing both platelet number and activity (Burstein *et al.*, 1996; Esmon, 2004). Moreover, endothelial cells and vascular smooth muscle cells are targets of IL-6 action, resulting in an increased expression of adhesion molecules and activation of the local renin–angiotensin system, which favours vascular wall inflammation and damage (Wassmann *et al.*, 2004).

In a study performed in the Framingham population, a polymorphism in the IL-6 gene promoter (–174 G/C, G = major allele) has been reported to modify the association of obesity with the development of insulin resistance and the risk of T2DM (Herbert *et al.*, 2005, 2006). On the one hand, the –174 GG genotype is associated with lower plasma glucose concentrations being protective against the onset of T2DM (Herbert *et al.*, 2005). However, weight gain induces a higher degree of insulin resistance in men with a –174 IL-6 CC genotype (Herbert *et al.*, 2006). These studies underscore the importance of gene–environment interactions in T2DM. In this context, men with the –174 IL-6 CC genotype may benefit especially from weight loss regimens to improve the risk of developing T2DM.

Plasminogen activator inhibitor-1

PAI-1 is the most important inhibitor of fibrinolysis and it is synthesized by vascular tissues, platelets, liver and visceral adipose tissue (De Taeye *et al.*, 2005). Elevations in plasma levels of PAI-1 are characteristic of obesity and contribute to the increased risk of atherothrombotic events in excess body weight and the metabolic syndrome (Sobel, 1999; Mertens *et al.*, 2006). Increased plasma PAI-1 concentrations are derived directly from cellular constituents of fat (adipocytes, stroma-vascular or adipose tissue matrix cells) or indirectly through the effects of other adipose-derived factors (TNF-α, Ang II, TGF-β, FFA) that stimulate local and systemic PAI-1 production (Fain *et al.*, 2004; De Taeye *et al.*, 2005). Obesity is also associated with increased circulating concentrations of the procoagulants fibrinogen, von Willebrand factor, factor VII and tissue factor. Many of the circulating cytokines elevated in obesity trigger an endothelial activation, which results in platelet aggregation and clot formation (Davi *et al.*, 2002; Gómez-Ambrosi *et al.*, 2002). Thus, the increase in clotting factor levels, together with platelet activation, constitute a procoagulant state, which contributes to atherogenesis via the deposition of platelets and fibrinous products in the developing plaques.

Adiponectin

Adiponectin (also known as Acrp30, AdipoQ, apM1 and gelatin-binding protein 28) is synthesized mainly by adipocytes and can be found in three oligomeric forms; namely as trimer, hexamer and high molecular weight molecules (Maeda *et al.*, 1996; Waki *et al.*, 2003). Adiponectin displays anti-diabetic and anti-atherogenic properties and is reduced in patients with obesity, T2DM and CAD (Arita *et al.*, 1999; Ouchi *et al.*, 1999; Hotta *et al.*, 2000). Two adiponectin receptors (AdipoR1 and AdipoR2) have been described (Yamauchi *et al.*, 2003). AdipoR1 is widely expressed in muscle, whereas AdipoR2 is expressed mainly in the liver. These receptors mediate the insulin-sensitizing action of adiponectin by increasing the activity of AMP kinase and peroxisome proliferator-activated receptor-α (PPARα) ligands, as well as fatty oxidation and glucose uptake. Adiponectin-deficient mice develop diet-induced insulin resistance on a high-fat, high-sucrose diet (Maeda *et al.*, 2002), while sustained peripheral expression of adiponectin decreases the development of diet-induced obesity and improves insulin sensitivity (Shklyaev *et al.*, 2003). Clinically, elevated adiponectin concentrations have been shown to be associated with higher insulin sensitivity and a reduced risk of T2DM (Hotta *et al.*, 2000, 2001; Spranger *et al.*, 2003). Thus, the development of interventions that increase adiponectin levels has been proposed as a target to improve insulin sensitivity and glucose tolerance, and probably coronary heart disease (CHD) (Bodary and Eitzman, 2006).

Adiponectin exerts an anti-inflammatory effect by downregulating the expression of adhesion molecules in endothelial cells, upregulating anti-inflammatory cytokines such as IL-10 and IL-1 receptor antagonist in monocytes and suppressing lipid accumulation and interferon-γ (IFN-γ) in macrophages. Most of the anti-inflammatory properties of adiponectin in endothelial cells and macrophages are mediated by the inhibition of the nuclear factor-kB (NF-κB) pathway. Likewise, adiponectin plays a protective role against atherosclerotic vascular alterations. Exogenous administration of adiponectin protects apolipoprotein E-deficient mice against the development of atherosclerosis. Adiponectin reduces the proliferation and migration of vascular smooth muscle cells by decreasing the effects of growth factors, such as platelet-derived growth factor (PDGF) and heparin-binding epidermal growth factor (HEGF). In this regard, adiponectin-deficient mice exhibit an exaggerated vascular remodelling response to injury and an impaired endothelium-dependent vasodilation on an atherogenic diet (Ouchi *et al.*, 2003b), increased leukocyte–endothelium adhesiveness (Ouedraogo *et al.*, 2007), increased neointimal hyperplasia after acute vascular injury (Kubota *et al.*, 2002; Matsuda *et al.*, 2002) and increased BP values compared with their wild-type littermates (Ohashi *et al.*, 2006). The beneficial effects of adiponectin on the endothelium are mediated by its ability to increase nitric oxide (NO) bioavailability (Li *et al.*, 2007).

In humans, hypoadiponectinaemia has been linked to endothelial dysfunction, CAD and stroke, with concentric hypertrophy and diastolic dysfunction commonly being observed in diabetes and other obesity-related disorders which are associated with decreased adiponectin concentrations (Kumada *et al.*, 2003; Shimabukuro *et al.*, 2003; Shibata *et al.*, 2004; Chen *et al.*, 2005). The local production of adiponectin by cardiomyocytes suggests an autocrine–paracrine effect (Piñeiro *et al.*, 2005). Adiponectin exerts its cardioprotective role modulating myocardial remodelling after ischaemic injury through AMPK and cyclo-oxygenase-2 (COX-2) (Ishikawa *et al.*, 2003; Shibata *et al.*, 2005), attenuating cardiac hypertrophy and interstitial fibrosis (Shibata *et al.*, 2004, 2007). Accordingly, high plasma adiponectin concentrations are associated with lower risk of acute coronary syndrome (Wolk *et al.*, 2007), infarction in men (Pischon *et al.*, 2004) and CHD in diabetic patients (Schulze *et al.*, 2005). However, other studies have found hyperadiponectinaemia in patients with chronic and congestive HF (Kistorp *et al.*, 2005; George *et al.*, 2006). Furthermore, a meta-analysis has reported that any association of adiponectin with CHD risk is comparatively moderate and requires further investigation (Sattar *et al.*, 2006).

Serum amyloid A

SAA is the major acute-reactant protein produced in the liver in response to infection, inflammation, injury and stress (Jensen and Whitehead, 1998; Uhlar and Whitehead, 1999). It has been reported that SAA can be more sensitive than CRP as an indicator of inflammation in some non-cardiovascular inflammatory conditions (Malle and De Beer, 1996). SAA is an apolipoprotein and a component of high-density lipoprotein (HDL) particles (Jensen and Whitehead, 1998). Increased concentrations of SAA are associated with an elevated risk of CVD (Morrow *et al.*, 2000; Jousilahti *et al.*, 2001; Delanghe *et al.*, 2002). During the non-acute-phase reaction, adipose tissue constitutes the major expression site of SAA, providing a direct link between adipose tissue mass and cardiovascular risk (Sjöholm *et al.*, 2005).

Adipocyte-derived SAA stimulates lipolysis in an autocrine way and, consequently, induces an increase in the release of FFA and a decrease in insulin sensitivity (Yang *et al.*, 2006). SAA also acts as a paracrine factor stimulating the secretion of proinflammatory cytokines (IL-6, IL-8, MCP-1) in adipose stromavascular cells. In addition, macrophages infiltrated in the adipose tissue may also constitute target cells for SAA action, further increasing the release of cytokines and chemokines. Finally, circulating SAA also stimulates the release of inflammatory cytokines from endothelial cells and monocytes, contributing to the infiltration of monocytes into the vasculature and to endothelial dysfunction, thus accelerating the development of atherosclerosis. In addition, the interaction of SAA with HDL further aggravates the atherosclerotic process, since SAA is incorporated into HDL particles and impairs its function. Taken together, SAA is a direct mediator of obesity-associated inflammation and its related cardiometabolic consequences. Importantly, weight loss reduces circulating SAA concentrations, which may mediate, in part, the improvements in systemic inflammation and cardiovascular risk associated with weight reduction (Gómez-Ambrosi *et al.*, 2006). Thus, SAA may be a valuable diagnostic and prognostic marker of obesity-associated CVD.

C-reactive protein

Circulating C-reactive protein (CRP) concentrations are strongly associated with obesity and obesity-related diseases, including insulin resistance, T2DM and hyperlipidaemia (Ouchi *et al.*, 2003a; Flórez *et al.*, 2006). In fact, obesity is the major determinant of elevated CRP levels in subjects with the metabolic syndrome (Aronson *et al.*, 2004). CRP is an acute-phase reactant produced by the liver and a well-known marker of chronic low-grade inflammation. It is a member of the pentraxin family that attaches damaged cells causing cell death through the activation of the complement cascade (Pepys and Hirschfield, 2003). Excess adiposity drives to an enhanced production of proinflammatory cytokines such as IL-6 and TNF-α, which, in turn, stimulate the hepatic production of CRP (Flórez *et al.*, 2006). Moreover, human adipose tissue reportedly produces CRP and an inverse relationship between CRP and adiponectin in both plasma and adipose tissue has been observed (Ouchi *et al.*, 2003a). Modest elevations in CRP are associated with the pathogenesis of atherosclerosis, with increased CRP concentrations being a risk factor for CHD. Thus, the measurement of CRP is recommended in some clinical settings to stratify CVD risk and to guide clinical management (Pearson *et al.*, 2003).

Osteopontin

Osteopontin, also known as early T-lymphocyte activation, secreted phosphoprotein-1 and bone sialoprotein-1, is a phosphoprotein identified originally in osteoblasts and osteoclasts that has been shown subsequently to be secreted by a wide variety of cells (Naldini *et al.*, 2006; Rangaswami *et al.*, 2006), including adipocytes (Gómez-Ambrosi *et al.*, 2007). Although associated initially with bone mineralization, it has been recognized that osteopontin also participates in wound healing and inflammation, as well as immunity. Osteopontin influences cardiovascular function, playing a role in atherosclerosis (Isoda *et al.*, 2003), LV hypertrophy (Graf *et al.*, 1997) and cardiac fibrosis (Lenga *et al.*, 2008), processes commonly associated with obesity. In this regard, circulating osteopontin levels reportedly are increased in obesity (Gómez-Ambrosi *et al.*, 2007). Interestingly, in the post-infarcted heart, osteopontin has been shown to operate coordinating the intracellular signals required to integrate myofibroblast proliferation, migration and extracellular matrix deposition with the recruitment of macrophages and initiation of collateral vessel formation, thus ensuring that the mechanical properties of the heart are not further compromised (Zahradka, 2008).

Hypertension

Obesity, in particular if accompanied by an increased visceral fat accumulation, is an independent risk factor for the development of hypertension. A prospective study performed in the Framingham population showed that overweight and obesity are associated with an increased relative risk for the onset of hypertension (Wilson *et al.*, 2002). It is well documented that BP increases with weight gain and decreases with weight loss. Alterations in sodium and water reabsorption have been shown to participate in the onset of obesity-associated hypertension (Krauss *et al.*, 1998). An increased arterial pressure is required to maintain sodium balance in obese subjects, indicating an impaired renal natriuresis (Hall, 1997). Both glomerular filtration rate and renal plasma flow are elevated in obesity, suggesting that impaired renal excretion is a consequence of increased renal tubular reabsorption. In addition, the stimulation of the sympathetic nervous system (SNS) found in obesity further worsens the renal tubular reabsorption and altered natriuresis (Esler, 2000).

Obesity-associated hypertension results from the complex interaction between haemodynamic and endocrine-metabolic factors. Among the latter, a central role has been attributed to insulin resistance, which characterizes both obesity and hypertension (Krauss *et al.*, 1998). In the past decade, growing evidence supports the contribution of adipose tissue-derived factors in BP homeostasis, thus improving our understanding of obesity-related hypertension.

Leptin

Leptin, the *OB* gene product, participates in the control of body weight by regulating food intake and energy expenditure (Frühbeck *et al.*, 2001). Leptin secretion is proportional to the amount of adipose tissue stores, with plasma concentrations being increased markedly in obese individuals (Considine *et al.*, 1996). Beyond its participation in the maintenance of energy balance, leptin contributes to the homeostasis of the vascular tone (Frühbeck, 2004). It is suggested that hyperleptinaemia plays an important role in the pathogenesis of obesity-associated hypertension (Agata *et al.*, 1997; Ren, 2004; Rahmouni *et al.*, 2005). Intracerebroventricular and intravenous administration of leptin reportedly increases mean arterial pressure (MAP) and HR, as well as sympathetic outflow to kidneys, adipose tissue, skeletal vasculature and adrenal medulla in rodents (Matsumura *et al.*, 2000). Leptin increases the vasomotor sympathetic activity through the activation of leptin receptors (OB-R) in the ventromedial and dorsomedial hypothalamic regions (Haynes *et al.*, 1997; Marsh *et al.*, 2003). Interestingly, administration of leptin is not always accompanied by changes in MAP and HR (Haynes *et al.*, 1997; Shek *et al.*, 1998; Frühbeck, 1999). The explanation for this apparent paradox is that, in addition to its central sympathoexcitatory action, leptin induces a depressor effect simultaneously in peripheral tissues. Leptin also has been shown to induce a depressor response attributable to the vasodilation of conduit and resistance vessels (Frühbeck, 1999; Lembo *et al.*, 2000; Beltowski *et al.*, 2006). In the aorta and coronary arteries, leptin reportedly induces vasodilation via NO (Kimura *et al.*, 2000; Lembo *et al.*, 2000; Knudson *et al.*, 2005), whereas the relaxation induced by the hormone in mesenteric arteries is mediated by the endothelium-derived hyperpolarizing factor (EDHF) (Lembo *et al.*, 2000; Gálvez *et al.*, 2006). Leptin also inhibits the Ang II-induced calcium increase and vasoconstriction in the smooth muscle layer of the aorta (Fortuño *et al.*, 2002; Rodríguez *et al.*, 2007b). A further mechanism whereby leptin decreases BP values is the induction of natriuresis and diuresis at the tubular level through NO-dependent mechanisms (Jackson and Li, 1997; Villarreal *et al.*, 1998, 2004). Finally, leptin reduces insulin secretion and improves insulin sensitivity in skeletal muscle and the liver (Zhao *et al.*, 1998, 2000; Yaspelkis *et al.*, 2001). Leptin, therefore, appears to have a dual effect on BP control with a pressor response attributable to sympathetic activation via the central nervous system and a depressor response attributable to a direct effect of leptin on peripheral tissues (Fig. 9.4).

Leptin has been found to be synthesized by cardiomyocytes and released to the coronary effluent, raising the possibility that cardiac leptin exerts direct physiological effects on the myocardium (Purdham *et al.*, 2004). In this sense, leptin has been shown to decrease the contractility of ventricular myocytes via NO (Nickola *et al.*, 2000) and to promote the hypertrophy of rat cardiomyocytes via activation of the mitogen-activated protein kinase cascade (Rajapurohitam *et al.*, 2006). Moreover, leptin exhibits a cardioprotective effect in myocardial ischaemia-reperfusion injury (Smith *et al.*, 2006). After a brief period of myocardial ischaemia, a rapid local inflammatory cascade takes place in the infarcted tissue after reperfusion. Since inflammation and vascularization play an important role in tissue healing, the proinflammatory properties, together with the angiogenic and wound-healing actions of leptin, may improve the infarcted tissue repair considerably (Bouloumié *et al.*, 1998; Otero *et al.*, 2005). Taken together, the local production of leptin and the presence of OB-R in cardiac myocytes indicate that this cytokine acts as an autocrine and paracrine agent in cardiac function regulation under both physiological and pathophysiological conditions.

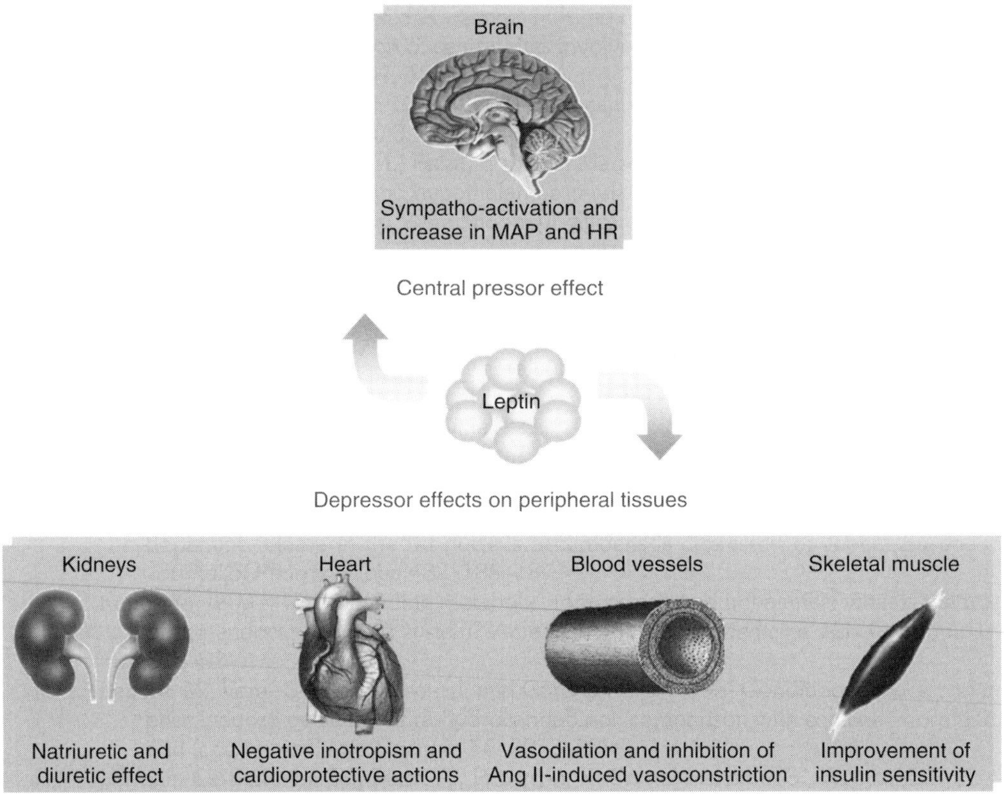

Fig. 9.4. Dual effects of leptin on blood pressure control. MAP, mean arterial pressure; HR, heart rate; Ang II, angiotensin II.

Leptin predicts the worsening of features of the metabolic syndrome independently of obesity (Franks *et al.*, 2005). Leptin levels are elevated in essential hypertension, suggesting a possible link between hyperleptinaemia and cardiovascular dysfunction in hypertension (Agata *et al.*, 1997). In this respect, it has been reported that the beneficial vascular, renal and cardiac responses induced by leptin are impaired in hypertensive rats (Villarreal *et al.*, 1998; Wold *et al.*, 2002; Gálvez *et al.*, 2006; Rodríguez *et al.*, 2006). Moreover, a strong positive correlation was found between hyperleptinaemia and tachycardia in mildly obese and mildly hypertensive patients (Narkiewicz *et al.*, 1999). Furthermore, hyperleptinaemia constitutes an independent risk marker for different cardiovascular events, such as chronic heart failure (CHF) or ischaemic and non-ischaemic stroke, indicating that leptin represents an important link between obesity and CVD (Schulze *et al.*, 2003; Söderberg *et al.*, 2003). Nevertheless, it has to be taken into consideration that the supposedly detrimental effects of leptin on cardiovascular homeostasis may only underlie a state of leptin-resistance, which has been shown clearly in obese subjects (Caro *et al.*, 1996; Rahmouni *et al.*, 2005).

Ghrelin

Ghrelin is a growth hormone (GH)-releasing peptide, isolated originally from the stomach, identified as an endogenous ligand for the GH secretagogue receptor (GHS-R) (Kojima *et al.*, 1999). Although gastric and intestinal ghrelin constitute the two major origins of this hormone (Kojima *et al.*, 1999), ghrelin is also synthesized to a lesser extent by adipose tissue (Knerr *et al.*, 2006). Two major forms of ghrelin are present in plasma and stomach: ghrelin, with an *n*-octanoyl group at the serine 3 residue, and desacyl-ghrelin, without the acylation (Hosoda *et al.*, 2000). Ghrelin acts on the pituitary and hypothalamus to stimulate GH release, food intake and weight gain (Tschöp *et al.*, 2000, 2001; Wren *et al.*, 2000, 2001; Wortley *et al.*, 2004; Ahima, 2006). The secretion of GH stimulated by ghrelin is independent of that evoked by the hypothalamic GH-releasing hormone (GHRH). GH and its mediator, insulin-like growth factor (IGF)-1, are anabolic hormones that are essential for myocardial development and performance. Ghrelin has cardiovascular effects through both GH-dependent and -independent mechanisms (Fig. 9.5).

Ghrelin acts on the neurones of the nucleus tractus solitarius to decrease MAP in rodents (Lin *et al.*, 2004; Tsubota *et al.*, 2005). Intravenous administration of ghrelin to healthy individuals and patients with CHF decreases MAP without changing HR and improves cardiac function by increasing stroke volume and the cardiac index (Nagaya *et al.*, 2001a,b). The beneficial haemodynamic effects of ghrelin in patients with CHF seem to be attributable to both an inotropism of GH and a fall in cardiac overload. The presence of GHS-R in cardiac ventricles provides evidence for direct cardiac effects of ghrelin (Iglesias *et al.*, 2004), which has been shown to be synthesized by cardiomyocytes and to operate as an endogenous cardioprotective factor protecting cardiomyocytes and endothelial cells against apoptosis through the activation of an intracellular survival pathway (Baldanzi *et al.*, 2002). Moreover, ghrelin administration after myocardial infarction has been shown to attenuate LV enlargement and myocardial fibrosis in rodents (Soeki *et al.*, 2008).

Based on the widespread expression of ghrelin and GHS-R in the human cardiovascular system, the possible participation of ghrelin in the paracrine regulation of the vascular tone was investigated further (Kleinz *et al.*, 2006). Intra-arterial infusion of ghrelin to healthy individuals induces vasodilation through GH/IGF-1-independent mechanisms (Okumura *et al.*, 2002). Moreover, ghrelin improves the endothelial dysfunction of patients with the metabolic syndrome by increasing NO biodisponibility (Tesauro *et al.*, 2005). In addition, both ghrelin and desacyl-ghrelin potently reverse endothelin-1-induced vasoconstriction, a peptide that is upregulated in atherosclerosis (Kleinz *et al.*, 2006). In this sense, ghrelin may play a modulatory role in atherosclerosis since this peptide also inhibits proinflammatory cytokine production, mononuclear cell binding and NF-κB activation in human endothelial cells *in vitro* and endotoxin-induced cytokine production *in vivo* (Li *et al.*, 2004). In summary, the low circulating ghrelin concentrations found in obese and hypertensive patients might be haemodynamically disadvantageous, due to the positive vascular and anti-inflammatory properties of ghrelin (Tschöp *et al.*, 2001; Yildiz *et al.*, 2004; Poykko *et al.*, 2005).

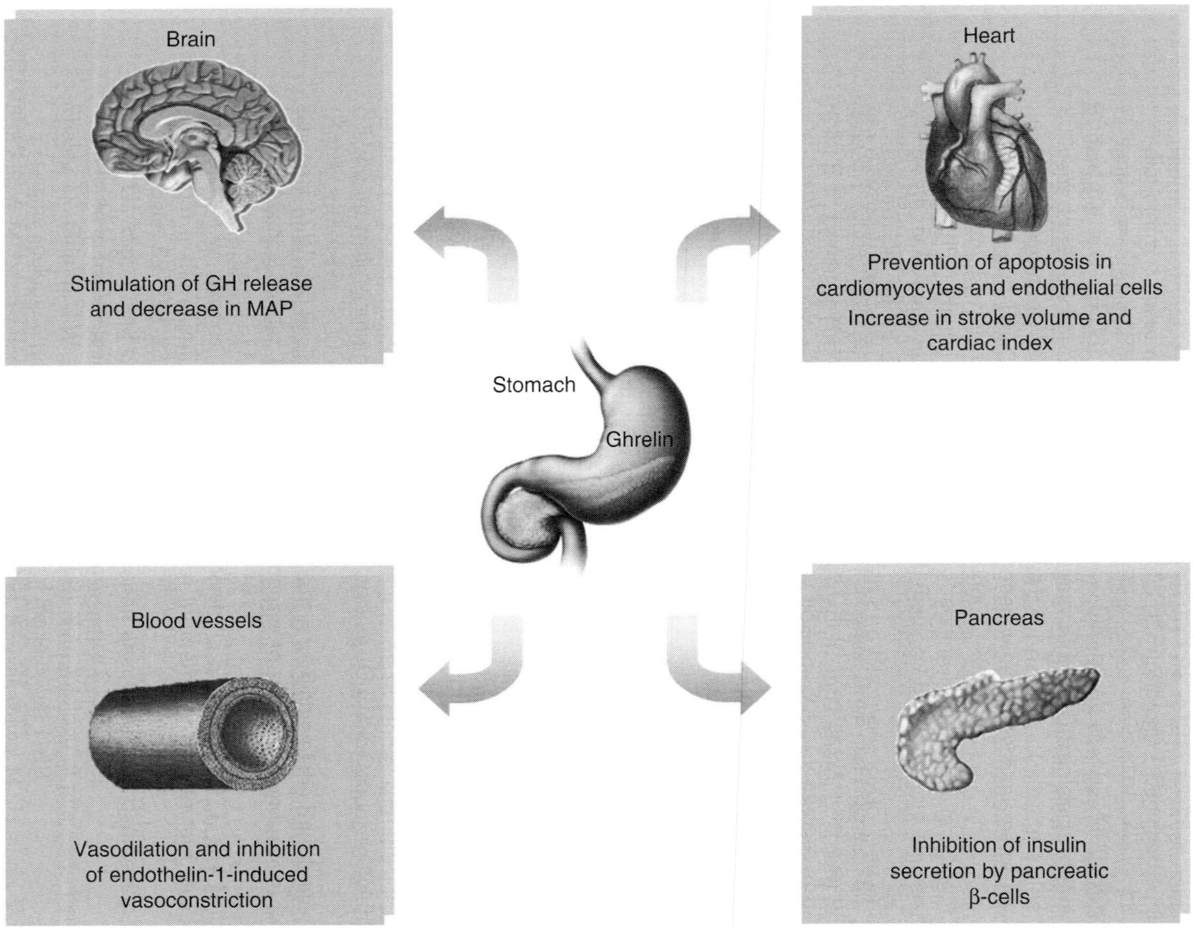

Fig. 9.5. Participation of ghrelin in cardiovascular homeostasis. GH, growth hormone; MAP, mean arterial pressure.

Angiotensinogen–Angiotensin II

Angiotensin (Ang) II is a well-known hypertensive hormone, derived from the precursor molecule angiotensinogen, which is cleaved by enzymes of the renin–angiotensin system (RAS) [renin, angiotensin-converting enzyme (ACE)], as well as the non-renin-angiotensin system (NRAS) (cathepsin D, cathepsin G, tonin, chymase) (Karlsson *et al.*, 1998). Human adipose tissue has been shown to express angiotensinogen and the enzymes required for its conversion to Ang II. Likewise, both Ang II receptor subtypes, AT_1 and AT_2, are expressed in the cell membrane of adipocytes (Crandall *et al.*, 1994). Obese individuals reportedly exhibit elevated circulating concentrations of ACE, angiotensinogen, renin and aldosterone and increased adipose tissue angiotensinogen expression (Engeli *et al.*, 2005). A 5% reduction in body weight leads to a meaningful reduction in the renin–angiotensin–aldosterone system in plasma and adipose tissue, contributing to systolic BP decrease. Collectively, these findings show that adipose tissue constitutes an important peripheral site of Ang II production and a target for this hypertensive hormone, suggesting the involvement of adipocyte-derived Ang II in obesity-associated hypertension (Kim and Moustaid-Moussa, 2000).

Apelin

Apelin was identified as the endogenous ligand of an orphan G protein-coupled receptor, the human APJ receptor (Tatemoto *et al.*, 1998). To date, four forms of the peptide have been isolated (apelin-12, 13, 17 and 36), each showing different receptor-binding capabilities. Similarities between the structure and anatomical distribution of apelin and its receptor and that of Ang II and the AT_1 receptor provide clues about its potential cardiovascular effects (Lee *et al.*, 2006). Apelin is expressed in rat and human adipocytes and is influenced markedly by the nutritional status, with its expression being reduced during fasting and increased by re-feeding (Boucher *et al.*, 2005). Insulin also influences the production of apelin in adipose tissue, upregulating its synthesis both *in vitro* and *in vivo*.

The cardiovascular system appears to be a primary target of apelin since APJ is expressed in the heart and the media layer of human coronary arteries, aorta and saphenous vein grafts. The intravenous administration of apelin to rats is followed by a decrease in MAP through NO-dependent mechanisms ranging from 5% for apelin-36 to 25% for apelin-12 (Tatemoto *et al.*, 2001). This hypotensive effect is accompanied by a slight increase in HR, which results from the baroreceptor reflex-mediated stimulation of the SNS. In this sense, it has been reported that apelin increases myocardial contractility in isolated perfused rat hearts (Szokodi *et al.*, 2002). Moreover, circulating apelin concentrations, atrial apelin and atrial and ventricular APJ expression are decreased markedly in patients with HF (Földes *et al.*, 2003). It has been reported recently that in ischaemic myocardium of isolated rat heart, apelin expression is upregulated but returns back to baseline values after reperfusion (Kleinz and Baxter, 2008). During the period of reduced apelin expression, administration of exogenous apelin-13 attenuated the ischaemic/reperfusion injury, reducing the infarct size.

In spite of increasing myocardial contractility (Szokodi *et al.*, 2002), apelin exerts a weak effect on cardiac output, probably because it induces vasodilation and reduces the preload. Apelin-12 has been shown to dilate peripheral veins more efficiently than the Ca^{2+}-antagonists, hydralazine or nitroglycerin (Cheng *et al.*, 2003). This hypotensive effect is accompanied by a slight increase in HR, which results from the baroreceptor reflex-mediated stimulation of the SNS. Moreover, APJ-deficient mice have been shown to increase the vasopressor response to the potent vasoconstrictor Ang II, suggesting that APJ might play a counter-regulatory role opposing the pressor action of Ang II (Ishida *et al.*, 2004). The evidence that apelin acts as a vasodilator as well as a cardioprotective factor and that the sensitivity to apelin might be altered in disease states makes the apelin–APJ system a promising therapeutic target.

Insulin resistance and type 2 diabetes mellitus

Insulin resistance is one of the core defects of the metabolic syndrome, lying at the centre of the pathogenesis of T2DM and the associated CVD risk. As mentioned before, the proinflammatory mediators released by adipose tissue (PAI-1, TNF-α, IL-6, resistin and others), together with hyposecretion of beneficial adipokines (such as adiponectin), exert a detrimental effect on vascular endothelial function, thereby increasing the CVD risk in the metabolic syndrome.

Resistin

Resistin, also known as Fizz3, is a member of a gene family that includes resistin-like molecule α (RELM-α), RELM-β and RELM-γ. In mice, resistin is produced mainly by adipocytes (Yang *et al.*, 2003). In humans, resistin is strongly expressed by macrophages and, in lesser amounts, by fat cells. According to its name, resistin was found to increase insulin resistance in obese mice (Steppan *et al.*, 2001). Moreover, treatment of a murine adipocyte cell line with thiazolidinedione (TZD), an anti-diabetic drug that increases insulin sensitivity via the stimulation of PPARγ, decreases resistin expression markedly in adipocytes. Despite the clear link between elevated serum resistin concentrations in obese mice, the association between circulating resistin levels and obesity in humans is more controversial. Several groups have described increased concentrations of resistin in human obesity (Azuma *et al.*, 2003; Degawa-Yamauchi *et al.*, 2003), while others report no differences (Lee *et al.*, 2003; Silha *et al.*, 2003; Heilbronn *et al.*, 2004). Studies of the association of plasma resistin levels with insulin resistance and T2DM have also yielded inconsistent results. In spite of the hyperresistinaemia found in diabetic animal models, several human studies reported no differences in circulating concentrations of resistin among normal subjects and patients with insulin resistance or T2DM (Lee *et al.*, 2003; Silha *et al.*, 2003; Heilbronn *et al.*, 2004; Iqbal *et al.*, 2005; Kusminski *et al.*, 2005), while some authors reported that subjects with T2DM exhibited higher resistin levels (McTernan *et al.*, 2003; Youn *et al.*, 2004). In addition, whereas murine models of insulin resistance show dramatic changes in resistin expression after treatment with PPARγ agonists, such agents have more modest effects in humans (Savage *et al.*, 2001). Taken together, the relation between obesity, adipose tissue resistin expression, systemic insulin resistance and amelioration of the latter by PPARγ agonists reported in mice may not translate completely to human pathophysiology.

Growing evidence links resistin with inflammation and CVD (Lehrke *et al.*, 2004; Gómez-Ambrosi and Frühbeck, 2005; Yaturu *et al.*, 2006). A significant association between hyperresistinaemia and proatherogenic inflammatory markers, unstable angina, congestive HF and coronary atherosclerosis has been shown (Gómez-Ambrosi and Frühbeck, 2001; Reilly *et al.*, 2005; Kunnari *et al.*, 2006; Lubos *et al.*, 2007; Norata *et al.*, 2007; Takeishi *et al.*, 2007). Macrophages infiltrating human atherosclerotic aneurysms have been shown to secrete resistin (Jung *et al.*, 2006). In turn, resistin stimulates the synthesis of proinflammatory cytokines such as TNF-α, IL-1, IL-6 and IL-12 through an NF-κB dependent pathway, upregulates the expression of adhesion molecules (VCAM1 and ICAM1) and promotes the release of endothelin-1 in the human endothelial cells (Tilg and Moschen, 2006). Interestingly, resistin also stimulates the synthesis of monocyte chemoattractant protein-1 (MCP-1) in the endothelium, which might perpetuate a vicious circle of macrophage recruitment and production of proinflammatory cytokines (Verma *et al.*, 2003). Furthermore, resistin induces endothelial dysfunction in isolated coronary artery rings (Dick *et al.*, 2006) and worsens cardiac ischaemia-reperfusion injury in isolated perfused rat hearts (Rothwell *et al.*, 2006). In this respect, patients with CAD exhibit a strong correlation between resistin levels and inflammatory markers, namely CRP and TNF-α (Yaturu *et al.*, 2006). Recently, resistin has been shown to be able to induce a selective vascular insulin resistance-impairing endothelial IRS-1 signalling pathway that leads to endothelial nitric oxide synthase (eNOS) activation and vasodilation (Gentile *et al.*, 2008). Although a clear-cut function for resistin in humans is still lacking, it may play a role in the progression from vascular inflammation to endothelial dysfunction and accelerate the eventual development of overt CVD (Gómez-Ambrosi and Frühbeck, 2005).

Visfatin

Visfatin, which was identified initially as pre-B-cell-colony-enhancing factor, is produced mainly by visceral adipose tissue of mice and humans (Fukuhara *et al.*, 2005). Acute and chronic administration of visfatin to mice reduces glycaemia without changes in insulin concentrations. Visfatin apparently was reported to bind to the insulin receptor at a different site from insulin to exert insulin-mimetic properties, such as the stimulation of glucose uptake and lipogenesis in 3T3-L1 adipocytes or L6 myocytes, and the suppression of glucose production by cultured hepatocytes (Fukuhara *et al.*, 2005). In addition, visfatin facilitates adipogenesis by stimulating markers of adipocyte differentiation, including PPARγ, fatty acid synthase, diacylglycerol acyltransferase or adiponectin (Fukuhara *et al.*, 2005). However, part of these findings are currently controversial, with the authors having been forced to retract some of their original conclusions (Fukuhara *et al.*, 2007).

To date, the relationship between visfatin, obesity and T2DM remains controversial. Increased plasma visfatin concentrations have been reported in patients with type 2 diabetes, gestational diabetes and obesity (Fukuhara *et al.*,

2005; Chen *et al.*, 2006; Krzyzanowska *et al.*, 2006), while other studies have found reduced visfatin levels in obesity and no correlation with insulin resistance (Pagano *et al.*, 2006). Interestingly, hyperglycaemia causes an increase in circulating visfatin concentrations, with this increase being more prominent as glucose intolerance worsens in patients with T2DM (Dogru *et al.*, 2007). Improvement of the glycaemic profile with either exercise training or weight loss lowers the elevated visfatin levels found in patients with obesity and type 1 diabetes mellitus (Haider *et al.*, 2006a,b). Thus, the increase in visfatin synthesis associated with obesity and diabetes may represent a compensatory mechanism to maintain normoglycaemia. In fact, TZD treatment in healthy volunteers increases the release of visfatin from adipose tissue, improving their insulin sensitivity, with FFA reportedly counteracting this effect. Recently, it has been observed that visfatin upregulates key molecules of the angiogenic process, such as matrix metalloproteinases (MMP) and vascular endothelial growth factor (VEGF) in human endothelial cells (Adya *et al.*, 2008), revealing a novel insight into the potential role of visfatin in CVD.

Retinol-binding protein 4

Retinol-binding protein 4 (RBP4) is the only specific transport protein for retinol (vitamin A) and its main function is to deliver retinol to tissues (Quadro *et al.*, 1999). Recently, it has been shown that this adipokine may contribute to the pathogenesis of T2DM. Transgenic overexpression of human RBP4 or injection of RBP4 to normal mice causes insulin resistance, whereas genetic deletion of *Rbp4* enhances insulin sensitivity (Yang *et al.*, 2005). Circulating RBP4 is increased substantially, not only in several murine models of obesity and insulin resistance but also in humans with these conditions (Yang *et al.*, 2005; Graham *et al.*, 2006). It has been proposed that adipocytes might detect the absence of glucose uptake by the glucose transporter, GLUT4, and in response, secrete adipokines such as RBP4 to restrict glucose uptake by skeletal muscle and increase hepatic glucose output via the induction of the expression of the gluconeogenic enzyme, phosphoenolpyruvate carboxykinase (PEPCK), thereby increasing glycaemia (Tamori *et al.*, 2006). However, in human obesity, the exact contribution of RBP4 has not been disentangled completely, with some studies observing normal concentrations in obese, insulin-resistant and also diabetic patients (Janke *et al.*, 2006; Broch *et al.*, 2007; Gómez-Ambrosi *et al.*, 2008). Further research is needed to unravel the involvement of RBP4 in the development of obesity-associated insulin resistance in humans.

Acylation-stimulating protein: C3, factor B and adipsin

Acylation-stimulating protein (ASP) was determined initially in human plasma and identified as a derivative of the third complement component (C3) (Cianflone *et al.*, 1989). ASP is a hormone produced by adipocytes through the interaction of C3 with factor B and adipsin in the alternative complement pathway; none the less, ASP potentially could be generated through the two other complement pathways, the classical and the lectin pathway (Cianflone *et al.*, 2003). ASP increases the esterification of FFA into triacylglycerol (TG) synthesis in fat-storing cells. This effect is achieved through the stimulation of glucose uptake (by enhancing the translocation of glucose transporters to the plasma membrane) and diacylglycerol transferase enzyme (DGAT), as well as the inhibition of hormone-sensitive lipase-mediated lipolysis (Cianflone *et al.*, 2003). ASP also participates in the regulation of glucose homeostasis, since this hormone increases glucose-stimulated insulin secretion via a direct action on pancreatic β-cells (Ahren *et al.*, 2003). Elevated plasma ASP, C3 and adipsin concentrations have been found in obesity, type 1 and T2DM (Koistinen *et al.*, 2001; Cianflone *et al.*, 2003). Although plasma ASP concentrations are correlated inversely with insulin sensitivity, this association is lost in T2DM (Koistinen *et al.*, 2001). Taken together, ASP is associated with whole-body glucose and lipid metabolism in healthy individuals, whereas metabolic disturbances in obesity and T2DM may overcome the regulatory role of ASP in lipid and glucose homeostasis.

Only C3 and, to a lesser extent, ASP have been examined with respect to CAD and dyslipidaemia. On the one hand, increased C3 levels are associated with hypertension and T2DM (with an additive effect) and C3 has been shown to be a powerful predictor of myocardial infarction (Muscari *et al.*, 1995). On the other hand, ASP is increased in subjects with CAD, especially in those with increased plasma TG and/or cholesterol, as characterized by increased plasma apolipoprotein B levels (Cianflone *et al.*, 1997). Growing evidence supports a role for ASP and C3 in adipose tissue function and maintenance of whole-body glucose homeostasis (Cianflone *et al.*, 2003). The aetiology of the links between ASP and C3 with T2DM and CAD and well-recognized risk factors such as insulin resistance and lipid profile has not been disentangled completely and future studies are required to unravel the exact role of these molecules in the ethiopathogenesis of CVD.

Concluding Remarks

Given the current prevalence of obesity and that this condition is a major modifiable contributor to CHD, a better understanding of the underlying mechanisms that relate fat mass to cardiovascular health is of paramount importance. Adipose tissue constitutes an important source of circulating mediators of inflammation that participate in the mechanism of cardiovascular injury and atherogenesis. Adipocytes and adipose tissue-embedded macrophages secrete proinflammatory cytokines (TNF-α, IL-6), acute-phase reactants (CRP, SAA), complement factors (adipsin and ASP), prothrombotic molecules (PAI-1, tissue factor), growth factors (cardiotrophin-1, EGF, FGF) and hormones implicated in the regulation of inflammation (leptin, resistin, osteopontin, adiponectin). In addition, increased adiposity is accompanied by a defective lipid partitioning that favours the development of CVD. Excessive fatty acid release in obesity leads to lipid deposition in muscle, liver, pancreas and heart. This ectopic lipid accumulation contributes to the development of insulin resistance, atherogenic dyslipidaemia and hyperinsulinaemia.

The exact participation of the complex network of bioactive mediators on vasoactivity and inflammation remains to be disentangled fully, in particular as regards gaining more insight into the mechanisms involved in the activation and integration of the diverse signalling pathways. Major advances in unravelling the molecular events underlying inflammation and atherogenesis are to be expected by focusing on how the known vasoactive factors are related to the more recently identified hormones, adipokines, receptors, channels and peptides such as obestatin, adrenomedullin, hypoxia-sensitive molecules, aquaporins, caveolins and caspases. Undoubtedly, given the adipose tissue's versatile and ever-expanding list of activities, additional and unexpected vasoactive peptides are sure to emerge. The intense ongoing epidemiological, interventional and molecular research warrants the incorporation of relevant and novel information in many different frontiers of our current cardiovascular knowledge.

References

Adya, R., Tan, B.K., Punn, A., Chen, J. and Randeva, H.S. (2008) Visfatin induces human endothelial VEGF and MMP-2/9 production via MAPK and PI3K/Akt signalling pathways: novel insights into visfatin-induced angiogenesis. *Cardiovascular Research* 78, 356–365.

Agata, J., Masuda, A., Takada, M., Higashiura, K., Murakami, H., Miyazaki, Y. and Shimamoto, K. (1997) High plasma immunoreactive leptin level in essential hypertension. *The American Journal of Hypertension* 10, 1171–1174.

Ahima, R. (2006) Ghrelin – a new player in glucose homeostasis? *Cell Metabolism* 3, 301–302.

Ahren, B., Havel, P., Pacini, G. and Cianflone, K. (2003) Acylation stimulating protein stimulates insulin secretion. *International Journal of Obesity and Related Metabolic Disorders* 27, 1037–1043.

Arita, Y., Kihara, S., Ouchi, N., Takahashi, M., Maeda, K., Miyagawa, J., Hotta, K., Shimomura, I., Nakamura, T., Miyaoka, K., Kuriyama, H., Nishida, M., Yamashita, S., Okubo, K., Matsubara, K., Muraguchi, M., Ohmoto, Y., Funahashi, T. and Matsuzawa, Y. (1999) Paradoxical decrease of an adipose-specific protein, adiponectin, in obesity. *Biochemical and Biophysical Research Communications* 257, 79–83.

Aronson, D., Bartha, P., Zinder, O., Kerner, A., Markiewicz, W., Avizohar, O., Brook, G.J. and Levy, Y. (2004) Obesity is the major determinant of elevated C-reactive protein in subjects with the metabolic syndrome. *International Journal of Obesity and Related Metabolic Disorders* 28, 674–679.

Azuma, K., Katsukawa, F., Oguchi, S., Murata, M., Yamazaki, H., Shimada, A. and Saruta, T. (2003) Correlation between serum resistin level and adiposity in obese individuals. *Obesity Research* 11, 997–1001.

Baldanzi, G., Filigheddu, N., Cutrupi, S., Catapano, F., Bonissoni, S., Fubini, A., Malan, D., Baj, G., Granata, R., Broglio, F., Papotti, M., Surico, N., Bussolino, F., Isgaard, J., Deghenghi, R., Sinigaglia, F., Prat, M., Muccioli, G., Ghigo, E. and Graziani, A. (2002) Ghrelin and des-acyl ghrelin inhibit cell death in cardiomyocytes and endothelial cells through ERK1/2 and PI 3-kinase/AKT. *The Journal of Cell Biology* 159, 1029–1037.

Beltowski, J., Wojcicka, G. and Jamroz-Wisniewska, A. (2006) Role of nitric oxide and endothelium-derived hyperpolarizing factor (EDHF) in the regulation of blood pressure by leptin in lean and obese rats. *Life Sciences* 79, 63–71.

Berg, A.H. and Scherer, P.E. (2005) Adipose tissue, inflammation, and cardiovascular disease. *Circulation Research* 96, 939–949.

Bodary, P. and Eitzman, D. (2006) Adiponectin: vascular protection from the fat? *Arteriosclerosis, Thrombosis, and Vascular Biology* 26, 235–236.

Boucher, J., Masri, B., Daviaud, D., Gesta, S., Guigne, C., Mazzucotelli, A., Castan-Laurell, I., Tack, I., Knibiehler, B., Carpene, C., Audigier, Y., Saulnier-Blache, J. and Valet, P. (2005) Apelin, a newly identified adipokine up-regulated by insulin and obesity. *Endocrinology* 146, 1764–1771.

Bouloumié, A., Drexler, H.C., Lafontan, M. and Busse, R. (1998) Leptin, the product of *Ob* gene, promotes angiogenesis. *Circulation Research* 83, 1059–1066.

Broch, M., Vendrell, J., Ricart, W., Richart, C. and Fernández-Real, J.M. (2007) Circulating retinol-binding protein-4, insulin sensitivity, insulin secretion, and insulin disposition index in obese and non-obese subjects. *Diabetes Care* 30, 1802–1806.

Burstein, S., Peng, J., Friese, P., Wolf, R., Harrison, P., Downs, T., Hamilton, K., Comp, P. and Dale, G. (1996) Cytokine-induced alteration of platelet and hemostatic function. *Stem Cells* 14 (Suppl. 1), 154–162.

Caro, J., Sinha, M., Kolaczynski, J., Zhang, P. and Considine, R. (1996) Leptin: the tale of an obesity gene. *Diabetes* 45, 1455–1462.

Chen, M., Tsai, J., Chung, F., Yang, S., Hsing, L., Shin, S. and Lee, Y. (2005) Hypoadiponectinemia is associated with ischemic cerebrovascular disease. *Arteriosclerosis, Thrombosis, and Vascular Biology* 25, 821–826.

Chen, M.P., Chung, F.M., Chang, D.M., Tsai, J.C., Huang, H.F., Shin, S.J. and Lee, Y.J. (2006) Elevated plasma level of visfatin/pre-B cell colony-enhancing factor in patients with type 2 diabetes mellitus. *Journal of Clinical Endocrinology and Metabolism* 91, 295–299.

Cheng, X., Cheng, X.S. and Pang, C.C. (2003) Venous dilator effect of apelin, an endogenous peptide ligand for the orphan APJ receptor, in conscious rats. *The European Journal of Pharmacology* 470, 171–175.

Cianflone, K.M., Sniderman, A.D., Walsh, M.J., Vu, H.T., Gagnon, J. and Rodríguez, M.A. (1989) Purification and characterization of acylation stimulating protein. *Journal of Biological Chemistry* 264, 426–430.

Cianflone, K., Zhang, X., Genest, J.J. and Sniderman, A. (1997) Plasma acylation-stimulating protein in coronary artery disease. *Arteriosclerosis, Thrombosis, and Vascular Biology* 17, 1239–1244.

Cianflone, K., Xia, Z. and Chen, L. (2003) Critical review of acylation-stimulating protein physiology in humans and rodents. *Biochimica et Biophysica Acta* 1609, 127–143.

Considine, R.V., Sinha, M.K., Heiman, M.L., Kriauciunas, A., Stephens, T.W., Nyce, M.R., Ohannesian, J.P., Marco, C.C., McKee, L.J., Bauer, T.L. and Caro, J.F. (1996) Serum immunoreactive-leptin concentrations in normal-weight and obese humans. *The New England Journal of Medicine* 334, 292–295.

Crandall, D., Herzlinger, H., Saunders, B., Armellino, D. and Kral, J. (1994) Distribution of angiotensin II receptors in rat and human adipocytes. *Journal of Lipid Research* 35, 1378–1385.

Davi, G., Guagnano, M., Ciabattoni, G., Basili, S., Falco, A., Marinopiccoli, M., Nutini, M., Sensi, S. and Patrono, C. (2002) Platelet activation in obese women: role of inflammation and oxidant stress. *The Journal of the American Medical Association* 288, 2008–2014.

Degawa-Yamauchi, M., Bovenkerk, J., Juliar, B., Watson, W., Kerr, K., Jones, R., Zhu, Q. and Considine, R. (2003) Serum resistin (FIZZ3) protein is increased in obese humans. *Journal of Clinical Endocrinology and Metabolism* 88, 5452–5455.

Delanghe, J.R., Langlois, M.R., De Bacquer, D., Mak, R., Capel, P., Van Renterghem, L. and De Backer, G. (2002) Discriminative value of serum amyloid A and other acute-phase proteins for coronary heart disease. *Atherosclerosis* 160, 471–476.

De Taeye, B., Smith, L.H. and Vaughan, D.E. (2005) Plasminogen activator inhibitor-1: a common denominator in obesity, diabetes and cardiovascular disease. *Current Opinion in Pharmacology* 5, 149–154.

Dick, G.M., Katz, P.S., Farias, M. 3rd, Morris, M., James, J., Knudson, J.D. and Tune, J.D. (2006) Resistin impairs endothelium-dependent dilation to bradykinin, but not acetylcholine, in the coronary circulation. *American Journal of Physiology – Heart and Circulatory Physiology* 291, H2997–3002.

Dogru, T., Sonmez, A., Tasci, I., Bozoglu, E., Yilmaz, M.I., Genc, H., Erdem, G., Gok, M., Bingol, N., Kilic, S., Ozgurtas, T. and Bingol, S. (2007) Plasma visfatin levels in patients with newly diagnosed and untreated type 2 diabetes mellitus and impaired glucose tolerance. *Diabetes Research and Clinical Practice* 76, 24–29.

Eckel, R.H. and Krauss, R.M. (1998) American Heart Association call to action: obesity as a major risk factor for coronary heart disease. AHA Nutrition Committee. *Circulation* 97, 2099–2100.

Eckel, R.H., Grundy, S.M. and Zimmet, P.Z. (2005) The metabolic syndrome. *Lancet* 365, 1415–1428.

Engeli, S., Bohnke, J., Gorzelniak, K., Janke, J., Schling, P., Bader, M., Luft, F. and Sharma, A. (2005) Weight loss and the renin–angiotensin–aldosterone system. *Hypertension* 45, 356–362.

Esler, M. (2000) The sympathetic system and hypertension. *American Journal of Hypertension* 13, 99S–105S.

Esmon, C. (2004) The impact of the inflammatory response on coagulation. *Thrombosis Research* 114, 321–327.

Fain, J., Madan, A., Hiler, M., Cheema, P. and Bahouth, S. (2004) Comparison of the release of adipokines by adipose tissue, adipose tissue matrix, and adipocytes from visceral and subcutaneous abdominal adipose tissues of obese humans. *Endocrinology* 145, 2273–2282.

Flórez, H., Castillo-Flórez, S., Méndez, A., Casanova-Romero, P., Larreal-Urdaneta, C., Lee, D. and Goldberg, R. (2006) C-reactive protein is elevated in obese patients with the metabolic syndrome. *Diabetes Research and Clinical Practice* 71, 92–100.

Földes, G., Horkay, F., Szokodi, I., Vuolteenaho, O., Ilves, M., Lindstedt, K.A., Mäyränpää, M., Sarman, B., Seres, L., Skoumal, R., Lako-Futo, Z., deChatel, R., Ruskoaho, H. and Toth, M. (2003) Circulating and cardiac levels of apelin, the novel ligand of the orphan receptor APJ, in patients with heart failure. *Biochemical and Biophysical Research Communications* 308, 480–485.

Fortuño, A., Rodríguez, A., Gómez-Ambrosi, J., Muñiz, P., Salvador, J., Díez, J. and Frühbeck, G. (2002) Leptin inhibits angiotensin II-induced intracellular calcium increase and vasoconstriction in rat aorta. *Endocrinology* 143, 3555–3560.

Franks, P., Brage, S., Luan, J., Ekelund, U., Rahman, M., Farooqi, I., Halsall, I., O'Rahilly, S. and Wareham, N. (2005) Leptin predicts a worsening of the features of the metabolic syndrome independently of obesity. *Obesity Research* 13, 1476–1484.

Frühbeck, G. (1999) Pivotal role of nitric oxide in the control of blood pressure after leptin administration. *Diabetes* 48, 903–908.

Frühbeck, G. (2004) The adipose tissue as a source of vasoactive factors. *Current Medicinal Chemistry – Cardiovascular and Hematological Agents* 2, 197–208.

Frühbeck, G., Gómez-Ambrosi, J., Muruzábal, F.J. and Burrell, M.A. (2001) The adipocyte: a model for integration of endocrine and metabolic signaling in energy metabolism regulation. *American Journal of Physiology – Endocrinology and Metabolism* 280, E827–847.

Fukuhara, A., Matsuda, M., Nishizawa, M., Segawa, K., Tanaka, M., Kishimoto, K., Matsuki, Y., Murakami, M., Ichisaka, T., Murakami, H., Watanabe, E., Takagi, T., Akiyoshi, M.,

Ohtsubo, T., Kihara, S., Yamashita, S., Makishima, M., Funahashi, T., Yamanaka, S., Hiramatsu, R., Matsuzawa, Y. and Shimomura, I. (2005) Visfatin: a protein secreted by visceral fat that mimics the effects of insulin. *Science* 307, 426–430.

Fukuhara, A., Matsuda, M., Nishizawa, M., Segawa, K., Tanaka, M., Kishimoto, K., Matsuki, Y., Murakami, M., Ichisaka, T., Murakami, H., Watanabe, E., Takagi, T., Akiyoshi, M., Ohtsubo, T., Kihara, S., Yamashita, S., Makishima, M., Funahashi, T., Yamanaka, S., Hiramatsu, R., Matsuzawa, Y. and Shimomura, I. (2007) Visfatin: a protein secreted by visceral fat that mimics the effects of insulin. Retraction. *Science* 318, 565.

Gálvez, B., Castro, J. de, Herold, D., Dubrovska, G., Arribas, S., González, M.C., Aránguez, I., Luft, F.C., Ramos, M.P., Gollasch, M. and Fernández Alfonso, M.S. (2006) Perivascular adipose tissue and mesenteric vascular function in spontaneously hypertensive rats. *Arteriosclerosis, Thrombosis, and Vascular Biology* 26, 1297–1302.

Gentile, M.T., Vecchione, C., Marino, G., Aretini, A., Di Pardo, A., Antenucci, G., Maffei, A., Cifelli, G., Iorio, L., Landolfi, A., Frati, G. and Lembo, G. (2008) Resistin impairs insulin-evoked vasodilation. *Diabetes* 57, 577–583.

George, J., Patal, S., Wexler, D., Sharabi, Y., Peleg, E., Kamari, Y., Grossman, E., Sheps, D., Keren, G. and Roth, A. (2006) Circulating adiponectin concentrations in patients with congestive heart failure. *Heart* 92, 1420–1424.

Gómez-Ambrosi, J. and Frühbeck, G. (2001) Do resistin and resistin-like molecules also link obesity to inflammatory diseases? *Annals of Internal Medicine* 135, 306–307.

Gómez-Ambrosi, J. and Frühbeck, G. (2005) Evidence for the involvement of resistin in inflammation and cardiovascular disease. *Current Diabetes Reviews* 1, 227–234.

Gómez-Ambrosi, J., Salvador, J., Páramo, J., Orbe, J., Irala, J. de, Díez-Caballero, A., Gil, M., Cienfuegos, J. and Frühbeck, G. (2002) Involvement of leptin in the association between percentage of body fat and cardiovascular risk factors. *Clinical Biochemistry* 35, 315–320.

Gómez-Ambrosi, J., Salvador, J., Rotellar, F., Silva, C., Catalán, V., Rodríguez, A., Gil, M.J. and Frühbeck, G. (2006) Increased serum amyloid A concentrations in morbid obesity decrease after gastric bypass. *Obesity Surgery* 16, 262–269.

Gómez-Ambrosi, J., Catalán, V., Ramírez, B., Rodríguez, A., Colina, I., Silva, C., Rotellar, F., Mugueta, C., Gil, M.J., Cienfuegos, J.A., Salvador, J. and Frühbeck, G. (2007) Plasma osteopontin levels and expression in adipose tissue are increased in obesity. *Journal of Clinical Endocrinology and Metabolism* 92, 3719–3727.

Gómez-Ambrosi, J., Rodríguez, A., Catalán, V., Ramírez, B., Silva, C., Rotellar, F., Gil, M.J., Salvador, J. and Frühbeck, G. (2008) Serum retinol-binding protein 4 is not increased in obesity or obesity-associated type 2 diabetes mellitus, but is reduced after relevant reductions in body fat following gastric bypass. *Clinical Endocrinology* 69, 208–215.

Goralski, K.B., McCarthy, T.C., Hanniman, E.A., Zabel, B.A., Butcher, E.C., Parlee, S.D., Muruganandan, S. and Sinal, C.J. (2007) Chemerin, a novel adipokine that regulates adipogenesis and adipocyte metabolism. *Journal of Biological Chemistry* 282, 28175–28188.

Graf, K., Do, Y.S., Ashizawa, N., Meehan, W.P., Giachelli, C.M., Marboe, C.C., Fleck, E. and Hsueh, W.A. (1997) Myocardial osteopontin expression is associated with left ventricular hypertrophy. *Circulation* 96, 3063–3071.

Graham, T., Yang, Q., Bluher, M., Hammarstedt, A., Ciaraldi, T., Henry, R., Wason, C., Oberbach, A., Jansson, P., Smith, U. and Kahn, B. (2006) Retinol-binding protein 4 and insulin resistance in lean, obese, and diabetic subjects. *New England Journal of Medicine* 354, 2552–2563.

Gualillo, O., González-Juanatey, J.R. and Lago, F. (2007) The emerging role of adipokines as mediators of cardiovascular function: physiologic and clinical perspectives. *Trends in Cardiovascular Medicine* 17, 275–283.

Haider, D.G., Pleiner, J., Francesconi, M., Wiesinger, G.F., Muller, M. and Wolzt, M. (2006a) Exercise training lowers plasma visfatin concentrations in patients with type 1 diabetes. *Journal of Clinical Endocrinology and Metabolism* 91, 4702–4704.

Haider, D.G., Schindler, K., Schaller, G., Prager, G., Wolzt, M. and Ludvik, B. (2006b) Increased plasma visfatin concentrations in morbidly obese subjects are reduced after gastric banding. *Journal of Clinical Endocrinology and Metabolism* 91, 1578–1581.

Hall, J. (1997) Mechanisms of abnormal renal sodium handling in obesity hypertension. *American Journal of Hypertension* 10, 49S–55S.

Haynes, W., Morgan, D., Walsh, S., Mark, A. and Sivitz, W. (1997) Receptor-mediated regional sympathetic nerve activation by leptin. *The Journal of Clinical Investigation* 100, 270–278.

Heilbronn, L., Rood, J., Janderova, L., Albu, J., Kelley, D., Ravussin, E. and Smith, S. (2004) Relationship between serum resistin concentrations and insulin resistance in non-obese, obese, and obese diabetic subjects. *Journal of Clinical Endocrinology and Metabolism* 89, 1844–1848.

Herbert, A., Liu, C., Karamohamed, S., Schiller, J., Liu, J., Yang, Q., Wilson, P., Cupples, L. and Meigs, J. (2005) The –174 IL-6 GG genotype is associated with a reduced risk of type 2 diabetes mellitus in a family sample from the National Heart, Lung and Blood Institute's Framingham Heart Study. *Diabetologia* 48, 1492–1495.

Herbert, A., Liu, C., Karamohamed, S., Liu, J., Manning, A., Fox, C., Meigs, J. and Cupples, L. (2006) BMI modifies associations of IL-6 genotypes with insulin resistance: the Framingham Study. *Obesity* 14, 1454–1461.

Hernández, R., Teruel, T., Alvaro, C. de and Lorenzo, M. (2004) Rosiglitazone ameliorates insulin resistance in brown adipocytes of Wistar rats by impairing TNF-alpha induction of p38 and p42/p44 mitogen-activated protein kinases. *Diabetologia* 47, 1615–1624.

Hosoda, H., Kojima, M., Matsuo, H. and Kangawa, K. (2000) Ghrelin and des-acyl ghrelin: two major forms of rat ghrelin peptide in gastrointestinal tissue. *Biochemical and Biophysical Research Communications* 279, 909–913.

Hotamisligil, G., Murray, D., Choy, L. and Spiegelman, B. (1994) Tumor necrosis factor alpha inhibits signaling from the insulin receptor. *The Proceedings of the National Academy of Science of the United States of America* 91, 4854–4858.

Hotamisligil, G., Arner, P., Caro, J., Atkinson, R. and Spiegelman, B. (1995) Increased adipose tissue expression of tumor necrosis factor-alpha in human obesity and insulin resistance. *The Journal of Clinical Investigation* 95, 2409–2415.

Hotta, K., Funahashi, T., Arita, Y., Takahashi, M., Matsuda, M., Okamoto, Y., Iwahashi, H., Kuriyama, H., Ouchi, N., Maeda, K., Nishida, M., Kihara, S., Sakai, N., Nakajima, T., Hasegawa, K., Muraguchi, M., Ohmoto, Y., Nakamura, T., Yamashita, S., Hanafusa, T. and Matsuzawa, Y. (2000) Plasma concentrations of a novel, adipose-specific protein, adiponectin, in type 2 diabetic patients. *Arteriosclerosis, Thrombosis, and Vascular Biology* 20, 1595–1599.

Hotta, K., Funahashi, T., Bodkin, N., Ortmeyer, H., Arita, Y., Hansen, B. and Matsusawa, Y. (2001) Circulating concentrations of the adipocyte protein adiponectin are decreased in parallel with reduced insulin sensitivity during the progression to type 2 diabetes in rhesus monkeys. *Diabetes* 50, 1126–1133.

Iglesias, M.J., Pineiro, R., Blanco, M., Gallego, R., Dieguez, C., Gualillo, O., Gonzalez-Juanatey, J.R. and Lago, F. (2004) Growth hormone releasing peptide (ghrelin) is synthesized and secreted by cardiomyocytes. *Cardiovascular Research* 62, 481–488.

Iqbal, N., Seshadri, P., Stern, L., Loh, J., Kundu, S., Jafar, T. and Samaha, F.F. (2005) Serum resistin is not associated with obesity or insulin resistance in humans. *The European Review for Medical and Pharmacological Sciences* 9, 161–165.

Ishida, J., Hashimoto, T., Hashimoto, Y., Nishiwaki, S., Iguchi, T., Harada, S., Sugaya, T., Matsuzaki, H., Yamamoto, R., Shiota, N., Okunishi, H., Kihara, M., Umemura, S., Sugiyama, F., Yagami, K., Kasuya, Y., Mochizuki, N. and Fukamizu, A. (2004) Regulatory roles for APJ, a seven-transmembrane receptor related to angiotensin-type 1 receptor in blood pressure *in vivo*. *Journal of Biological Chemistry* 279, 26274–26279.

Ishikawa, Y., Akasaka, Y., Ishii, T., Yoda-Murakami, M., Choi-Miura, N.H., Tomita, M., Ito, K., Zhang, L., Akishima, Y., Ishihara, M., Muramatsu, M. and Taniyama, M. (2003) Changes in the distribution pattern of gelatin-binding protein of 28 kDa (adiponectin) in myocardial remodelling after ischaemic injury. *Histopathology* 42, 43–52.

Isoda, K., Kamezawa, Y., Ayaori, M., Kusuhara, M., Tada, N. and Ohsuzu, F. (2003) Osteopontin transgenic mice fed a high-cholesterol diet develop early fatty-streak lesions. *Circulation* 107, 679–681.

Jackson, E. and Li, P. (1997) Human leptin has natriuretic activity in the rat. *American Journal of Physiology* 272, F333–F338.

Janke, J., Engeli, S., Boschmann, M., Adams, F., Bohnke, J., Luft, F.C., Sharma, A.M. and Jordan, J. (2006) Retinol-binding protein 4 in human obesity. *Diabetes* 55, 2805–2810.

Jensen, L.E. and Whitehead, A.S. (1998) Regulation of serum amyloid A protein expression during the acute-phase response. *Biochemical Journal* 334 (Pt 3), 489–503.

Jousilahti, P., Salomaa, V., Rasi, V., Vahtera, E. and Palosuo, T. (2001) The association of c-reactive protein, serum amyloid a and fibrinogen with prevalent coronary heart disease – baseline findings of the PAIS project. *Atherosclerosis* 156, 451–456.

Jung, H.S., Park, K.H., Cho, Y.M., Chung, S.S., Cho, H.J., Cho, S.Y., Kim, S.J., Kim, S.Y., Lee, H.K. and Park, K.S. (2006) Resistin is secreted from macrophages in atheromas and promotes atherosclerosis. *Cardiovascular Research* 69, 76–85.

Karlsson, C., Lindell, K., Ottosson, M., Sjostrom, L., Carlsson, B. and Carlsson, L.M. (1998) Human adipose tissue expresses angiotensinogen and enzymes required for its conversion to angiotensin II. *Journal of Clinical Endocrinology and Metabolism* 83, 3925–3929.

Kim, S. and Moustaid-Moussa, N. (2000) Secretory, endocrine and autocrine/paracrine function of the adipocyte. *Journal of Nutrition* 130, 3110S–3115S.

Kimura, K., Tsuda, K., Baba, A., Kawabe, T., Boh-oka, S., Ibata, M., Moriwaki, C., Hano, T. and Nishio, I. (2000) Involvement of nitric oxide in endothelium-dependent arterial relaxation by leptin. *Biochemical and Biophysical Research Communications* 273, 745–749.

Kistorp, C., Faber, J., Galatius, S., Gustafsson, F., Frystyk, J., Flyvbjerg, A. and Hildebrandt, P. (2005) Plasma adiponectin, body mass index, and mortality in patients with chronic heart failure. *Circulation* 112, 1756–1762.

Klein, J., Perwitz, N., Kraus, D. and Fasshauer, M. (2006) Adipose tissue as source and target for novel therapies. *Trends in Endocrinology and Metabolism* 17, 26–32.

Kleinz, M.J. and Baxter, G.F. (2008) Apelin reduces myocardial reperfusion injury independently of PI3K/Akt and P70S6 kinase. *Regulatory Peptides* 146, 271–277.

Kleinz, M., Maguire, J., Skepper, J. and Davenport, A. (2006) Functional and immunocytochemical evidence for a role of ghrelin and des-octanoyl ghrelin in the regulation of vascular tone in man. *Cardiovascular Research* 69, 227–235.

Knerr, I., Herzog, D., Rauh, M., Rascher, W. and Horbach, T. (2006) Leptin and ghrelin expression in adipose tissues and serum levels in gastric banding patients. *The European Journal of Clinical Investigation* 36, 389–394.

Knudson, J.D., Dincer, U.D., Dick, G.M., Shibata, H., Akahane, R., Saito, M. and Tune, J.D. (2005) Leptin resistance extends to the coronary vasculature in prediabetic dogs and provides a protective adaptation against endothelial dysfunction. *American Journal of Physiology – Heart and Circulatory Physiology* 289, H1038–1046.

Koistinen, H., Vidal, H., Karonen, S., Dusserre, E., Vallier, P., Koivisto, V. and Ebeling, P. (2001) Plasma acylation stimulating protein concentration and subcutaneous adipose tissue C3 mRNA expression in non-diabetic and type 2 diabetic men. *Arteriosclerosis, Thrombosis, and Vascular Biology* 21, 1034–1039.

Kojima, M., Hosoda, H., Date, Y., Nakazato, M., Matsuo, H. and Kangawa, K. (1999) Ghrelin is a growth-hormone-releasing acylated peptide from stomach. *Nature* 402, 656–660.

Kopelman, P.G. (2000) Obesity as a medical problem. *Nature* 404, 635–643.

Krauss, R., Winston, M., Fletcher, B. and Grundy, S. (1998) Obesity: impact on cardiovascular disease. *Circulation* 98, 1472–1476.

Krzyzanowska, K., Krugluger, W., Mittermayer, F., Rahman, R., Haider, D., Shnawa, N. and Schernthaner, G. (2006) Increased visfatin concentrations in women with gestational diabetes mellitus. *Clinical Science* 110, 605–609.

Kubota, N., Terauchi, Y., Yamauchi, T., Kubota, T., Moroi, M., Matsui, J., Eto, K., Yamashita, T., Kamon, J., Satoh, H., Yano, W., Froguel, P., Nagai, R., Kimura, S., Kadowaki, T. and Noda, T. (2002) Disruption of adiponectin causes insulin resistance and neointimal formation. *Journal of Biological Chemistry* 277, 25863–25866.

Kuk, J.L., Katzmarzyk, P.T., Nichaman, M.Z., Church, T.S., Blair, S.N. and Ross, R. (2006) Visceral fat is an independent predictor of all-cause mortality in men. *Obesity* 14, 336–341.

Kumada, M., Kihara, S., Sumitsuji, S., Kawamoto, T., Matsumoto, S., Ouchi, N., Arita, Y., Okamoto, Y., Shimomura, I., Hiraoka, H., Nakamura, T., Funahashi, T. and Matsuzawa, Y. (2003) Association of hypoadiponectinemia with coronary artery disease in men. *Arteriosclerosis, Thrombosis, and Vascular Biology* 23, 85–89.

Kunnari, A., Ukkola, O., Paivansalo, M. and Kesaniemi, Y.A. (2006) High plasma resistin level is associated with enhanced highly sensitive C-reactive protein and leukocytes. *Journal of Clinical Endocrinology and Metabolism* 91, 2755–2760.

Kusminski, C.M., McTernan, P.G. and Kumar, S. (2005) Role of resistin in obesity, insulin resistance and Type II diabetes. *Clinical Science* 109, 243–256.

Lee, D.K., George, S.R. and O'Dowd, B.F. (2006) Unravelling the roles of the apelin system: prospective therapeutic applications in heart failure and obesity. *Trends in Pharmaceutical Sciences* 27, 190–194.

Lee, J., Chan, J., Yiannakouris, N., Kontogianni, M., Estrada, E., Seip, R., Orlova, C. and Mantzoros, C. (2003) Circulating resistin levels are not associated with obesity or insulin resistance in humans and are not regulated by fasting or leptin administration: cross-sectional and interventional studies in normal, insulin-resistant, and diabetic subjects. *Journal of Clinical Endocrinology and Metabolism* 88, 4848–4856.

Lehrke, M., Reilly, M., Millington, S., Iqbal, N., Rader, D. and Lazar, M. (2004) An inflammatory cascade leading to hyperresistinemia in humans. *PLoS Medicine* 1, e45.

Lembo, G., Vecchione, C., Fratta, L., Marino, G., Trimarco, V., d'Amati, G. and Trimarco, B. (2000) Leptin induces direct vasodilation through distinct endothelial mechanisms. *Diabetes* 49, 293–297.

Lenga, Y., Koh, A., Perera, A.S., McCulloch, C.A., Sodek, J. and Zohar, R. (2008) Osteopontin expression is required for myofibroblast differentiation. *Circulation Research* 102, 319–327.

Li, R., Wang, W.Q., Zhang, H., Yang, X., Fan, Q., Christopher, T.A., Lopez, B.L., Tao, L., Goldstein, B.J., Gao, F. and Ma, X.L. (2007) Adiponectin improves endothelial function in hyperlipidemic rats by reducing oxidative/nitrative stress and differential regulation of eNOS/iNOS activity. *American Journal of Physiology – Endocrinology and Metabolism* 293, E1703–1708.

Li, W., Gavrila, D., Liu, X., Wang, L., Gunnlaugsson, S., Stoll, L., McCormick, M., Sigmund, C., Tang, C. and Weintraub, N. (2004) Ghrelin inhibits proinflammatory responses and nuclear factor-kappaB activation in human endothelial cells. *Circulation* 109, 2221–2226.

Lin, Y., Matsumura, K., Fukuhara, M., Kagiyama, S., Fujii, K. and Iida, M. (2004) Ghrelin acts at the nucleus of the solitary tract to decrease arterial pressure in rats. *Hypertension* 43, 977–982.

Lubos, E., Messow, C.M., Schnabel, R., Rupprecht, H.J., Espinola-Klein, C., Bickel, C., Peetz, D., Post, F., Lackner, K.J., Tiret, L., Munzel, T. and Blankenberg, S. (2007) Resistin, acute coronary syndrome and prognosis results from the AtheroGene study. *Atherosclerosis* 193, 121–128.

McTernan, P., Fisher, F., Valsamakis, G., Chetty, R., Harte, A., McTernan, C., Clark, P., Smith, S., Barnett, A. and Kumar, S. (2003) Resistin and type 2 diabetes: regulation of resistin expression by insulin and rosiglitazone and the effects of recombinant resistin on lipid and glucose metabolism in human differentiated adipocytes. *Journal of Clinical Endocrinology and Metabolism* 88, 6098–6106.

Maeda, K., Okubo, K., Shimomura, I., Funahashi, T., Matsuzawa, Y. and Matsubara, K. (1996) cDNA cloning and expression of a novel adipose specific collagen-like factor, apM1 (AdiPose Most abundant Gene transcript 1). *Biochemical and Biophysical Research Communications* 221, 286–289.

Maeda, N., Shimomura, I., Kishida, K., Nishizawa, H., Matsuda, M., Nagaretani, H., Furuyama, N., Kondo, H., Takahashi, M., Arita, Y., Komuro, R., Ouchi, N., Kihara, S., Tochino, Y., Okutomi, K., Horie, M., Takeda, S., Aoyama, T., Funahashi, T. and Matsuzawa, Y. (2002) Diet-induced insulin resistance in mice lacking adiponectin/ACRP30. *Nature Medicine* 8, 731–737.

Malle, E. and De Beer, F.C. (1996) Human serum amyloid A (SAA) protein: a prominent acute-phase reactant for clinical practice. *European Journal of Clinical Investigation* 26, 427–435.

Marsh, A., Fontes, M., Killinger, S., Pawlak, D., Polson, J. and Dampney, R. (2003) Cardiovascular responses evoked by leptin acting on neurons in the ventromedial and dorsomedial hypothalamus. *Hypertension* 42, 488–493.

Matsuda, M., Shimomura, I., Sata, M., Arita, Y., Nishida, M., Maeda, N., Kumada, M., Okamoto, Y., Nagaretani, H., Nishizawa, H., Kishida, K., Komuro, R., Ouchi, N., Kihara, S., Nagai, R., Funahashi, T. and Matsuzawa, Y. (2002) Role of adiponectin in preventing vascular stenosis. The missing link of adipo-vascular axis. *Journal of Biological Chemistry* 277, 37487–37491.

Matsumura, K., Abe, I., Tsuchihashi, T. and Fujishima, M. (2000) Central effects of leptin on cardiovascular and neurohormonal responses in conscious rabbits. *American Journal of Physiology – Regulatory, Integrative and Comparative Physiology* 278, R1314–1320.

Mertens, I., Verrijken, A., Michiels, J., van der Planken, M., Ruige, J. and van Gaal, L. (2006) Among inflammation and coagulation markers, PAI-1 is a true component of the metabolic syndrome. *International Journal of Obesity* 30, 1308–1314.

Mohamed-Ali, V., Goodrick, S., Rawesh, A., Katz, D.R., Miles, J.M., Yudkin, J.S., Klein, S. and Coppack, S.W. (1997) Subcutaneous adipose tissue releases interleukin-6, but not tumor necrosis factor-alpha, *in vivo*. *Journal of Clinical Endocrinology and Metabolism* 82, 4196–4200.

Moller, D. (2000) Potential role of TNF-alpha in the pathogenesis of insulin resistance and type 2 diabetes. *Trends in Endocrinology and Metabolism* 11, 212–217.

Morrow, D.A., Rifai, N., Antman, E.M., Weiner, D.L., McCabe, C.H., Cannon, C.P. and Braunwald, E. (2000) Serum amyloid A predicts early mortality in acute coronary syndromes: A TIMI 11A substudy. *Journal of the American College of Cardiology* 35, 358–362.

Muscari, A., Bozzoli, C., Puddu, G., Sangiorgi, Z., Dormi, A., Rovinetti, C., Descovich, G. and Puddu, P. (1995) Association of serum C3 levels with the risk of myocardial infarction. *American Journal of Medicine* 98, 357–364.

Nagaya, N., Kojima, M., Uematsu, M., Yamagishi, M., Hosoda, H., Oya, H., Hayashi, Y. and Kangawa, K. (2001a) Hemodynamic and hormonal effects of human ghrelin in healthy volunteers. *American Journal of Physiology – Regulatory, Integrative and Comparative Physiology* 280, R1483–1487.

Nagaya, N., Miyatake, K., Uematsu, M., Oya, H., Shimizu, W., Hosoda, H., Kojima, M., Nakanishi, N., Mori, H. and Kangawa, K. (2001b) Hemodynamic, renal, and hormonal effects of ghrelin infusion in patients with chronic heart failure. *Journal of Clinical Endocrinology and Metabolism* 86, 5854–5859.

Naldini, A., Leali, D., Pucci, A., Morena, E., Carraro, F., Nico, B., Ribatti, D. and Presta, M. (2006) Cutting edge: IL-1beta mediates the proangiogenic activity of osteopontin-activated human monocytes. *The Journal of Immunology* 177, 4267–4270.

Narkiewicz, K., Somers, V., Mos, L., Kato, M., Accurso, V. and Palatini, P. (1999) An independent relationship between plasma leptin and heart rate in untreated patients with essential hypertension. *Journal of Hypertension* 17, 245–249.

Natal, C., Fortuno, M.A., Restituto, P., Bazan, A., Colina, I., Diez, J. and Varo, N. (2008) Cardiotrophin-1 is expressed in adipose tissue and upregulated in the metabolic syndrome. *American Journal of Physiology – Endocrinology and Metabolism* 294, E52–60.

Nickola, M.W., Wold, L.E., Colligan, P.B., Wang, G.J., Samson, W.K. and Ren, J. (2000) Leptin attenuates cardiac contraction in rat ventricular myocytes. Role of NO. *Hypertension* 36, 501–505.

Norata, G.D., Ongari, M., Garlaschelli, K., Raselli, S., Grigore, L. and Catapano, A.L. (2007) Plasma resistin levels correlate with determinants of the metabolic syndrome. *European Journal of Endocrinology* 156, 279–284.

Ohashi, K., Kihara, S., Ouchi, N., Kumada, M., Fujita, K., Hiuge, A., Hibuse, T., Ryo, M., Nishizawa, H., Maeda, N., Maeda, K., Shibata, R., Walsh, K., Funahashi, T. and Shimomura, I. (2006) Adiponectin replenishment ameliorates obesity-related hypertension. *Hypertension* 47, 1108–1116.

Okumura, H., Nagaya, N., Enomoto, M., Nakagawa, E., Oya, H. and Kangawa, K. (2002) Vasodilatory effect of ghrelin, an endogenous peptide from the stomach. *Journal of Cardiovascular Pharmacology* 39, 779–783.

Oral, H., Dorn, G. and Mann, D. (1997) Sphingosine mediates the immediate negative inotropic effects of tumor necrosis factor-alpha in the adult mammalian cardiac myocyte. *Journal of Biological Chemistry* 272, 4836–4842.

Otero, M., Lago, R., Lago, F., Casanueva, F.F., Diéguez, C., Gómez-Reino, J.J. and Gualillo, O. (2005) Leptin, from fat to inflammation: old questions and new insights. *FEBS Letters* 579, 295–301.

Ouchi, N., Kihara, S., Arita, Y., Maeda, K., Kuriyama, H., Okamoto, Y., Hotta, K., Nishida, M., Takahashi, M., Nakamura, T., Yamashita, S., Funahashi, T. and Matsuzawa, Y. (1999) Novel modulator for endothelial adhesion molecules: adipocyte-derived plasma protein adiponectin. *Circulation* 100, 2473–2476.

Ouchi, N., Kihara, S., Funahashi, T., Nakamura, T., Nishida, M., Kumada, M., Okamoto, Y., Ohashi, K., Nagaretani, H., Kishida, K., Nishizawa, H., Maeda, N., Kobayashi, H., Hiraoka, H. and Matsuzawa, Y. (2003a) Reciprocal association of C-reactive protein with adiponectin in blood stream and adipose tissue. *Circulation* 107, 671–674.

Ouchi, N., Ohishi, M., Kihara, S., Funahashi, T., Nakamura, T., Nagaretani, H., Kumada, M., Ohashi, K., Okamoto, Y., Nishizawa, H., Kishida, K., Maeda, N., Nagasawa, A., Kobayashi, H., Hiraoka, H., Komai, N., Kaibe, M., Rakugi, H., Ogihara, T. and

Matsuzawa, Y. (2003b) Association of hypoadiponectinemia with impaired vasoreactivity. *Hypertension* 42, 231–234.

Ouedraogo, R., Gong, Y., Berzins, B., Wu, X., Mahadev, K., Hough, K., Chan, L., Goldstein, B.J. and Scalia, R. (2007) Adiponectin deficiency increases leukocyte-endothelium interactions via upregulation of endothelial cell adhesion molecules *in vivo*. *The Journal of Clinical Investigation* 117, 1718–1726.

Pagano, C., Pilon, C., Olivieri, M., Mason, P., Fabris, R., Serra, R., Milan, G., Rossato, M., Federspil, G. and Vettor, R. (2006) Reduced plasma visfatin/pre-B cell colony-enhancing factor in obesity is not related to insulin resistance in humans. *Journal of Clinical Endocrinology and Metabolism* 91, 3165–3170.

Pearson, T.A., Mensah, G.A., Alexander, R.W., Anderson, J.L., Cannon, R.O. 3rd, Criqui, M., Fadl, Y.Y., Fortmann, S.P., Hong, Y., Myers, G.L., Rifai, N., Smith, S.C. Jr, Taubert, K., Tracy, R.P. and Vinicor, F. (2003) Markers of inflammation and cardiovascular disease: application to clinical and public health practice. A statement for healthcare professionals from the Centers for Disease Control and Prevention and the American Heart Association. *Circulation* 107, 499–511.

Pepys, M.B. and Hirschfield, G.M. (2003) C-reactive protein: a critical update. *The Journal of Clinical Investigation* 111, 1805–1812.

Piñeiro, R., Iglesias, M.J., Gallego, R., Raghay, K., Eiras, S., Rubio, J., Diéguez, C., Gualillo, O., González-Juanatey, J.R. and Lago, F. (2005) Adiponectin is synthesized and secreted by human and murine cardiomyocytes. *FEBS Letters* 579, 5163–5169.

Pischon, T., Girman, C.J., Hotamisligil, G.S., Rifai, N., Hu, F.B. and Rimm, E.B. (2004) Plasma adiponectin levels and risk of myocardial infarction in men. *The Journal of the American Medical Association* 291, 1730–1737.

Poykko, S., Ukkola, O., Kauma, H., Kellokoski, E., Horkko, S. and Kesaniemi, Y. (2005) The negative association between plasma ghrelin and IGF-I is modified by obesity, insulin resistance and type 2 diabetes. *Diabetologia* 48, 309–316.

Purdham, D.M., Zou, M.X., Rajapurohitam, V. and Karmazyn, M. (2004) Rat heart is a site of leptin production and action. *American Journal of Physiology – Heart and Circulatory Physiology* 287, H2877–2884.

Quadro, L., Blaner, W.S., Salchow, D.J., Vogel, S., Piantedosi, R., Gouras, P., Freeman, S., Cosma, M.P., Colantuoni, V. and Gottesman, M.E. (1999) Impaired retinal function and vitamin A availability in mice lacking retinol-binding protein. *The EMBO Journal* 18, 4633–4644.

Rahmouni, K., Morgan, D.A., Morgan, G.M., Mark, A.L. and Haynes, W.G. (2005) Role of selective leptin resistance in diet-induced obesity hypertension. *Diabetes* 54, 2012–2018.

Rajapurohitam, V., Javadov, S., Purdham, D.M., Kirshenbaum, L.A. and Karmazyn, M. (2006) An autocrine role for leptin in mediating the cardiomyocyte hypertrophic effects of angiotensin II and endothelin-1. *The Journal of Molecular and Cellular Cardiology* 41, 265–274.

Rangaswami, H., Bulbule, A. and Kundu, G.C. (2006) Osteopontin: role in cell signaling and cancer progression. *Trends in Cell Biology* 16, 79–87.

Reilly, M.P., Lehrke, M., Wolfe, M.L., Rohatgi, A., Lazar, M.A. and Rader, D.J. (2005) Resistin is an inflammatory marker of atherosclerosis in humans. *Circulation* 111, 932–939.

Ren, J. (2004) Leptin and hyperleptinemia – from friend to foe for cardiovascular function. *Journal of Endocrinology* 181, 1–10.

Rodríguez, A., Frühbeck, G., Gómez-Ambrosi, J., Catalán, V., Sáinz, N., Díez, J., Zalba, G. and Fortuño, A. (2006) The inhibitory effect of leptin on angiotensin II-induced vasoconstriction is blunted in spontaneously hypertensive rats. *Journal of Hypertension* 24, 1589–1597.

Rodríguez, A., Catalán, V., Gómez-Ambrosi, J. and Frühbeck, G. (2007a) Visceral and subcutaneous adiposity: are both potential therapeutic targets for tackling the metabolic syndrome? *Current Pharmaceutical Design* 13, 2169–2175.

Rodríguez, A., Fortuño, A., Gómez-Ambrosi, J., Zalba, G., Díez, J. and Frühbeck, G. (2007b) The inhibitory effect of leptin on angiotensin II-induced vasoconstriction in vascular smooth muscle cells is mediated via a nitric oxide-dependent mechanism. *Endocrinology* 148, 324–331.

Rothwell, S.E., Richards, A.M. and Pemberton, C.J. (2006) Resistin worsens cardiac ischaemia-reperfusion injury. *Biochemical and Biophysical Research Communications* 349, 400–407.

Sattar, N., Wannamethee, G., Sarwar, N., Tchernova, J., Cherry, L., Wallace, A., Danesh, J. and Whincup, P. (2006) Adiponectin and coronary heart disease: a prospective study and meta-analysis. *Circulation* 114, 623–629.

Savage, D., Sewter, C., Klenk, E., Segal, D., Vidal-Puig, A., Considine, R. and O'Rahilly, S. (2001) Resistin/Fizz3 expression in relation to obesity and peroxisome proliferator-activated receptor-gamma action in humans. *Diabetes* 50, 2199–2202.

Schulze, P., Kratzsch, J., Linke, A., Schoene, N., Adams, V., Gielen, S., Erbs, S., Moebius-Winkler, S. and Schuler, G. (2003) Elevated serum levels of leptin and soluble leptin receptor in patients with advanced chronic heart failure. *European Journal of Heart Failure* 5, 33–40.

Schulze, M.B., Shai, I., Rimm, E.B., Li, T., Rifai, N. and Hu, F.B. (2005) Adiponectin and future coronary heart disease events among men with type 2 diabetes. *Diabetes* 54, 534–539.

Sharma, A.M. (2006) The obese patient with diabetes mellitus: from research targets to treatment options. *The American Journal of Medicine* 119, S17–S23.

Shek, E., Brands, M. and Hall, J. (1998) Chronic leptin infusion increases arterial pressure. *Hypertension* 31, 409–414.

Shibata, R., Ouchi, N., Ito, M., Kihara, S., Shiojima, I., Pimentel, D.R., Kumada, M., Sato, K., Schiekofer, S., Ohashi, K., Funahashi, T., Colucci, W.S. and Walsh, K. (2004) Adiponectin-mediated modulation of hypertrophic signals in the heart. *Nature Medicine* 10, 1384–1389.

Shibata, R., Sato, K., Pimentel, D.R., Takemura, Y., Kihara, S., Ohashi, K., Funahashi, T., Ouchi, N. and Walsh, K. (2005) Adiponectin protects against myocardial ischemia-reperfusion injury through AMPK- and COX-2-dependent mechanisms. *Nature Medicine* 11, 1096–1103.

Shibata, R., Izumiya, Y., Sato, K., Papanicolaou, K., Kihara, S., Colucci, W.S., Sam, F., Ouchi, N. and Walsh, K. (2007) Adiponectin protects against the development of systolic dysfunction following myocardial infarction. *The Journal of Molecular and Cellular Cardiology* 42, 1065–1074.

Shimabukuro, M., Higa, N., Asahi, T., Oshiro, Y., Takasu, N., Tagawa, T., Ueda, S., Shimomura, I., Funahashi, T. and Matsuzawa, Y. (2003) Hypoadiponectinemia is closely linked to endothelial dysfunction in man. *Journal of Clinical Endocrinology and Metabolism* 88, 3236–3240.

Shklyaev, S., Aslanidi, G., Tennant, M., Prima, V., Kohlbrenner, E., Kroutov, V., Campbell-Thompson, M., Crawford, J., Shek, E.W., Scarpace, P.J. and Zolotukhin, S. (2003) Sustained peripheral expression of transgene adiponectin offsets the development of diet-induced obesity in rats. *The Proceedings of the National Academy of Science of the United States of America* 100, 14217–14222.

Silha, J., Krsek, M., Skrha, J., Sucharda, P., Nyomba, B. and Murphy, L. (2003) Plasma resistin, adiponectin and leptin levels in lean and obese subjects: correlations with insulin resistance. *European Journal of Endocrinology* 149, 331–335.

Sjöholm, K., Palming, J., Olofsson, L., Gummesson, A., Svensson, P., Lystig, T., Jennische, E., Brandberg, J., Torgerson, J., Carlsson, B. and Carlsson, L. (2005) A microarray search for genes predominantly expressed in human omental adipocytes: adipose tissue as a major production site of serum amyloid A. *Journal of Clinical Endocrinology and Metabolism* 90, 2233–2239.

Smith, C.C., Mocanu, M.M., Davidson, S.M., Wynne, A.M., Simpkin, J.C. and Yellon, D.M. (2006) Leptin, the obesity-associated hormone, exhibits direct cardioprotective effects. *British Journal of Pharmacology* 149, 5–13.

Smith, S.C. Jr, Blair, S.N., Bonow, R.O., Brass, L.M., Cerqueira, M.D., Dracup, K., Fuster, V., Gotto, A., Grundy, S.M., Miller, N.H., Jacobs, A., Jones, D., Krauss, R.M., Mosca, L., Ockene, I., Pasternak, R.C., Pearson, T., Pfeffer, M.A., Starke, R.D. and Taubert, K.A. (2001) AHA/ACC Guidelines for preventing heart attack and death in patients with atherosclerotic cardiovascular disease: 2001 update. A statement for healthcare professionals from the American Heart Association and the American College of Cardiology. *Journal of the American College of Cardiology* 38, 1581–1583.

Sobel, B. (1999) Increased plasminogen activator inhibitor-1 and vasculopathy. A reconcilable paradox. *Circulation* 99, 2496–2498.

Söderberg, S., Stegmayr, B., Ahlbeck-Glader, C., Slunga-Birgander, L., Ahren, B. and Olsson, T. (2003) High leptin levels are associated with stroke. *Cerebrovascular Diseases* 15, 63–69.

Soeki, T., Kishimoto, I., Schwenke, D.O., Tokudome, T., Horio, T., Yoshida, M., Hosoda, H. and Kangawa, K. (2008) Ghrelin suppresses cardiac sympathetic activity and prevents early left ventricular remodeling in rats with myocardial infarction. *American Journal of Physiology – Heart and Circulatory Physiology* 294, H426–432.

Spranger, J., Kroke, A., Mohlig, M., Bergmann, M., Ristow, M., Boeing, H. and Pfeiffer, A. (2003) Adiponectin and protection against type 2 diabetes mellitus. *Lancet* 361, 226–228.

Steppan, C.M., Bailey, S.T., Bhat, S., Brown, E.J., Banerjee, R.R., Wright, C.M., Patel, H.R., Ahima, R.S. and Lazar, M.A. (2001) The hormone resistin links obesity to diabetes. *Nature* 409, 307–312.

Sun, M., Dawood, F., Wen, W., Chen, M., Dixon, I., Kirshenbaum, L. and Liu, P. (2004) Excessive tumor necrosis factor activation after infarction contributes to susceptibility of myocardial rupture and left ventricular dysfunction. *Circulation* 110, 3221–3228.

Szokodi, I., Tavi, P., Foldes, G., Voutilainen-Myllyla, S., Ilves, M., Tokola, H., Pikkarainen, S., Piuhola, J., Rysa, J., Toth, M. and Ruskoaho, H. (2002) Apelin, the novel endogenous ligand of the orphan receptor APJ, regulates cardiac contractility. *Circulation Research* 91, 434–440.

Takeishi, Y., Niizeki, T., Arimoto, T., Nozaki, N., Hirono, O., Nitobe, J., Watanabe, T., Takabatake, N. and Kubota, I. (2007) Serum resistin is associated with high risk in patients with congestive heart failure – a novel link between metabolic signals and heart failure. *Circulation Journal* 71, 460–464.

Tamori, Y., Sakaue, H. and Kasuga, M. (2006) RBP4, an unexpected adipokine. *Nature Medicine* 12, 30–31.

Tatemoto, K., Hosoya, M., Habata, Y., Fujii, R., Kakegawa, T., Zou, M.X., Kawamata, Y., Fukusumi, S., Hinuma, S., Kitada, C., Kurokawa, T., Onda, H. and Fujino, M. (1998) Isolation and characterization of a novel endogenous peptide ligand for the human APJ receptor. *Biochemical and Biophysical Research Communications* 251, 471–476.

Tatemoto, K., Takayama, K., Zou, M.X., Kumaki, I., Zhang, W., Kumano, K. and Fujimiya, M. (2001) The novel peptide apelin lowers blood pressure via a nitric oxide-dependent mechanism. *Regulatory Peptides* 99, 87–92.

Tesauro, M., Schinzari, F., Iantorno, M., Rizza, S., Melina, D., Lauro, D. and Cardillo, C. (2005) Ghrelin improves endothelial function in patients with metabolic syndrome. *Circulation* 112, 2986–2992.

Tilg, H. and Moschen, A.R. (2006) Adipocytokines: mediators linking adipose tissue, inflammation and immunity. *Nature Reviews Immunology* 6, 772–783.

Tschöp, M., Smiley, D.L. and Heiman, M.L. (2000) Ghrelin induces adiposity in rodents. *Nature* 407, 908–913.

Tschöp, M., Weyer, C., Tataranni, P.A., Devanarayan, V., Ravussin, E. and Heiman, M.L. (2001) Circulating ghrelin levels are decreased in human obesity. *Diabetes* 50, 707–709.

Tsubota, Y., Owada-Makabe, K., Yukawa, K. and Maeda, M. (2005) Hypotensive effect of des-acyl ghrelin at nucleus tractus solitarii of rat. *Neuroreport* 16, 163–166.

Uhlar, C.M. and Whitehead, A.S. (1999) Serum amyloid A, the major vertebrate acute-phase reactant. *European Journal of Biochemistry* 265, 501–523.

Verma, S., Li, S.H., Wang, C.H., Fedak, P.W., Li, R.K., Weisel, R.D. and Mickle, D.A. (2003) Resistin promotes endothelial cell activation: further evidence of adipokine-endothelial interaction. *Circulation* 108, 736–740.

Villarreal, D., Reams, G., Freeman, R. and Taraben, A. (1998) Renal effects of leptin in normotensive, hypertensive, and obese rats. *American Journal of Physiology* 275, R2056–R2060.

Villarreal, D., Reams, G., Samar, H., Spear, R. and Freeman, R. (2004) Effects of chronic nitric oxide inhibition on the renal excretory response to leptin. *Obesity Research* 12, 1006–1010.

Waki, H., Yamauchi, T., Kamon, J., Ito, Y., Uchida, S., Kita, S., Hara, K., Hada, Y., Vasseur, F., Froguel, P., Kimura, S., Nagai, R. and Kadowaki, T. (2003) Impaired multimerization of human adiponectin mutants associated with diabetes. Molecular structure and multimer formation of adiponectin. *Journal of Biological Chemistry* 278, 40352–40363.

Wassmann, S., Stumpf, M., Strehlow, K., Schmid, A., Schieffer, B., Bohm, M. and Nickenig, G. (2004) Interleukin-6 induces oxidative stress and endothelial dysfunction by overexpression of the angiotensin II type 1 receptor. *Circulation Research* 94, 534–541.

Weisberg, S.P., McCann, D., Desai, M., Rosenbaum, M., Leibel, R.L. and Ferrante, A.W. Jr (2003) Obesity is associated with macrophage accumulation in adipose tissue. *The Journal of Clinical Investigation* 112, 1796–1808.

Wilson, P., D'Agostino, R., Sullivan, L., Parise, H. and Kannel, W. (2002) Overweight and obesity as determinants of cardiovascular risk: the Framingham experience. *Archives of Internal Medicine* 162, 1867–1872.

Wisse, B.E. (2004) The inflammatory syndrome: the role of adipose tissue cytokines in metabolic disorders linked to obesity. *Journal of the American Society of Nephrology* 15, 2792–2800.

Wold, L., Relling, D., Duan, J., Norby, F. and Ren, J. (2002) Abrogated leptin-induced cardiac contractile response in ventricular myocytes under spontaneous hypertension: role of Jak/STAT pathway. *Hypertension* 39, 69–74.

Wolk, R., Berger, P., Lennon, R.J., Brilakis, E.S., Davison, D.E. and Somers, V.K. (2007) Association between plasma adiponectin levels and unstable coronary syndromes. *European Heart Journal* 28, 292–298.

Wortley, K.E., Anderson, K.D., Garcia, K., Murray, J.D., Malinova, L., Liu, R., Moncrieffe, M., Thabet, K., Cox, H.J., Yancopoulos, G.D., Wiegand, S.J. and Sleeman, M.W. (2004) Genetic deletion of ghrelin does not decrease food intake but influences metabolic fuel preference. *Proceedings of the National Academy of Sciences of the United States of America* 101, 8227–8232.

Wren, A.M., Small, C.J., Ward, H.L., Murphy, K.G., Dakin, C.L., Taheri, S., Kennedy, A.R., Roberts, G.H., Morgan, D.G., Ghatei, M.A. and Bloom, S.R. (2000) The novel hypothalamic peptide ghrelin stimulates food intake and growth hormone secretion. *Endocrinology* 141, 4325–4328.

Wren, A.M., Small, C.J., Abbott, C.R., Dhillo, W.S., Seal, L.J., Cohen, M.A., Batterham, R.L., Taheri, S., Stanley, S.A., Ghatei, M.A. and Bloom, S.R. (2001) Ghrelin causes hyperphagia and obesity in rats. *Diabetes* 50, 2540–2547.

Yamauchi, T., Kamon, J., Ito, Y., Tsuchida, A., Yokomizo, T., Kita, S., Sugiyama, T., Miyagishi, M., Hara, K., Tsunoda, M., Murakami, K., Ohteki, T., Uchida, S., Takekawa, S., Waki, H., Tsuno, N.H., Shibata, Y., Terauchi, Y., Froguel, P., Tobe, K., Koyasu, S., Taira, K., Kitamura, T., Shimizu, T., Nagai, R. and Kadowaki, T. (2003) Cloning of adiponectin receptors that mediate antidiabetic metabolic effects. *Nature* 423, 762–769.

Yang, Q., Graham, T., Mody, N., Preitner, F., Peroni, O., Zabolotny, J., Kotani, K., Quadro, L. and Kahn, B. (2005) Serum retinol binding protein 4 contributes to insulin resistance in obesity and type 2 diabetes. *Nature* 436, 356–362.

Yang, R., Lee, M., Hu, H., Pollin, T., Ryan, A., Nicklas, B., Snitker, S., Horenstein, R., Hull, K., Goldberg, N., Goldberg, A., Shuldiner, A., Fried, S. and Gong, D. (2006) Acute-phase serum amyloid A: an inflammatory adipokine and potential link between obesity and its metabolic complications. *PLoS Medicine* 3, e287.

Yang, R.Z., Huang, Q., Xu, A., McLenithan, J.C., Eisen, J.A., Shuldiner, A.R., Alkan, S. and Gong, D.W. (2003) Comparative studies of resistin expression and phylogenomics in human and mouse. *Biochemical and Biophysical Research Communications* 310, 927–935.

Yaspelkis, B., Davis, J., Saberi, M., Smith, T., Jazayeri, R., Singh, M., Fernández, V., Trevino, B., Chinookoswong, N., Wang, J., Shi, Z. and Levin, N. (2001) Leptin administration improves skeletal muscle insulin responsiveness in diet-induced insulin-resistant rats. *American Journal of Physiology – Endocrinology and Metabolism* 280, E130–E142.

Yaturu, S., Daberry, R., Rains, J. and Jain, S. (2006) Resistin and adiponectin levels in subjects with coronary artery disease and type 2 diabetes. *Cytokine* 34, 219–223.

Yildiz, B., Suchard, M., Wong, M., McCann, S. and Licinio, J. (2004) Alterations in the dynamics of circulating ghrelin, adiponectin, and leptin in human obesity. *Proceedings of the National Academy of Sciences of the United States of America* 101, 10434–10439.

Youn, B., Yu, K., Park, H., Lee, N., Min, S., Youn, M., Cho, Y., Park, Y., Kim, S., Lee, H. and Park, K. (2004) Plasma resistin concentrations measured by enzyme-linked immunosorbent assay using a newly developed monoclonal antibody are elevated in individuals with type 2 diabetes mellitus. *Journal of Clinical Endocrinology and Metabolism* 89, 150–156.

Yusuf, S., Hawken, S., Ounpuu, S., Bautista, L., Franzosi, M.G., Commerford, P., Lang, C.C., Rumboldt, Z., Onen, C.L., Lisheng, L., Tanomsup, S., Wangai, P. Jr, Razak, F., Sharma, A.M. and Anand, S.S. (2005) Obesity and the risk of myocardial infarction in 27,000 participants from 52 countries: a case-control study. *Lancet* 366, 1640–1649.

Zahradka, P. (2008) Novel role for osteopontin in cardiac fibrosis. *Circulation Research* 102, 270–272.

Zhao, A., Bornfeldt, K. and Beavo, J. (1998) Leptin inhibits insulin secretion by activation of phosphodiesterase 3B. *The Journal of Clinical Investigation* 102, 869–873.

Zhao, A., Shinohara, M., Huang, D., Shimizu, M., Eldar-Finkelman, H., Krebs, E., Beavo, J. and Bornfeldt, K. (2000) Leptin induces insulin-like signaling that antagonizes cAMP elevation by glucagon in hepatocytes. *Journal of Biological Chemistry* 275, 11348–11354.

10 Hierarchy of Neural Pathways Controlling Energy Homeostasis

ALFONSO ABIZAID[1] AND TAMAS L. HORVATH[2]

[1]*Institute for Neuroscience, Carleton University, Canada;* [2]*Department of Neurobiology, Yale University School of Medicine, USA*

Introduction

The past 20 years have witnessed tremendous advances in the understanding of the central mechanisms regulating food intake and energy balance, possibly in response to the accelerated increase in the incidence of obesity worldwide. This renewed interest, as well as drastic improvements in the tools that are now currently available to neuroscientists, has yielded a great deal of insight into the mechanisms by which the brain regulates metabolic function and volitional aspects of feeding in response to metabolic signals like leptin, insulin and ghrelin. Among these mechanisms are the complex intracellular signals elicited by these hormones in neurones. Moreover, these signals produce and modulate the metabolism of the cell at the level of the mitochondria and, finally, they promote plastic changes that alter the synaptic circuitry in a number of circuits and ultimately affect cellular, physiological and behavioural responses in defence of energy homeostasis (Abizaid *et al.*, 2006a; Horvath, 2005, 2006a,b; Gao and Horvath, 2007a,b, 2008). This chapter provides a synopsis of these advances, leading to the idea of synaptic plasticity as an important factor in the regulation of food intake and energy homeostasis.

Hypothalamic Homeostatic Circuits: The Renaissance

It is now well established that the hypothalamus plays a critical role in the regulation of energy balance. This was first suspected in the last century after descriptions of obesity in patients with hypothalamic tumours (Brobeck, 1946) but, at the time, it was thought that the pituitary gland regulated most endocrine functions and that alterations of the pituitary lead to metabolic disorders (Brobeck, 1946). Confirmation of the hypothalamus as important for regulation of food intake and energy balance was obtained from animal studies using brain lesions

 Peptides in Energy Balance and Obesity (ed. G. Frühbeck)

of hypothalamic structures (Hetherington and Ranson, 1940, 1942; Brobeck *et al.*, 1943). In essence, evidence obtained from both the clinical descriptions in tumour patients, and from the lesion work, showed that gross damage to mediobasal hypothalamic areas, in particular the ventromedial hypothalamic nucleus (VMH), was clearly associated with increased food intake, morbid obesity and insulin resistance, while damage to more lateral hypothalamic structures was associated with anorexia and adipsia (Anand and Brobeck, 1951). In turn, electrical stimulation of the VMH resulted in decreased feeding, whereas stimulation of the lateral hypothalamic region increased appetite (Coons and Cruce, 1968; Valenstein *et al.*, 1968; Valenstein and Mittleman, 1984). Taken together, these data suggested that the mediobasal hypothalamus was a satiety centre and that the lateral hypothalamus was an orexigenic centre (Grossman, 1979; Elmquist *et al.*, 1999).

This dual centre hypothesis dominated the field for several decades until a number of studies began to trickle data showing that neither the VMH and adjacent structures were solely satiety centres, nor was the lateral hypothalamus involved uniquely in appetite (Valenstein and Mittleman, 1984; Stellar and Stellar, 1985). For example, it was found that knife cuts that separated the ventral from the lateral hypothalamus without damage to the VMH were sufficient to cause hypothalamic obesity (Albert and Storlien, 1969). Similarly, vagotomy appeared to ameliorate obesity caused by VMH destruction (Inoue and Bray, 1977; Bray *et al.*, 1981). Finally, destruction of dopaminergic fibres of the medial forebrain bundle, which course through the lateral hypothalamus, resulted in animals that showed similar anorexic and adipsic symptoms as animals with lesions to the lateral hypothalamus (Zigmond and Stricker, 1972). Indeed, it seemed that disconnections of pathways coursing through these regions were as effective in inducing obesity or anorexia as the lesions themselves. For many years, the study of ingestive behaviour and obesity focused on exploring the relative contribution of different neurotransmitter systems on the regulation of energy balance.

While a tremendous amount of data was obtained during this time, the discovery of neuropeptide Y (NPY) and leptin can be regarded as the most important discoveries in the past 25 years. First, NPY, a 36-amino acid peptide homologue of the pancreatic polypeptide family (Tatemoto *et al.*, 1982), was found to be produced within the brain primarily (although not uniquely) in the arcuate nucleus (ARC), a hypothalamic nucleus ventral to the VMH previously implicated in the regulation of body weight and energy balance (Nemeroff *et al.*, 1978). When injected into the ventricles of rats or within other hypothalamic nuclei, NPY elicited food intake potently (Nemeroff *et al.*, 1978; Clark *et al.*, 1984; Stanley and Leibowitz, 1984, 1985; Stanley *et al.*, 1985). Moreover, NPY synthesis and content within the ARC was elevated in fasted and in genetically obese animals (Sahu *et al.*, 1988; Sanacora *et al.*, 1990). NPY infusions also increased fat deposition and decreased brown fat thermogenesis and oxygen consumption, suggesting that NPY was not only an orexigenic peptide but also one important in the regulation of metabolism (Billington *et al.*, 1991; Walker and Romsos, 1993).

In 1994, Friedman's group cloned the gene that produced leptin, a peptide hormone produced in adipocytes, and that was mutated in the *ob/ob* line of genetically obese mice (Zhang *et al.*, 1994). Treatment with leptin reversed the phenotypic abnormalities seen in *ob/ob* mice and was also effective in reducing body weight and food intake, while increasing energy expenditure in normal animals (Campfield *et al.*, 1995; Halaas *et al.*, 1995; Pelleymounter *et al.*, 1995). A second line of genetically obese and diabetic mice, known as the *db/db* mice, was soon after found to be the result of a deletion of the gene encoding the long form of the leptin receptor (ObRb) (Maffei *et al.*, 1995; Tartaglia *et al.*, 1995; Chen *et al.*, 1996). Finally, it was established that leptin targeted NPY neurones within the ARC to produce these dramatic changes in metabolism (Stephens *et al.*, 1995). These groundbreaking discoveries laid the foundation of what could be termed as a renaissance in the study of neural control of obesity and energy balance. Reports of other peptides with either anorexic or orexigenic properties began to appear routinely in high impact journals, and continue to make headlines.

Because the ARC contains the largest concentration of cells that produce NPY and has the densest concentration of leptin-sensitive neurones in the brain, it is accepted generally that this region is key to the regulation of energy balance (Fig. 10.1). This is supported by the fact that, in addition to NPY, the ARC also contains a second set of neurones that produce α-melanocyte-stimulating hormone (α-MSH), an anorectic peptide formed from the cleavage of the pro-opiomelanocortin (POMC) protein (Cone *et al.*, 2001). This protein acts on melanocortin receptors types 3 and 4 (MC3/4, respectively) present in various hypothalamic nuclei to reduce food intake and energy expenditure in a manner similar to leptin (Boston *et al.*, 1997). Moreover, the pharmacological blockade of MC3/4 receptors, or the deletion of the gene encoding the MC4 receptor, results in obesity and leptin resistance in rodents and primates (Fan *et al.*, 1997; Huszar *et al.*, 1997; Ollmann *et al.*, 1997). In addition, NPY neurones produce a second orexigenic peptide, the agouti-related peptide (AgRP), an endogenous antagonist to the MC3/4 receptor (Ollmann *et al.*, 1997). This peptide, like NPY, stimulates food intake dramatically, but the increase in food intake produced by this peptide is long lasting, an effect that is still not well understood (Hagan *et al.*, 2000). Similarly, POMC cells also synthesize a second anorexic peptide, the cocaine- and amphetamine-related transcript (CART) (Elias *et al.*, 2001). The relative contribution of CART versus α-MSH in the regulation of food intake and energy expenditure remains unexplained. What is known is that both NPY/AgRP and POMC/CART neurones within the ARC appear to modulate food intake primarily via their output targets (Fig. 10.1). Both POMC and NPY cells have a widespread projection field that has been implicated in a variety of physiological and behavioural events that include reproduction, water balance, body temperature and energy balance. The main output of both NPY/AgRP and POMC/CART cells appears to be the hypothalamic paraventricular nucleus (PVN), where NPY, α-MSH and AgRP have strong effects on food intake and body temperature. These cells, however, target other hypothalamic nuclei like the VMH, dorsomedial hypothalamus (DMH) and lateral hypothalamus (LH), among others, potentially to modulate food intake and energy expenditure, and the relative contribution of these nuclei to produce the orexigenic or anorexic effects of these peptides continues to be investigated (Kalra *et al.*, 1999; Elmquist, 2001; Leibowitz and

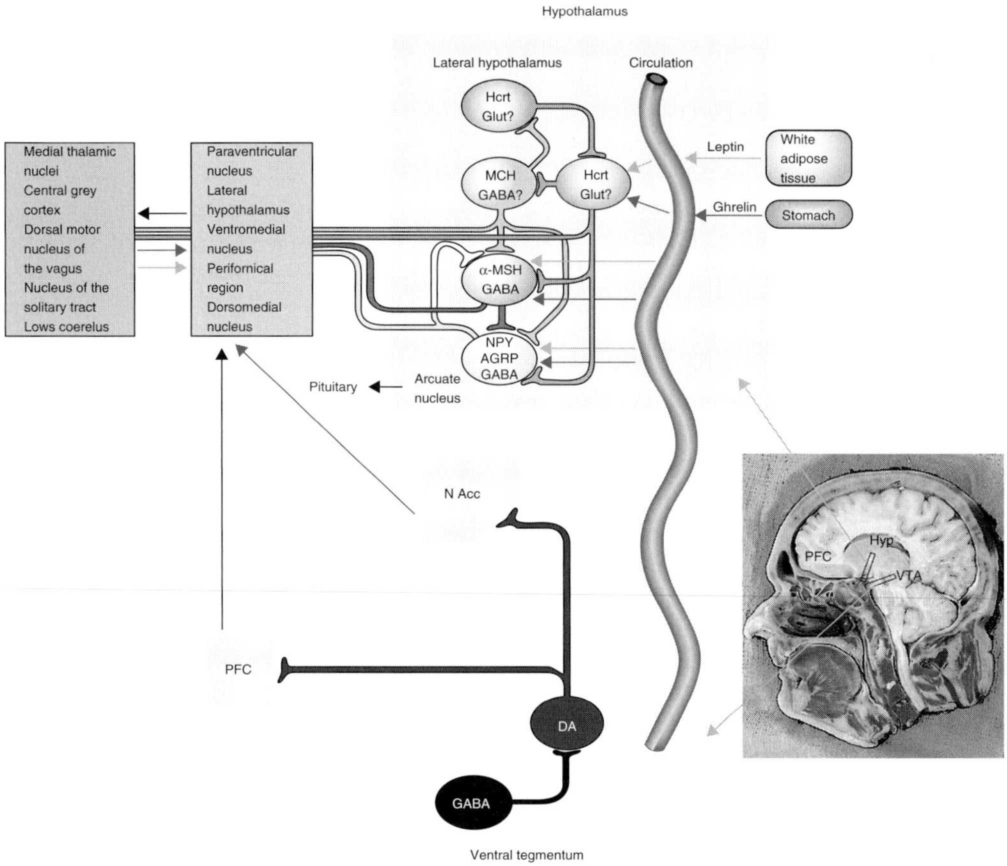

Fig. 10.1. Parallel system regulating food intake and energy balance. The upper level within the figure describes the complex hypothalamic peptidergic circuits involved in the regulation of energy balance. Neurones of the melanocortin system, NPY/AgRP and α-MSH cells in the ARC, act at various hypothalamic sites to modulate food intake and energy expenditure. Lateral hypothalamic neurones producing hypocretin/orexin and MCH modulate cells in the melanocortin system, as well as target extrahypothalamic sites to increase food intake. The bottom level of the panel includes the mesolimbic dopaminergic system that has been shown to be implicated in motivational behavioural patterns related to energy balance. Because all of these systems are sensitive to metabolic signals like leptin, ghrelin, insulin and glucose, it is likely that these systems are activated in parallel by metabolic signals to alter physiological and behavioural responses according to the energetic state of the organism. Hcrt, hypocretin/orexin; Glut, glutamate; MCH, melanin-concentrating hormone; GABA, γ-aminobutyric acid; α-MSH, α-melanocyte-stimulating hormone; NPY, neuropeptide Y; AgRP, agouti-related protein; N Acc, nucleus accumbens; PFC, prefrontal cortex; DA, dopamine.

Hoebel, 2004; Gao and Horvath, 2008a,b). Finally, NPY/AgRP neurones of the ARC appear to synapse onto neighbouring POMC/CART cells to inhibit them using γ-aminobutyric acid (GABA) as a neurotransmitter (Horvath *et al.*, 1992; Pu *et al.*, 1999; Horvath, 2005).

While the lateral hypothalamus previously had been described as a 'hunger' centre, it was not until the end of the 1990s that two orexigenic peptides, hypocretin/orexin and melanin-concentrating hormone (MCH), were identified and localized within this area (Qu *et al.*, 1996; Sakurai *et al.*, 1998; Ohno and Sakurai, 2008). Interestingly, both hypocretin/orexin and MCH increase food intake via different mechanisms. In the case of hypocretin/orexin neurones, their role in the regulation of food intake has been questioned, given that their effects on food intake are short-lived (Edwards *et al.*, 1999) and that *ob/ob* and *db/db* mice show lower levels of hypocretin/orexin mRNA and peptide content than their wild-type littermates (Stricker-Krongrad *et al.*, 2002). Nevertheless, mice with genetic deletion of the gene encoding the prepro-orexin peptide are hypophagic (Hara *et al.*, 2001). Hypocretin/orexin cells send projections to the ARC where they synapse on to NPY/AgRP cells, which, in turn, project back to hypocretin/orexin cells (Horvath *et al.*, 1999a). This particular circuit has been shown to play an important role in hypocretin/orexin-induced food intake (Elias *et al.*, 1998; Horvath *et al.*, 1999a; Yamanaka *et al.*, 2000; Muroya *et al.*, 2004; López *et al.*, 2007). Moreover, the presence of receptors for signals like leptin and ghrelin, as well as changes in electrophysiological activity of hypocretin/orexin neurones in response to these signals, demonstrates that hypocretin/orexin cells can be modified directly by peripheral signals (Horvath *et al.*, 1999a). Moreover, hypocretins/orexins play a crucial role in activating arousal circuits in response to energetic challenges, resulting in food-seeking behaviours and food-anticipatory behaviours (Sakurai, 2003, 2005; Boutrel and de Lecea, 2008).

In contrast, the role of MCH hypothalamic neurones in the regulation of energy balance appears to be more straightforward. For example, *ob/ob*, *db/db* mice have high levels of MCH expression in the hypothalamus and MCH transgenic mice are overweight and gain more weight under a high-fat diet (Qu *et al.*, 1996; Ludwig *et al.*, 2001). In contrast, MCH or MCH receptor knockout mice are leaner, eat less and have increased metabolism than their wild-type littermates (Shimada *et al.*, 1998; Marsh *et al.*, 2002). Interestingly, α-MSH/POMC cells inhibit the activity of MCH neurones and thus prevent increases in food intake (Ludwig *et al.*, 1998; Tritos *et al.*, 1998). Given the widespread distribution of both hypocretin/orexin and MCH projections (Elias *et al.*, 1998), it has been suggested that most aspects of food intake and energy regulation could be modulated by the interaction between these two cell groups at these target sites (Berthoud, 2002, 2004) and, given their close proximity and synaptic interconnections, perhaps by modulating each other's cellular activity reciprocally (Gao and van den Pol, 2001; Guan *et al.*, 2002; Li *et al.*, 2002b).

The list of peripheral factors that, like leptin, target the ARC to modulate energy balance has also grown (Woods *et al.*, 1998). Metabolic signals such as glucose availability, insulin, cholecystokinin (CCK), pancreatic polypeptides (PP and PYY) and ghrelin have, among others, all been found to modulate NPY and POMC in the ARC to alter food intake and metabolism. Of these, ghrelin has received special attention given that, in contrast to the other peptides, ghrelin acts in NPY cells within the ARC to increase food intake, adiposity and the secretion of growth hormone (Kojima *et al.*, 1999; Tschöp *et al.*, 2000; Horvath *et al.*, 2001). Although ghrelin is produced primarily in the stomach (Kojima *et al.*,

1999; van der Lely *et al.*, 2004), a subset of ghrelin-secreting neurones has been identified in the dorsal portion of the ARC and in the spaces that surround different hypothalamic nuclei implicated in the regulation of energy balance (Kojima *et al.*, 1999; Cowley *et al.*, 2003a). The role of these neurones remains to be fully determined but, anatomically, it appears that these cells integrate metabolic and circadian outputs to regulate energy balance (Cowley *et al.*, 2003a).

We are then left with a model where metabolic signals that monitor energetic state, like leptin, ghrelin, insulin and PYY, target the hypothalamus, and particularly the ARC, to modulate the activity of NPY/AgRP and POMC neurones. The activation of these neurones by 'satiety' signals leads to a reduction in NPY/AgRP and an increase in the release of α-MSH from POMC neurones. Consequently, α-MSH binds to MC3/4 receptors in MCH cells in the lateral hypothalamus to reduce food intake and with thyroid hormone and corticotropin-releasing hormones (THS and CRH) in the PVN to increase energy expenditure. In contrast, hunger signals like a reduction in glucose availability or increased circulating ghrelin will lead to increases in ARC nucleus NPY release that inhibit POMC, THS and CRH and stimulate the secretion of hypocretin/orexin and MCH from the LH ultimately to increase food intake and reduce metabolic rate. The ARC appears to be, therefore, a brain nucleus orchestrating brain responses to changes in energy demands (Cone *et al.*, 2001).

Tools for the Study of Feeding Circuits

In addition to improved lesion techniques and increased availability of agonists or antagonists that target different neuropeptide receptors specifically, the molecular biology and molecular genetics revolution has proven pivotal for the unveiling of feeding circuits (Horvath, 2005). Molecular biological techniques have revealed that the ObRb leptin receptor belongs to the same family (gp130) of receptors associated with cytokines such as the interleukins (Tartaglia *et al.*, 1995). Activation of this receptor by leptin can achieve gene transcription by at least three signalling cascades that include the activation of the JAK2/STAT3, the ERK/MAP kinase and the phosphoinositol 3-kinase (PI3K) pathways (Vaisse *et al.*, 1996; Bjørbaek *et al.*, 2001; Sahu, 2003; Frühbeck, 2006). Much attention has been focused on the ability of leptin to activate STAT3, which, in turn, will act as a transcription factor for several genes that include the suppressor of cytokine signalling 3 (SOCS3) gene, an intracellular protein that prevents further activation of the ObRb (Bjørbaek *et al.*, 1998, 2000). The pivotal role of STAT3 as a transcription factor that mediates the effects of leptin on energy balance has been highlighted recently by the generation of mice with targeted deletions to different sites for STAT3 phosphorylation, rendering animals with deficient STAT3 signalling. These mice are severely obese and insulin resistant, and show high expression of NPY and AgRP and diminished expression of POMC in the ARC (Bates *et al.*, 2003, 2004; Gao *et al.*, 2004). Several knockout mice lines have underlined the importance of the melanocortin system in the regulation of leptin's effects and in energy balance in general. Thus, targeted deletions to the genes that encode α-MSH, MC4 receptor and the specific deletion of the ObRb in POMC neurones also result in obese, hyperinsulinaemic and leptin-resistant mice (Huszar *et al.*, 1997; Butler and Cone, 2003; Balthasar *et al.*, 2004). Moreover, naturally occurring mutations of the Ob and α-MSH genes also produce the same symptoms in humans (Barsh *et al.*, 2000).

In contrast, deletions to the genes that encode NPY, ghrelin, or the active form of the ghrelin receptor (growth hormone secretagogue receptor 1a, or GHS-R 1a) result in few phenotypic abnormalities (Erickson *et al.*, 1997; Palmiter *et al.*, 1998; Sun *et al.*, 2004; Wortley *et al.*, 2004). Nevertheless, NPY/leptin double knockout animals show decreased food intake, body weight and adiposity in comparison to the regular leptin-deficient (*ob/ob*) mice (Palmiter *et al.*, 1998), and ghrelin-deficient animals appear to be slightly resistant to diet-induced obesity (Wortley *et al.*, 2004). Physiological responses of NPY-, ghrelin- and GHS-R-deficient animals remain to be fully determined. In any event, there are a variety of mutations that lead to a lean phenotype (i.e. MCH KO mice) and some, as in the dopamine-deficient mice, become completely aphagic, needing dopamine replacement to continue eating (Szczypka *et al.*, 1999a,b). The relative contribution of these genes in the regulation of hypothalamic homeostatic circuits is a matter of continuous research efforts.

Finally, the development of reporter genes that can be used as tags has become a welcome addition to the study of hypothalamic circuits. For example, the gene that encodes the green fluorescent protein (GFP), a protein that is produced in a specific species of jellyfish, has been tagged on to the promoters of several of the peptides implicated in energy regulation. These gene 'knock-ins' have enabled the visualization of cells that synthesize neuropeptides such as NPY and POMC, or neurotransmitters like GABA that are difficult to visualize using immunocytochemical techniques. The use of mice with specific insertions of the GFP gene has proven invaluable to the study of anatomical and physiological properties of specific hypothalamic neuropeptides. For instance, Cowley and colleagues used mice with the GFP gene inserted in the POMC promoter to unveil the electrophysiological properties of POMC neurones in response to signals like leptin, NPY, ghrelin and PYY (Cowley *et al.*, 2001, 2003b). To determine the mechanisms by which different metabolic signals and neurotransmitters act on NPY and POMC cells, mice with the GFP gene inserted in the NPY and POMC promoters have been used (Roseberry *et al.*, 2004). In collaboration with Friedman's laboratory, we have used these mice lines crossbred with *ob/ob* mice to describe the dynamic synaptic remodelling that occurs in both POMC and NPY cells in response to leptin and ghrelin, and which may be critical for the regulation of energy balance as described below.

Synaptic Plasticity and Energy Balance

The concept of homeostasis implies that physiological events in all organisms necessitate a degree of plasticity or flexibility to allow for constant dynamic changes to achieve balance. Within the brain, this plasticity is afforded by systems that can change in response to given stimuli and that rearrange in ways that allow for more efficient responses to future stimuli. In contrast to old dogma, it is now well accepted that connections between cells within the adult brain are capable to change in response to a variety of stimuli and that these changes play an important role in critical brain functions as learning, memory and motivated behaviour. Such changes are referred to as synaptic plasticity.

Within the hypothalamus, synaptic changes have been implicated in a variety of processes that include osmoregulation, lactation, circadian rhythmicity and reproductive function (Guldner and Ingham, 1979; Olmos *et al.*, 1989; Nishikawa *et al.*, 1995; Flanagan-Cato, 2000; Theodosis and Poulain, 2001; Langle *et al.*, 2002; Hayashi *et al.*, 2004). Interestingly, proteins that are commonly found in the developing brain and that are associated with the formation of new synapses are expressed selectively in the hypothalamus of adult organisms, and particularly in the ARC (Garcia-Segura *et al.*, 1994). Interestingly, ultrastructural studies of the ARC revealed that synaptic remodelling occurred on cells within this region across the oestrus cycle in female rats (Olmos *et al.*, 1989). It was also revealed that this effect was produced by oestrogen and that, in addition to rats, it was also observable across the reproductive cycle of non-human primates (Garcia-Segura *et al.*, 1994; Naftolin *et al.*, 2007). The ARC contains both oestrogen receptor alpha and beta subtypes, yet the effects of oestrogen on ARC nucleus cells can occur within minutes of the presence of oestrogen in the media and mimic those elicited in cells by growth factors (Garcia-Segura *et al.*, 1996). While these studies were correlated with the onset and termination of the preovulatory luteinizing hormone surge, it has become clear that these changes might mediate the metabolic effects of oestrogen.

Coinciding with these data, researchers soon discovered that leptin, like oestrogen, targeted hypothalamic and extrahypothalamic structures that demonstrated a high degree of synaptic remodelling, including the ARC, VMH and hippocampus (Huang *et al.*, 1996; Hakansson *et al.*, 1998; Shioda *et al.*, 1998; Gao and Horvath, 2008b). Within the hippocampus, it has been demonstrated that leptin can lower the threshold for the induction of long-term potentiation (LTP) after activation of the *N*-methyl-d-aspartate (NMDA) subtype of glutamate receptors (Shanley *et al.*, 2001; Li *et al.*, 2002a; Harvey and Ashford, 2003). Because LTP is thought to result from synaptic changes, these data suggest that leptin can induce synaptic remodelling to increase sensitivity to excitatory stimulation.

Taken together, this information made it plausible that leptin, like oestrogen, could target the ARC and other structures to modulate energy balance by actually remodelling inputs to the different cell groups in the ARC. In a collaboration with Friedman, our laboratory engaged in a project examining the effects of leptin on the number and type of synapses contacting both POMC and NPY neurones (Pinto *et al.*, 2004). To do this, mice in which the gene encoding the GFP protein was inserted in the promoter for either NPY or POMC were crossbred with heterozygous leptin-deficient (*ob/ob*) mice, to produce *ob/ob* GFP transgenic mice. Electron microscopic examination determined that NPY cells in the ARC of *ob/ob* mice had more synapses than NPY cells of wild-type mice. Surprisingly, POMC neurones of *ob/ob* mice had a lower number of synapses than those of wild-type mice. Nevertheless, synapses on to POMC cells of *ob/ob* mice were predominantly putative inhibitory (symmetric), whereas NPY cells of *ob/ob* mice exhibited primarily putative excitatory (asymmetric). These data were consistent with electrophysiological recordings showing that the frequency of spontaneous inhibitory postsynaptic currents (sIPSCs) on to POMC cells of *ob/ob* mice was higher than that on POMC cells of wild-type mice, with no significant differences in the frequency of spontaneous excitatory postsynaptic currents (sEPSCs) on these cells. In contrast, the frequency of sEPSCs was increased and that of sIPSCs was decreased in NPY neurones of *ob/ob* mice compared to NPY neurones of wild-type mice. Finally, leptin administration to *ob/ob* mice restored the balance of excitatory and inhibitory synapses rapidly to the levels observed in untreated wild-type mice, whereas ghrelin treatment to wild-type mice had just the opposite effect. The outcome of these experiments provided anatomical and electrophysiological evidence of a dynamic model of energy regulation in which hypothalamic neurones were in a constant 'tug of war' between inhibitory and excitatory synapses and where peripheral signals like leptin, ghrelin and oestrogen shifted the balance ultimately to increase or decrease food intake, providing for a dynamic framework we have termed the 'floating blueprint' (Horvath and Diano, 2004).

Plasticity and Mitochondrial UCP2

The plastic nature of ARC nucleus cells, and indeed that of any system that is capable of actual architectural remodelling, may involve high energy expenditure, which may be reflected in the activity as well as in the proliferation of the mitochondria. The mitochondria are involved in the generation of cellular metabolism and optimal mitochondrial functioning determines the fate of individual cells (Balaban *et al.*, 2005). Increased mitochondrial activity may, however, also result in the generation of free radicals that can lead to cellular stress and degeneration (Balaban *et al.*, 2005). It has been suggested that uncoupling proteins (UCPs) are capable of preventing cell damage by dissociating the production of energy in the form of ATP and the resultant high levels of free radicals by regulating the proton leak from the inner membrane of the mitochondria (Bechmann *et al.*, 2002; Paradis *et al.*, 2003).

Of the different UCPs identified, UCP2 has been shown to play an important role in neuroprotection (Conti *et al.*, 2005; Kim-Han and Dugan, 2005) and may, as has been suggested previously, play a role in neurotransmission (Horvath *et al.*, 1999a; Fuxe *et al.*, 2005). This is, indeed, the case in the mammalian hypothalamus, where UCP2 is expressed constitutively (Richard *et al.*, 1998; Horvath *et al.*, 1999b; Diano *et al.*, 2000). Within the ARC, UCP2 appears to be present in NPY/AgRP-producing cells, as well as in oestrogen- and leptin-sensitive cells, which could also be POMC-secreting neurones (Horvath *et al.*, 1999b). The role of UCP2 in these systems remains to be fully determined, although it has been suggested that locally produced active thyroid hormone (T3) activates UCP2 in NPY/AgRP cells, a response that may be critical to activate these cells during negative energy balance (Horvath and Diano, 2004). A role for UCP2 in obesity continues to be considered, although UCP2 knockout mice do not seem to be obese (Paradis *et al.*, 2003). Nevertheless, spontaneously obese yellow agouti mice have a leaner phenotype when cross-bred with mice that overexpress the human form of UCP2 (hUCP2) (Horvath *et al.*, 2003). Interestingly, although these mice are heavier than their wild-type littermates at the age of 3 months, they appear to have less body fat. As they age, hUCP2 transgenics do not continue to gain weight and, by the age of 10 months, they are leaner than their wild-type counterparts (Horvath *et al.*, 2003). UCP2 is involved in protecting ARC cells from free radical damage that results from the high metabolic rate of these cells. As animals age, uncoupling mechanisms that include the induction of UCP2 and the production of new mitochondria may become deficient, leading to alterations in cell function and, ultimately, obesity. Finally, a role for UCP2 in sustaining synaptic plasticity in the ARC has also been put forward. Dendritic mitochondria have been implicated directly in the generation and maintenance of new synapses following hippocampal stimulation. In general, it appears that increases in the number of mitochondria present in dendrites is related directly to the number of synapses that are formed (Mattson and Liu, 2003; Ben-Shachar and Laifenfeld, 2004; Li *et al.*, 2004). Induction of UCP2 also increases the number of mitochondria in hippocampal cells (Diano *et al.*, 2003), modulating synaptic remodelling through the elevation in the number of mitochondria. UCPs present in selected neurones do not operate as constitutive uncouplers. However, they can be activated by free radicals and free fatty acids and their activity has a profound influence on neuronal function. By regulating mitochondrial biogenesis, calcium flux, free radical production and local temperature, neuronal UCPs can influence neurotransmission, synaptic plasticity and neurodegenerative processes directly (Andrews *et al.*, 2005). Insights into the regulation and function of these proteins offer unsuspected avenues for a better understanding of synaptic transmission and neurodegeneration.

Parallel Systems Regulating Food Intake and Body Weight

While it appears that the hypothalamus, and in particular, the ARC, are key regions regulating energy balance, previous and emerging data demonstrate the existence of other circuits that, when activated, modulate food intake and body weight (Woods *et al.*, 2000; Grill and Kaplan, 2002; Saper *et al.*, 2002; Berthoud, 2002, 2004). The importance of these circuits has often been overshadowed by the attention paid to hypothalamic homeostatic circuits, yet their study may prove to be more relevant to human obesity (Berthoud, 2002, 2004). In addition, these systems often are viewed as either secondary or connected in series with the hypothalamus; that is, they only function once the hypothalamus has been activated. Although these systems cannot be considered fully homeostatic, they may be activated in parallel with and/or perhaps recruit homeostatic centres to modulate the ingestion of food. In addition, activation of these pathways may override regulatory signals from hypothalamic homeostatic centres to either increase or decrease appetite. For example, it is well established that rats whose brainstem is isolated continue to regulate the food they consume, and even show affective responses to palatable foods (Grill and Kaplan, 2001). Corticolimbic pathways are capable of integrating sensory inputs and produce cognitive as well as affective representations, which are stored and used for making decisions, and lesions to various corticolimbic regions result in obesity (Stein *et al.*, 1990; Carr and Wolinsky, 1994; Tataranni *et al.*, 1999). Feeding is also associated with motivational mechanisms, the 'liking' and 'wanting', which are required for the behavioural responses that are necessary to seek and obtain food (Berridge, 1996, 2004). These mechanisms are commonly associated with midbrain and forebrain centres that regulate arousal, locomotor activity, mood and reward (Adamantidis and de Lecea, 2008). Reward pathways in particular have received special attention, given the universality of food as a natural reinforcer (see Chapter 11).

Dopamine produced in cells within the midbrain ventral tegmental area (VTA) is released into several forebrain structures like the hippocampus, ventral striatum and prefrontal cortex, and this release is commonly associated with the experience or the expectation of reward (Wise and Rompre, 1989; Wise, 2004a,b). Within the ventral striatum, dopamine release into the nucleus accumbens has been implicated in the rewarding aspects of food, sex and drugs of abuse (Mitchell and Gratton, 1994; Kelley, 1999). Interestingly, genetic deletion of dopamine suppresses food intake markedly, in a manner that is similar to that of lesions of the lateral hypothalamus (Szczypka *et al.*, 1999a,b). Numerous studies suggest that hypothalamic peptides like NPY, α-MSH, AgRP, orexin and MCH play an important role in modulating the activity of dopaminergic cells targeting the nucleus accumbens (Kelley, 2004b; Palmiter, 2007). The idea is that the ARC funnels metabolic information from signals like leptin or ghrelin, to modulate the activity of the mesolimbic dopaminergic system via direct projections to the nucleus accumbens, or indirectly through the activation of hypocretin/orexin or MCH cells that also project to both the VTA and nucleus accumbens (Berthoud, 2002; Kelley, 2004b). Emerging evidence, however, supports the notion that at least the VTA is sensitive to leptin, insulin and ghrelin, and that the activity of dopaminergic cells within the VTA can be modulated by these signals (Guan *et al.*, 1997; Figlewicz *et al.*, 2003; Abizaid *et al.*, 2006b). Interestingly, it has been shown that calorie-rich nutrients can influence directly brain reward circuits that control food intake independently of palatability or functional taste transduction by activating the mesolimbic dopamine–accumbal pathway (de Araujo *et al.*, 2008; Andrews and Horvath, 2008).

In this context, a widespread dysfunction in mechanisms regulating dopamine release in obesity has been put forward. Electrically evoked dopamine release has been shown to be attenuated in obesity-prone rats, not only in the nucleus accumbens but also in additional terminal sites of dopamine neurones, such as the accumbens shell, dorsal striatum and medial prefrontal cortex (Geiger *et al.*, 2008). Moreover, dopamine impairment in these rats was apparent at birth and associated with changes in expression of several factors regulating dopamine synthesis and release, such as vesicular monoamine transporter-2, tyrosine hydroxylase, dopamine transporter and dopamine receptor-2 short-form. Taken together, these results suggest that an attenuated central dopamine system would reduce the hedonic response associated with feeding and induce compensatory hyperphagia, leading to obesity.

For decades, endogenous dopaminergic and opioid systems have been considered the most important systems in mediating brain reward processes.

However, recent evidence suggests that the endogenous cannabinoid (endocannabinoid) system also exerts a relevant role in signalling of rewarding events (Solinas *et al.*, 2008). Several findings support this notion. First, cannabinoid-1 (CB1) receptors are found in brain areas involved in reward processes, such as the dopaminergic mesolimbic system. Second, activation of CB1 receptors by plant-derived, synthetic or endogenous CB1 receptor agonists stimulates dopaminergic neurotransmission, produces rewarding effects and increases rewarding effects of abused drugs and food. Third, pharmacological or genetic blockade of CB1 receptors prevents activation of dopaminergic neurotransmission by several addictive drugs and reduces the rewarding effects of food and these drugs. Fourth, brain levels of the endocannabinoids, anandamide and 2-arachidonoylglycerol, are altered by activation of reward processes. The intrinsic activity of the endocannabinoid system, however, does not appear to play a facilitatory role in brain stimulation reward, and some evidence suggests it may even oppose it (Solinas *et al.*, 2008). The influence of the endocannabinoid system on brain reward processes may depend on the degree of activation of the different brain areas involved and might represent a mechanism for fine-tuning dopaminergic activity. Thus, the endocannabinoid system appears to play a major role in modulating reward processes, although the involvement of the various elements of this pathway may differ, depending on the type of reward studied. Further research may reveal that, in contrast to the funnel hypothesis, metabolic signals may act directly on reward systems to modulate motivational aspects of feeding in tandem with homeostatic systems to increase or reduce food intake (see Chapter 11).

Future Considerations

We believe that the ability of the ARC to rewire dynamically in response to ever-changing signals is necessary for cells within this nucleus to modulate energy balance efficiently. Interestingly, synaptic plasticity also appears to be an important feature in extrahypothalamic circuits affecting food intake. For instance, synaptic rearrangement within the VTA and nucleus accumbens has been implicated in the mechanisms that lead to addiction to substances like opioids, cocaine and amphetamine (Kelley, 2004a; Robinson and Kolb, 2004). Within the VTA, the crosstalk between astrocytes and dopaminergic neurones is important in the sensitization to amphetamine (Flores and Stewart, 2000a,b). Chronic cocaine stimulation leads to long-lasting changes in gene expression within the nucleus accumbens that may reflect permanent changes in the inputs to cells within this region (Chao and Nestler, 2004). We know that, in addition to targeting the ARC to modulate homeostatic pathways, leptin and ghrelin reach cells in the VTA, where they may also alter their synaptic inputs to enhance or decrease their activity. Further research in this area will lead to a better understanding of the mechanisms that cause food cravings and those that increase or decrease the incentive value of palatable foods. They may also lead to insight in the study of eating disorders like obesity and anorexia nervosa and lead ultimately to more efficient treatments for these disorders.

References

Abizaid, A., Gao, Q. and Horvath, T.L. (2006a) Thoughts for food: brain mechanisms and peripheral energy balance. *Neuron* 51, 691–702.

Abizaid, A., Liu, Z.W., Andrews, Z.B., Shanabrough, M., Borok, E., Elsworth, J.D., Roth, R.H., Sleeman, M.W., Picciotto, M.R., Tschöp, M.H., Gao, X.B. and Horvath, T.L. (2006b) Ghrelin modulates the activity and synaptic input organization of midbrain dopamine neurons while promoting appetite. *Journal of Clinical Investigation* 116, 3229–3239.

Adamantidis, A. and Lecea, L. de (2008) Physiological arousal: a role for hypothalamic systems. *Cellular and Molecular Life Sciences* 65, 1475–1488.

Albert, D.J. and Storlien, L.H. (1969) Hyperphagia in rats with cuts between the ventromedial and lateral hypothalamus. *Science* 165, 599–600.

Anand, B.K. and Brobeck, J.R. (1951) Localization of a feeding center in the hypothalamus of the rat. *Proceedings for the Society of Experimental Biology and Medicine* 77, 323–324.

Andrews, Z.B. and Horvath, T.L. (2008) Tasteless food reward. *Neuron* 57, 806–808.

Andrews, Z.B., Diano, S. and Horvath, T.L. (2005) Mitochondrial uncoupling proteins in the CNS: in support of function and survival. *Nature Reviews Neuroscience* 6, 829–840.

Araujo, I.E. de, Oliveira-Maia, A.J., Sotnikova, T.D., Gainetdinov, R.R., Caron, M.G., Nicolelis, M.A. and Simon, S.A. (2008) Food reward in the absence of taste receptor signaling. *Neuron* 57, 930–941.

Balaban, R.S., Nemoto, S. and Finkel, T. (2005) Mitochondria, oxidants, and aging. *Cell* 120, 483–495.

Balthasar, N., Coppari, R., McMinn, J., Liu, S.M., Lee, C.E., Tang, V., Kenny, C.D., McGovern, R.A., Chua, S.C. Jr, Elmquist, J.K. and Lowell, B.B. (2004) Leptin receptor signaling in POMC neurons is required for normal body weight homeostasis. *Neuron* 42, 983–991.

Barsh, G.S., Farooqi, I.S. and O'Rahilly, S. (2000) Genetics of body-weight regulation. *Nature* 404, 644–651.

Bates, S.H., Stearns, W.H., Dundon, T.A., Schubert, M., Tso, A.W., Wang, Y., Banks, A.S., Lavery, H.J., Haq, A.K., Maratos-Flier, E., Neel, B.G., Schwartz, M.W. and Myers, M.G. Jr (2003) STAT3 signalling is required for leptin regulation of energy balance but not reproduction. *Nature* 421, 856–859.

Bates, S.H., Dundon, T.A., Seifert, M., Carlson, M., Maratos-Flier, E. and Myers, M.G. Jr (2004) LRb-STAT3 signaling is required for the neuroendocrine regulation of energy expenditure by leptin. *Diabetes* 53, 3067–3073.

Bechmann, I., Diano, S., Warden, C.H., Bartfai, T., Nitsch, R. and Horvath, T.L. (2002) Brain mitochondrial uncoupling protein 2 (UCP2): a protective stress signal in neuronal injury. *Biochemical Pharmacology* 64, 363–367.

Ben-Shachar, D. and Laifenfeld, D. (2004) Mitochondria, synaptic plasticity, and schizophrenia. *International Reviews of Neurobiology* 59, 273–296.

Berridge, K.C. (1996) Food reward: brain substrates of wanting and liking. *Neuroscience Biobehavior Reviews* 20, 1–25.

Berridge, K.C. (2004) Motivation concepts in behavioral neuroscience. *Physiology and Behavior* 81, 179–209.

Berthoud, H.R. (2002) Multiple neural systems controlling food intake and body weight. *Neuroscience Biobehavior Reviews* 26, 393–428.

Berthoud, H.R. (2004) Mind versus metabolism in the control of food intake and energy balance. *Physiology and Behavior* 81, 781–793.

Billington, C.J., Briggs, J.E., Grace, M. and Levine, A.S. (1991) Effects of intracerebroventricular injection of neuropeptide Y on energy metabolism. *American Journal of Physiology – Regulatory, Integrative and Comparative Physiology* 260, R321–327.

Bjørbaek, C., Elmquist, J.K., Frantz, J.D., Shoelson, S.E. and Flier, J.S. (1998) Identification of SOCS-3 as a potential mediator of central leptin resistance. *Molecular Cell* 1, 619–625.

Bjørbaek, C., Lavery, H.J., Bates, S.H., Olson, R.K., Davis, S.M., Flier, J.S. and Myers, M.G. Jr (2000) SOCS3 mediates feedback inhibition of the leptin receptor via Tyr985. *American Journal of Biological Chemistry* 275, 40649–40657.

Bjørbaek, C., Buchholz, R.M., Davis, S.M., Bates, S.H., Pierroz, D.D., Gu, H., Neel, B.G., Myers, M.G. Jr and Flier, J.S. (2001) Divergent roles of SHP-2 in ERK activation by leptin receptors. *American Journal of Biological Chemistry* 276, 4747–4755.

Boston, B.A., Blaydon, K.M., Varnerin, J. and Cone, R.D. (1997) Independent and additive effects of central POMC and leptin pathways on murine obesity. *Science* 278, 1641–1644.

Boutrel, B. and Lecea, L. de (2008) Addiction and arousal: the hypocretin connection. *Physiology and Behavior* 93, 947–951.

Bray, G.A., Inoue, S. and Nishizawa, Y. (1981) Hypothalamic obesity. The autonomic hypothesis and the lateral hypothalamus. *Diabetologia* 20 (Suppl.) 366–377.

Brobeck, J.R. (1946) Mechanisms of the development of obesity in animals with hypothalamic lesions. *Physiological Reviews* 26, 541–559.

Brobeck, J.R., Tepperman, J. and Long, C.N.H. (1943) Experimental hypothalamic hyperphagia in the albino rat. *Yale Journal of Biology and Medicine* 15, 831–853.

Butler, A.A. and Cone, R.D. (2003) Knockout studies defining different roles for melanocortin receptors in energy homeostasis. *Annals of the New York Academy of Sciences* 994, 240–245.

Campfield, L.A., Smith, F.J., Guisez, Y., Devos, R. and Burn, P. (1995) Recombinant mouse OB protein: evidence for a peripheral signal linking adiposity and central neural networks. *Science* 269, 546–549.

Carr, K.D. and Wolinsky, T.D. (1994) Regulation of feeding by multiple opioid receptors in cingulate cortex; follow-up to an *in vivo* autoradiographic study. *Neuropeptides* 26, 207–213.

Chao, J. and Nestler, E.J. (2004) Molecular neurobiology of drug addiction. *Annual Review of Medicine* 55, 113–132.

Chen, H., Charlat, O., Tartaglia, L.A., Woolf, E.A., Weng, X., Ellis, S.J., Lakey, N.D., Culpepper, J., Moore, K.J., Breitbart, R.E., Duyk, G.M., Tepper, R.I. and Morgenstern, J.P. (1996) Evidence that the diabetes gene encodes the leptin receptor: identification of a mutation in the leptin receptor gene in *db*/*db* mice. *Cell* 84, 491–495.

Clark, J.T., Kalra, P.S., Crowley, W.R. and Kalra, S.P. (1984) Neuropeptide Y and human pancreatic polypeptide stimulate feeding behavior in rats. *Endocrinology,* 115, 427–429.

Cone, R.D., Cowley, M.A., Butler, A.A., Fan, W., Marks, D.L. and Low, M.J. (2001) The arcuate nucleus as a conduit for diverse signals relevant to energy homeostasis. *International Journal of Obesity* 25 (Suppl. 5), S63–67.

Conti, B., Sugama, S., Lucero, J., Winsky-Sommerer, R., Wirz, S.A., Maher, P., Andrews, Z., Barr, A.M., Morale, M.C., Paneda, C., Pemberton, J., Gaidarova, S., Behrens, M.M., Beal, F., Sanna, P.P., Horvath, T. and Bartfai, T. (2005) Uncoupling protein 2 protects dopaminergic neurons from acute 1,2,3,6-methyl-phenyl-tetrahydropyridine toxicity. *Journal of Neurochemistry* 93, 493–501.

Coons, E.E. and Cruce, J.A.F. (1968) Lateral hypothalamus: food and current intensityin maintaining self-stimulation of hunger. *Science* 159, 1117–1119.

Cowley, M.A., Smart, J.L., Rubinstein, M., Cerdán, M.G., Diano, S., Horvath, T.L., Cone, R.D. and Low, M.J. (2001) Leptin activates anorexigenic POMC neurons through a neural network in the arcuate nucleus. *Nature* 411, 480–484.

Cowley, M.A., Smith, R.G., Diano, S., Tschöp, M., Pronchuk, N., Grove, K.L., Strasburger, C.J., Bidlingmaier, M., Esterman, M., Heiman, M.L., Garcia-Segura, L.M., Nillni, E.A., Mendez, P., Low, M.J., Sotonyi, P., Friedman, J.M., Liu, H., Pinto, S., Colmers, W.F., Cone, R.D. and Horvath, T.L. (2003a) The distribution and mechanism of action of ghrelin in the CNS demonstrates a novel hypothalamic circuit regulating energy homeostasis. *Neuron* 37, 649–661.

Cowley, M.A., Cone, R.D., Enriori, P., Louiselle, I., Williams, S.M. and Evans, A.E. (2003b) Electrophysiological actions of peripheral hormones on melanocortin neurons. *Annals of the New York Academy of Sciences* 994, 175–186.

Diano, S., Urbanski, H.F., Horvath, B., Bechmann, I., Kagiya, A., Nemeth, G., Naftolin, F., Warden, C.H. and Horvath, T.L. (2000) Mitochondrial uncoupling protein 2 (UCP2) in the non-human primate brain and pituitary. *Endocrinology* 141, 4226–4238.

Diano, S., Matthews, R.T., Patrylo, P., Yang, L., Beal, M.F., Barnstable, C.J. and Horvath, T.L. (2003) Uncoupling protein 2 prevents neuronal death including that occurring during seizures: a mechanism for preconditioning. *Endocrinology* 144, 5014–5021.

Edwards, C.M., Abusnana, S., Sunter, D., Murphy, K.G., Ghatei, M.A. and Bloom, S.R. (1999) The effect of the orexins on food intake: comparison with neuropeptide Y, melanin-concentrating hormone and galanin. *Journal of Endocrinology* 160, R7–12.

Elias, C.F., Saper, C.B., Maratos-Flier, E., Tritos, N.A., Lee, C., Kelly, J., Tatro, J.B., Hoffman, G.E., Ollmann, M.M., Barsh, G.S., Sakurai, T., Yanagisawa, M. and Elmquist, J.K. (1998) Chemically defined projections linking the mediobasal hypothalamus and the lateral hypothalamic area. *Journal of Comparative Neurology* 402, 442–459.

Elias, C.F., Lee, C.E., Kelly, J.F., Ahima, R.S., Kuhar, M., Saper, C.B. and Elmquist, J.K. (2001) Characterization of CART neurons in the rat and human hypothalamus. *Journal of Comparative Neurology* 432, 1–19.

Elmquist, J.K. (2001) Hypothalamic pathways underlying the endocrine, autonomic, and behavioral effects of leptin. *Physiology and Behavior* 74, 703–708.

Elmquist, J.K., Elias, C.F. and Saper, C.B. (1999) From lesions to leptin: hypothalamic control of food intake and body weight. *Neuron* 22, 221–232.

Erickson, J.C., Ahima, R.S., Hollopeter, G., Flier, J.S. and Palmiter, R.D. (1997) Endocrine function of neuropeptide Y knockout mice. *Regulatory Peptides* 70, 199–202.

Fan, W., Boston, B.A., Kesterson, R.A., Hruby, V.J. and Cone, R.D. (1997) Role of melanocortinergic neurons in feeding and the agouti obesity syndrome. *Nature* 385, 165–168.

Figlewicz, D.P., Evans, S.B., Murphy, J., Hoen, M. and Baskin, D.G. (2003) Expression of receptors for insulin and leptin in the ventral tegmental area/substantia nigra (VTA/SN) of the rat. *Brain Research* 964, 107–115.

Flanagan-Cato, L.M. (2000) Estrogen-induced remodeling of hypothalamic neural circuitry. *Frontiers in Neuroendocrinology* 21, 309–329.

Flores, C. and Stewart, J. (2000a) Basic fibroblast growth factor as a mediator of the effects of glutamate in the development of long-lasting sensitization to stimulant drugs: studies in the rat. *Psychopharmacology* 151, 152–165.

Flores, C. and Stewart, J. (2000b) Changes in astrocytic basic fibroblast growth factor expression during and after prolonged exposure to escalating doses of amphetamine. *Neuroscience* 98, 287–293.

Frühbeck, G. (2006) Intracellular signalling pathways activated by leptin. *Biochemical Journal* 393, 7–20.

Fuxe, K., Rivera, A., Jacobsen, K.X., Hoistad, M., Leo, G., Horvath, T.L., Staines, W., De la Calle, A. and Agnati, L.F. (2005) Dynamics of volume transmission in the brain. Focus on catecholamine and opioid peptide communication and the role of uncoupling protein 2. *Journal of Neural Transmission* 112, 65–76.

Gao, Q. and Horvath, T.L. (2007) Neurobiology of feeding and energy expenditure. *Annual Reviews in Neuroscience* 30, 367–398.

Gao, Q. and Horvath, T.L. (2008a) Neuronal control of energy homeostasis. *FEBS Letters* 582, 132–141.

Gao, Q. and Horvath, T.L. (2008b) Cross-talk between estrogen and leptin signaling in the hypothalamus. *American Journal of Physiology – Endocrinology and Metabolism* 294, E817–E826.

Gao, Q., Wolfgang, M.J., Neschen, S., Morino, K., Horvath, T.L., Shulman, G.I. and Fu, X.Y. (2004) Disruption of neural signal transducer and activator of transcription 3 causes obesity, diabetes, infertility, and thermal dysregulation. *Proceedings of the National Academy of Sciences of the United States of America* 101, 4661–4666.

Gao, X.B. and Pol, A.N. van den (2001) Melanin concentrating hormone depresses synaptic activity of glutamate and GABA neurons from rat lateral hypothalamus. *Journal of Physiology* 533, 237–252.

Garcia-Segura, L.M., Chowen, J.A., Parducz, A. and Naftolin, F. (1994) Gonadal hormones as promoters of structural synaptic plasticity: cellular mechanisms. *Progress in Neurobiology* 44, 279–307.

Garcia-Segura, L.M., Chowen, J.A., Duenas, M., Parducz, A. and Naftolin, F. (1996) Gonadal steroids and astroglial plasticity. *Cellular and Molecular Neurobiology* 16, 225–237.

Geiger, B.M., Behr, G.G., Frank, L.E., Caldera-Siu, A.D., Beinfeld, M.C., Kokkotou, E.G. and Pothos, E.N. (2008) Evidence for defective mesolimbic dopamine exocytosis in obesity-prone rats. *FASEB Journal* 22, 2740–2746.

Grill, H.J. and Kaplan, J.M. (2001) Interoceptive and integrative contributions of forebrain and brainstem to energy balance control. *International Journal of Obesity* 25 (Suppl. 5), S73–77.

Grill, H.J. and Kaplan, J.M. (2002) The neuroanatomical axis for control of energy balance. *Frontiers in Neuroendocrinology* 23, 2–40.

Grossman, S.P. (1979) The biology of motivation. *Annual Reviews of Psychology* 30, 209–242.

Guan, J.L., Uehara, K., Lu, S., Wang, Q.P., Funahashi, H., Sakurai, T., Yanagizawa, M. and Shioda, S. (2002) Reciprocal synaptic relationships between orexin- and melanin-concentrating hormone-containing neurons in the rat lateral hypothalamus: a novel circuit implicated in feeding regulation. *International Journal of Obesity* 26, 1523–1532.

Guan, X.M., Yu, H., Palyha, O.C., McKee, K.K., Feighner, S.D., Sirinathsinghji, D.J., Smith, R.G., Van der Ploeg, L.H. and Howard, A.D. (1997) Distribution of mRNA encoding the growth hormone secretagogue receptor in brain and peripheral tissues. *Brain Research – Molecular Brain Research* 48, 23–29.

Guldner, F.H. and Ingham, C.A. (1979) Plasticity in synaptic appositions of optic nerve afferents under different lighting conditions. *Neuroscience Letters* 14, 235–240.

Hagan, M.M., Rushing, P.A., Pritchard, L.M., Schwartz, M.W., Strack, A.M., Van Der Ploeg, L.H., Woods, S.C. and Seeley, R.J. (2000) Long-term orexigenic effects of AgRP-(83–132) involve mechanisms other than melanocortin receptor blockade. *American Journal of Physiology – Regulatory, Integrative and Comparative Physiology* 279, R47–52.

Hakansson, M.L., Brown, H., Ghilardi, N., Skoda, R.C. and Meister, B. (1998) Leptin receptor immunoreactivity in chemically defined target neurons of the hypothalamus. *Journal of Neuroscience* 18, 559–572.

Halaas, J.L., Gajiwala, K.S., Maffei, M., Cohen, S.L., Chait, B.T., Rabinowitz, D., Lallone, R.L., Burley, S.K. and Friedman, J.M. (1995) Weight-reducing effects of the plasma protein encoded by the obese gene. *Science* 269, 543–546.

Hara, J., Beuckmann, C.T., Nambu, T., Willie, J.T., Chemelli, R.M., Sinton, C.M., Sugiyama, F., Yagami, K., Goto, K., Yanagisawa, M. and Sakurai, T. (2001) Genetic ablation of orexin neurons in mice results in narcolepsy, hypophagia, and obesity. *Neuron* 30, 345–354.

Harvey, J. and Ashford, M.L. (2003) Leptin in the CNS: much more than a satiety signal. *Neuropharmacology* 44, 845–854.

Hayashi, M.L., Choi, S.Y., Rao, B.S., Jung, H.Y., Lee, H.K., Zhang, D., Chattarji, S., Kirkwood, A. and Tonegawa, S. (2004) Altered cortical synaptic morphology and impaired memory consolidation in forebrain- specific dominant-negative PAK transgenic mice. *Neuron* 42, 773–787.

Hetherington, A.W. and Ranson, S.W. (1940) Hypothalamic lesions and adipocity in the rat. *Anatomical Record* 78, 149–172.

Hetherington, A.W. and Ranson, S.W. (1942) The relation of various hypothalamic lesions to adiposity in the rat. *Journal of Comparative Neurology* 76, 475–499.

Horvath, T.L. (2005) The hardship of obesity: a soft-wired hypothalamus. *Nature Neuroscience* 8, 561–565.

Horvath, T.L. (2006a) Synaptic plasticity in energy balance regulation. *Obesity* 14 (Suppl, 5), 228S–233S.

Horvath, T.L. (2006b) Synaptic plasticity mediating leptin's effect on metabolism. *Progress in Brain Research* 153, 47–55.

Horvath, T.L. and Diano, S. (2004) The floating blueprint of hypothalamic feeding circuits. *Nature Reviews Neurosciences* 5, 662–667.

Horvath, T.L., Naftolin, F., Kalra, S.P. and Leranth, C. (1992) Neuropeptide-Y innervation of beta-endorphin-containing cells in the rat mediobasal hypothalamus: a light and electron microscopic double immunostaining analysis. *Endocrinology* 131, 2461–2467.

Horvath, T.L., Diano, S. and Pol, A.N. van den (1999a) Synaptic interaction between hypocretin (orexin) and neuropeptide Y cells in the rodent and primate hypothalamus: a novel circuit implicated in metabolic and endocrine regulations. *Journal of Neuroscience* 19, 1072–1087.

Horvath, T.L., Warden, C.H., Hajos, M., Lombardi, A., Goglia, F. and Diano, S. (1999b) Brain uncoupling protein 2: uncoupled neuronal mitochondria predict thermal synapses in homeostatic centers. *Journal of Neuroscience* 19, 10417–10427.

Horvath, T.L., Diano, S., Sotonyi, P., Heiman, M. and Tschöp, M. (2001) Minireview: ghrelin and the regulation of energy balance – a hypothalamic perspective. *Endocrinology* 142, 4163–4169.

Horvath, T.L., Diano, S., Miyamoto, S., Barry, S., Gatti, S., Alberati, D., Livak, F., Lombardi, A., Moreno, M., Goglia, F., Mor, G., Hamilton, J., Kachinskas, D., Horwitz, B. and Warden, C.H. (2003) Uncoupling proteins-2 and -3 influence obesity and inflammation in transgenic mice. *International Journal of Obesity* 27, 433–442.

Huang, X.F., Koutcherov, I., Lin, S., Wang, H.Q. and Storlien, L. (1996) Localization of leptin receptor mRNA expression in mouse brain. *Neuroreport* 7, 2635–2638.

Huszar, D., Lynch, C.A., Fairchild-Huntress, V., Dunmore, J.H., Fang, Q., Berkemeier, L.R., Gu, W., Kesterson, R.A., Boston, B.A., Cone, R.D., Smith, F.J., Campfield, L.A., Burn, P. and Lee, F. (1997) Targeted disruption of the melanocortin-4 receptor results in obesity in mice. *Cell* 88, 131–141.

Inoue, S. and Bray, G.A. (1977) The effects of subdiaphragmatic vagotomy in rats with ventromedial hypothalamic obesity. *Endocrinology* 100, 108–114.

Kalra, S.P., Dube, M.G., Pu, S., Xu, B., Horvath, T.L. and Kalra, P.S. (1999) Interacting appetite-regulating pathways in the hypothalamic regulation of body weight. *Endocrine Reviews* 20, 68–100.

Kelley, A.E. (1999) Functional specificity of ventral striatal compartments in appetitive behaviors. *Annals of the New York Academy of Sciences* 877, 71–90.

Kelley, A.E. (2004a) Memory and addiction: shared neural circuitry and molecular mechanisms. *Neuron* 44, 161–179.

Kelley, A.E. (2004b) Ventral striatal control of appetitive motivation: role in ingestive behavior and reward-related learning. *Neuroscience Biobehavioral Reviews* 27, 765–776.

Kim-Han, J.S. and Dugan, L.L. (2005) Mitochondrial uncoupling proteins in the central nervous system. *Antioxidants and Redox Signaling* 7, 1173–1181.

Kojima, M., Hosoda, H., Date, Y., Nakazato, M., Matsuo, H. and Kangawa, K. (1999) Ghrelin is a growth-hormone-releasing acylated peptide from stomach. *Nature* 402, 656–660.

Langle, S.L., Poulain, D.A. and Theodosis, D.T. (2002) Neuronal-glial remodeling: a structural basis for neuronal-glial interactions in the adult hypothalamus. *Journal of Physiology* 96, 169–175.

Leibowitz, S.F. and Hoebel, B.G. (2004) Behavioral neuroscience and obesity. In: Bray, G.A. and Bouchard, C. (eds) *Handbook of Obesity. Etiology and Pathophysiology*. Marcel Dekker, New York.

Lely, A.J. van der, Tschöp, M., Heiman, M.L. and Ghigo, E. (2004) Biological, physiological, pathophysiological, and pharmacological aspects of ghrelin. *Endocrine Reviews* 25, 426–457.

Li, X.L., Aou, S., Oomura, Y., Hori, N., Fukunaga, K. and Hori, T. (2002a) Impairment of long-term potentiation and spatial memory in leptin receptor-deficient rodents. *Neuroscience* 113, 607–615.

Li, Y., Gao, X.B., Sakurai, T. and Pol, A.N. van den (2002b) Hypocretin/orexin excites hypocretin neurons via a local glutamate neuron-A potential mechanism for orchestrating the hypothalamic arousal system. *Neuron* 36, 1169–1181.

Li, Z., Okamoto, K., Hayashi, Y. and Sheng, M. (2004) The importance of dendritic mitochondria in the morphogenesis and plasticity of spines and synapses. *Cell* 119, 873–887.

López, M., Lage, R., Tung, Y.C., Challis, B.G., Varela, L., Virtue, S., O'Rahilly, S., Vidal-Puig, A., Diéguez, C. and Coll, A.P. (2007) Orexin expression is regulated by alpha-melanocyte-stimulating hormone. *Journal of Neuroendocrinology* 19, 703–707.

Ludwig, D.S., Mountjoy, K.G., Tatro, J.B., Gillette, J.A., Frederich, R.C., Flier, J.S. and Maratos-Flier, E. (1998) Melanin-concentrating hormone: a functional melanocortin antagonist in the hypothalamus. *American Journal of Physiology – Endocrinology and Metabolism* 274, E627–633.

Ludwig, D.S., Tritos, N.A., Mastaitis, J.W., Kulkarni, R., Kokkotou, E., Elmquist, J., Lowell, B., Flier, J.S. and Maratos-Flier, E. (2001) Melanin-concentrating hormone overexpression in transgenic mice leads to obesity and insulin resistance. *Journal of Clinical Investigation* 107, 379–386.

Maffei, M., Fei, H., Lee, G.H., Dani, C., Leroy, P., Zhang, Y., Proenca, R., Negrel, R., Ailhaud, G. and Friedman, J.M. (1995) Increased expression in adipocytes of ob RNA in mice with lesions of the hypothalamus and with mutations at the db locus. *Proceedings of the National Academy of Sciences of the United States of America* 92, 6957–6960.

Marsh, D.J., Weingarth, D.T., Novi, D.E., Chen, H.Y., Trumbauer, M.E., Chen, A.S., Guan, X.M., Jiang, M.M., Feng, Y., Camacho, R.E., Shen, Z., Frazier, E.G., Yu, H.,

Metzger, J.M., Kuca, S.J., Shearman, L.P., Gopal-Truter, S., MacNeil, D.J., Strack, A.M., MacIntyre, D.E., Van der Ploeg, L.H. and Qian, S. (2002) Melanin-concentrating hormone 1 receptor-deficient mice are lean, hyperactive, and hyperphagic and have altered metabolism. *Proceedings of the National Academy of Sciences of the United States of America* 99, 3240–3245.

Mattson, M.P. and Liu, D. (2003) Mitochondrial potassium channels and uncoupling proteins in synaptic plasticity and neuronal cell death. *Biochemical Biophysical Research Communications* 304, 539–549.

Mitchell, J.B. and Gratton, A. (1994) Involvement of mesolimbic dopamine neurons in sexual behaviors: implications for the neurobiology of motivation. *Reviews in Neuroscience* 5, 317–329.

Muroya, S., Funahashi, H., Yamanaka, A., Kohno, D., Uramura, K., Nambu, T., Shibahara, M., Kuramochi, M., Takigawa, M., Yanagisawa, M., Sakurai, T., Shioda, S. and Yada, T. (2004) Orexins (hypocretins) directly interact with neuropeptide Y, POMC and glucose-responsive neurons to regulate Ca 2+ signaling in a reciprocal manner to leptin: orexigenic neuronal pathways in the mediobasal hypothalamus. *European Journal of Neuroscience* 19, 1524–1534.

Naftolin, F., Garcia-Segura, L.M., Horvath, T.L., Zsarnovszky, A., Demir, N., Fadiel, A., Leranth, C., Vondracek-Klepper, S., Lewis, C., Chang, A. and Parducz, A. (2007) Estrogen-induced hypothalamic synaptic plasticity and pituitary sensitization in the control of the estrogen-induced gonadotrophin surge. *Reproductive Science* 14, 101–116.

Nemeroff, C.B., Lipton, M.A. and Kizer, J.S. (1978) Models of neuroendocrine regulation: use of monosodium glutamate as an investigational tool. *Developmental Neuroscience* 1, 102–109.

Nishikawa, Y., Shibata, S. and Watanabe, S. (1995) Circadian changes in long-term potentiation of rat suprachiasmatic field potentials elicited by optic nerve stimulation *in vitro*. *Brain Research* 695, 158–162.

Ohno, K. and Sakurai, T. (2008) Orexin neuronal circuitry: role in the regulation of sleep and wakefulness. *Frontiers in Neuroendocrinology* 29, 70–87.

Ollmann, M.M., Wilson, B.D., Yang, Y.K., Kerns, J.A., Chen, Y., Gantz, I. and Barsh, G.S. (1997) Antagonism of central melanocortin receptors *in vitro* and *in vivo* by agouti-related protein. *Science* 278, 135–138.

Olmos, G., Naftolin, F., Perez, J., Tranque, P.A. and Garcia-Segura, L.M. (1989) Synaptic remodeling in the rat arcuate nucleus during the estrous cycle. *Neuroscience* 32, 663–667.

Palmiter, R.D. (2007) Is dopamine a physiologically relevant mediator of feeding behavior? *Trends in Neuroscience* 30, 375–381.

Palmiter, R.D., Erickson, J.C., Hollopeter, G., Baraban, S.C. and Schwartz, M.W. (1998) Life without neuropeptide Y. *Recent Progress in Hormone Research* 53, 163–199.

Paradis, E., Clavel, S., Bouillaud, F., Ricquier, D. and Richard, D. (2003) Uncoupling protein 2: a novel player in neuroprotection. *Trends in Molecular Medicine* 9, 522–525.

Pelleymounter, M.A., Cullen, M.J., Baker, M.B., Hecht, R., Winters, D., Boone, T. and Collins, F. (1995) Effects of the obese gene product on body weight regulation in *ob/ob* mice. *Science* 269, 540–543.

Pinto, S., Roseberry, A.G., Liu, H., Diano, S., Shanabrough, M., Cai, X., Friedman, J.M. and Horvath, T.L. (2004) Rapid rewiring of arcuate nucleus feeding circuits by leptin. *Science* 304, 110–115.

Pu, S., Jain, M.R., Horvath, T.L., Diano, S., Kalra, P.S. and Kalra, S.P. (1999) Interactions between neuropeptide Y and gamma-aminobutyric acid in stimulation of feeding: a morphological and pharmacological analysis. *Endocrinology* 140, 933–940.

Qu, D., Ludwig, D.S., Gammeltoft, S., Piper, M., Pelleymounter, M.A., Cullen, M.J., Mathes, W.F., Przypek, R., Kanarek, R. and Maratos-Flier, E. (1996) A role for melanin-concentrating hormone in the central regulation of feeding behaviour. *Nature* 380, 243–247.

Richard, D., Rivest, R., Huang, Q., Bouillaud, F., Sanchis, D., Champigny, O. and Ricquier, D. (1998) Distribution of the uncoupling protein 2 mRNA in the mouse brain. *Journal of Comparative Neurology* 397, 549–560.

Robinson, T.E. and Kolb, B. (2004) Structural plasticity associated with exposure to drugs of abuse. *Neuropharmacology* 47 (Suppl. 1), 33–46.

Roseberry, A.G., Liu, H., Jackson, A.C., Cai, X. and Friedman, J.M. (2004) Neuropeptide Y-mediated inhibition of proopiomelanocortin neurons in the arcuate nucleus shows enhanced desensitization in *ob/ob* mice. *Neuron* 41, 711–722.

Sahu, A. (2003) Leptin signaling in the hypothalamus: emphasis on energy homeostasis and leptin resistance. *Frontiers in Neuroendocrinology* 24, 225–253.

Sahu, A., Kalra, P.S. and Kalra, S.P. (1988) Food deprivation and ingestion induce reciprocal changes in neuropeptide Y concentrations in the paraventricular nucleus. *Peptides* 9, 83–86.

Sakurai, T. (2003) Orexin: a link between energy homeostasis and adaptive behaviour. *Current Opinion in Clinical Nutrition and Metabolic Care* 6, 353–360.

Sakurai, T. (2005) Reverse pharmacology of orexin: from an orphan GPCR to integrative physiology. *Regulatory Peptides* 126, 3–10.

Sakurai, T., Amemiya, A., Ishii, M., Matsuzaki, I., Chemelli, R.M., Tanaka, H., Williams, S.C., Richardson, J.A., Kozlowski, G.P., Wilson, S., Arch, J.R., Buckingham, R.E., Haynes, A.C., Carr, S.A., Annan, R.S., McNulty, D.E., Liu, W.S., Terrett, J.A., Elshourbagy, N.A., Bergsma, D.J. and Yanagisawa, M. (1998) Orexins and orexin receptors: a family of hypothalamic neuropeptides and G protein-coupled receptors that regulate feeding behavior. *Cell* 92, 573–585.

Sanacora, G., Kershaw, M., Finkelstein, J.A. and White, J.D. (1990) Increased hypothalamic content of preproneuropeptide Y messenger ribonucleic acid in genetically obese Zucker rats and its regulation by food deprivation. *Endocrinology* 127, 730–737.

Saper, C.B., Chou, T.C. and Elmquist, J.K. (2002) The need to feed: homeostatic and hedonic control of eating. *Neuron* 36, 199–211.

Shanley, L.J., Irving, A.J. and Harvey, J. (2001) Leptin enhances NMDA receptor function and modulates hippocampal synaptic plasticity. *Journal of Neuroscience* 21, RC1–6.

Shimada, M., Tritos, N.A., Lowell, B.B., Flier, J.S. and Maratos-Flier, E. (1998) Mice lacking melanin-concentrating hormone are hypophagic and lean. *Nature* 396, 670–674.

Shioda, S., Funahashi, H., Nakajo, S., Yada, T., Maruta, O. and Nakai, Y. (1998) Immunohistochemical localization of leptin receptor in the rat brain. *Neuroscience Letters* 243, 41–44.

Solinas, M., Goldberg, S.R. and Piomelli, D. (2008) The endocannabinoid system in brain reward processes. *British Journal of Pharmacology* 154, 369–383.

Stanley, B.G. and Leibowitz, S.F. (1984) Neuropeptide Y: stimulation of feeding and drinking by injection into the paraventricular nucleus. *Life Sciences* 35, 2635–2642.

Stanley, B.G. and Leibowitz, S.F. (1985) Neuropeptide Y injected in the paraventricular hypothalamus: a powerful stimulant of feeding behavior. *Proceedings of the National Academy of Sciences of the United States of America* 82, 3940–3943.

Stanley, B.G., Chin, A.S. and Leibowitz, S.F. (1985) Feeding and drinking elicited by central injection of neuropeptide Y: evidence for a hypothalamic site(s) of action. *Brain Research Bulletin* 14, 521–524.

Stein, E.A., Carr, K.D. and Simon, E.J. (1990) Brain stimulation-induced feeding alters regional opioid receptor binding in the rat: an *in vivo* autoradiographic study. *Brain Research* 533, 213–222.

Stellar, J.R. and Stellar, E. (1985) *The Neurobiology of Motivation and Reward*. Springer-Verlag, New York.

Stephens, T.W., Basinski, M., Bristow, P.K., Bue-Valleskey, J.M., Burgett, S.G., Craft, L., Hale, J., Hoffmann, J., Hsiung, H.M., Kriauciunas, A., MacKellar, W., Rosteck, P.R. Jr, Schoner, B., Smith, D., Tinsley, F.C., Zhang, X.-Y. and Heiman, M. (1995) The role of neuropeptide Y in the antiobesity action of the obese gene product. *Nature* 377, 530–532.

Stricker-Krongrad, A., Richy, S. and Beck, B. (2002) Orexins/hypocretins in the *ob/ob* mouse: hypothalamic gene expression, peptide content and metabolic effects. *Regulatory Peptides* 104, 11–20.

Sun, Y., Wang, P., Zheng, H. and Smith, R.G. (2004) Ghrelin stimulation of growth hormone release and appetite is mediated through the growth hormone secretagogue receptor. *Proceedings of the National Academy of Sciences of the United States of America* 101, 4679–4684.

Szczypka, M.S., Mandel, R.J., Donahue, B.A., Snyder, R.O., Leff, S.E. and Palmiter, R.D. (1999a) Viral gene delivery selectively restores feeding and prevents lethality of dopamine-deficient mice. *Neuron* 22, 167–178.

Szczypka, M.S., Rainey, M.A., Kim, D.S., Alaynick, W.A., Marck, B.T., Matsumoto, A.M. and Palmiter, R.D. (1999b) Feeding behavior in dopamine-deficient mice. *Proceedings of the National Academy of Sciences of the United States of America* 96, 12138–12143.

Tartaglia, L.A., Dembski, M., Weng, X., Deng, N., Culpepper, J., Devos, R., Richards, G.J., Campfield, L.A., Clark, F.T., Deeds, J., Muir, C., Sanker, S., Moriarty, A., Moore, K.J., Smutko, J.S., Mays, G.G., Wool, E.A., Monroe, C.A. and Tepper, R.I. (1995) Identification and expression cloning of a leptin receptor, OB-R. *Cell* 83, 1263–1271.

Tataranni, P.A., Gautier, J.F., Chen, K., Uecker, A., Bandy, D., Salbe, A.D., Pratley, R.E., Lawson, M., Reiman, E.M. and Ravussin E. (1999) Neuroanatomical correlates of hunger and satiation in humans using positron emission tomography. *Proceedings of the National Academy of Sciences of the United States of America* 96, 4569–4574.

Tatemoto, K., Carlquist, M. and Mutt, V. (1982) Neuropeptide Y – a novel brain peptide with structural similarities to peptide YY and pancreatic polypeptide. *Nature* 296, 659–660.

Theodosis, D.T. and Poulain, D.A. (2001) Maternity leads to morphological synaptic plasticity in the oxytocin system. *Progress in Brain Research* 133, 49–58.

Tritos, N.A., Vicent, D., Gillette, J., Ludwig, D.S., Flier, E.S. and Maratos-Flier, E. (1998) Functional interactions between melanin-concentrating hormone, neuropeptide Y, and anorectic neuropeptides in the rat hypothalamus. *Diabetes* 47, 1687–1692.

Tschöp, M., Smiley, D.L. and Heiman, M.L. (2000) Ghrelin induces adiposity in rodents. *Nature* 407, 908–913.

Vaisse, C., Halaas, J.L., Horvath, C.M., Darnell, J.E. Jr, Stoffel, M. and Friedman, J.M. (1996) Leptin activation of Stat3 in the hypothalamus of wild-type and *ob/ob* mice but not *db/db* mice. *Nature Genetics* 14, 95–97.

Valenstein, E.S., Cox, V.C. and Kakolewski, J.W. (1968) Modification of motivated behavior elicited by electrical stimulation of the lateral hypothalamus. *Science* 159, 1119–1121.

Valenstein, E.S. and Mittleman, G. (1984) Ingestive behavior evoked by hypothalamic stimulation and schedule-induced polydipsia are related. *Science* 224, 415–417.

Walker, H.C. and Romsos, D.R. (1993) Similar effects of NPY on energy metabolism and on plasma insulin in adrenalectomized *ob/ob* and lean mice. *American Journal of Physiology – Endocrinology and Metabolism* 264 (2 Pt 1), E226–230.

Wise, R.A. (2004a) Dopamine and food reward: back to the elements. *American Journal of Physiology – Regulatory, Integrative and Comparative Physiology* 286, R13.

Wise, R.A. (2004b) Dopamine, learning and motivation. *Nature Reviews Neuroscience* 5, 483–494.

Wise, R.A. and Rompre, P.P. (1989) Brain dopamine and reward. *Annual Review of Psychology* 40, 191–225.

Woods, S.C., Seeley, R.J., Porte, D. Jr and Schwartz, M.W. (1998) Signals that regulate food intake and energy homeostasis. *Science* 280, 1378–1383.

Woods, S.C., Schwartz, M.W., Baskin, D.G. and Seeley, R.J. (2000) Food intake and the regulation of body weight. *Annual Review of Psychology* 51, 255–277.

Wortley, K.E., Anderson, K.D., Garcia, K., Murray, J.D., Malinova, L., Liu, R., Moncrieffe, M., Thabet, K., Cox, H.J., Yancopoulos, G.D., Wiegand, S.J. and Sleeman, M.W. (2004) Genetic deletion of ghrelin does not decrease food intake but influences metabolic fuel preference. *Proceedings of the National Academy of Sciences of the United States of America* 101, 8227–8232.

Yamanaka, A., Kunii, K., Nambu, T., Tsujino, N., Sakai, A., Matsuzaki, I., Miwa, Y., Goto, K. and Sakurai, T. (2000) Orexin-induced food intake involves neuropeptide Y pathway. *Brain Research* 859, 404–409.

Zhang, Y., Proenca, R., Maffei, M., Barone, M., Leopold, L. and Friedman, J.M. (1994) Positional cloning of the mouse obese gene and its human homologue. *Nature* 372, 425–432.

Zigmond, M.J. and Stricker, E.M. (1972) Deficits in feeding behavior after intraventricular injection of 6-hydroxydopamine in rats. *Science* 177, 1211–1214.

11 Energy Regulatory Signals and Food Reward

Dianne Figlewicz Lattemann,[1,2] Nicole M. Sanders[1,2] and Alfred J. Sipols[3]

[1]VA Puget Sound Health Care System (151), Seattle, USA; [2]Department of Psychiatry and Behavioral Sciences, University of Washington, USA; [3]Department of Medicine, University of Latvia, Latvia

Introduction

Obesity is recognized as a significant risk factor for diabetes, cardiovascular disease and several cancers, as well as shortened lifespan (Mokdad *et al.*, 2001; Hill *et al.*, 2003; Pi-Sunyer, 2003). Several chapters in this book discuss the role of specific central nervous system (CNS) neuronal pathways and neurotransmitters, in the maintenance of caloric intake to satisfy physiological needs (see Chapters 1–3 and 10). The original model of a negative feedback loop between the brain and energy regulatory signals (circulating factors such as insulin and leptin, whose concentrations reflect the size of adipose stores and which signal this information to the CNS) has received substantial experimental support (Woods *et al.*, 1979; Baskin *et al.*, 1999). While there also is support for the concept that feeding behaviour can be modified by the rewarding aspects of food, the concept that the perceived rewarding value of food in turn may be regulated is still somewhat novel. The purpose of this chapter is to summarize current knowledge regarding CNS mediation of food reward and the evidence in support of its potential regulation or modulation, including discussion of the neurochemical and neuroanatomical substrates for crosstalk between the CNS energy regulatory and the CNS reward circuitry.

Functional Definition of Reward

Although the meaning of 'reward' has been controversial among psychologists (Wise and Hoffman, 1992; Robbins and Everitt, 1996; Berridge and Robinson, 1998; Wise, 2002), for the purposes of this chapter, we will define 'reward' functionally: the 'rewarding' aspect of food is gauged by the function of its being sought out and consumed, be it in an animal experiment or a free-choice setting for humans. How the 'final bottom line' of food intake is arrived at in a free-choice situation is the complex result of numerous factors, only one of which is true caloric need. Thus, we need to understand, at anatomical and functional levels, how caloric need, food palatability and emotional/learned aspects of food experience are interwoven and how they might be manipulated either behaviourally or through the use of pharmacotherapeutics to impact total, chronic caloric intake in a meaningful way.

Both the nutritional status of an animal and its environment – type of food, obstacles to obtain food – modulate the rewarding or motivational value of food (Wilson *et al.*, 1995; Levine and Billington, 1997; Salamone *et al.*, 2003). How or whether this modified value of the food impacts on the energy balance regulatory circuitry of the CNS remains essentially unknown. One reason, perhaps, that there are so few data currently available to address this issue is that, historically, studies of CNS reward function and studies of regulation of food intake have, for the most part, focused on anatomically distinct circuitry and have used behavioural paradigms that might not be appropriate for asking or answering questions about food reward.

Identification of Anatomical Targets

Historically, studies evaluating the physiological defence of caloric intake by the CNS have focused on the medial hypothalamus as a major anatomical target (Williams *et al.*, 2001; Saper *et al.*, 2002) and have evaluated the actions of hormones and neurotransmitters experimentally using highly controlled and stimulus-deprived environments for these tests. While this experimental approach has been necessary and is correct for these sorts of studies, the applicability of the findings to human eating behaviour has been challenged. Hill *et al.* (2003) have pointed out that the current obesity epidemic may be ascribed to an environment of convenient and economically affordable food that is both highly palatable and high in caloric density and fat content. If, in fact, the medial hypothalamic circuitry acts as the final common arbiter of caloric intake, then why does caloric intake (and in adults, body adiposity) not remain constant and appropriate in the face of whatever foodstuffs are available? One response to this query is that the data we have gleaned regarding the calorie regulatory circuitry of the hypothalamus have been obtained in circumstances where there are no environmental challenges or choices, and probably limited activation of the CNS reward/motivational circuitry. That is, the majority of data on food intake regulation by energy regulatory signals have been collected from animals feeding in their home cages on commercial rodent chow, a bland, monotonous and relatively low-fat diet which is presented in abundance; thus, the animal needs minimal engagement of motor systems in order to eat and generally can expect that there will be as much to eat as it wants. This situation offers an almost perfect model for studying the regulation of caloric need by neural and endocrine factors. However, in 1988 a key study by Bray's group made it clear that the function of the CNS–energy regulatory signal feedback loop could be altered by an environmental intervention, i.e. changing the fat content of the diet. They demonstrated that rats fed a high-fat diet lost less body weight in response to a direct CNS infusion of insulin than rats maintained on standard laboratory chow (Arase *et al.*, 1988). This finding subsequently was replicated, showing that the effect was 'dose-dependent' on the concentration of fat in the diet (Chavez *et al.*, 1996). While the precise mechanism(s) of this effect remains unclear, these studies made the point that diet composition has a major impact on the function of the calorie regulatory CNS circuitry.

Historically, the analysis of the CNS and reward has focused anatomically on the lateral hypothalamic area (LHA) and midbrain dopaminergic cell bodies and their projection sites, and functionally on paradigms such as brain self-stimulation or self-administration of various neurally active substances (Olds, 1962; Wise, 1988). Not surprisingly, it has become appreciated that additional CNS sites have a role in mediating the rewarding aspects of stimuli. As an anatomical basis for potential crosstalk between the energy regulatory circuitry and the reward circuitry, it must be appreciated that the medial hypothalamic nuclei are connected extensively with the CNS regions that mediate reward and motivation. For example, the LHA is a major relay area for projections from the mediobasal hypothalamus and thus could serve as a critical integrator for signals from both the reward circuitry and calorie regulatory circuitry. The limbic reward system can be defined functionally as those CNS structures that mediate the rewarding, reinforcing and emotional aspects of stimuli. From an anatomical perspective, there is a general consensus that the LHA, amygdala, select regions of the cerebral cortex, the ventral tegmental area (VTA) and ventral striatum or nucleus accumbens (NAc) are components of this circuitry (DeOlmos and Heimer, 1999; Everitt *et al.*, 1999). Reciprocal synaptic connections exist between the amygdala and cortex and between the NAc and cortex, and there are substantial efferent projections from the amygdala to the hypothalamus and the VTA/substantia nigra pars compacta (SNc). There appear to be limited forebrain inputs directly to the paraventricular nucleus of the hypothalamus (PVN), although PVN efferent projections to the LHA are abundant. Rather, the arcuate nucleus of the mediobasal hypothalamus appears to receive critical limbic input projections from the LHA containing the feeding-stimulatory peptide, orexin, as well as input from the central nucleus of the amygdala. In turn, there are direct projections from the arcuate nucleus to the LHA. The central nucleus of the amygdala also projects to the LHA, and LHA and amygdala receive direct taste inputs from the nucleus of the solitary tract (NTS, the critical primary–secondary relay site for the taste pathway). Other relevant synaptic connections include reciprocal projections from the NAc to the VTA and projections from the NAc to the LHA. For more detailed anatomical discussion, the reader is referred to key reviews (Berthoud, 2002; Kelley *et al.*, 2002; Solinas *et al.*, 2008).

The collection of mesocorticolimbic (VTA) dopamine (DA) neurones, which project to the ventral striatum or NAc and to the prefrontal cortex, has been viewed as a central neuroanatomical substrate for reward and motivation (Ikemoto and Panksepp, 1996; McBride *et al.*, 1999; Palmiter, 2007). Activation of VTA DA neurones, and release of DA within the NAc, have long been viewed as indicative of reward enhancement. What does activation of the VTA/NAc pathway reflect in terms of food reward? This remains a topic of lively debate (Hoebel *et al.*, 1989; Schultz, 2002, 2004; Salamone *et al.*, 2003); Berridge (1996) has demonstrated that this activity is not correlated with enhanced hedonic value of food, as evaluated in the 'taste reactivity' paradigm, and thus dopaminergic activation does not reflect an increase in the animal's 'liking' of the food. Rather, Berridge proposed that mesocorticolimbic DA activity reflects an increase in the 'incentive salience' of a stimulus, including food. This property can be modulated by the nutritional status of an animal. In this schema, with food deprivation, the food stimulus would be more relevant and more motivating. It has been documented that with repeated training to gain access to a diet in a defined physical environment, initial exposure leads to increased release of DA in the NAc shell, whereas subsequent exposure leads to either no increase of DA (Richardson and Gratton, 1996); an increase of DA in anticipation of the presentation of food (Kiyatkin, 1995); increased release of DA within a different part of the NAc (Bassareo and DiChiara, 1997); and sustained release of DA in the prefrontal cortex (Bassareo and DiChiara, 1997, 1999). Although interpretation of these findings remains controversial, the concept that the contextual stimuli themselves (odour or visual cues) become salient and can elicit DA release with repeated exposure to food in the same context seems experimentally validated. However, if DA release specifically in the NAc shell reflects 'reward', the habitual presentation of the same food could be predicted to result in a loss of its primary reward value.

Experimental Paradigms

Most current studies of candidate endocrine or neural factors for the regulation of food intake include an evaluation of these elements in association with chronic consumption of a high-fat diet and, perhaps, diet-induced obesity. While this experimental approach may be valuable for understanding the impact of diet composition on the efficacy of putative energy regulatory factors, it may not be a meaningful diet manipulation for evaluating the effect of diet composition on the reward circuitry function. As summarized above, one might speculate that there is very limited NAc DA release when rats eat commercial rat chow in the habitual home cage environment and, after the initial exposures, there probably would be limited DA release to a high-fat diet if that were the only food available. In addition to the possibility that chronic, non-contingent exposure to any diet might neutralize its primary rewarding properties, studies from the scientific literature on drug abuse have highlighted the point that there are differences in CNS DA release, neuroendocrine response and drug-seeking behaviour between rats that receive a drug such as cocaine passively (non-contingently) and rats that have the opportunity to self-administer it (Wilson *et al.*, 1994; Markou *et al.*, 1999; Gallici *et al.*, 2000). This effect also might hold true for diets: that is, the consequences of diet choice on VTA/NAc activity (or any other component of the reward circuitry) may be different from those that occur when there is no choice, regardless of the diet composition. Studies that evaluate CNS neurochemistry, where feeding is only one of several simultaneous activities in a slightly more complex environment (e.g. Nonogaki *et al.*, 2003), represent an experimental approach that should be more applicable to human feeding.

Perhaps the strongest clue to common functional links between the reward circuitry and the energy regulatory circuitry is the observation that fasting or food restriction have marked behavioural and neurochemical effects on both. The function of the medial hypothalamic energy regulatory circuitry in the context of starvation has been reviewed elsewhere (Schwartz and Seeley, 1997). Fasting or food restriction also activates, or enhances the activation of, the reward circuitry, as evaluated in several different behavioural paradigms (Shizgal *et al.*, 2001; Carr, 2002). In one of the most striking illustrations of this effect, Carroll and Meisch (1984) studied rats allowed to self-administer a threshold dose of cocaine, which caused the release of DA in the NAc. Rats were fasted or fed on an alternating day schedule prior to having access to the cocaine. When tested on days after they were fed overnight, they self-administered almost no cocaine. When tested on days after they had been fasted overnight, rats self-administered cocaine robustly. This result demonstrates that the reward circuitry in the CNS is rapidly responsive to changes in metabolic status. The finding has been replicated with food restriction rather than fasting and has been observed in a somewhat different self-administration paradigm: food restriction will enhance the propensity to drug-taking relapse in rats that have extinguished drug self-administration (Shalev *et al.*, 2002).

A second behavioural task, the conditioned place preference (CPP) paradigm, assesses the strength of a learned association between the perceived reward value of a stimulus (such as a drug treatment or food) and the location in which the animal receives the stimulus (Bardo and Bevins, 2000). The strength of the conditioning can be sensitive to the nutritional status of the animal. Conditioning of a place preference by cocaine is enhanced by food restriction (Bell *et al.*, 1997). Place preference also can be conditioned by food and, perhaps not surprisingly, the strength of the CPP is enhanced by food restriction (Swerdlow *et al.*, 1983; Papp, 1988; Agmo *et al.*, 1995; Lepore *et al.*, 1995; Figlewicz *et al.*, 2001). Place preference conditioning by food is dependent on intact dopaminergic activation, as development of the CPP is blocked by administration of DA receptor antagonists during training sessions (Agmo *et al.*, 1995; Figlewicz *et al.*, 2001).

Given the central role of the VTA DA neurones in the reward circuitry, it has been hypothesized that the effect of food restriction is due to enhanced activation of these neurones. It was shown that food-restricted rats trained to drink a palatable liquid food had greater DA release in the NAc than free-feeding rats (Wilson *et al.*, 1995). One question, then, is whether these DA neurones are a target for neural or endocrine factors that change in association with fasting and food restriction. The study by Carroll and Meisch (1984) suggests that neural or humoral factors modulating these phenomena must be able to change with a time course of several hours. Adrenal glucocorticoid levels are elevated with fasting. Evidence has been provided that glucocorticoids can facilitate DA release and DA-mediated behaviours (Marinelli and Piazza, 2002). Additionally, both insulin and leptin levels decrease rapidly in association with food restriction or fasting (Havel, 2000) and inhibit performance in food reward behavioural tasks that are DA-dependent (e.g. Shalev *et al.*, 2001; Figlewicz *et al.*, 2004). Insulin exerts effects at the level of DA reuptake (Figlewicz, 2003); *in vivo*, intracerebroventricular

(ICV) insulin increases steady-state mRNA levels of the DA reuptake transporter (DAT); *in vitro*, insulin administration (at a physiological concentration) facilitates striatal DA reuptake, which should, in turn, curtail dopaminergic signalling. The functional consequence of decreased DA signalling should be that insulin decreases the rewarding aspect of stimuli. Consistent with the possibility that modulation of DAT function has behavioural consequences, Pecina *et al.* (2003) demonstrated that DAT-knockdown mice ate more and had decreased latency to onset for eating food treats. Krugel *et al.* (2003) have shown that ICV leptin administration decreases both baseline and food-stimulated NAc DA release. Receptors for both insulin and leptin have been identified on the VTA DA cell bodies (Figlewicz *et al.*, 2003), supporting the possibility that insulin and leptin may act directly at the VTA and/or on NAc nerve terminals to blunt dopaminergic activity and its contribution to food reward. Collectively, studies of the effects of glucocorticoids, insulin and leptin support the conclusion that a neuroendocrine milieu exists in fasted animals that would bias them towards enhanced dopaminergic function.

The Opioid System

In addition to the mesocorticolimbic dopaminergic system, other neurotransmitter systems (e.g. GABAergic, cholinergic) have a role in the VTA/NAc reward circuitry. Brain opioid systems have served as a focus for investigation, as endogenous opioid neural networks appear to play a role in the regulation of food intake, food hedonics and food choice (Glass *et al.*, 1999; Levine *et al.*, 2003). Mu, kappa and delta opioids and opiatergic agonists stimulate feeding independently of the nutritional status of the animal when administered into multiple CNS sites. Conversely, opiate antagonists decrease feeding. Although experimental evidence demonstrates that DA and opioids play somewhat different roles in the mediation of food reward, the neuroanatomical circuitry that is implicated in opioid effects overlaps significantly with the VTA/NAc reward circuitry; opioidergic activation may mediate the hedonic valuation of foods, whereas activation of the VTA/NAc mediates the rewarding (motivating, reinforcing, incentive salient) properties of food. The potential interaction of opioidergic and dopaminergic systems seems obvious, as one would predict that a 'more pleasing' food would be more rewarding. A compelling reason for targeting the opioids for continued investigation by basic scientists is the observation that endogenous opioids may play a role in hedonic valuation of foods in human subjects. In one report, the opiate antagonist, nalmefene, decreased fat and protein intake from a standardized buffet meal in non-obese subjects (Yeomans *et al.*, 1990). Both taste preferences for, and intake of, sweet high-fat foods (such as cookies or chocolate) are decreased by treatment with the opioid antagonist, naloxone, in binge eaters but not in non-binge eaters (Drewnowski *et al.*, 1992, 1995). This finding suggests that some endogenous opioid systems may be (more) active in association with food binges.

The role of the CNS opioid systems in mediating food intake suggests that the opioids act to sustain, rather than initiate, feeding; that stimulation of all major subclasses of opioid receptor (mu, delta, kappa) can result in enhanced food intake; and that activation of these receptor populations can stimulate intake of preferred food (Morley *et al.*, 1983; Levine and Billington, 1990, 2004; Cooper and Kirkham, 1993; Bodnar *et al.*, 1995; Gosnell and Levine, 1996). Opioids do not enhance the actual sensory properties of palatable foods. Rather, exogenous opioids appear to enhance the hedonic value or perceived palatability of food (Pecina and Berridge, 2000). Mu opioid activation within the VTA can stimulate short-term intake of palatable food in non-deprived rats (Figlewicz, 2003) and can enhance motivated work (i.e. running) to obtain food (Noel and Wise, 1995), whereas mu-opioid receptor knockout mice exhibit diminished food-anticipatory activity (Kas *et al.*, 2004). The effect of the non-selective opiate agonist, butorphanol, was evaluated in the 'progressive ratios' paradigm (Rudski *et al.*, 1994). In this paradigm, rats receive a food reward after pressing a lever for a successively increasing number of times within a session. Thus, initially one lever press may result in a food reward, but the rat then has to press more often (i.e. work harder) for each successive food reward. Butorphanol enhanced responding for food pellets and, conversely, the opiate antagonist, naloxone, decreased this responding. The effect of naloxone could be reversed partially by food restriction, demonstrating that – similar to the VTA dopaminergic system – there is an interaction between nutritional status and this component of the reward circuitry. Likewise, B-endorphin knockout mice show decreased responding for regular chow, or sweet or fat chow (Low *et al.*, 2003); a similar result is observed in enkephalin knockout mice (Hayward *et al.*, 2002).

The influence of opioids to stimulate intake of preferred foods has been demonstrated in several studies; some examples follow. Naloxone specifically can inhibit intake of preferred sweetened chow versus regular chow (Levine *et al.*, 1995). Naloxone also decreases intake of the individually preferred diet when rats are offered high-fat and high-carbohydrate diets (Glass *et al.*, 1996). The longer-acting opiate antagonist, naltrexone, blocks the reinstatement of sucrose-diet feeding in rats that had prior access to a sucrose diet and then were returned to a less preferred diet (Levine *et al.*, 2002). Endogenous opioid-mediated preference for high-fat diets may have a genetic basis, since the administration of the dynorphin antagonist, norbinaltorphimine, has been shown to decrease intake of high-fat diet preferentially in Osborne–Mendel rats, but does not do so in non-fat-preferring Sprague–Dawley rats, when the rats are given a choice between a high-fat and a high-carbohydrate diet (Ookuma *et al.*, 1998). Thus, the CNS opioids play a potentially unique role in feeding behaviour, mediating the expression of food preferences when food choices are available.

Discrete opioid receptor populations within several CNS areas mediate the feeding effects of endogenous or exogenous opiate peptides. Mapping of c-fos in response to peripheral administration of butorphanol has revealed neuronal activation in the PVN, the NTS and the central nucleus of the amygdala (CeA) (Kim *et al.*, 2001). Opioid receptors in the NTS and PVN appear to be important in the total amount of energy content consumed, while opioid receptors in the CeA may be more important in affecting consumption influenced by the sensory or hedonic value of ingestates (Glass *et al.*, 1999). Thus, administration of naltrexone into the PVN suppresses both preferred and non-preferred diets, but administration of naltrexone into the CeA decreases intake of the preferred diet only (Glass *et al.*, 2000). Naltrexone administration into the NTS decreases deprivation-induced feeding and can decrease body weight (Kotz *et al.*, 1997). Functional links appear to exist between the opioid receptor populations of the NTS and CeA, since naltrexone administration into the NTS also increases dynorphin mRNA expression in the amygdala (Glass *et al.*, 2002). Although these observations have not yet been tied to behaviour, the efficacy of an opioid antagonist, in this case at the molecular level, reveals neural connections that intrinsically must be active. One speculative scenario would be that NTS opioidergic neurones facilitate deprivation-induced feeding and connection with the amygdala would enhance the rewarding aspect of the food in this condition. This finding is consistent with other studies implicating the amygdala in feeding (Petrovich *et al.*, 2002). Opiate administration directly into the CeA stimulates feeding (Gosnell *et al.*, 1986; Levine *et al.*, 2004). The NAc is another direct limbic target for opioids, with opiates stimulating feeding when administered directly into the NAc (e.g. Zhang *et al.*, 1998; Will *et al.*, 2003). Finally, mu opioids induce feeding when given into the VTA and expression of the behaviour is dependent on DA release in the NAc (MacDonald *et al.*, 2004). A limited number of observations suggest that there is interaction between these opioid receptor populations and intrinsic or extrinsic energy regulatory signals; insulin reverses feeding stimulated by intra-VTA administration of the mu opioid agonist DAMGO ([D-Ala2,*N*-MePhe4,Gly5-ol]-enkephalin) (Figlewicz, 2003) and ICV insulin reverses dynorphin agonist-stimulated feeding (Sipols *et al.*, 2002). The opioid, nociceptin, may enhance or sustain feeding by interacting with feeding-termination neuropeptide pathways such as α-melanocyte-stimulating hormone (MSH), oxytocin, or corticotrophin-releasing hormone (CRH) (Olszewski and Levine, 2004).

The Endocannabinoid System

Recent evidence supports the role of the endogenous cannabinoids (endocannabinoids, or ECs), anandamide and 2-arachidonoyl glycerol (2-AG), in food intake (Berry and Mechoulam, 2002; Fride, 2002; Harrold and Williams, 2003; Kirkham and Williams, 2004; DiMarzo, 2008; Solinas *et al.*, 2008). One conclusion that may be made with the current status of knowledge is that the ECs may be at the interface of CNS energy regulatory systems and the reward system (see, for example, Ravinet Trillou *et al.*, 2003; Kunos, 2007; Bellocchio *et al.*, 2008). There is some evidence demonstrating interaction with CNS leptin effects (Jo *et al.*, 2005; Farooqi *et al.*, 2007; Kunos, 2007) and also interaction with the mesocorticolimbic DA system (Geiger *et al.*, 2008). Genetic models of leptin (*ob/ob* mouse) or leptin receptor (Zucker *fa/fa* rat; *db/db* mouse) deficiency have higher hypothalamic levels of the EC, 2-AG (Thiemann *et al.*, 2008). Exogenous (ip) administration of leptin decreases ECs in the hypothalamus but not cerebellum of rats. However, ECs were not measured in the limbic circuitry in that study (DiMarzo *et al.*, 2001), although EC content is higher in limbic circuitry (brainstem, striatum and hippocampus) than in the diencephalon (Fride, 2002). Measurements in normal rats demonstrate increases in limbic content of both anandamide and 2-AG with food deprivation, with a modest increase of 2-AG in the hypothalamus (and no change of hypothalamic anandamide content). Further, direct administration of 2-AG into the NAc shell stimulates feeding significantly (Kirkham *et al*., 2002). Endocannabinoid type 1 (CB1) receptor content, assessed by quantitative autoradiography, is decreased in the hippocampus and NAc of rats fed a palatable diet for 10 weeks and the energy intake from the diet is correlated inversely with CB1 receptor density, interpreted as enhanced receptor–ligand interaction, i.e. increased EC activity (Harrold *et al*., 2002).

Protocols evaluating motivation or reward (as opposed to free-feeding measurements) certainly implicate ECs in feeding (Solinas *et al*., 2006, 2008; Mathes *et al*., 2008). Thus, the CB1 antagonist, SR141716 (rimonabant), decreases CPP by food (Chaperon *et al*., 1998), as well as self-administration of food (Arnone *et al*., 1997). It has been demonstrated that relapse to food intake is enhanced by a D3 agonist and this effect is blocked by the CB1 receptor antagonist, suggesting some synergy between the EC and DA pathways (Duarte *et al*., 2004; Thanos *et al*., 2008). Further, 'progressive ratio' performance for sucrose self-administration is decreased in CB1 knockout mice (Sanchis-Segura *et al*., 2004). Finally, LH-stimulation-induced feeding is increased by (exogenous) tetrahydrocannabinol (Trojniar and Wise, 1991) and decreased by CB1 antagonism (Deroche-Gamonet *et al*., 2001). Together, these studies provide evidence for a role of the ECs in the motivational aspect(s) of feeding.

In addition to the studies examining the potential leptin–EC connection, the EC interaction with ghrelin also has been addressed (Cummings and Shannon, 2003). It has been shown that a dose of CB1 antagonist, which does not decrease food intake on its own, reverses feeding induced by ghrelin administration into the PVN (Tucci *et al*., 2004). Furthermore, ghrelin binds to neurones of the VTA, where it triggers increased DA neuronal activity, synapse formation and DA turnover in the NAc in a growth hormone secretagogue 1 receptor-dependent way (Abizaid *et al*., 2006). Direct VTA administration of ghrelin also triggered feeding, while intra-VTA delivery of a selective growth hormone secretagogue 1 receptor antagonist blocked the orexigenic effect of circulating ghrelin and blunted rebound feeding following fasting. In addition, ghrelin- and ghrelin receptor (GHSR)-deficient mice showed attenuated feeding responses to restricted feeding schedules. Taken together, these data suggest that the mesolimbic reward circuitry is targeted by peripheral ghrelin to influence physiological mechanisms related to feeding.

As discussed above, a central neuroanatomical substrate for coordinating both reward inputs and energy circuitry inputs may be the LHA. The anatomical basis for this concept is well established, as the LHA receives direct and indirect limbic inputs and direct projections from the arcuate nucleus of the hypothalamus (which is a major target for candidate adiposity signals), as well as numerous intrahypothalamic and neuroendocrine inputs (Elias *et al*., 1998). Studies dating from the 1960s have demonstrated the capacity of animals to self-stimulate their brains electrically when electrodes are placed within specific sites of the LHA. This behavioural paradigm has been interpreted as representing 'activation of the reward circuitry within the CNS' (Wise, 1996). While the physiological meaning of this behavioural paradigm may be open to question, it is clear that the choice to self-stimulate the brain electrically in lieu of pursuing other activities or other stimuli reflects access to some powerfully motivating neural circuitry. Thus, studies of LHA self-stimulation, interpreted cautiously, may provide insights into the modulation of the CNS reward circuitry. Relevant to the present discussion are the facts that the self-stimulation activity can be enhanced by complete or partial food deprivation and that the threshold current necessary to sustain this behaviour is decreased in association with food deprivation (Carr, 1996). The question of which neurochemical or neuroendocrine substrate(s) mediate(s) this phenomenon has been pursued, with some interesting results. Enhanced opioidergic activity may be the intrinsic neurochemical change that mediates the shift in current threshold, since administration of naltrexone into the lateral ventricles can reverse the effect of fasting on threshold current shift within individual rats (Carr and Wolinsky, 1993).

Shizgal *et al.* (2001) have localized some of these food restriction-sensitive sites to the perifornical area of the LHA, a region that contains an abundance of neurones that synthesize melanin-concentrating hormone (MCH), an orexigenic neuropeptide (Broberger *et al.*, 1998). The LHA also contains neurones that co-synthesize dynorphin and orexin (Chou *et al.*, 2001). Both of these neuronal phenotypes might be altered in their activity in response to food restriction, as they receive projections from arcuate nucleus neurones that are a critical component of the hypothalamic calorie regulatory circuitry. Recent studies by Berthoud and colleagues demonstrate that in association with (NAc-induced) feeding, there is enhanced neuronal activation in some LHA neurones expressing orexin (Zheng *et al.*, 2003). Finally, we have localized receptors for both insulin and leptin within the LHA (Figlewicz, 2003) and have postulated that the LHA may be a direct target for these peptides after they are transported into the CNS. Collectively, this evidence suggests that the LHA is responsive to both signals of nutritional status and CNS inputs reflecting reward.

Other Elements

Other forebrain structures are implicated in both limbic function and food reward/motivation. The subthalamic nucleus has been shown to play a critical role in modulating food-related motivation (Baunez *et al.*, 2002). This nucleus is a component of the basal ganglia circuitry, within a functional loop that includes the NAc and ventral pallidum. The subthalamic nucleus is connected to the prefrontal cortex through a two-synapse pathway (subthalamic nucleus–substantia nigra–cortex), as well as through the pallidal loop (Maurice *et al.*, 1999). Bilateral lesion of the subthalamic nucleus results in an increased rate of eating food pellets, an increase in performance in the 'progressive ratios' paradigm and increased reinforcing properties of food-associated stimuli (Baunez *et al.*, 2002). These effects were situation-dependent and, therefore, not due to a non-specific enhancement of motor responding. Additionally, specific subcomponents of the cerebral cortex are involved integrally in taste recognition, taste memory and valuation and executive function in initiating ingestive decisions based on visual and olfactory cues (Rolls, 2004). Primary taste cortex (i.e. agranular insular cortex) has efferent connections to the orbitofrontal cortex (OFC) and the other major limbic areas, including NAc, LH and amygdala. Additionally, there are direct projections to the NTS and autonomic motor CNS structures. The OFC receives multimodal inputs including gustatory, olfactory, visual and somatosensory information. For example, some OFC neurones respond to the oral texture of fat (Rolls *et al.*, 1999). Outputs from this region of the cortex project to the striatum, the ventral midbrain and the sympathetic nervous system (Berthoud, 2002). In rats, electrical stimulation of the OFC initiates feeding (Bielajew and Trzcinska, 1994) and infusion of various neuropeptides or neurotransmitters into the OFC can alter respiratory quotient and thermogenesis (McGregor *et al.*, 1990a,b; Westerhaus and Loewy, 2001).

Information Derived from Imaging Techniques

Recent advances in imaging technologies, most notably functional magnetic resonance imaging (fMRI) and positron emission tomography (PET), highlight the need for delineating a potential relationship between the energy regulatory circuitry and limbic forebrain and cortex. These studies not only have confirmed previously identified CNS loci of appetitive and ingestive behaviour in animals, but have expanded greatly our knowledge of the roles that reward and palatability play in human ingestion and have identified their neural substrates (see Chapter 12). Previous studies in non-human primates indicated that neurones within the amygdala, OFC and hypothalamus responded to visual presentations of food (Rolls, 1994). Killgore *et al.* (2003) tested the cerebral responses (blood oxygen level-dependent activation) of normal-weight adult women to colour photographs of foods having different caloric content. Whereas photographs of all foods were associated with bilateral amygdala and ventromedial prefrontal cortex (especially OFC) activation in comparison to photographs of non-edible dining-related items, only the high-calorie foods resulted in activation within the medial and dorsolateral prefrontal cortex, thalamus, hypothalamus, corpus callosum and cerebellum. In contrast, low-calorie foods resulted in smaller regions of activation within the medial OFC, primary gustatory/somatosensory cortex and superior, middle and medial temporal regions of the brain. These findings suggest that the amygdala and OFC are involved in non-specific food recognition without regard to caloric value, whereas the medial prefrontal cortex, thalamus and hypothalamus may be central to the rewarding and motivating aspects of food stimuli.

A number of human imaging studies involving taste and ingestion similarly have extended our knowledge of neural substrates involved with food affinity. In a study of brain activity in response to glucose taste, Frank *et al.* (2003) used fMRI in a double-blind protocol in which they observed increased right medial OFC activation in healthy normal-weight adult women, relative to the response to artificial saliva. The marked OFC activation suggests that tastes with greater hedonic or emotional value are represented preferentially in this brain region. Gottfried *et al.* (2003) used fMRI to study hungry volunteers who were first presented with picture–odour pairings and then fed a meal specific to the presented odour until satiated, simultaneously lessening the subject's reported value of the picture. This devaluation was associated with strikingly decreased neural responses in the left dorsomedial amygdala and OFC to the picture after the meal in comparison to those just before the meal, and modest decreases in the ventral striatum, insula and cingulate. A speculative interpretation of this study is that satiety signals generated during feeding contribute to the decrease in the value of a food-related sensory cue, whether the cue is primary or secondary to a learned association. OFC activation has been observed to be low in obese (relative to lean) men in response to a satiating meal (Gautier *et al*., 2000). This latter study strongly suggests a role of the OFC in human energy balance.

Studies investigating localization of neurotransmitter signalling during meal consumption have also yielded noteworthy insights concerning food reward and reinforcement. Small *et al*. (2003) used labelled raclopride PET scanning following a 16-h fast and a favourite meal in normal volunteers to measure regional DA binding. Reduced DA binding was observed in the full versus hungry state in the dorsal striatum, indicating DA release on food consumption. The reduction in raclopride binding was correlated with meal pleasantness, but not with hunger before eating or satiety following the meal, indicating that the amount of dorsal striatal DA released correlates with pleasure. Using similar imaging methodology, Volkow *et al*. (2003) correlated DA release with eating behaviour survey results and food stimulation (smell and taste) in normal volunteers. Eating restraint scores were correlated positively with DA release to food stimulation and emotionality scores were correlated negatively with baseline D_2 receptors – all in the dorsal, not ventral, striatum. Hence, dorsal striatal DA is likely implicated in at least two different neurobiologic aspects of eating behaviour.

One obvious clinical application of these new imaging techniques is the assessment of neural substrates of body weight dysregulation and eating disorders. In a study using PET scanning to investigate brain DA involvement in pathologically obese individuals, Wang *et al*. (2001) found that D_2 receptor availability in these subjects was correlated inversely with body weight (in contrast to normal-weight controls), suggesting that decreased brain DA activity in the obese may well predispose them to excessive food intake (i.e. the greater the BMI, the fewer the DA receptors). The same investigators (Volkow *et al*., 2002) found an increase in dorsal striatal extracellular DA following non-consumed food display in normal-weight fasting subjects, further implicating the dorsal striatum as the neural substrate in the incentive properties of ingestion. Recently, the associations between striatal D_2 receptors and prefrontal metabolism in obese subjects suggest that decreases in striatal D_2 receptors may contribute to overeating via their modulation of striatal prefrontal pathways that participate in inhibitory control and salience attribution (Volkow *et al*., 2008). The association between striatal D_2 receptors and metabolism in somatosensory cortices (regions that process palatability) may underlie one of the mechanisms through which DA regulates the reinforcing properties of food.

Relation to Reinforcement of Addictive Behaviour

It is becoming abundantly clear that the reward circuitry involved in feeding appetitive behaviours is also associated intimately with, if not identical to, the neural substrate involved in reinforcement of many addictive behaviours. For example, assessing 90 female eating disorder patients and 115 healthy female controls, Shinohara *et al.* (2004) found that binge eating was associated with a defect (short allele) in the DA transporter gene similar to that observed in nicotine, cocaine and alcohol abuse, suggesting that dysregulation of DA reuptake which results in overstimulation of DA receptors may be a mechanism common to binging, obesity and substance abuse (Boutrel, 2008; Davis *et al.*, 2008). Furthermore, the same reductions in brain DA receptors observed by Wang *et al.* (2001) in the obese have also been reported in abusers of cocaine, methamphetamine, alcohol and heroin (Wang *et al.*, 2002). Davis *et al.* (2003) tested sensitivity to reward (STR) using food reward in normal, overweight and obese women. STR was correlated positively with measures of emotional overeating and overweight women (BMI > 25 kg/m^2) scored higher on STR than normal-weight women. However, obese women (BMI > 30 kg/m^2) were more anhedonic (i.e. lower on the STR scale) than overweight women. These findings suggest that activity of reward circuits in the brain may be implicated in the initial stages of intake-driven obesity but, on achieving frank obesity, some neuroadaptation to brain reward circuit overactivity may occur.

The valence of various rewards of different modalities (e.g. food and drugs) and how these rewards interact using at least part of the same neural substrate has also been the subject of numerous studies. In a comparison of carbohydrate snack (CHO) preference to money rewards during nicotine deprivation in female smokers, Spring *et al.* (2003) found that abstinent smokers worked harder for CHO rewards relative to money in comparison to non-smokers. They also worked harder for CHO during nicotine deprivation than when smoking, indicating that deprivation of one reward may increase the reinforcing value of the other. Whether there is an overlap in the neural circuitry of these rewards of differing modality remains to be addressed. However, the genetic basis of such interaction was the subject of a study by Lerman *et al.* (2004), who observed the effect of bupropion (a DA reuptake blocker) and the DA D_2 receptor gene (DRD2) on food reward in smokers. Subjects underwent a test of food reward before bupropion or abstinence and again after 3 weeks of bupropion or 1 week of abstinence. It was found that DRD2 A1 allele carriers exhibited greater food reward value after abstinence, which was attenuated by bupropion. Higher food reward levels predicted a 6-month weight increase in the placebo but not the bupropion group. Since the A1 allele renders D_2 receptors less able to bind DA, lowered neuronal DA activity in the face of reinforcement (nicotine) deprivation is thought to increase the rewarding value of food, leading to increased body weight. By promoting DA activity, bupropion could well decrease the value of food reward, hence lessen hedonically-driven overeating and subsequent weight gain. In a more general study, Hodgkins *et al.* (2004) investigated the relationship between drug abstinence and body weight change in adolescents in a treatment facility, finding a significant body weight and BMI increase with abstinence. These results suggest that patients seeking drug treatment may be substituting food for their drug of choice, leading to obesity. It then should follow that individuals who rely on food's rewarding aspects to maintain signalling in neural reward circuits should not need to use drugs for such signalling; hence, exceptional drug use ought not to be a problem in an obese population. This was borne out in a study by Kleiner *et al.* (2004), who found a significant inverse relationship between BMI and alcohol use in obese females, concluding that overeating may compete with alcohol for brain reward sites, making alcohol less reinforcing.

The collective point of these studies is that the multisensory experience of feeding and food choice in humans strongly activates the limbic forebrain, and activation patterns differ depending on the nutritional status (physiology) and degree of obesity (pathophysiology) of human subjects. These findings not only support conclusions obtained from animal studies but also shed light on which neural substrates should be investigated further in animal studies.

Crosstalk Between the Energy Regulatory and Reward Circuits

How does the energy regulatory circuitry communicate with the reward circuitry? A large cast of neuropeptide and neuroendocrine players has been identified in roles contributing to the regulation of food intake and body weight (Beck, 2000; Ahima and Osei, 2001). For many of these molecules, the locations of either neuronal cells of origin, or receptor-containing neuronal populations, have begun to be identified. Having this knowledge, it should be possible to test the effects of these signalling molecules in behavioural paradigms of reward and motivation. One might begin with the admittedly simple hypothesis that energy regulatory signals which are known to decrease feeding may decrease reward values of foods (or, conversely, orexigenic signals might enhance reward values of foods). In this context, whether leptin, the orexigenic neuropeptide Y (NPY) or the anorectic peptide corticotropin-releasing factor can modulate specifically food restriction-sensitive LHA sites of self-stimulation has been addressed (Shizgal *et al.*, 2001). Whereas leptin modulates LHA stimulation reward (Fulton *et al.*, 2000, 2004), NPY does not and CRH affects only LHA sites that are insensitive to chronic food restriction (Fulton *et al.*, 2002a,b). Likewise, it has been demonstrated that the food restriction-induced changes in LHA self-stimulation can be reversed by ICV insulin administration (Carr *et al.*, 2000). These findings argue for the usefulness of this experimental approach. However, the focus of much of the work to date has been on evaluating the effects of food deprivation or food restriction. Given the reasonable perspective that, in fact, the calorie regulatory circuitry has evolved to defend caloric intake rather than to inhibit it (Pi-Sunyer, 2003; Schwartz *et al.*, 2003), one may not be able to make inferences for overeating based on differences observed with feeding versus fasting. Not only is this issue crucial for those who study the calorie regulatory circuitry, it is also probably an important issue for the study of mechanisms underlying the overingestion of calories. The point here is that the delineation of mechanisms underlying food reward that is responsive to food restriction may not be useful for understanding motivation that is enhanced purely in response to palatable food. Thus, while food restriction is a relatively simple experimental model that has been studied extensively, its relevance for questions of overfeeding may be limited.

How does the reward circuitry talk to the energy regulatory circuitry? Given the central anatomical placement of the medial hypothalamus in the reward circuitry (and 'the energy regulatory circuitry' as defined here), the possibilities for direct and indirect inputs to the medial hypothalamus are numerous. One obvious interface is the LHA, with its abundant primary and secondary projections from the medial hypothalamus and its many links to other limbic structures and taste inputs. Thus, inputs from components of the reward circuitry such as the NAc or amygdala might synergize with orexigenic inputs from the medial hypothalamus when a palatable food is offered. These inputs (such as the NAc/LHA neuronal activation described above) might be substantial enough to result in feeding, even when medial hypothalamic orexigenic input is decreased (e.g. dessert at the end of a meal).

The energy regulatory circuitry has been clearly shown to be linked intricately with the reward circuitry. So, it is a logical, rather than a radical, proposition that energy regulatory signals communicate directly with the reward circuitry and vice versa. The concept has been put forward that the energy regulatory circuitry is part of a negative feedback loop which includes the generation of peripheral signals that reflect body adipose stores, and these signals act primarily at the medial hypothalamus to regulate the efferent components of this feedback loop, specifically food intake and energy balance. However, CNS anatomy suggests that the reward circuitry ultimately should not be viewed as functionally separate from the energy regulatory circuitry but as part of the loop. Inputs from the reward circuitry may not be just 'modulatory input' but are undoubtedly one critical component of the total CNS network that regulates food intake. The potential clinical and public health significance of the holistic understanding of how the food reward circuitry functions warrants future investigation.

Acknowledgements

Dianne Figlewicz Lattemann is supported by the Merit Review Programme of the Department of Veterans Affairs and NIH Grant RO1-DK40963. Alfred J. Sipols is supported by the University of Latvia, Riga, Latvia and Nicole M. Sanders is supported by a Transition Faculty Grant from the American Diabetes Association.

References

Abizaid, A., Liu, Z.W., Andrews, Z.B., Shanabrough, M., Borok, E., Elsworth, J.D., Roth, R.H., Sleeman, M.W., Picciotto, M.R., Tschöp, M.H., Gao, X.B. and Horvath, T.L. (2006) Ghrelin modulates the activity and synaptic input organization of midbrain dopamine neurons while promoting appetite. *Journal of Clinical Investigation* 116, 3229–3239.

Agmo, A., Galvan, A. and Talamantes, B. (1995) Reward and reinforcement produced by drinking sucrose: two processes that may depend on different neurotransmitters. *Pharmacology, Biochemistry and Behavior* 52, 403–414.

Ahima, R.S. and Osei, S.Y. (2001) Molecular regulation of eating behavior: new insights and prospects for therapeutic strategies. *Trends in Molecular Medicine* 7, 205–213.

Arase, K., Fisler, J.S., Shargill, N.S., York, D.A. and Bray, G.A. (1988) Intracerebroventricular infusions of 3-OHB and insulin in a rat model of dietary obesity. *American*

Journal of Physiology – Regulatory, Integrative and Comparative Physiology 255, R974–R981.

Arnone, M., Maruani, J., Chaperon, F., Thiebot, M.H., Poncelet, M., Soubrie, P. and LeFur, G. (1997) Selective inhibition of sucrose and ethanol intake by SR141716, an antagonist of central cannabinoid (CB1) receptors. *Psychopharmacology* 132, 104–106.

Bardo, M.T. and Bevins, R.A. (2000) Conditioned place preference: what does it add to our preclinical understanding of drug reward? *Psychopharmacology* 153, 31–43.

Baskin, D.G., Figlewicz Lattemann, D., Seeley, R.J., Woods, S.C., Porte, D. Jr and Schwartz, M.W. (1999) Insulin and leptin: dual adiposity signals to the brain for the regulation of food intake and body weight. *Brain Research* 848, 114–123.

Bassareo, V. and DiChiara, G. (1997) Differential influence of associative and non-associative learning mechanisms on the responsiveness of prefrontal and accumbal dopamine transmission to food stimulu in rats fed *ad libitum*. *Journal of Neuroscience* 17, 851–861.

Bassareo, V. and DiChiara, G. (1999) Modulation of feeding-induced activation of mesolimbic dopamine transmission by appetitive stimuli and its relation to motivational state. *European Journal of Neuroscience* 11, 4389–4397.

Baunez, C., Amalric, M. and Robbins, T.W. (2002) Enhanced food-related motivation after bilateral lesions of the subthalamic nucleus. *Journal of Neuroscience* 22, 562–568.

Beck, B. (2000) Neuropeptides and obesity. *Nutrition* 16, 916–923.

Bell, S.M., Stewart, R.B., Thompson, S.C. and Meisch, R.A. (1997) Food deprivation increases cocaine-induced conditioned place preference and locomotor activity in rats. *Psychopharmacology* 131, 1–8.

Bellocchio, L., Cervino, C., Pasquali, R. and Pagotto, U. (2008) The endocannabinoid system and energy metabolism. *Journal of Neuroendocrinology* 20, 850–857.

Berridge, K.C. (1996) Food reward: brain substrates of wanting and liking. *Neuroscience and Biobehavioral Reviews* 28, 309–369.

Berridge, K.C. and Robinson, T.E. (1998) What is the role of dopamine in reward: hedonic impact, reward learning, or incentive salience? *Brain Research Reviews* 28, 309–369.

Berry, E.M. and Mechoulam, R. (2002) Tetrahydrocannabinol and endocannabinoids in feeding and appetite. *Pharmacological Therapeutics* 95, 185–190.

Berthoud, H.R. (2002) Multiple neural systems controlling food intake and body weight. *Neuroscience and Biobehavioral Reviews* 26, 393–428.

Bielajew, C. and Trzcinska, M. (1994) Characteristics of stimulation-induced feeding sites in the sulcal prefrontal cortex. *Behavior Brain Research* 61, 29–35.

Bodnar, R.J., Glass, M.J., Ragnauth, A. and Cooper, M.L. (1995) General, mu and kappa opioid antagonists in the nucleus accumbens alter food intake under deprivation, glucoprivic and palatable conditions. *Brain Research* 700, 205–212.

Boutrel, B. (2008) A neuropeptide-centric view of psychostimulant addiction. *British Journal of Pharmacology* 154, 343–357.

Broberger, C., DeLecea, L., Sutcliffe, J.G. and Hokfelt, T. (1998) Hypocretin/orexin- and melanin-concentrating hormone-expressing cells form distinct populations in the rodent lateral hypothalamus: relationship to the neuropeptide Y and agouti gene-related protein systems. *Journal of Comparative Neurology* 402, 460–474.

Carr, K.D. (1996) Feeding, drug abuse, and the sensitization of reward by metabolic need. *Neurochemical Research* 21, 1455–1467.

Carr, K.D. (2002) Augmentation of drug reward by chronic food restriction: behavioral evidence and underlying mechanisms. *Physiology and Behavior* 76, 353–364.

Carr, K.D. and Wolinsky, T.D. (1993) Chronic food restriction and weight loss produce opioid facilitation of perifornical hypothalamic self-stimulation. *Brain Research* 607, 141–148.

Carr, K.D., Kim, G.Y. and Cabeza de Vaca, S. (2000) Hypoinsulinemia may mediate the lowering of self-stimulation thresholds by food restriction and streptozotocin-induced diabetes. *Brain Research* 863, 160–168.

Carroll, M.E. and Meisch, R.A. (1984) Increased drug-reinforced behavior due to food deprivation. *Advances in Behavioral Pharmacology* 4, 47–88.

Chaperon, F., Soubrie, P., Puech, A.J. and Thiebot, M.H. (1998) Involvement of central cannabinoid (CB1) receptors in the establishment of place conditioning in rats. *Psychopharmacology* 135, 324–332.

Chavez, M., Riedy, C.A., Van Dijk, G.V. and Woods, S.C. (1996) Central insulin and macronutrient intake in the rat. *American Journal of Physiology – Regulatory, Integrative and Comparative Physiology* 271, R727–R731.

Chou, T.C., Lee, C.E., Lu, J., Elmquist, J.K., Hara, J., Willie, J.T., Beuckmann, C.T., Chemelli, R.M., Sakurai, T., Yanagisawa, M., Saper, C.B. and Scammel, T.E. (2001) Orexin (hypocretin) neurons contain dynorphin. *Journal of Neuroscience* 21, RC168.

Cooper, S. and Kirkham, T. (1993) Opioid mechanisms in the control of food consumption and taste preferences. In: Herz, A. (ed.) *Handbook of Experimental Pharmacology*. Springer, Berlin, pp. 239–262.

Cummings, D.E. and Shannon, M.H. (2003) Roles for ghrelin in the regulation of appetite and body weight. *Archives of Surgery* 138, 389–396.

Davis, C., Strachan, S. and Berkson, M. (2003) Sensitivity to reward: implications for overeating and overweight. *Appetite* 42, 131–138.

Davis, C., Levitan, R.D., Kaplan, A.S., Carter, J., Reid, C., Curtis, C., Patte, K., Hwang, R. and Kennedy, J.L. (2008) Reward sensitivity and the D2 dopamine receptor gene: a case-control study of binge eating disorder. *Progress in Neuropsychopharmacology and Biological Psychiatry* 32, 620–628.

DeOlmos, J.S. and Heimer, L. (1999) The concepts of the ventral striatopallidal system and extended amygdala. *Annals of the New York Academy of Sciences* 877, 1–32.

Deroche-Gamonet, V., LeMoal, M., Piazza, P.V. and Soubrie, P. (2001) SR141716, a CB1 receptor antagonist, decreases the sensitivity to the reinforcing effects of electrical brain stimulation in rats. *Psychopharmacology* 157, 254–259.

DiMarzo, V. (2008) Targeting the endocannabinoid system: to enhance or reduce? *Nature Reviews Drug Discovery* 7, 438–455.

DiMarzo, V., Goparaju, S.K., Wang, L., Liu, J., Batkal, S., Jaral, Z., Fezza, F., Miura, G.I., Palmiter, R.D., Sugiura, T. and Kunos, G. (2001) Leptin-regulated endocannabinoids are involved in maintaining food intake. *Nature* 410, 822–825.

Drewnowski, A., Krahn, D.D., Demitrack, M.A., Nairn, K. and Gosnell, B.A. (1992) Taste responses and preferences for sweet high fat foods: evidence of opioid involvement. *Physiology and Behavior* 51, 371–379.

Drewnowski, A., Krahn, D.D., Demitrack, M.A., Nairn, K. and Gosnell, B.A. (1995) Naloxone, an opiate blocker, reduces the consumption of sweet high fat foods in obese and lean female binge eaters. *American Journal of Clinical Nutrition* 61, 1206–1212.

Duarte, C., Alonso, R., Bichet, N., Cohen, C., Soubrie, P. and Thiebot, M.-H. (2004) Blockade by the cannabinoid CB1 receptor antagonist, Rimonabant (SR141716), of the potentiation by quinelorane of food-primed reinstatement of food-seeking behavior. *Neuropsychopharmacology* 29, 911–920.

Elias, C.F., Saper, C.B., Maratos-Flier, E., Tritos, N.A., Lee, C., Kelly, J., Tatro, J.B., Hoffman, G.E., Ollman, M.M., Barsh, G.S., Sakurai, T., Yanagisawa, M. and Elmquist,

J.K. (1998) Chemically defined projections linking the mediobasal hypothalamus and the lateral hypothalamic area. *Journal of Comparative Neurology* 402, 442–459.

Everitt, B.J., Parkinson, J.A., Olmstead, M.C., Arroyo, M., Robledo, P. and Robbins, T.W. (1999) Associative processes in addiction and reward. The role of amygdala–ventral striatal subsystems. *Annals of the New York Academy of Sciences* 877, 412–438.

Farooqi, I.S., Bullmore, E., Keogh, J., Gillard, J., O'Rahilly,. S. and Fletcher, P.C. (2007) Leptin regulates striatal regions and human eating behavior. *Science* 317, 1355.

Figlewicz, D.P. (2003) Adiposity signals and food reward: expanding the CNS roles of insulin and leptin. *American Journal of Physiology – Regulatory, Integrative and Comparative Physiology* 284, R882–R892.

Figlewicz, D.P., Higgins, M.S., Ng-Evans, S.B. and Havel, P.J. (2001) Leptin reverses sucrose-conditioned place preference in food-restricted rats. *Physiology and Behavior* 73, 229–234.

Figlewicz, D.P., Evans, S.B., Murphy, J., Hoen, M. and Baskin, D.G. (2003). Expression of receptors for insulin and leptin in the ventral tegmental area/substantia nigra (VTA/SN) of the rat. *Brain Research* 964, 107–115.

Figlewicz, D.P., Bennett, J., Evans, S.B., Kaiyala, K., Sipols, A.J. and Benoit, S.B. (2004) Intraventricular insulin and leptin reverse place preference conditioned with high-fat diet in rats. *Behavioral Neuroscience* 118, 479–487.

Frank, G.K., Kaye, W.H., Carter, C.S., Brooks, S., May, C., Fissell, K. and Stenger, V.A. (2003) The evaluation of brain activity in response to taste stimuli – a pilot study and method for central taste activation as assessed by event-related fMRI. *Journal of Neuroscience Methods* 131, 99–105.

Fride, E. (2002) Endocannabinoids in the central nervous system – an overview. *Prostaglandins, Leukotrienes, and Essential Fatty Acids* 66, 221–233.

Fulton, S., Woodside, B. and Shizgal, P. (2000) Modulation of brain reward circuitry by leptin. *Science* 287, 125–128.

Fulton, S., Richard, D., Woodside, B. and Shizgal, P. (2002a) Interaction of CRH and energy balance in the modulation of brain stimulation reward. *Behavioral Neuroscience* 116, 651–659.

Fulton, S., Woodside, B. and Shizgal, P. (2002b) Does neuropeptide Y contribute to the modulation of brain stimulation reward by chronic food restriction? *Behavior Brain Research* 134, 157–164.

Fulton, S., Richard, D., Woodside, B. and Shizgal, P. (2004) Food restriction and leptin impact brain reward circuitry in lean and obese Zucker rats. *Behavior Brain Research* 155, 319–329.

Gallici, R., Pechnick, R.N., Poland, R.E. and France, C.P. (2000) Comparison of non-contingent vs. contingent cocaine administration on plasma corticosterone levels in rats. *European Journal of Pharmacology* 387, 59–62.

Gautier, J.F., Chen, K., Salbe, A.D., Bandy, D., Pratley, R.E., Heiman, M., Ravussin, E., Reiman, E.M. and Tataranni, P.A. (2000) Differential brain responses to satiation in obese and lean men. *Diabetes* 49, 838–846.

Geiger, B.M., Behr, G.G., Frank, L.E., Caldera-Siu, A.D., Beinfeld, M.C., Kokkotou, E.G. and Pothos, E.N. (2008) Evidence for defective mesolimbic dopamine exocytosis in obesity-prone rats. *FASEB Journal* 22, 2740–2746.

Glass, M.J., Grace, M., Cleary, J.P., Billington, C.J. and Levine, A.S. (1996) Potency of naloxone's anorectic effect in rats is dependent on diet preference. *American Journal of Physiology – Regulatory, Integrative and Comparative Physiology* 271, R217–R221.

Glass, M.J., Billington, C.J. and Levine, A.S. (1999) Opioids and food intake: distributed functional neural pathways? *Neuropeptides* 33, 360–368.

Glass, M.J., Billington, C.J. and Levine, A.S. (2000) Naltrexone administered to central nucleus of amygdala or PVN: neural dissociation of diet and energy. *American Journal of Physiology – Regulatory, Integrative and Comparative Physiology* 279, R86–R92.

Glass, M.J., Briggs, J.E., Billington, C.J., Kotz, C.M. and Levine, A.S. (2002) Opioid receptor blockade in the rat nucleus solitarius alters amygdala dynorphin gene expression. *American Journal of Physiology – Regulatory, Integrative and Comparative Physiology* 283, R161–R167.

Gosnell, B.A. and Levine, A.S. (1996) The stimulation of ingestive behavior by preferential and selective opioid agonists. In: Cooper, S.J. and Clifton, P.J. (eds) *Drug Receptor Subtypes and Ingestive Behavior*. Academic Press, London, pp. 147–166.

Gosnell, B.A., Morley, J.E. and Levine, A.S. (1986) Opioid-induced feeding: localization of brain sensitive sites. *Brain Research* 369, 177–184.

Gottfried, J.A., O'Doherty, J. and Dolan, R.J. (2003) Encoding predictive reward value in human amygdala and orbitofrontal cortex. *Science* 301, 1104–1107.

Harrold, J.A. and Williams, G. (2003) The cannabinoid system: a role in both the homeostatic and hedonic control of eating? *British Journal of Nutrition* 90, 729–734.

Harrold, J.A., Elliott, J.C., King, P.J., Widdowson, P.S. and Williams, G. (2002) Down-regulation of cannabinoid-1 (CB-1) receptors in specific extrahypothalamic regions of rats with dietary obesity: a role for endogenous cannabinoids in driving appetite for palatable food? *Brain Research* 952, 232–238.

Havel, P.J. (2000) Role of adipose tissue in body-weight regulation: mechanisms regulating leptin production and energy balance. *Proceedings of the Nutrition Society* 59, 359–371.

Hayward, M.D., Pintar, J.E. and Low, M.J. (2002) Selective reward deficit in mice lacking B-endorphin and enkephalin. *Journal of Neuroscience* 22, 8251–8258.

Hill, J.O., Wyatt, H.R., Reed, G.W. and Peters, J.C. (2003) Obesity and the environment: where do we go from here? *Science* 299, 853–855.

Hodgkins, C., Cahill, K.S., Seraphine, A.E., Frost-Pineda, K. and Gold, M.S. (2004) Adolescent drug addiction treatment and weight gain. *Journal of Addictive Diseases* 23, 55–65.

Hoebel, B., Hernandez, L., Schwartz, D.H., Mark, G.P. and Hunter, G.A. (1989) Microdialysis studies of brain norepinephine, serotonin, and dopamine release during ingestive behavior. Theoretical and clinical implications. *Annals of the New York Academy of Sciences* 575, 171–191.

Ikemoto, S. and Panksepp, J. (1996) Dissociations between appetitive and consummatory responses by pharmacological manipulations of reward-relevant brain regions. *Behavioral Neuroscience* 100, 331–345.

Jo, Y.H., Chen, Y.J., Chua, S.C. Jr, Talmage, D.A. and Role, L.W. (2005) Integration of endocannabinoid and leptin signaling in an appetite-related neural circuit. *Neuron* 48, 1055–1066.

Kas, M.J.H., van den Bos, R., Baars, A.M., Lubbers, M., Lesscher, H.M.B., Hillebrand, J.J.G., Schuller, A.G., Pintar, J.E. and Spruijt, B.M. (2004) Mu-opioid receptor knockout mice show diminished food-anticipatory activity. *European Journal of Neuroscience* 20, 1624–1632.

Kelley, A.E., Bakshi, V.P., Haber, S.N., Steininger, T.L., Will, M.J. and Zhang, M. (2002) Opioid modulation of taste hedonics within the ventral striatum. *Physiology and Behavior* 76, 365–377.

Killgore, W.D.S., Young, A.D., Femia, L.A., Bogorodzki, P., Rogowska, J. and Yurgelun-Todd, D.A. (2003) Cortical and limbic activation during viewing of high- versus low-calorie foods. *NeuroImage* 19, 1381–1394.

Kim, E.M., Shi, Q., Olszewski, P.K., Grace, M.K., O'Hare, E., Billington, C.J. and Levine, A.S. (2001) Identification of central sites involved in butorphanol induced feeding in rats. *Brain Research* 907, 125–129.

Kirkham, T.C. and Williams, C.M. (2004) Endocannabinoid receptor antagonists. *Treatments in Endocrinology* 3, 1–16.

Kirkham, T.C., Williams, C.M., Fezza, F. and Di Marzo, V. (2002) Endocannabinoid levels in rat limbic forebrain and hypothalamus in relation to fasting, feeding and satiation: stimulation of eating by 2-arachidonoyl glycerol. *British Journal of Pharmacology* 136, 550–557.

Kiyatkin, A.E. (1995) Functional significance of mesolimbic dopamine. *Neuroscience and Biobehavioral Reviews* 19, 573–598.

Kleiner, K.D., Gold, M.S., Frost-Pineda, K., Lenz-Brunsman, B., Perri, M.G. and Jacobs, W.S. (2004) Body mass index and alcohol use. *Journal of Addictive Diseases* 23, 105–118.

Kotz, C.M., Billington, C.J. and Levine, A.S. (1997) Opioids in the nucleus of the solitary tract are involved in feeding in the rat. *American Journal of Physiology – Regulatory, Integrative and Comparative Physiology* 41, R1028–R1032.

Krugel, U., Schraft, T., Kittner, H., Kiess, W. and Illes, P. (2003) Basal and feeding-evoked dopamine release in the rat nucleus accumbens is depressed by leptin. *European Journal of Pharmacology* 482, 185–187.

Kunos, G. (2007) Understanding metabolic homeostasis and imbalance: what is the role of the endocannabinoid system? *American Journal of Medicine* 120 (9 Suppl. 1), S18–S24.

Lepore, M., Vorel, S.R., Lowinson, J. and Gardner, E.L. (1995) Conditioned place preference induced by delta9-tetra-hydrocannabinol: comparison with cocaine, morphine, and food reward. *Life Sciences* 56, 2073–2080.

Lerman, C., Berrettini, W., Pinto, A., Patterson, F., Crystal-Mansour, S., Wileyto, E.P., Restine, S., Leonard, D.G.B., Shields, P.G. and Epstein, L.H. (2004) Changes in food reward following smoking cessation; a pharmacogenetic investigation. *Psychopharmacology* 174, 571–577.

Levine, A.S. and Billington, C.J. (1990) Opioids: are they regulators of feeding? *Annals of the New York Academy of Sciences* 575, 209–220.

Levine, A.S. and Billington, C.J. (1997) Why do we eat? A neural systems approach. *Annal Reviews of Nutrition* 17, 597–619.

Levine, A.S. and Billington, C.J. (2004) Opioids as agents of reward-related feeding: a consideration of the evidence. *Physiology and Behavior* 82, 57–61.

Levine, A.S., Weldon, D.T., Grace, M., Cleary, J.P. and Billington, C.J. (1995) Naloxone blocks that portion of feeding driven by sweet tastes in food restricted rats. *American Journal of Physiology – Regulatory, Integrative and Comparative Physiology* 268, R248–R252.

Levine, A.S., Grace, M.K., Cleary, J.P. and Billington, C.J. (2002) Naltrexone infusion inhibits the development of preference for a high sucrose diet. *American Journal of Physiology – Regulatory, Integrative and Comparative Physiology* 283, R1149–R1154.

Levine, A.S., Kotz, C.M. and Gosnell, B.A. (2003) Sugars and fats: the neurobiology of preference. *Journal of Nutrition* 133, 831S–834S.

Levine, A.S., Olszewski, P.K., Mullett, M.A., Pomonis, J.D., Grace, M.K., Kotz, C.M. and Billington, C.J. (2004) Intra-amygdalar injection of DAMGO: effects on c-Fos levels in brain sites associated with feeding behavior. *Brain Research* 1015, 9–14.

Low, M.J., Hayward, M.D., Appleyard, S.M. and Rubinstein, M. (2003) State-dependent modulation of feeding behavior by proopiomelanocortin-derived beta-endorphin. *Annals of the New York Academy of Sciences* 994, 192–201.

McBride, W.J., Murphy, J.M. and Ikemoto, S. (1999) Localization of brain reinforcement mechanisms: intracranial self-administration and intracranial place-conditioning studies. *Behavioral Brain Research* 101, 129–152.

MacDonald, A.F., Billington, C.J. and Levine, A.S. (2004) Alterations in food intake by opioid and dopamine signaling pathways between the ventral tegmental area and the shell of the nucleus accumbens. *Brain Research* 1018, 78–85.

McGregor, I.S., Menendez, J.A. and Atrens, D.M. (1990a) Metabolic effects obtained from excitatory amino acid stimulation of the sulcal prefrontal cortex. *Brain Research* 529, 1–6.

McGregor, I.S., Menendez, J.A. and Atrens, D.M. (1990b) Metabolic effects of neuropeptide Y injected into the sulcal prefrontal cortex. *Brain Research Bulletin* 24, 363–367.

Marinelli, M. and Piazza, P.V. (2002) Interaction between glucocorticoid hormones, stress, and psychostimulant drugs. *European Journal of Neuroscience* 16, 387–394.

Markou, A., Arroyo, M. and Everitt, B.J. (1999) Effects of contingent and non-contingent cocaine on drug-seeking behavior measured using a second-order schedule of cocaine reinforcement in rats. *Neuropsychopharmacology* 20, 542–555.

Mathes, C.M., Ferrara, M. and Rowland, N.E. (2008) Cannabinoid CB1 receptor antagonists reduce caloric intake by decreasing palatable diet selection in a novel dessert protocol in female rats. *American Journal of Physiology – Regulatory, Integrative and Comparative Physiology* 295, R67–R75.

Maurice, N., Deniau, J.M., Glowinski, J. and Thierry, A.M. (1999) Relationships between the prefrontal cortex and the basal ganglia in the rat: physiology of the cortico-nigral circuits. *Journal of Neuroscience* 19, 4674–4681.

Mokdad, A., Bowman, B., Ford, E., Vinicor, R., Marks, J. and Koplan, J. (2001) The continuing epidemics of obesity and diabetes in the United States. *Journal of the American Medical Association* 286, 1195–1200.

Morley, J.E., Levine, A.S., Yim, G.K.W. and Lowy, M.T. (1983) Opioid modulation of appetite. *Neuroscience and Biobehavioral Reviews* 7, 399–410.

Noel, M.B. and Wise, R.A. (1995) Ventral tegmental injections of a selective mu or delta opioid enhance feeding in food-deprived rats. *Brain Research* 673, 304–312.

Nonogaki, K., Abdallah, L., Goulding, E.H., Bonasera, S.J. and Tecott, L.H. (2003) Hyperactivity and reduced energy cost of physical activity in serotonin 5-HT(2C) receptor mutant mice. *Diabetes* 52, 315–320.

Olds, J. (1962) Hypothalamic substrate of reward. *Physiological Reviews* 42, 554–604.

Olszewski, P.K. and Levine, A.S. (2004) Minireview: characterization of influence of central nociceptin/orphanin FQ on consummatory behavior. *Endocrinology* 145, 2627–2632.

Ookuma, K., Barton, C., York, D.A. and Bray, G.A. (1998) Differential response to kappa-opioidergic agents in dietary fat selection between Osborne–Mendel and S5B/P1 rats. *Peptides* 19, 141–147.

Palmiter, R.D. (2007) Is dopamine a physiologically relevant mediator of feeding behavior? *Trends in Neuroscience* 30, 375–381.

Papp, M. (1988) Different effects of short- and long-term treatment with imipramine on the apomorphine- and food-induced place preference conditioning in rats. *Pharmacology, Biochemistry and Behavior* 30, 889–893.

Pecina, S. and Berridge, K.C. (2000) Opioid eating site in nucleus accumbens shell mediates food intake and hedonic 'liking': map based on micro-injection Fos plumes. *Brain Research* 863, 71–86.

Pecina, S., Cagniard, B., Berridge, K.C., Aldridge, J.W. and Zhuang, X. (2003) Hyperdopaminergic mutant mice have higher 'wanting' but not 'liking' for sweet rewards. *Journal of Neuroscience* 23, 9395–9402.

Petrovich, G.D., Setlow, B., Holland, P.C. and Gallagher, M. (2002) Amygdalo-hypothalamic circuit allows learned cues to override satiety and promote eating. *Journal of Neuroscience* 22, 8748–8753.

Pi-Sunyer, X. (2003) A clinical view of the obesity problem. *Science* 299, 859–860.

Ravinet Trillou, C., Arnone, M., Delgorge, C., Gonalons, N., Keane, P., Maffrand, J.-P. and Soubrie, P. (2003) Anti-obesity effect of SR141716, a CB1 receptor antagonist, in diet-induced obese mice. *American Journal of Physiology – Regulatory, Integrative and Comparative Physiology* 284, R345–353.

Richardson, N.R. and Gratton, A. (1996) Behavior-relevant changes in nucleus accumbens dopamine transmission elicited by food reinforcement: an electrochemical study in rat. *Journal of Neuroscience* 24, 8160–8169.

Robbins, T.W. and Everitt, B.J. (1996) Neurobehavioural mechanisms of reward and motivation. *Current Opinion in Neurobiology* 6, 228–236.

Rolls, E.T. (1994) Neural processing related to feeding in primates. In: Legg, C.R. and Booth, D.A. (eds) *Appetite: Neural and Behavioral Bases.* Oxford University Press, Oxford, UK, pp. 11–53.

Rolls, E.T. (2004) The functions of the orbitofrontal cortex. *Brain and Cognition* 55, 11–29.

Rolls, E.T., Critchley, H.D., Browning, A.S., Hernadi, I. and Lenard, L. (1999) Responses to the sensory properties of fat of neurons in the primate orbitofrontal cortex. *Journal of Neuroscience* 19, 1532–1540.

Rudski, J.M., Billington, C.J. and Levine, A.S. (1994) Butorphanol increases food reinforcement operant responding in satiated rats. *Pharmacology, Biochemistry and Behavior* 49, 843–847.

Salamone, J.D., Correa, M., Mingote, S. and Weber, S.M. (2003) Nucleus accumbens dopamine and the regulation of effort in food-seeking behavior: implications for studies of natural motivation, psychiatry, and drug abuse. *Journal of Pharmacology and Experimental Therapeutics* 305, 1–8.

Sanchis-Segura, C., Cline, B.H., Marsicano G., Lutz, B. and Spanagel, R. (2004) Reduced sensitivity to reward in CB1 knockout mice. *Psychopharmacology* 176, 223–232.

Saper, C.B., Chou, T.C. and Elmquist, J.K. (2002) The need to feed: homeostatic and hedonic control of eating. *Neuron* 36, 199–211.

Schultz, W. (2002) Getting formal with dopamine and reward. *Neuron* 36, 241–263.

Schultz, W. (2004) Neural coding of basic reward terms of animal learning theory, game theory, microeconomics and behavioural ecology. *Current Opinion in Neurobiology* 14, 139–147.

Schwartz, M.W. and Seeley, R.J. (1997) Seminars in medicine of the Beth Israel Deaconess Medical Center. Neuroendocrine responses to starvation and weight loss. *New England Journal of Medicine* 336, 1802–1811.

Schwartz, M.W., Woods, S.C., Seeley, R.J., Barsh, G.S., Baskin, D.G. and Leibel, R.L. (2003) Is the energy homeostasis system inherently biased toward weight gain? *Diabetes* 52, 232–238.

Shalev, U., Yap, J. and Shaham, Y. (2001) Leptin attenuates food deprivation-induced relapse to heroin seeking. *Journal of Neuroscience* 21, RC129.

Shalev, U., Grimm, J.W. and Shaham, Y. (2002) Neurobiology of relapse to heroin and cocaine seeking: a review. *Pharmacological Reviews* 54, 1–42.

Shinohara, M., Mizushima, H., Hirano, M., Shioe, K., Nakazawa, M., Hiejima, Y., Ono, Y. and Kanba, S. (2004) Eating disorders with binge-eating behaviour are associated with the s allele of the 3′-UTR VNTR polymorphism of the dopamine transporter gene. *Journal of Psychological Neurosciences* 29, 134–137.

Shizgal, P., Fulton, S. and Woodside, B. (2001) Brain reward circuitry and the regulation of energy balance. *International Journal of Obesity* 25 (Suppl. 5), S17–S21.

Sipols, A.J., Bayer, J., Bennett, R. and Figlewicz, D.P. (2002) Intraventricular insulin decreases kappa opioid-mediated sucrose intake in rats. *Peptides* 23, 2181–2187.

Small, D.M., Jones-Gotman, M. and Dagher, A. (2003) Feeding-induced dopamine release in dorsal striatum correlates with meal pleasantness ratings in healthy human volunteers. *NeuroImage* 19, 1709–1715.

Solinas, M., Justinova, Z., Goldberg, S.R. and Tanda, G. (2006) Anandamide administration alone and after inhibition of fatty acid amide hydrolase (FAAH) increases dopamine levels in the nucleus accumbens shell in rats. *Journal of Neurochemistry* 98, 408–419.

Solinas, M., Goldberg, S.R. and Piomelli, D. (2008) The endocannabinoid system in brain reward processes. *British Journal of Pharmacology* 154, 369–383.

Spring, B., Pagoto, S., McChargue, D., Hedeker, D. and Werth, J. (2003) Altered reward value of carbohydrate snacks for female smokers withdrawn from nicotine. *Pharmacology, Biochemistry and Behavior* 76, 351–360.

Swerdlow, N.R., Vander Kooy, D., Koob, G.F. and Wenger, J.R. (1983) Cholecystokinin produces conditioned place-aversions, not place preferences, in food-deprived rats: evidence against involvement in satiety. *Life Sciences* 32, 2087–2093.

Thanos, P.K., Michaelides, M., Ho, C.W., Wang, G.J., Newman, A.H., Heidbreder, C.A., Ashby, C.R. Jr, Gardner, E.L. and Volkow, N.D. (2008) The effects of two highly selective dopamine D3 receptor antagonists (SB-277011A and NGB-2904) on food self-administration in a rodent model of obesity. *Pharmacology, Biochemistry, and Behavior* 89, 499–507.

Thiemann, G., van der Stelt, M., Petrosino, S., Molleman, A., Di Marzo, V. and Hasenöhrl, R.U. (2008) The role of the CB1 cannabinoid receptor and its endogenous ligands, anandamide and 2-arachidonoylglycerol, in amphetamine-induced behavioural sensitization. *Behavioral Brain Research* 187, 289–296.

Trojniar, W. and Wise, R.A. (1991) Facilitory effect of delta 9-tetrahydrocannabinol on hypothalamically-induced feeding. *Psychopharmacology* 103, 172–176.

Tucci, S.A., Rogers, E.K., Korbonits, M. and Kirkham, T.C. (2004) The cannabinoid CB1 receptor antagonist SR141716 blocks the orexigenic effects of intrahypothalamic ghrelin. *British Journal of Pharmacology* 143, 520–523.

Volkow, N.D., Wang, G.-J., Fowler, J.S., Logan, J., Jayne, M., Franceschi, D., Wong, C., Gatley, S.J., Gifford, A.N., Ding, Y.S. and Pappas, N. (2002) 'Non-hedonic' food motivation in humans involves dopamine in the dorsal striatum and methylphenidate amplifies this effect. *Synapse* 44, 175–180.

Volkow, N.D., Wang, G.-J., Maynard, L., Jayne, M., Fowler, J.S., Zhu, W., Logan, J., Gatley, S.J., Ding, Y.-S., Wong, C. and Pappas, N. (2003) Brain dopamine is associated with eating behaviors in humans. *International Journal of Eating Disorders* 33, 136–142.

Volkow, N.D., Wang, G.J., Telang, F., Fowler, J.S., Thanos, P.K., Logan, J., Alexoff, D., Ding, Y.S. and Wong, C. (2008) Low dopamine striatal D2 receptors are associated with prefrontal metabolism in obese subjects: possible contributing factors. *NeuroImage* 42, 1537–1543.

Wang, G.-J., Volkow, N.D., Logan, J., Pappas, N.R., Wong, C.T., Zhu, W., Netusil, L. and Fowler, J.S. (2001) Brain dopamine and obesity. *Lancet* 357, 354–357.

Wang, G.-J., Volkow, N.D. and Fowler, J.S. (2002) The role of dopamine in motivation for food in humans: implications for obesity. *Expert Opinion on Therapeutic Targets* 6, 601–609.

Westerhaus, M.J. and Loewy, A.D. (2001) Central representation of the sympathetic nervous system in the cerebral cortex. *Brain Research* 903, 117–127.

Will, M.J., Franzblau, E.B. and Kelley, A.E. (2003) Nucleus accumbens μ-opioids regulate intake of a high-fat diet via activation of a distributed brain network. *Journal of Neuroscience* 23, 2882–2888.

Williams, G., Bing, C., Cai, X.J., Harrold, J.A., King, P.J. and Liu, X.H. (2001) The hypothalamus and the control of energy homeostasis: different circuits, different purposes. *Physiology and Behavior* 74, 683–701.

Wilson, C., Nomikos, G.G., Collu, M. and Fibiger, H.C. (1995) Dopaminergic correlates of motivated behavior: importance of drive. *Journal of Neuroscience* 15, 5169–5178.

Wilson, J.M., Nobrega, J.N., Corrigal, W.A., Coen, K.M. and Kish, S.J. (1994) Amygdala dopamine levels are markedly elevated after self – but not passive – administration of cocaine. *Brain Research* 668, 39–45.

Wise, R.A. (1988) Psychomotor stimulant properties of addictive drugs. *Annals of the New York Academy of Sciences* 537, 228–234.

Wise, R.A. (1996) Addictive drugs and brain stimulation reward. *Annual Review of Neuroscience* 19, 319–340.

Wise, R.A. (2002) Brain reward circuitry: insights from unsensed incentives. *Neuron* 36, 229–240.

Wise, R.A. and Hoffman, D.C. (1992) Localization of drug reward mechanisms by intracranial injections. *Synapse* 10, 247–263.

Woods, S.C., Lotter, E.C., McKay, L.D. and Porte, D. Jr (1979) Chronic intracerebroventricular infusion of insulin reduces food intake and body weight of baboons. *Nature* 282, 503–505.

Yeomans, M.R., Wright, P., Macleod, H.A. and Critchley, J.A.J.H. (1990) Effect of nalmefene on feeding in humans. Dissociation of hunger and palatability. *Psychopharmacology* 100, 426–432.

Zhang, M., Gosnell, B.A. and Kelley, A.E. (1998) Intake of high fat food is selectively enhanced by mu opioid receptor stimulation within the nucleus accumbens. *Journal of Pharmacology and Experimental Therapeutics* 285, 908–914.

Zheng, H., Corkern, M., Stoyanova, I., Patterson, L.M., Tian, R. and Berthoud, H.-R. (2003) Appetite-inducing accumbens manipulation activates hypothalamic orexin neurons and inhibits POMC neurons. *American Journal of Physiology – Regulatory, Integrative and Comparative Physiology* 284, R1436–R1444.

12 Embracing Complexity: The Emergence of Functional Neuroimaging and Other Methodologies to Study the Role of the Human Brain in the Pathophysiology of Obesity

P. Antonio Tataranni, Nicola Pannacciulli, Duc Son NT Le and Angelo Del Parigi

National Institute of Diabetes and Digestive and Kidney Diseases, National Institute of Health, Department of Health and Human Services, Phoenix, USA

Obesity: Metabolic or Neurological Disease?

It is inherently difficult to study the aetiology of obesity in humans. The methodologies available to measure various components of daily energy metabolism are either barely precise (energy expenditure) or profoundly inaccurate (energy intake) and thus unable to detect the small differences in energy balance that, when chronically sustained, are likely to be responsible for the development of obesity in the majority of people.

For the past 30 years, regulation of energy expenditure has been a dominant topic in human obesity research. The majority of studies have been conducted under the controlled, artificial conditions of a metabolic study unit and often in the resting state. It is only in the past decade, with the advent of the doubly-labelled water technique (Schoeller and Taylor, 1987), that a limited number of studies have started to provide information on energy expenditure in individuals who are unencumbered by the confines of the laboratory setting. However, to date, these areas of research have not produced convincing evidence that abnormal regulation of energy expenditure is either a common or a major risk factor for weight gain in humans.

The study of molecular mechanisms and resulting behaviours that underlie energy intake in humans has been even less conclusive. First of all, there is surprisingly sparse experimental proof that people who are at risk of obesity eat in excess of their daily energy requirements (Stunkard *et al.*, 1999; Tataranni *et al.*,

1999). Secondly, there is even uncertainty about the macronutrient preferences that may predispose to weight gain.

The principal risk for weight gain is mainly the consequence of eating in excess of daily energy expenditure (Tataranni, 2000). Consequently, the brain, which controls eating behaviour, must play a major role in the aetiology of the disease. This conclusion is based not only on deductive reasoning, but also on the fact that complex neuropeptidergic pathways control energy balance. This finding supports that body fat content is, at least in part, under non-conscious homeostatic control in the hypothalamus (Gao and Horvath, 2007). However, animals and humans seldom eat in response to acute changes in energy balance. Eating is not a simple, stereotypical behaviour. It requires a set of tasks to be carried out by the central and peripheral nervous systems to coordinate the initiation of a meal episode, procurement of food, consumption of the procured food and termination of the meal (Berthoud and Morrison, 2008). Most of these tasks are behaviours learned after weaning. Accordingly, there is now universal recognition that the hypothalamus is not likely to be the only, or even the major, compartment of the brain involved in the control of eating behaviour.

Thus, obesity, once the prototypical metabolic disorder, is also being recognized increasingly as a neurological disease due to inherited and localized neurochemical defects (Bray, 2004). By subscribing to the notion that the brain plays a critical role in the control of energy homeostasis, and ultimately the genesis of obesity, one must acknowledge that the greatest challenge, following the identification of the genetic make-up of obese individuals, will be to understand how these molecular defects work together to alter the neurophysiology of those regions of the brain that control energy balance.

Functional Neuroimaging and the Study of Human Eating Behaviour

The investigation of the complex interplay among brain regions involved in both the homeostatic and hedonic regulation of eating behaviour has been addressed by a relatively new research tool, referred to as functional neuroimaging (FN), which allows for *in vivo* whole-brain evaluation of the neuronal response to several processes, including food-related stimuli. FN encompasses a number of non-invasive brain imaging techniques and measurements of local neuronal activity that explore patterns of brain activation associated with cognitive and other behavioural processes. Developed primarily to study the functional architecture of the normal living brain, neuroimaging is being used increasingly to study neurological and psychiatric disorders (Rolls, 2007). Among FN techniques, positron emission tomography (PET) and functional magnetic resonance imaging (fMRI) have emerged as new tools to search for regions of the brain that are involved in the regulation of eating behaviours and may be relevant to the pathophysiology of obesity (Tataranni and Del Parigi, 2003).

The use of FN provides the fundamental advantage of allowing investigation of the whole brain, thus making it possible to study the entire system rather than restricting the investigation to preselected regions of interest. Obviously, imaging of the human brain also has significant intrinsic limitations, which have been discussed extensively elsewhere (Reiman *et al.*, 1997). While different kinds of functional imaging systems have been improved in terms of spatial resolution (i.e. their ability to distinguish signals that are close together), temporal resolution (i.e. their ability to characterize and contrast signals over short durations) and sensitivity (i.e. their ability to detect processes that occur in minute concentrations), the scale of certain brain events is still beyond the capability of most FN techniques. Furthermore, while the range of neurochemical processes that can be assessed in the living human brain by these techniques is growing, it still remains limited. Finally, while FN provides information about the neuroanatomical correlates of normal and abnormal behaviours, it does not indicate the extent to which they are necessary or sufficient for the behaviour of interest.

There is More to it than just the Hypothalamus: The Discovery of Putative Orexigenic and Anorexigenic Brain Networks

Food intake is not influenced solely by physiological signals for hunger and satiety but is an essential human activity which relies on the interaction between homeostatic regulation and hedonic pleasure. While the unconscious homeostatic need to eat aimed at ensuring adequate nutrition plays an important role in the regulation of food intake and body weight, one of the most potent drives for feeding is its rewarding nature (see Chapter 11).

Non-conscious regulation of energy intake: the role of the hypothalamus

Although the role of the hypothalamus in the non-conscious regulation of energy homeostasis has long been established, remarkable progress in our understanding of the neurobiological complexity of the hypothalamic pathways involved in the regulation of body weight has been seen in recent years (Gao and Horvath, 2008). Studies in mice, genetically engineered to suppress or overexpress specific gene products, indicate that food intake and body weight regulation depend on the balance between orexigenic (neuropeptide Y [NPY], agouti-related peptide [AgRP], melanin-concentrating hormone [MCH], orexins and ghrelin) and anorexigenic (pro-opiomelanocortin/α-melanocyte-stimulating hormone [POMC/α-MSH], cocaine- and amphetamine-related transcript [CART]) neurotransmitters and their receptors. These central pathways are, in turn, modulated by peripheral hormones secreted by the stomach (e.g. ghrelin, cholecystokinin [CCK]), intestine (e.g. peptide YY [PYY]), pancreas (e.g. insulin) and adipose tissue (e.g. leptin).

FN studies have contributed only minimally to translating this neurochemical complexity from animal to human neurophysiology. Indeed, the hypothalamus is a small organ located deep in the brain, lining the lateral walls of the third ventricle, and surrounded by a very rich vascular network. For these reasons, the study of the hypothalamus using FN techniques poses some technical problems, mostly related to spatial resolution and accuracy of the image deformation algorithms.

Despite these limitations, functional imaging of the hypothalamic response to caloric intake in humans has been pursued in a number of studies. Liu *et al.* (2000) reported that, 7–12 min after the administration of a glucose load, a profound and sustained (approximately 10 min) decrease in neuronal activity was detectable by fMRI in two distinct regions of the hypothalamus (possibly corresponding to the ventromedial nucleus and paraventricular nucleus). This was confirmed in a study from Smeets *et al.* (2005), who reported a significant decrease of hypothalamic fMRI signal, which was dose-dependent (i.e. the larger the glucose load, the more pronounced the signal reduction) and possibly related to decreased neuronal activity of glucose-sensitive neurones located in the lateral hypothalamic area. Moreover, the time course of the decrease in neuronal activity of the hypothalamus following glucose ingestion (characterized by a rapid onset within a few minutes followed by a prolonged duration of over 30 min) suggests a possible association with changes in circulating insulin concentrations. This is in agreement with a similar decrease in neuronal activity in the hypothalamic region observed in response to liquid food (Tataranni *et al.*, 1999; Gautier *et al.*, 2000), but not after solid food (Small *et al.*, 2001), using measurements of local neuronal activity by PET. Importantly, a similar response also can be observed in the hypothalamus of Sprague–Dawley rats (Mahankali *et al.*, 2000). The exact neurophysiological mechanisms underlying this finding are unclear. It is possible that the meal inhibits hypothalamic neuronal activity directly, which may be elevated in a state of hunger. Alternatively, the meal may activate inhibitory pathways (prefrontocortical hypothalamic pathways), which, in turn, suppress the neuronal activity of the hypothalamus (Del Parigi *et al.*, 2002a). It is also unclear whether it is a component of the meal itself (glucose or other macronutrients) or the physiological response to the meal (insulin, other gut hormones, or autonomic nervous system afferent signals) that mediates the observed hypothalamic response. More importantly, from these initial studies, it is not possible to resolve if these responses relate primarily to the role of the hypothalamus in the regulation of glucose metabolism or energy homeostasis.

Reward as a drive for feeding: the role of extrahypothalamic brain regions

While the subconscious control of energy homeostasis by these highly organized orexigenic and anorexigenic hypothalamic pathways plays an important role in the regulation of food intake and body weight, there is general agreement that much remains to be understood about the conscious control of eating behaviour. For example, we know that the relative abundance of food and its reward value often override the physiologic signals of hunger and satiety (Grill *et al.*, 2007; Berthoud and Morrison, 2008). On the other hand, the rewarding effects of food are modulated by the internal state, such that a food that is pleasurable when one is hungry may be unpleasant after satiety.

Psychophysiology of food reward

The psychophysiology of food reward is comprised of at least two phases: anticipation and consumption (Finlayson *et al.*, 2007; Dillon *et al.*, 2008). Anticipation is elicited usually by the presentation of a sensory cue, which reliably signals the forthcoming delivery of a rewarding stimulus. Once food reaches the mouth, it is smelled and tasted before it is ingested. The neuroanatomical correlates of taste and olfaction, which are known to be primary reinforcers of food intake, have been fairly well described (Zald and Pardo, 2000). However, it remains unclear as to where in the brain the interaction between taste and olfaction takes place to produce the perception of flavour (Small *et al.*, 2004), which ultimately provides the true integrated sensory assessment of the chemical/biological suitability of a certain food for ingestion. As eating is often driven by the hedonic value of food, the brain response to the affective component of taste and olfaction may contribute to what and how much we eat. Predictably, the hedonic aspects of chemosensory stimulation are represented mainly in limbic and paralimbic areas (Zald and Pardo, 1997; O'Doherty *et al.*, 2001).

Taste also can affect the reward value of food during ingestion. The phenomenon called 'sensory-specific satiety' describes the decreasing firing rate (habituation) of specific sensory neurones in response to the prolonged application of the same, but not a novel, stimulus (Rolls, 2007). Sensory-specific satiety is thought to modulate food consumption in rodents, non-human primates and humans (Critchley and Rolls, 1996; Rolls *et al.*, 1996; Kringelbach *et al.*, 2003). Indeed, an inverse relationship between orbitofrontal neuronal activity and subjective pleasantness rating when a liquid food is eaten to satiety has been reported in humans (Kringelbach *et al.*, 2003; Burke *et al.*, 2008). In addition, a study aimed at assessing the change in brain activity associated with a transition from a state of high motivation to eat food (chocolate) to a state of aversion (Small *et al.*, 2001) revealed a shift in brain structures recruited selectively to respond to this stimulus (i.e. from the primary gustatory cortex, striatum, midbrain, subcallosal region and caudomedial orbitofrontal cortex to the parahippocampal gyrus, caudolateral orbitofrontal and prefrontal cortex). These findings provide evidence for an additional role of brain dopaminergic pathways in the regulation of food intake.

The role of dopamine in the regulation of eating behaviour

Involvement of dopamine (DA) in eating behaviour has been strongly suggested by both animal and human studies. Indeed, neurotoxin-induced degeneration of the dopamine system provokes aphagia in rats, causing them to die from starvation (Ungerstedt, 1971). Similarly, administration of dopamine antagonists in rats impairs the reinforcing and rewarding value of food, thus reducing food intake (Wise *et al.*, 1978).

DA action in the control of feeding behaviour is mediated by different receptor subtypes. D_1-like receptors, including D_1 and D_3 subtypes, are related to the regulation of meal size and duration, whereas D_2-like receptors, including D_2, D_4 and D_5 subtypes, are related to the regulation of the rate of feeding (Wang *et al.*, 2002a). Consistently, DA seems to exert its effects on different features of food intake regulation in a site-specific manner. In fact, its action in the hypothalamus, where DA inhibits expression and activity of orexigenic NPY and stimulates expression of anorectic peptide precursor POMC, is likely to be related to regulation of the duration of eating episodes (Palmiter, 2007).

By contrast, mesolimbic DA is thought to participate in food intake regulation by modulating motivational and rewarding processes. Animal studies have shown previously that DA antagonism in the dorsal striatum produces a steady decline in food intake and reward, which is reversed by restoration of DA production (Szczypka *et al.*, 2001). In accordance with these findings from animal models, FN studies in humans have demonstrated that both overall neuronal activity (Small *et al.*, 2001) and DA release (Small *et al.*, 2003) in the dorsal striatum while eating a favourite meal are associated strongly with the perceived pleasantness of the meal, thus confirming that DA release in the dorsal striatum is related to food reward. Similarly, extracellular DA changes with food stimulation in the dorsal striatum have been reported to correlate with restraint and emotionality (Small *et al.*, 2003; Volkow *et al.*, 2003), thus suggesting that striatal DA is involved with motivational and emotional variables regulating eating behaviour. In summary, both animal and human studies point to a central role for brain DA, particularly in the dorsal striatum, in the regulation of food intake by modulating appetitive motivational processes. The association between low striatal D_2 receptors in obesity in somatosensory cortices (i.e. regions that process palatability) may underlie one of the mechanisms through which DA regulates the reinforcing properties of food (Volkow *et al.*, 2008).

Putative brain networks integrating unconscious and conscious regulation of energy intake

One of the most significant contributions of FN studies thus far has been the identification of extrahypothalamic orexigenic and anorexigenic brain networks and the initial exploration of the interaction between hunger, satiety and reward responses. For example, several regions of the brain which previously have been demonstrated to participate in motivational, decisional and rewarding processing, have been associated also with hunger or satiety and, therefore, they are likely to play a role in the conscious hedonic control of food intake as a stimulus for pleasure and reward. Specifically, we have demonstrated that hunger, as elicited by a 36-h fast, is associated with increased neuronal activity in the hypothalamus, insular, orbitofrontal and anterior cingulate cortices, striatum, hippocampal and parahippocampal formations, precuneus, thalamus and cerebellum (Tataranni *et al.*, 1999). In contrast, satiety, as induced by the administration of a liquid meal providing 50% of the daily resting metabolic rate, is associated with increased neuronal activity in the dorsolateral and ventromedial prefrontal cortices (Tataranni *et al.*, 1999). Additional studies are needed to determine the extent to which each region participates in conscious or non-conscious information processing, hedonic or non-hedonic contributions to the regulation of food intake, or aspects of hunger and satiety not involved in the regulation of food intake.

By using a different paradigm based on the visual presentation of images of preferred food items before and after eating to satiety, the participation of such brain areas in a network underlying feeding behaviour has been further confirmed (Hinton *et al.*, 2004). Indeed, areas of increased activation, as elicited by

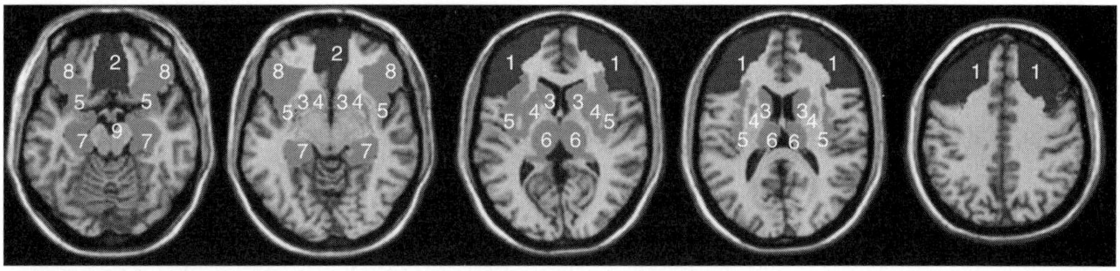

Anorexigenic domain	References	Orexigenic domain	References
Dorsolateral prefrontal cortex (1)	a, b, c	Dorsal striatum (caudate, putamen) (3,4)	a, b, c
Ventromedial prefrontal cortex (2)	a, b, c	Insular cortex (5)	a, b, c
		Thalamus (6)	a, b
		Hippocamus and parahippocampal gyrus (7)	a
		Medial orbitofrontal cortex (8)	c
		Hypothalamus (9)	a, b

a, Tataranni *et al.*, 1999
b, Small *et al.*, 2001
c, Hinton *et al.*, 2004

Fig. 12.1. Neuroanatomical correlates of hunger and satiety.

the visual task, included the hypothalamus, amygdala, striatum and insular and anterior cingulate cortices during the state of hunger and lateral prefrontal and temporal cortices during the state of satiety.

The engagement of these limbic, neocortical and chemosensory areas in orexigenic and anorexigenic brain networks has also been confirmed by a study demonstrating that eating a preferred food (i.e. chocolate) beyond satiety causes a shift in brain structures recruited selectively to respond to this stimulus. Specifically, limbic/paralimbic areas (including subcallosal region, caudomedial oribitofrontal cortex, insula/operculum, striatum and midbrain) are activated mainly when the motivation to eat is high, whereas other regions (including prefrontal areas, but also parahippocampal gyrus and caudolateral orbitofrontal cortex) are engaged when the motivational state is low due to satiety (Small *et al.*, 2001).

In summary, the study of the human brain response to the ingestion of liquid (Tataranni *et al.*, 1999) and solid food (Small *et al.*, 2001) and to the sight of food images before and after eating to satiety (Hinton *et al.*, 2004) has revealed the existence of an orexigenic domain (represented mainly by limbic and paralimbic areas, including the orbitofrontal and insular cortices, anterior cingulate and hypothalamic region) and a satiety domain (represented almost exclusively by prefrontal areas, Fig. 12.1). These findings also reveal that the functional neuroanatomy of hunger/satiety overlaps partially with the functional neuroanatomy of reward.

Atypical Responses in Obesity and Other Eating Disorders

Atypical neuronal responses to food-related stimuli of some brain areas involved in the regulation of eating behaviour have been reported in FN studies. Interestingly, it has been shown in both fMRI and PET studies that the increase in hypothalamic activity elicited by the state of hunger is significantly higher in obese subjects compared to lean individuals (Gautier *et al*., 2000). Several other subcortical and cortical brain regions, in addition to the hypothalamus, have been shown to be engaged abnormally in obese individuals (Del Parigi *et al*., 2002a). In PET studies, it has been reported that obese subjects have abnormal brain responses to both hunger (after a 36-h fast) and satiety (after ingesting a satiating liquid meal). In response to hunger versus satiety, obese individuals showed a greater neuronal activation in several limbic/paralimbic areas, including the insula, hippocampus and orbitofrontal cortex compared to lean individuals. In response to satiety versus hunger, greater activation was observed in the dorso- and ventrolateral prefrontal areas in obese subjects compared to lean individuals. These responses generally were consistent in men and women (Del Parigi *et al*., 2002a).

Obese subjects also have higher resting metabolic activity (as measured by [^{18}F]fluorodeoxyglucose PET) in the vicinity of the post-central area on the lateral surface of the parietal cortex bilaterally, which may suggest a higher sensitivity to food palatability (i.e. taste, consistency) in obesity (Wang *et al*., 2002b; Small *et al*., 2003). In contrast, Karhunen *et al*. (1997) reported previously that obese subjects had lower baseline cerebral blood flow (CBF), a marker of local neuronal activity (as evaluated by ^{99}mTc-ethyl-cysteine-dimer single photon emission computed tomography [SPECT]), in both parietal and temporal cortices compared to normal-weight individuals. They also suggested that the higher increase in CBF of these cortical areas in response to food exposure in obese subjects might well be due to an initially lower neuronal activity at rest (Karhunen *et al*., 2000).

Moreover, it has been suggested that a dysfunction of the dopaminergic brain may have a role in the pathophysiology of obesity. In particular, it has been suggested that abnormalities in dopaminergic transmission can be evidenced in obese individuals even when the brain is idling (Wang *et al*., 2001). Using PET and ^{11}C-raclopride, a specific ligand for the dopamine type 2 receptor (DRD2), they found that the availability of DRD2 is decreased in the dorsal striatum of obese individuals, a neurochemical feature associated previously with other reward deficiency syndromes (Wang *et al*., 2001). In our PET studies (Tataranni *et al*., 1999; Gautier *et al*., 2000), where individuals had fasted for 36 h, there were no changes in neuronal activity in the striatum after tasting a liquid meal, but we observed a decrease after ingesting a satiating amount of the same meal. This was not different between obese and normal-weight individuals (Del Parigi *et al*., 2003). Interestingly, other regions receiving dopaminergic afferents through the mesolimbic and mesocortical pathways have been reported to have an atypical neuronal response to food stimuli in obesity. We have reported previously on the abnormal response of the insula of obese people after tasting the liquid meal. We also observed larger decreases of neuronal activity in the orbitofrontal cortex and anteromedial temporal lobe in obese compared to lean individuals who had just eaten a satiating amount of liquid meal (Gautier *et al.*, 2000). It is important to point out that other systems, such as the opioidergic and serotoninergic systems, are also thought to play a role in the neurophysiology of reward (Saper *et al.*, 2002; Halford *et al.*, 2005; Olszewski and Levine, 2007).

FN studies have also provided evidence for a neuronal disturbance in eating disorders (Table 12.1). Remarkably, an abnormal response of prefrontal regions to food-related stimulation has been reported consistently in bulimia nervosa (BN) and binge-eating disorder (BED), as well as in anorexia nervosa (AN), suggesting that these regional changes may play a common functional role (Kaye, 2008). In particular, in an fMRI study, women with BN responded to the sight of food with increased activation of medial prefrontal and anterior cingulate cortices and decreased activation of anterior and lateral prefrontal and temporal cortices compared to control subjects (Uher *et al.*, 2004). A greater increase of CBF in the left than the right hemisphere, especially in the frontal and prefrontal regions, in women affected by BED compared to control subjects had been demonstrated previously by SPECT (Karhunen *et al.*, 2000). In addition, the neuronal activity of the left frontal and prefrontal regions was associated strongly with an increase in the feeling of hunger during the exposure to food (Karhunen *et al.*, 2000).

These cortical areas participate in a neuronal network involved in emotional processing and implicated in the cognitive-emotional features of mood disorders, including obsessive-compulsive and affective disorders (Drevets, 2001). As noted below, additional studies are needed to determine whether the cognitive-emotional processes subserved by these regions have a central, causative role in the development of eating disorders or are consequences of eating disorders (e.g. increased discomfort in response to food). Finally, a defect in serotoninergic neurotransmission persisting after recovery seems likely to be a trait-related disturbance that may play a primary aetiological role in the development of all major eating disorders (Olszewski and Levine, 2007; Kaye, 2008).

Cause or Consequence

Simply identifying functional abnormalities in the brain of obese subjects does not prove that these alterations cause the disease. Indeed, the search for the causes of obesity in humans has yielded definitive results only for some rare and severe monogenic forms, accounting altogether for not more than 5% of the prevalence. This is likely because idiopathic obesity, like other complex diseases, is due not to a single genetic mutation but to multiple allelic defects which determine susceptibility to environmental factors. Individuals who carry only one or some of these alleles still may not develop the disease because they either lack another allele (gene–gene interaction) or are not exposed to the precipitating environment (gene–environment interaction). Known monogenic forms of obesity result from mutations in the genes encoding either for leptin, leptin receptor, POMC, prohormone convertase 1, or the melanocortin 4 receptor (Farooqi, 2008). The latter is, by far, the most common monogenic form of obesity. Proteins encoded by all these genes are involved critically in central homeostatic

Table 12.1. Reported alterations of brain activity in the resting state and in response to food-related stimuli in eating disorders.

Anorexia nervosa (AN)	Ref.	Bulimia nervosa (BN)	Ref.	Binge-eating disorder (BED)	Ref.
Basal neural activity: $\downarrow$ medial prefrontal cortex $\downarrow$ anterior cingulate cortex $\downarrow$ temporo-parietal regions	a, b, c	*Basal neural activity:* $\downarrow$ parietal cortex $\uparrow$ inferior frontal cortex $\uparrow$ temporal cortex	a, d	*Neural response to the sight of food:* $\uparrow$ frontal – prefrontal regions (left > right)	g
Neural response to food intake (d) or the sight of food (e): $\uparrow$ medial prefrontal $\uparrow$ anterior cingulate cortex	d, e	*Neural response to the sight of food:* $\downarrow$ inferior frontal $\downarrow$ temporal cortex $\uparrow$ medial prefrontal cortex $\uparrow$ anterior cingulate cortex	f		
Activation of lateral prefrontal cortex in response to food-related stimuli: BN < AN			f		

Note: a, Delvenne *et al.*, 1999. b, Naruo *et al.*, 2001. c, Takano *et al.*, 2001. d, Nozoe *et al.*, 1995. e, Ellison *et al.*, 1998. f, Uher *et al.*, 2004. g, Karhunen *et al.*, 2000.

pathways, located mainly in the hypothalamus and brainstem (Gao and Horvath, 2007, 2008).

On the other hand, as discussed previously, current evidence indicates that the involvement of the central nervous system in common forms of obesity spans beyond non-conscious homeostatic circuits and includes cortical and subcortical regions implicated in hedonic and cognitive processing. Studies in post-obese subjects (i.e. individuals who achieved and maintain a normal body weight despite a past history of severe obesity and who are at high risk of relapse) are informative for the identification of phenotypic characteristics that precede and possibly cause the development of obesity (Wing and Hill, 2001). In fact, by identifying obese-like abnormalities in the brain response of post-obese individuals to the taste and consumption of a liquid meal, we have recognized putative markers of an increased risk of obesity. We found that, in obese and post-obese individuals, the middle insular cortex and posterior hippocampus responded similarly to the taste and consumption of a satiating meal, respectively (Del Parigi *et al.*, 2004). While this observation requires confirmation in longitudinal studies, these findings are consistent with the hypothesis that predisposition to obesity involves areas of the brain that control complex aspects of eating behaviour, including anticipation and reward, food-sensory perception and autonomic control of digestion (insular cortex), as well as enteroception and learning/memory (hippocampus) (Del Parigi *et al.*, 2004).

What Does All This Mean? A Hypothetical Model

The brain response to the ingestion of a meal in normal-weight individuals, as defined by FN studies, suggests the presence of an orexigenic domain (represented mainly by limbic and paralimbic areas including the orbitofrontal and insular cortices, anterior cingulate and hypothalamic region) and a satiety domain (represented almost exclusively by prefrontal areas) (Tataranni *et al.*, 1999; Small *et al.*, 2001; Del Parigi *et al.*, 2002a,b, 2004; Hinton *et al.*, 2004). Based on prior evidence, we have proposed a model in which the prefrontal cortex signals satiety by sending inhibitory inputs to the limbic/paralimbic areas, thus suppressing hunger (Del Parigi *et al.*, 2002a, 2005; Tataranni and Del Parigi, 2003). There is no easy explanation as to why the prefrontal cortex and some of the limbic/paralimbic areas show greater changes in obese versus lean individuals (Gautier *et al.*, 2000, 2001). However, if our model is correct, it is possible that the prefrontal cortex may be working harder to suppress chronically hyperactive orexigenic areas in obese individuals (Del Parigi *et al.*, 2002a). Alternatively, resistance of the hypothalamus to the inhibitory effects of the prefrontal cortex may also be playing a role (Fig. 12.2).

According to our theoretical model, among the paralimbic areas, the insular cortex and hippocampus may play a special role as putative markers of an increased risk of obesity (Del Parigi *et al.*, 2004). In particular, the role of the insular cortex in the context of the control of eating behaviour deserves to be emphasized. The insula represents an important relay of the neuronal network connecting the hypothalamus, orbitofrontal cortex and limbic system. Moreover,

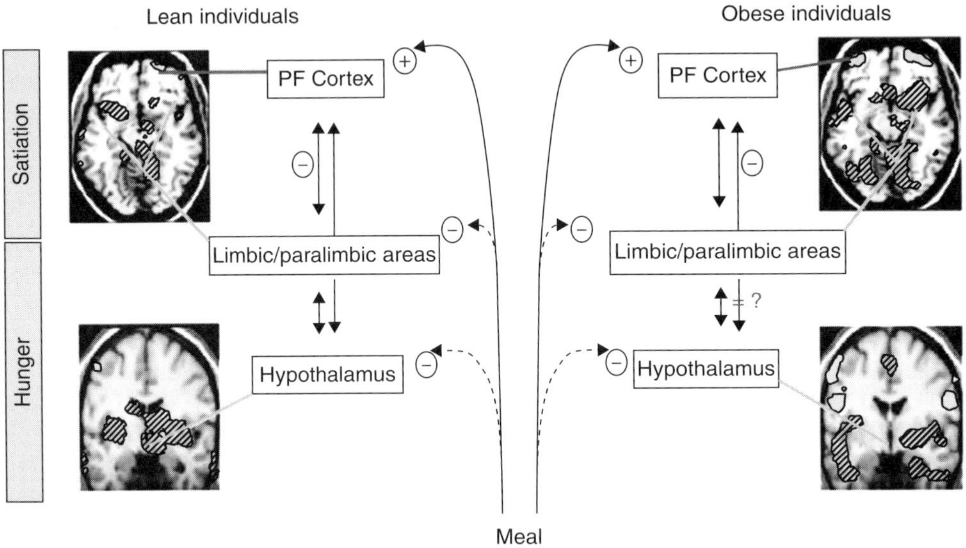

Fig. 12.2. Hypothetical model recapitulating differences in the brain response to a meal in obese and lean individuals. Brain images are from 11 lean and 11 obese men. Lower images are coronal sections through the hypothalamic region. Upper images are horizontal sections at 4 mm below a horizontal plane between the anterior and posterior commissure (co-ordinates from the Talairach and Tournoux brain atlas). The right hemisphere for each brain picture is on the reader's right. Brain regions with significant increases in regional cerebral blood flow in response to the meal are shown in outline, while decreases are outlined and hatched. This model predicts that the prefrontal (PF) cortex signals satiety by sending inhibitory inputs to the limbic/paralimbic areas, thus suppressing hunger. In obese individuals, the PF cortex may be working harder to suppress chronically hyperactive orexigenic areas. Alternatively, resistance of the hypothalamus to the inhibitory effects of the PF cortex may also be playing a role. Because neuroimaging allows investigation of the whole brain, without restriction to preselected regions, and because data analysis strategies are being developed that may permit better definition of the interplay between the hypothalamus, limbic/paralimbic areas and PF cortex, the hypothetical model proposed is likely to be improved significantly in the coming years, which should result in a better understanding of how the biological need and the pleasure of eating are integrated in the brain and contribute to normal and abnormal eating behaviour.

the insular cortex is known to contribute to the autonomic response to emotional states, and insular stimulation elicits gastrointestinal responses. In addition, the insular cortex is also a visceral sensory area and evidence suggests that it monitors distressing and potentially dangerous internal sensations (Augustine, 1996). Finally, we found a negative association between changes in plasma insulin concentrations after the administration of a satiating meal and the insular neuronal activity (Tataranni *et al.*, 1999; Del Parigi *et al.*, 2002a). Therefore, we postulate that this region is affected selectively by the metabolic, hormonal or neuronal events that signal the threatening state of hunger and that insulin may have a modulatory effect on this response.

Finally, it has been suggested repeatedly that control of energy homeostasis is biased inherently towards weight gain by more robust responses to energy restriction than to energy surpluses (Tataranni *et al.*, 1999; Schwartz *et al.*, 2003). Consistent with this possibility, maps of the human brain response to a meal show activation of a much larger number of brain areas in response to hunger than in response to satiety (Tataranni *et al.*, 1999).

Beyond Functional Neuroimaging

The use of FN in obesity research is a relatively recent development and it is, therefore, important to understand that this field is in its infancy. These types of studies have provided the first *in vivo* images of the human hypothalamic response to nutritional stimuli and revealed the complexity of the human brain response to hunger and satiety. Finding that the pattern of brain activity characterizes the response to a specific stimulus provided a foundation for investigating the neurofunctional features of normal and abnormal eating behaviour. A promising step in this direction was the demonstration that, in obese individuals, the decrease in hypothalamic activity following a meal was reduced significantly compared to lean individuals. Whether this, in turn, explains the associated differential responses in limbic/paralimbic areas and prefrontal cortex remains to be further examined.

The challenges in this line of research are daunting and meaningful progress is likely to depend as much on intelligent, hypothesis-driven study designs as it is on methodological and technical advances. For example, there are no easy methods to screen for psychopathological phenotypes underlying obesity (our current ability to define hyperphagia in a measurable way relies mostly on the observation of its main effect, i.e. weight gain). Consequently, it is not possible to identify individuals at risk for obesity based on an easy test. Despite these difficulties, metabolic risk factors for the development of obesity have been identified in carefully conducted longitudinal studies. Therefore, it is reasonable to believe that neurofunctional risk factors (and hopefully their neurochemical underpinnings) can be identified in the same way (Del Parigi *et al.*, 2004).

Emerging methodological developments may help researchers determine the extent to which functional alterations in obese patients are attributable to abnormalities in brain structure and how different brain regions work in concert in the neuronal network underlying normal and abnormal eating behaviours. One such method, voxel-based morphometry (VBM), is a fully-automated, model-free whole-brain technique detecting regionally specific differences in brain tissue composition (i.e. grey, white and CSF partitions) of stereologically normalized brains on a voxel-by-voxel basis (Ashburner and Friston, 2001). VBM and other voxel-based image-analysis algorithms circumvent problems inherent to classical region of interest (ROI)-based approaches, such as subjectivity of anatomical boundaries of some regions and rater-dependence of the measurements (Pruessner *et al.*, 2000; Good *et al.*, 2001). In addition, this technique has been cross-validated with both ROI measurements and functional data in a number of studies (Sowell *et al.*, 1999; Maguire *et al.*, 2000; Good *et al.*, 2001) and has made it possible to detect regional differences associated with some disease conditions which would have been difficult or unlikely to be detected by traditional ROI-based approaches (Suzuki *et al.*, 2005).

In a VBM study of 24 obese subjects versus 36 lean controls (Pannacciulli *et al.*, 2006), we found that the group of obese individuals had significantly lower grey matter density in the post-central gyrus, frontal operculum, middle frontal gyrus, putamen and cerebellum after adjustment for sex, age, handedness, global grey matter density and multiple comparisons. Moreover, BMI was associated negatively with grey matter density of the left post-central gyrus in obese but not lean subjects. This study identified structural brain differences in human obesity in several brain areas involved previously in the regulation of taste, reward, behavioural control and satiety. These alterations may either precede obesity, representing a neural marker of increased propensity to gaining weight, or occur as a consequence of obesity, indicating that the brain also is affected by increased adiposity. Interestingly, leptin has been shown recently to reverse weight loss-induced changes in regional neural activity responses to visual food stimuli (Rosenbaum *et al.*, 2008). Specifically, following a 10% weight loss leptin-reversible increases in neural activity in response to visual food cues were observed in the brainstem, culmen, parahippocampal gyrus, inferior and middle frontal gyri, middle temporal gyrus and lingual gyrus. Leptin-reversible decreases in activity in response to food cues in the hypothalamus, cingulate gyrus and middle frontal gyrus were also found (Rosenbaum *et al.*, 2008). These imaging findings warrant further studies to evaluate the possible biological and behavioural correlates of these alterations, their cytoarchitectural basis as well as their role in the natural history of obesity and weight maintenance.

Data analysis strategies are also being developed to explore the existence of separate functional networks giving rise to the patterns of brain activation in response to specific stimuli and how these networks are related to behavioural or biological measures. The scaled subprofile model (SSM) analysis (Alexander *et al.*, 1999), for example, may permit us to understand better the interplay between the hypothalamus, limbic/paralimbic areas and the prefrontal cortex, for which we have only a hypothetical model at this time.

In addition, researchers continue to develop new image analysis methods to study the functions of small regions, perhaps even as small as hypothalamic subnuclei. For instance, researchers have developed ingenious data analysis methods based on reducing the three-dimensional structure of brain regions as complex as the hippocampus to a single plane and conducting the final analysis of its subregions on a 'flat map' (Zeineh *et al.*, 2003). As another example, while high-strength magnetic fields (9.4 T) allow detection of blood oxygen level-dependent (BOLD)-fMRI signals with isotropic voxels on the order of 100 μm in experimental animals (Kim *et al.*, 2000), the ability to localize a neuronal response precisely may be limited by the distance between the activated neurones and the related vascular response detected by the BOLD signal. Confining the MRI field-of-view to a preselected region, it may be possible to use MRI with increasingly high field strength to provide information with a comparably high spatial resolution. A spatial resolution of this magnitude, when available for human studies, may allow imaging the activity of groups of neurones within a selected brain region.

Researchers also continue to develop, test and apply new image-acquisition methods (e.g. new radiotracer techniques using PET or SPECT) to allow the non-invasive assessment of additional molecular events. Much research is being carried out to develop tracers that can be used to study individual monoaminergic pathways that are obviously of potential interest to the field of obesity research. Among them, ^{123}I-*N*-methyl-2 beta-carbomethoxy-3beta-(4-iodophenyl) tropane can be used to image the density of dopamine (DAT) and serotonin transporter receptors (SERT) by SPECT (Emond *et al.*, 1997). Several radioligands are being developed to image SERT, norepinephrine transport (NET) and DAT (Wellsow *et al.*, 2002) using PET. Benzodiazepine receptor ligands (^{123}I-Iomazenil for SPECT and ^{11}C-flumazenil for PET) (Mountz *et al.*, 2002) will be used to investigate how some aspects of gamma-amino butyric acid (GABA) signalling are related to hunger and satiety.

As noted previously, PET can also provide indirect information about state-dependent alterations in the synaptic availability of dopamine, serotonin and certain other neurotransmitters. Magnetic resonance spectroscopy (MRS) is another imaging technique that provides information about certain metabolic processes in the living human brain, such as the concentration of *N*-acetyl aspartic acid (a putative marker of viable neurones), glucose transport in the grey and white matter, the neuronal tricarboxylic acid cycle (TCA) and ketone oxidation (de Graaf *et al.*, 2001; Pan *et al.*, 2002). Although MRS is limited in its sensitivity to detect processes that occur in small concentrations and in the spatial resolution required to distinguish subtle changes from noise, researchers continue to address these methodological challenges (Perez *et al.*, 2002; de Graaf *et al.*, 2003), providing new opportunities in the study of obesity.

Like other research strategies, imaging studies are likely to benefit from complementary research methods to elucidate more fully the role and underlying molecular processes involved in the regulation of food intake and their alterations in obesity. Complementary approaches include, but are not limited to: (i) lesion, stimulation and neural tract tracing studies in laboratory animals, which can help determine the extent to which brain regions in an imaging study are necessary or sufficient for performance of a task; (ii) reversible 'lesion' studies (e.g. using transcranial magnetic stimulation (TMS) for the same purpose in human studies); (iii) the further development and use of electrophysiological recording techniques (e.g. event-related potentials and magnetoencephalography), which have the temporal resolution needed to help determine the sequence in which implicated brain regions are activated; (iv) transcriptomic and proteomic studies using post-mortem brain tissue from the brain regions implicated in imaging studies to assess the molecular processes involved in the normal and abnormal regulation of food intake; and (v) a host of additional approaches (e.g. genetic and pharmacological) that are likely to result from the use of imaging methods in conjunction with these complementary approaches.

These complementary research methods will also help to determine the extent to which the neuroanatomical correlates of normal and abnormal responses to specific stimuli identified by FN techniques are necessary for the behaviour of interest. For instance, the above-cited TMS could be used to excite or inhibit temporally a region of the brain implicated in FN studies and determine the extent to which this experimental effect (e.g. regional brain inhibition) alters the behaviour of interest (Pascual-Leone *et al.*, 2000).

Another attractive way to investigate the functional molecular machinery of the living brain is the assessment of *in vivo* gene expression. *In vivo* hybridization techniques using antisense DNA probes with high specific radioactivity for SPECT or PET are being used in experimental animals (Lee *et al.*, 2002) and are providing important insights when combined with adenoviral vector-mediated expression, as has already been shown for DRD2 in the rat striatum (Umegaki *et al.*, 2002). How this will translate into human research is difficult to predict at this point. In the meantime, while waiting anxiously for some of these technical advancements, we have designed our research programme to combine FN with more classical cellular/molecular biology-based approaches. We hope that this strategy will allow us to unravel further the putative common neurochemical abnormalities underlying the aetiology of human obesity.

Conclusions

In conclusion, functional neuroimaging allows exploration of patterns of brain activation associated with subconscious and conscious (perceptual, emotional and cognitive) mental processes. Because regulation of eating behaviour spans the range of non-conscious (homeostatic) and conscious (hedonic) events, FN provides an increasingly important tool for investigating how different regions of the brain work in concert to orchestrate normal eating behaviours and how they conspire to produce obesity and other eating disorders. The extent to which the promise of neuroimaging in the study of obesity is realized depends on the ability of researchers to address methodological challenges associated with this technologically advancing field, their ability to address experimental challenges involved in the generation and testing of specific hypotheses and their ability to capitalize on methodological approaches that complement the role of imaging in the study of the normal and abnormal regulation of food intake.

Acknowledgements

The authors wish to thank Dr Clifton Bogardus, Dr Joy C. Bunt, Dr Eric M. Reiman and Dr Arline D. Salbe for their helpful comments and careful review of the manuscript.

References

Alexander, G.E., Mentis, M.J., Van Horn, J.D., Grady, C.L., Berman, K.F., Furey, M.L., Pietrini, P., Rapoport, S.I., Schapiro, M.B. and Moeller, J.R. (1999) Individual differences in PET activation of object perception and attention systems predict face matching accuracy. *Neuroreport* 10, 1965–1971.

Ashburner, J. and Friston, K.J. (2001) Why voxel-based morphometry should be used. *Neuroimage* 14,1238–1243.

Augustine, J.R. (1996) Circuitry and functional aspects of the insular lobe in primates including humans. *Brain Research Reviews* 22, 229–244.

Berthoud, H.R. and Morrison, C. (2008) The brain, appetite, and obesity. *Annual Review of Psychology* 59, 55–92.

Bray, G.A. (2004) Obesity is a chronic, relapsing neurochemical disease. *International Journal of Obesity* 28, 34–38.

Burke, K.A., Franz, T.M., Miller, D.N. and Schoenbaum, G. (2008) The role of the orbitofrontal cortex in the pursuit of happiness and more specific rewards. *Nature* 454, 340–344.

Critchley, H.D. and Rolls, E.T. (1996) Hunger and satiety modify the responses of olfactory and visual neurons in the primate orbitofrontal cortex. *Journal of Neurophysiology* 75, 1673–1686.

Del Parigi, A., Gautier, J.F., Chen, K., Salbe, A.D., Ravussin, E., Reiman, E. and Tataranni, P.A. (2002a) Neuroimaging and obesity: mapping the brain research responses to hunger and satiation in humans using positron emission tomography. *Annals of the New York Academy of Sciences* 967, 389–397.

Del Parigi, A., Chen, K., Gautier, J.F., Salbe, A.D., Pratley, R.E., Ravussin, E., Reiman, E.M. and Tataranni, P.A. (2002b) Sex differences in the human brain's response to hunger and satiation. *American Journal of Clinical Nutrition* 75, 1017–1022.

Del Parigi, A., Chen, K., Salbe, A.D., Reiman, E.M. and Tataranni, P.A. (2003) Are we addicted to food? *Obesity Research* 11, 493–495.

Del Parigi, A., Chen, K., Salbe, A.D., Hill, J.O., Wing, R.R., Reiman, E.M. and Tataranni, P.A. (2004) Persistence of abnormal neural responses to a meal in postobese individuals. *International Journal of Obesity* 28, 370–377.

Del Parigi, A., Pannacciulli, N., Le, D.N. and Tataranni, P.A. (2005) In pursuit of neural risk factors for weight gain in humans. *Neurobiology of Aging* 26 (Suppl. 1), 50–55.

Delvenne, V., Goldman, S., De Maertelaer, V. and Lotstra, F. (1999) Brain glucose metabolism in eating disorders assessed by positron emission tomography. *International Journal of Eating Disorders* 25, 29–37.

Dillon, D.G., Holmes, A.J., Jahn, A.L., Bogdan, R., Wald, L.L. and Pizzagalli, D.A. (2008) Dissociation of neural regions associated with anticipatory versus consummatory phases of incentive processing. *Psychophysiology* 45, 36–49.

Drevets, W.C. (2001) Neuroimaging and neuropathological studies of depression: implications for the cognitive-emotional features of mood disorders. *Current Opinion in Neurobiology* 11, 240–249.

Ellison, Z., Foong, J., Howard, R., Bullmore, E., Williams, S. and Treasure, J. (1998) Functional anatomy of calorie fear in anorexia nervosa. *Lancet* 352, 1192.

Emond, P., Chalon, S., Garreau, L., Dognon, A.M., Bodard, S., Frangin, Y., Baulieu, J.L., Besnard, J.C. and Guilloteau, D. (1997) A new iodinated tropane derivative (beta-CDIT) for *in vivo* dopamine transporter exploration: comparison with beta-CIT. *Synapse* 26, 72–80.

Farooqi, I.S. (2008) Monogenic human obesity. *Frontiers of Hormone Research* 36, 1–11.

Finlayson, G., King, N. and Blundell, J.E. (2007) Liking vs. wanting food: importance for human appetite control and weight regulation. *Neuroscience and Biobehavioral Reviews* 31, 987–1002.

Gao, Q. and Horvath, T.L. (2007) Neurobiology of feeding and energy expenditure. *Annual Review of Neuroscience* 30, 367–398.

Gao, Q. and Horvath, T.L. (2008) Neuronal control of energy homeostasis. *FEBS Letters* 582, 132–141.

Gautier, J.F., Chen, K., Salbe, A.D., Bandy, D., Pratley, R.E., Heiman, M., Ravussin, E., Reiman, E.M. and Tataranni, P.A. (2000) Differential brain research responses to satiation in obese and lean men. *Diabetes* 49, 838–846.

Gautier, J.F., Del Parigi, A., Chen, K., Salbe, A.D., Bandy, D., Pratley, R.E., Ravussin, E., Reiman, E.M. and Tataranni, P.A. (2001) Effect of satiation on brain activity in obese and lean women. *Obesity Research* 9, 676–684.

Good, C.D., Johnsrude, I., Ashburner, J., Henson, R.N., Friston, K.J. and Frackowiak, R.S. (2001) Cerebral asymmetry and the effects of sex and handedness on brain structure: a voxel-based morphometric analysis of 465 normal adult human brains. *Neuroimage* 14, 685–700.

Graaf, R.A. de, Pan, J.W., Telang, F., Lee, J.H., Brown, P., Novotny, E.J., Hetherington, H.P. and Rothman, D.L. (2001) Differentiation of glucose transport in human brain gray and white matter. *Journal of Cerebral Blood Flow and Metabolism* 21,483–492.

Graaf, R.A. de, Brown, P.B., Mason, G.F., Rothman, D.L. and Behar, K.L. (2003) Detection of [1,6-13C2]-glucose metabolism in rat brain by *in vivo* 1H-[13C]-NMR spectroscopy. *Magnetic Resonanace in Medicine* 49, 37–46.

Grill, H.J., Skibicka, K.P. and Hayes, M.R. (2007) Imaging obesity: fMRI, food reward, and feeding. *Cell Metabolism* 6, 423–425.

Halford, J.C., Harrold, J.A., Lawton, C.L. and Blundell, J.E. (2005) Serotonin (5-HT) drugs: effects on appetite expression and use for the treatment of obesity. *Current Drug Targets* 6, 201–213.

Hinton, E.C., Parkinson, J.A., Holland, A.J., Arana, F.S., Roberts, A.C. and Owen, A.M. (2004) Neural contributions to the motivational control of appetite in humans. *European Journal of Neuroscience* 20, 1411–1418.

Karhunen, L.J., Lappalainen, R.I., Vanninen, E.J., Kuikka, J.T. and Uusitupa, M.I. (1997) Regional cerebral blood flow during food exposure in obese and normal-weight women. *Brain* 120 Part 9, 1675–1684.

Karhunen, L.J., Vanninen, E.J., Kuikka, J.T., Lappalainen, R.I., Tiihonen, J. and Uusitupa, M.I. (2000) Regional cerebral blood flow during exposure to food in obese binge eating women. *Psychiatry Research* 99, 29–42.

Kaye, W. (2008) Neurobiology of anorexia and bulimia nervosa. *Physiology and Behavior* 94, 121–135.

Kim, D.S., Duong, T,Q. and Kim, S.G. (2000) High-resolution mapping of iso-orientation columns by fMRI. *Nature Neuroscience* 3, 164–169.

Kringelbach, M.L., O'Doherty, J., Rolls, E.T. and Andrews, C. (2003) Activation of the human orbitofrontal cortex to a liquid food stimulus is correlated with its subjective pleasantness. *Cerebral Cortex* 13, 1064–1071.

Lee, H.J., Boado, R.J., Braasch, D.A., Corey, D.R. and Pardridge, W.M. (2002) Imaging gene expression in the brain *in vivo* in a transgenic mouse model of Huntington's disease with an antisense radiopharmaceutical and drug-targeting technology. *Journal of Nuclear Medicine* 43, 948–956.

Liu, Y., Gao, J.H., Liu, H.L. and Fox, P.T. (2000) The temporal response of the brain after eating revealed by functional MRI. *Nature* 405, 1058–1062.

Maguire, E.A., Gadian, D.G., Johnsrude, I.S., Good, C.D., Ashburner, J., Frackowiak, R.S. and Frith, C.D. (2000) Navigation-related structural change in the hippocampi of taxi drivers. *Proceedings of the National Academy of Sciences of the United States of America* 97, 4398–4403.

Mahankali, S., Liu, Y., Pu, Y., Wang, J., Chen, C.W., Fox, P.T. and Gao, J.H. (2000) *In vivo* fMRI demonstration of hypothalamic function following intraperitoneal glucose administration in a rat model. *Magnetic Resonance in Medicine* 43, 155–159.

Mountz, J.D., Hsu, H.C., Wu, Q., Liu, H.G., Zhang, H.G. and Mountz, J.M. (2002) Molecular imaging: new applications for biochemistry. *Journal of Cellular Biochemistry* 39 (Suppl.), 162–171.

Naruo, T., Nakabeppu, Y., Deguchi, D., Nagai, N., Tsutsui, J., Nakajo, M. and Nozoe, S. (2001) Decreases in blood perfusion of the anterior cingulate gyri in anorexia nervosa restricters assessed by SPECT image analysis. *BMC Psychiatry* 1, 2.

Nozoe, S., Naruo, T., Yonekura, R., Nakabeppu, Y., Soejima, Y., Nagai, N., Nakajo, M. and Tanaka, H. (1995) Comparison of regional cerebral blood flow in patients with eating disorders. *Brain Research Bulletin* 36, 251–255.

O'Doherty, J., Rolls, E.T., Francis, S., Bowtell, R. and McGlone, F. (2001) Representation of pleasant and aversive taste in the human brain. *Journal of Neurophysiology* 85, 1315–1321.

Olszewski, P.K. and Levine, A.S. (2007) Central opioids and consumption of sweet tastants: when reward outweighs homeostasis. *Physiology and Behavior* 91, 506–512.

Palmiter, R.D. (2007) Is dopamine a physiologically relevant mediator of feeding behavior? *Trends in Neuroscience* 30, 375–381.

Pan, J.W., Graaf, R.A. de, Petersen, K.F., Shulman, G.I., Hetherington, H.P. and Rothman, D.L. (2002) [2,4-13 C2]-beta-Hydroxybutyrate metabolism in human brain. *Journal of Cerebral Blood Flow and Metabolism* 22, 890–898.

Pannacciulli, N., Del Parigi, A., Chen, K., Le, D.S.N.T., Reiman, E.M. and Tataranni, P.A. (2006) Brain abnormalities in human obesity: a voxel-based morphometric study. *Neuroimage* 31, 1419–1425.

Pascual-Leone, A., Walsh, V. and Rothwell, J. (2000) Transcranial magnetic stimulation in cognitive neuroscience – virtual lesion, chronometry, and functional connectivity. *Current Opinion in Neurobiology* 10, 232–237.

Perez, J.M., Josephson, L., O'Loughlin, T., Hogemann, D. and Weissleder, R. (2002) Magnetic relaxation switches capable of sensing molecular interactions. *Nature Biotechnology* 20, 816–820.

Pruessner, J.C., Li, L.M., Serles, W., Pruessner, M., Collins, D.L., Kabani, N., Lupien, S. and Evans, A.C. (2000) Volumetry of hippocampus and amygdala with high-resolution MRI and three-dimensional analysis software: minimizing the discrepancies between laboratories. *Cerebral Cortex* 10, 433–442.

Reiman, E.M., Lane, R.D., Ahern, G.L., Schwartz, G.E, Davidson, R.J., Friston, K.J., Yun, L.S. and Chen, K. (1997) Neuroanatomical correlates of externally and internally generated human emotion. *American Journal of Psychiatry* 154, 918–925.

Rolls, E.T. (2007) Sensory processing in the brain related to the control of food intake. *Proceedings of the Nutrition Society* 66, 96–112.

Rolls, E.T., Critchley, H.D. and Treves, A. (1996) Representation of olfactory information in the primate orbitofrontal cortex. *Journal of Neurophysiology* 75, 1982–1996.

Rosenbaum, M., Sy, M., Pavlovich, K., Leibel, R.L. and Hirsch, J. (2008) Leptin reverses weight loss-induced changes in regional neural activity responses to visual food stimuli. *Journal of Clinical Investigation* 118, 2583–2591.

Saper, C.B., Chou, T.C. and Elmquist, J.K. (2002) The need to feed: homeostatic and hedonic control of eating. *Neuron* 36, 199–211.

Schoeller, D.A. and Taylor, P.B. (1987) Precision of the doubly labelled water method using the two-point calculation. *Human Nutrition, Clinical Nutrition* 41, 215–223.

Schwartz, M.W., Woods, S.C., Seeley, R.J., Barsh, G.S., Baskin, D.G. and Leibel, R.L. (2003) Is the energy homeostasis system inherently biased toward weight gain? *Diabetes* 52, 232–238.

Small, D.M., Zatorre, R.J., Dagher, A., Evans, A.C. and Jones-Gotman, M. (2001) Changes in brain activity related to eating chocolate: from pleasure to aversion. *Brain* 124, 1720–1733.

Small, D.M., Jones-Gotman, M. and Dagher, A. (2003) Feeding-induced dopamine release in dorsal striatum correlates with meal pleasantness ratings in healthy human volunteers. *Neuroimage* 19, 1709–1715.

Small, D.M., Voss, J., Mak, Y.E., Simmons, K.B., Parrish, T. and Gitelman, D. (2004) Experience-dependent neural integration of taste and smell in the human brain. *Journal of Neurophysiology* 92, 1892–1903.

Smeets, P.A., Graaf, C. de, Stafleu, A., Osch, M.J. van and Grond, J. van der (2005) Functional MRI of human hypothalamic responses following glucose ingestion. *Neuroimage* 24, 363–368.

Sowell, E.R., Thompson, P.M., Holmes, C.J., Batth, R., Jernigan, T.L. and Toga, A.W. (1999) Localizing age-related changes in brain structure between childhood and adolescence using statistical parametric mapping. *Neuroimage* 9, 587–597.

Stunkard, A.J., Berkowitz, R.I., Stallings, V.A. and Schoeller, D.A. (1999) Energy intake, not energy output, is a determinant of body size in infants. *American Journal of Clinical Nutrition* 69, 524–530.

Suzuki, M., Hagino, H., Nohara, S., Zhou, S.Y., Kawasaki, Y., Takahashi, T., Matsui, M., Seto, H., Ono, T. and Kurachi, M. (2005) Male-specific volume expansion of the human hippocampus during adolescence. *Cerebral Cortex* 15, 187–193.

Szczypka, M.S., Kwok, K., Brot, M.D., Marck, B.T., Matsumoto, A.M., Donahue, B.A. and Palmiter, R.D. (2001) Dopamine production in the caudate putamen restores feeding in dopamine-deficient mice. *Neuron* 30, 819–828.

Takano, A., Shiga, T., Kitagawa, N., Koyama, T., Katoh, C., Tsukamoto, E. and Tamaki, N. (2001) Abnormal neuronal network in anorexia nervosa studied with I-123-IMP SPECT. *Psychiatry Research* 107, 45–50.

Tataranni, P.A. (2000) Mechanisms of weight gain in humans. *European Review for Medical and Pharmacological Sciences* 4, 1–7.

Tataranni, P.A. and Del Parigi, A. (2003) Functional neuroimaging: a new generation of human brain studies in obesity research. *Obesity Reviews* 4, 229–238.

Tataranni, P.A., Gautier, J.F., Chen, K., Uecker, A., Bandy, D., Salbe, A.D., Pratley, R.E., Lawson, M., Reiman, E.M. and Ravussin, E. (1999) Neuroanatomical correlates of hunger and satiation in humans using positron emission tomography. *Proceedings of the National Academy of Sciences of the United States of America* 96, 4569–4574.

Uher, R., Murphy, T., Brammer, M.J., Dalgleish, T., Phillips, M.L., Ng, V.W., Andrew, C.M., Williams, S.C., Campbell, I.C. and Treasure, J. (2004) Medial prefrontal cortex activity associated with symptom provocation in eating disorders. *American Journal of Psychiatry* 161, 1238–1246.

Umegaki, H., Ishiwata, K., Ogawa, O., Ingram, D.K., Roth, G.S., Yoshimura, J., Oda, K., Matsui-Hirai, H., Ikari, H., Iguchi, A. and Senda, M. (2002) *In vivo* assessment of adenoviral vector-mediated gene expression of dopamine D(2) receptors in the rat striatum by positron emission tomography. *Synapse* 43, 195–200.

Ungerstedt, U. (1971) Adipsia and aphagia after 6-hydroxydopamine induced degeneration of the nigro-striatal dopamine system. *Acta Physiologica Scandinava* 367 Supplement, 95–122.

Volkow, N.D., Wang, G.J., Maynard, L., Jayne, M., Fowler, J.S., Zhu, W., Logan, J., Gatley, S.J., Ding, Y.S., Wong, C. and Pappas, N. (2003) Brain dopamine is associated with eating behaviors in humans. *International Journal of Eating Disorders* 33, 136–142.

Volkow, N.D., Wang, G.J., Telang, F., Fowler, J.S., Thanos, P.K., Logan, J., Alexoff, D., Ding, Y.S., Wong, C., Ma, Y. and Pradhan, K. (2008) Low dopamine striatal D2 receptors are associated with prefrontal metabolism in obese subjects: possible contributing factors. *Neuroimage* 42, 1537–1543.

Wang, G.J., Volkow, N.D., Logan, J., Pappas, N.R., Wong, C.T., Zhu, W., Netusil, N. and Fowler, J.S. (2001) Brain dopamine and obesity. *Lancet* 357, 354–357.

Wang, G.J., Volkow, N.D. and Fowler, J.S. (2002a) The role of dopamine in motivation for food in humans: implications for obesity. *Expert Opinion in Therapeutic Targets* 6, 601–609.

Wang, G.J., Volkow, N.D., Felder, C., Fowler, J.S., Levy, A.V., Pappas, N.R., Wong, C.T., Zhu, W. and Netusil, N. (2002b) Enhanced resting activity of the oral somatosensory cortex in obese subjects. *Neuroreport* 13, 1151–1155.

Wellsow, J., Kovar, K.A. and Machulla, H.J. (2002) Molecular modeling of potential new and selective PET radiotracers for the serotonin transporter. Positron Emission Tomography. *Journal of Pharmacy and Pharmacological Sciences* 5, 245–257.

Wing, R.R. and Hill, J.O. (2001) Successful weight loss maintenance. *Annual Review of Nutrition* 21, 323–341.

Wise, R.A., Spindler, J., deWit, H. and Gerberg, G.J. (1978) Neuroleptic-induced 'anhedonia' in rats: pimozide blocks reward quality of food. *Science* 201, 262–264.

Zald, D.H. and Pardo, J.V. (1997) Emotion, olfaction, and the human amygdala: amygdala activation during aversive olfactory stimulation. *Proceedings of the National Academy of Sciences of the United States of America* 94, 4119–4124.

Zald, D.H. and Pardo, J.V. (2000) Functional neuroimaging of the olfactory system in humans. *International Journal of Psychophysiology* 36, 165–181.

Zeineh, M.M., Engel, S.A., Thompson, P.M. and Bookheimer, S.Y. (2003) Dynamics of the hippocampus during encoding and retrieval of face-name pairs. *Science* 299, 577–580.

13 Overview of the Integrative Physiology of Adipose Tissue in Energy Homeostasis

ISABELLE DUGAIL AND MICHELE GUERRE-MILLO

INSERM, Paris, France; Centre de Recherche des Cordeliers, Université Pierre et Marie Curie, France; Université Paris Descartes, France

Introduction

In mammals, energy storage and release is crucial for survival in the face of intermittent food availability. The adipose tissue fulfils this need through its unique capacity to store energy in the form of lipids (see Chapters 5 and 6). The adipocyte lipid droplet represents a specialized reservoir of triglycerides from which fatty acids can be mobilized and exported for use in other tissues. Multiple hormonal, metabolic and neural signals tightly control the pathways of fat accretion (lipogenesis) and fat mobilization (lipolysis), to adjust lipid storage to whole-body energy balance (Guerre-Millo, 2004; Ahima, 2006; Frühbeck, 2006; Coll *et al.*, 2007). In the long run, the amounts of triglycerides stored within adipocytes reflect the net balance between caloric intake and energy expenditure.

Adipose tissue is now recognized as playing a central role in energy homeostasis, not only as a reserve of energetic substrates but also through the secretion of a number of adipokines, principal among these being leptin. Produced in proportion to body fat mass, leptin signals to the brain the level of triglyceride stores, allowing feedback changes in food intake to promote stable adipose tissue weight. This function is crucial to prevent ectopic lipid accumulation in other organs, where they can exert toxic effects (Garg, 2004; Ahima, 2006; Coll *et al.*, 2007). The most frequent derangement of this signalling network results in situations where energy intake exceeds energy expenditure. In this case, the amounts of adipose cell triglycerides rise in the expanding lipid droplet. Initially adaptive, this process becomes deleterious as it increases in degree and duration, leading to fat cell hypertrophy and obesity. Interestingly, adipocyte size has been shown to be an important determinant of adipokine secretion, with a preferential expression of proinflammatory factors with increasing adipocyte size (Skurk *et al.*, 2007). Recently, it has been observed that the number of fat cells stays constant in adulthood in both lean and obese individuals, even after marked weight loss

(Spalding *et al.*, 2008). Thus, defects in adipocyte metabolism and secretory function define the pathophysiology of adipose tissue in obesity.

In past years, several transgenic mice with targeted alterations in adipose-specific gene expression have been created. These models are useful to elucidate the role of selected proteins in adipose tissue biology and their implication in the pathogenesis of obesity. Besides, they provide new insights into the pathophysiological importance of adipose tissue in whole-body energy homeostasis. The scope of this chapter is to review some of these transgenic mouse models, to illustrate how adipose tissue influences the level of fat storage through its own gene products. The observation that adipose tissue is infiltrated with macrophages in obesity has led to the hypothesis that adipose gene expression is influenced through paracrine pathways involving resident macrophages. This finding is discussed briefly in the last part of the chapter.

Adipose Cell Metabolism

Adipose cell size, and in turn most of the adipose tissue mass, is determined by the relative rates of triglyceride synthesis and breakdown. Transgenic mice have been created with genetic alterations in these two opposite pathways, with both expected and unanticipated outcomes. A consistent observation is that altering fuel metabolism in adipocytes influences whole-body homeostasis, at least in part, through adipose-secreted factors.

Anabolic pathways

The GLUT4 isoform of glucose transporter is expressed in insulin-sensitive tissues, including skeletal muscles and adipose tissue. In the fat cell, GLUT4 provides glucose for *de novo* fatty acid synthesis and for the formation of the glycerol backbone of triacylglycerol. Mice were generated that overexpressed GLUT4 in their adipose tissue (Shepherd *et al.*, 1993). As expected, these mice developed excess fat mass, due to enhanced rates of glucose utilization preferentially for *de novo* fatty acid synthesis in adipocytes (Tozzo *et al.*, 1995). Surprisingly, however, adipose tissue growth was hyperplastic rather than hypertrophic in this model, suggesting enhanced preadipocyte recruitment through an unknown mechanism. When the GLUT4 gene was deleted specifically in adipose tissue, the overall adiposity of the mice was not altered (Abel *et al.*, 2001). This phenotype suggested that another glucose transporter, likely the ubiquitously expressed GLUT1 isoform, substituted for GLUT4 to allow glucose entry in the adipose cells. The ablation of GLUT4 specifically in adipose tissue was not anticipated to have a major impact on glucose homeostasis, since adipose tissue contributed little to whole-body glucose utilization as compared to skeletal muscles. However, adipose-specific GLUT4-null mice exhibited a 50% reduction in whole-body glucose uptake. This unexpected phenotype was accounted for by a severe reduction of insulin-stimulated glucose utilization in muscles. Moreover, the effect of insulin to suppress hepatic glucose production was also impaired in these mice. These findings led to the hypothesis that a factor, either missing or overproduced in GLUT4-null adipocytes, was able to influence insulin action in the liver and muscles. Using DNA microarray technology, this factor was identified later as retinol-binding protein 4 (RBP4), whose adipose expression was increased in adipose-specific GLUT4-null mice and which caused insulin resistance in normal mice (Yang *et al.*, 2005). Thus, although not essential for adipose tissue growth, GLUT4 appears critical for normal adipose cell biology and whole-body glucose homeostasis. Therefore, a potential therapeutic approach reducing adipose cell GLUT4 expression could be more harmful than beneficial in obesity. Serum RBP4 concentrations have been described to be elevated in humans with obesity and type 2 diabetes mellitus (T2DM) (Cho *et al.*, 2006; Graham *et al.*, 2006). However, the true contribution of RBP4 to human obesity has not been clarified completely, with some studies observing normal concentrations of this protein (Janke *et al.*, 2006; Broch *et al.*, 2007; Gómez-Ambrosi *et al.*, 2008). Further research is needed to unravel the involvement of RBP4 in the development of obesity-associated insulin resistance in humans.

Diacylglycerol acyltransferase 1 (DGAT1) catalyses the final step in the triacylglycerol synthetic pathway. This enzyme was considered necessary for adipose tissue formation. Unexpectedly, invalidation of the DGAT1 gene in mice produced healthy animals, with only a twofold reduction in the weight of fat pads (Smith *et al.*, 2000). This phenotype revealed that alternative pathways for triacylglycerol synthesis could be used and paved the way for the identification of the DGAT2 gene (Cases *et al.*, 2001). A striking feature of the DGAT1-null mice phenotype is their ability to resist obesity when fed a high-fat diet and to keep their body weight at the level of chow-fed animals. Resistance to diet-induced obesity was not due to fat malabsorption, but rather to high rates of energy expenditure. The underlying mechanisms still remain to be fully clarified, but could implicate a factor or factors released by adipose tissue. Indeed, transplantation of DGAT1-null adipose tissue conferred partial obesity resistance in wild-type recipient mice (Chen *et al.*, 2003). One candidate factor is adiponectin, whose expression is higher in the adipose tissue of DGAT1-null mice as compared to wild-type controls. Besides being an important determinant of whole-body insulin sensitivity (Guerre-Millo, 2008), adiponectin may act centrally to decrease body weight by stimulating energy expenditure (Qi *et al.*, 2004). Thus, according to these studies, pharmacological inhibition of DGAT1 may represent a therapeutic strategy for human obesity without major side effects (Chen and Farese, 2005).

Caveolins form a gene family of three members (caveolin-1, 2 and 3) that encode membrane-associated proteins. Adipocytes express particularly high levels of caveolins and are enriched in caveolae, flask-shaped invaginations of the plasma membrane whose formation is driven by caveolins. To assess the physiological importance of these structural proteins, caveolin-1-null mice were generated, which were viable and presented no major developmental abnormalities (Drab *et al.*, 2001; Razani *et al.*, 2001). As expected, these mice lack any detectable caveolae. The metabolic phenotype of caveolin-1-null mice revealed an unsuspected role for caveolins in lipid storage, with caveolin-1-null mice being lean and resistant to diet-induced obesity (Razani *et al.*, 2002). This phenotype is not related to altered nutrient absorption or decreased food intake. It has been suggested that caveolin-1 acts as a molecular scaffold required for stabilization of the insulin receptor in adipocytes. Thus, the lean phenotype observed in the caveolin-1 knockout mice could be due, at least in part, to a reduction of insulin-stimulated lipogenesis (Cohen *et al.*, 2003). Interestingly, caveolin-1 gene expression is increased in obese leptin-deficient mice, as reported in a genome-wide-scale analysis of adipose tissue gene expression (Soukas *et al.*, 2000; Catalán *et al.*, 2008). It is also possible that caveolins are required for proper lipid storage in adipocytes. Indeed, the lipid droplets isolated from adipocytes of caveolin-1-null mice display a dramatic reduction of cholesterol content (Le Lay *et al.*, 2006). Another important adipose-related phenotypic trait of these mice is their reduced levels of circulating adiponectin (Razani *et al.*, 2002). Low adiponectinaemia is usually found in obese insulin-resistant rodents or humans. Accordingly, caveolin-1-null mice are insulin-resistant, but the reason why adiponectin production is reduced still remains unclear. Surprisingly, expression of caveolin-1 in human adipose tissue has been shown recently to be upregulated in obesity and obesity-associated T2DM and related to inflammation (Catalán *et al.*, 2008). These observations suggest that caveolins are implicated in the pathways leading to fat accretion in adipose cells and, as such, represent novel antiobesity targets. It is noteworthy that, in humans, a homozygous mutation in the caveolin-1 gene was reported recently in a patient with Berardinelli–Seip congenital lipodystrophy, a finding that further illustrates the importance of caveolin expression for normal lipid storage in humans, as well as in rodents (Kim *et al.*, 2008).

Catabolic pathways

Stimulation of lipolysis by cAMP in adipocytes has been known for decades (see Chapter 6). Phosphorylation of the hormone sensitive lipase (HSL) by cAMP-dependent protein kinase A (PKA) was established originally as the rate-limiting step in this reaction. However, subsequently, new molecular participants in adipocyte lipolysis have been discovered. Perilipin A, which belongs to the PAT protein family, is present in adipose cells, tightly associated at the surface of the lipid droplet (Londos *et al.*, 2005). This feature appeared to be crucial to protect triglycerides from uncontrolled hydrolysis by cytoplasmic lipases. Upon lipolytic stimulation (e.g. catecholamines), perilipin is phosphorylated by PKA and modified in a manner that facilitates the access of phosphorylated HSL to its substrate. Invalidation of the perilipin gene in mice has further established the role of perilipin *in vivo* (Martinez-Botas *et al.*, 2000; Tansey *et al.*, 2001). Although consuming equal, or even more, food than their wild-type littermates, perilipin-null mice exhibited a reduced adipose tissue mass and resistance to diet-induced or genetic obesity. Low adiposity was attributed to elevated rates of basal lipolysis, due to the loss of the protective effect of perilipin on lipid droplets. Consequently, perilipin-null mice showed an increased tendency to develop glucose intolerance and peripheral insulin resistance, a likely consequence of the unblunted release of free fatty acids from adipocytes. This phenotype contrasted with the lack of alteration of lipolysis observed in mice with disruption of the HSL gene, which, in turn, displayed no major change in adiposity (Osuga *et al.*, 2000; Wang *et al.*, 2001). This suggested that another lipase was able to substitute for the absence of HSL. Subsequently, the cloning of the adipose triglyceride lipase (ATGL), also named desnutrin and iPLA2ζ, identified the enzyme responsible for the residual lipolytic activity in HSL knockout mice (Jenkins *et al.*, 2004; Villena *et al.*, 2004; Zimmermann *et al.*, 2004). The generation of ATGL-null rodents revealed the rate-limiting importance of this lipase for triglyceride breakdown (Haemmerle *et al.*, 2006). Unlike HSL-deficient mice, ATGL-knockout mice show defective lipolysis and increased adipose tissue mass. However, due to a reduced fatty acid mobilization, the use of glucose as a metabolic fuel is enhanced, resulting in increased glucose tolerance and insulin sensitivity in this murine model.

A surprising phenotypic feature of perilipin-deficient mice is that their plasma leptin concentration is greater than that expected for their low adiposity (Tansey *et al.*, 2001). In many different physiological or pathological situations, a strong relationship between leptin secretion and the size of adipose lipid stores has been established, which underlies the concept of the 'adipostat' (see Chapter 5). An alteration in this relationship is indicative of a disruption of the mechanisms by which fat cells detect their lipid stores. In this sense, the implication of perilipin in such a process was unexpected. Interestingly, this suggests that the structure of the lipid droplet, of which perilipin is a major protein component, plays a central role in adipocyte lipid sensing. Alternatively, the sensing mechanisms may involve lipid species present in the lipid droplet interacting with perilipin. Such a role has been proposed for cholesterol, whose intracellular localization has been shown to vary with respect to fat cell size and to influence various aspects of adipocyte metabolism (Le Lay *et al.*, 2001). Thus, agents that could inactivate perilipin may prove useful as antiobesity drugs, although with a potential increased risk of alterations in glucose homeostasis and leptin regulation.

Adipose Cell Signalling

Among the multitude of factors that impinge on adipose cell functions, glucocorticoids and insulin have a key role. In fact, in normal-weight individuals, adipocytes are among the cells most sensitive to the effects of insulin. With regards to glucocorticoids, it is noteworthy that the distribution of glucocorticoid receptors is not uniform among fat depots, resulting in regionally distinct hormonal sensitivity. The phenotypic analysis of several transgenic mice, where these hormonal pathways were altered specifically in adipose tissue, provides insight into their implication in the control of body fat mass and energy homeostasis.

Glucocorticoids

Exposure to excessive glucocorticoids favours fat accretion mostly in visceral adipose tissue, as observed in patients with Cushing's syndrome. Increased glucocorticoid sensitivity in mesenteric adipose cells has been attributed to their higher receptor density than in adipocytes of other locations. The effects of glucocorticoids also depend on their intracellular conversion into an active form by the 11β-hydroxysteroid dehydrogenase type-1 (11β HSD-1). The importance of local glucocorticoid action in the biology of adipose tissue has been demonstrated in two 'opposite' transgenic mouse models. In the first model, glucocorticoid activation was enhanced by overexpression of 11β HSD-1 specifically in adipose tissue (Masuzaki *et al.*, 2001). In the second model, glucocorticoid action was reduced through ectopic adipose expression of the 11β HSD-2 isoform, known to inactivate corticosterone in the kidneys (Kershaw *et al.*, 2005). The phenotypic comparison of these transgenic mice indicates that activation of glucocorticoids in the adipose tissue is sufficient to promote an increased body fat mass, with a specific accumulation of visceral fat. As part of their obesity-prone phenotype, mice overexpressing 11β HSD-1 exhibited an increased food intake. By contrast, mice with glucocorticoid inactivation, which were resistant to weight gain on a high-fat diet, exhibited a reduced food intake and an increased energy expenditure. Interestingly, leptin was excluded as a determinant of food intake in both types of transgenic mice, suggesting the implication of other adipocyte-secreted factors. These elegant studies raise the possibility that reduction of active glucocorticoids in adipose tissue, potentially through adipose tissue-specific inhibition of 11β HSD-1, represents a target to improve energy balance and promote resistance to diet-induced obesity.

Insulin

Insulin exerts a lipogenic action by stimulating glucose transport and lipogenesis while inhibiting lipolysis in adipocytes. Not unexpectedly, the fat-specific disruption of the insulin receptor gene in mice, known as FIRKO mutants (fat-specific insulin receptor knockout mice), resulted in low fat mass and protection against obesity (Blüher *et al.*, 2002). The most striking feature of the FIRKO mice is their ability to maintain low adiposity with increasing age and to exhibit an unusually long lifespan (Blüher *et al.*, 2003). This occurred in the face of normal food intake, demonstrating the beneficial effect of reduced adiposity as a key contributor to longevity, even without caloric restriction. Interestingly, a similar phenotype where leanness was associated with extended lifespan was reported in mice genetically engineered to have their C/EBPα gene replaced with a second copy of C/EBPβ (Chiu *et al.*, 2004). In this model, protection against fat accumulation, despite increased food intake, was attributed to the upregulation of genes encoding factors involved in energy dissipation and mitochondrial uncoupling in adipose tissue. Such a metabolic feature is observed in distinct transgenic mouse models involving the manipulation of nuclear receptors or their co-factors, although without information regarding the lifespan of mice (Wang *et al.*, 2003; Leonardsson *et al.*, 2004). The acute ectopic expression of uncoupling protein-1 (UCP1) in epididymal adipose tissue of diet-induced obese mice was shown to reduce adipocyte size, improve glucose tolerance and decrease food intake (Yamada *et al.*, 2006). The adipose-derived signals, which might contribute to the beneficial effects of leanness on whole-body homeostasis and longevity, still need to be elucidated completely.

A number of studies have pointed out the role of sirtuins in mediating the life-extending effects of caloric restriction in yeast, *Caenorhabditis elegans* and *Drosophila* (Guarente and Picard, 2005). Sirtuin1 (Sirt1) is the mammalian homologue of the SIR2 protein, a NAD-dependent histone deacetylase that can act through the silencing of gene expression (Imai *et al.*, 2000). In mice lacking one Sirt1 allele, the mobilization of fatty acids on fasting is compromised. Conversely, overexpression of Sirt1 in cultured adipose cells attenuates adipogenesis and triggers lipolysis and loss of fat. At the molecular level, Sirt1 targets the lipogenic transcription factor PPARγ and represses its transcriptional activity (Picard *et al.*, 2004). In transgenic mice, a moderate overexpression of Sirt1 does not reduce adipose tissue mass significantly but protects against high-fat diet-induced metabolic damage, as evidenced by a lower lipid-induced inflammation in the liver, along with better glucose tolerance and complete protection from hepatic steatosis (Pfluger *et al.*, 2008). In addition, caloric restriction increases Sirt1 expression in a variety of rat tissues, including the adipose tissue (Cohen *et al.*, 2004). Thus, Sirt1 has been proposed as a molecular link connecting caloric restriction to reduced fat accretion and longevity. These observations further support the contribution of low adiposity in determining lifespan in rodents (Guarente and Picard, 2005).

Adipose Tissue-derived Factors

Through a wide range of secreted factors, adipose tissue represents a secretory and endocrine organ highly integrated into whole-body homeostasis. Factors secreted by adipose tissue establish to a network of communication with other tissues and organs and also influence lipid storage and mobilization within the adipose tissue itself. The local role of adipose-derived factors has been analysed in transgenic mouse models, through the manipulation of their own gene or alterations of their respective receptors.

Leptin

The well-known weight-reducing effects of leptin are mediated mainly through the central nervous system and this was demonstrated unequivocally by the observation that selective deletion of leptin receptors in neurones results in obesity (Cohen *et al.*, 2001). In another study, adipocyte-selective reduction of the leptin receptors has been induced by antisense RNA in mice (Huan *et al.*, 2003). Despite normal expression levels of leptin receptors in the hypothalamus and normal food intake, these mice exhibited an increased adiposity and body weight gain in response to high-fat feeding. In a reverse model, adenovirus-induced hyperleptinaemia led to profound morphological and molecular changes in adipose tissue, including increased mitochondriogenesis and upregulation of energy-dissipating genes (Orci *et al.*, 2004). Thus, a local role for leptin might entail the maintenance of an appropriate level of fat oxidation in adipocytes. Since intact leptin signalling in adipose cells appears to be required for the maintenance of normal adipose tissue mass (Orci *et al.*, 2004), this raises the possibility that, in addition to central leptin resistance leading to hyperphagia, peripheral leptin resistance at the adipocyte level favours fat accretion in the obese state.

Angiotensinogen

Adipose tissue has been implicated as an important source of extra-hepatic angiotensinogen (Cassis *et al.*, 1988; Frederich *et al.*, 1992). Angiotensinogen is the precursor of angiotensin II, a well-characterized peptide, which plays a critical role in the regulation of blood pressure. The physiological importance of adipose-produced angiotensinogen was brought to light by astute genetic manipulations in mice. In wild-type mice, targeted overexpression of angiotensinogen in adipose tissue induced hypertension and, rather unexpectedly, increased adiposity (Massiera *et al.*, 2001a). Conversely, angiotensinogen-deficient mice were hypotensive and gained less weight than wild-type mice, despite exhibiting similar food intakes (Massiera *et al.*, 2001b). Remarkably, the re-expression of angiotensinogen specifically in adipose tissue was sufficient to restore normal blood pressure and adipose tissue mass in angiotensinogen-null mice. These genetically modified animal models revealed that angiotensinogen production by adipose tissue influenced not only blood pressure, but also adipose tissue growth. Several reports suggest the existence of a functional renin–angiotensin system in the adipose tissue. Thus, angiotensinogen is converted locally into angiotensin II, which has been shown to act as a potent trophic factor in adipose tissue development (Saint-Marc *et al.*, 2001). Angiotensin receptors exist in the form of two major subtypes, AT1R and AT2R. The latter appears to mediate lipogenic effects of angiotensin II, as suggested by the phenotype of ATR2-null mice, which display small size adipose cells and are protected against nutritional obesity (Yvan-Charvet *et al.*, 2005). Further studies are needed to test whether specific ATR2 inhibitors might be therapeutically helpful in reducing adiposity.

PAI-1

Plasminogen activator inhibitor-1 (PAI-1), a member of the serine protease inhibitor family, is synthesized by a variety of cells including adipocytes. In humans and in animal models of obesity, adipose PAI-1 gene expression and plasma levels increase with body fat mass (Alessi *et al.*, 2000). To test whether PAI-1 contributes to the development of obesity, transgenic mice have been created through genetic knockout or overexpression of PAI-1, with controversial results to some extent. A report indicates that PAI-1 deficient mice are protected against diet-induced obesity, due to increased metabolic rates with no change in food intake (Ma *et al.*, 2004). Accordingly, disruption of the PAI-1 gene reduced adiposity in leptin-deficient obese mice (Schafer *et al.*, 2001). In contrast, other studies suggest that PAI-1 deficiency has no beneficial effects on obesity and that overexpression of PAI-1 specifically in adipose tissue attenuates nutritionally induced obesity (Morange *et al.*, 2000; Lijnen *et al.*, 2003). Thus, whether and how PAI-1 influences adipose tissue growth still needs to be unravelled completely. Locally, PAI-1 may play a role in pericellular proteolysis, a process central to tissue remodelling and angiogenesis. Several adipose-derived factors share the capacity to affect adipose vasculature (Bouloumie *et al.*, 2001). Besides PAI-1, further vasoactive factors include leptin (Bouloumie *et al.*, 1998), vascular endothelial growth factor (VEGF), matrix metalloproteinases (Bouloumie *et al.*, 2001), a matricellular protein named SPARC (Tartare-Deckert *et al.*, 2001) and the angiopoietin-like protein, Angptl 4 (Kersten *et al.*, 2000; Yoon *et al.*, 2000) (see Chapter 9). Based on the observation that induction of apoptosis in the adipose vascular system reduces adipose mass and normalizes metabolism in obese mice (Kolonin *et al.*, 2004), such adipose proteins, including PAI-1, emerge as potentially useful targets to affect vascular supply to adipose tissue.

Acylation-stimulating protein

Adipose tissue is known to produce multiple proteins of the alternate complement pathway. Adipsin is a secreted serine protease related to complement factor D, which was discovered originally as an adipose differentiation-dependent factor (Cook *et al.*, 1987). Later, adipose tissue was shown to produce the acylation-stimulating protein (ASP), which derives from the interaction of complement C3, factor B and adipsin (Cianflone *et al.*, 2003). ASP appears to be inactive as an immune modulator (see Chapter 8) and its best recognized bioactivity is to stimulate triglyceride storage in adipocytes. At the cellular level, ASP stimulates glucose uptake, increases the activity of DGAT and inhibits HSL and lipolysis. An orphan G protein-coupled receptor (C5L2) expressed in human adipose tissue has been proposed to be the receptor responsible for the metabolic actions of ASP (Kalant *et al.*, 2003). Mice in which the C3 gene has been deleted represent a model of ASP deficiency. The C3/ASP-deficient animals display a substantial reduction of white adipose tissue mass, both on a standard and a high-fat diet, despite higher food intake than wild-type mice (Murray *et al.*, 2000). Not unexpectedly, this phenotype resembles that of DGAT-null mice. Moreover, cross-breeding of C3/ASP knockout animals with leptin-deficient mice results in reduced adiposity and increased energy expenditure (Xia *et al.*, 2002). Thus, reducing the production of ASP or developing ASP receptor antagonists represents potential approaches for treating obesity.

Macrophage Infiltration in the Adipose Tissue

In past years, obesity has emerged as a chronic low-grade inflammatory condition (see Chapter 8). Adipose tissue itself produces inflammation-related proteins, whose expression is dependent of fat mass (Clement *et al.*, 2004). Moreover, several reports in human and animal models of obesity have revealed the presence of macrophages within the adipose tissue (Weisberg *et al.*, 2003; Xu *et al.*,

2003; Curat *et al.*, 2004; Cancello *et al.*, 2005). Preadipocytes can be converted into macrophages in an inflammatory environment (Charriere *et al.*, 2003). Alternatively, bioactive molecules released by adipose cells in response to sheer size stress or oxidative damage might increase the chemotaxis of blood monocytes, similar to what occurs during atherosclerosis (Wellen and Hotamisligil, 2003). Whatever the mechanisms of their recruitment, the presence of macrophages is not inconsequential for the biology of adipose tissue. Indeed, macrophages release proinflammatory cytokines, including TNFα, IL-1, IL-10 and IL-6. In adipocytes, TNFα and IL-6 interact negatively with insulin signalling (Hotamisligil *et al.*, 1994; Lagathu *et al.*, 2003) and also interfere with the adipose tissue secretory function. TNFα is a key determinant of PAI-1 production (Samad *et al.*, 1999), while both cytokines inhibit adiponectin production (Fasshauer *et al.*, 2002, 2003). In addition, infusion of TNFα has been shown to elicit major changes in adipocyte gene expression, repressing genes involved in fat accretion (Ruan *et al.*, 2002). As such, the release of TNFα by macrophages might actually protect adipocytes from increasing excessively in size, although with unwanted consequences related to an elevated fatty acid release and an altered adipose tissue secretion profile. Gaining insight into the interactions between

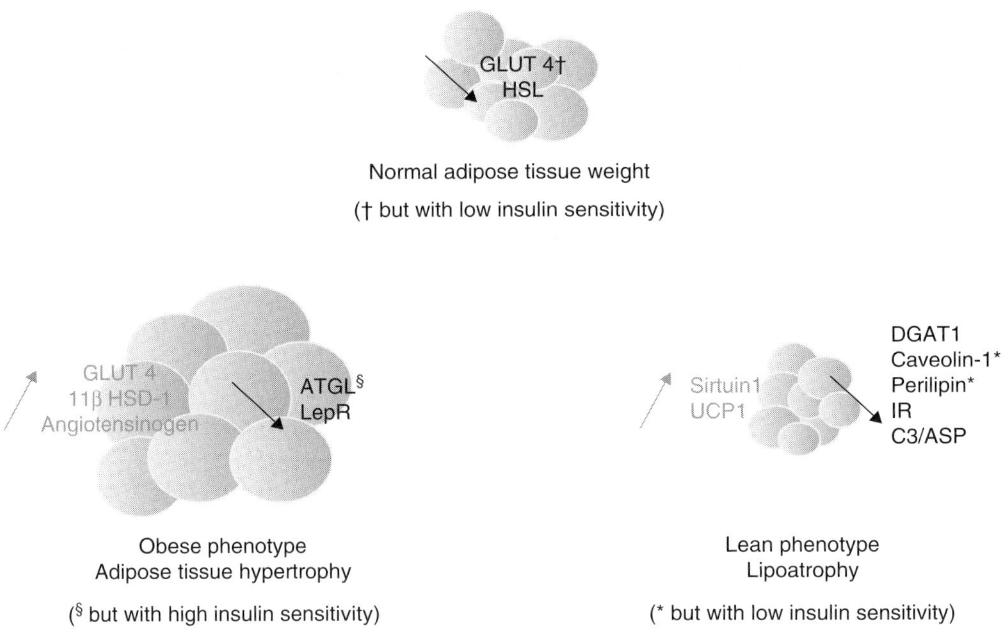

Fig. 13.1. Overexpression (up arrows) or invalidation (down arrows) of selected genes specifically in adipose tissue influence energy homeostasis and adipose tissue mass in transgenic mice, with sometimes unexpected effects on insulin sensitivity, as indicated. GLUT 4, glucose transporter isoform 4; HSL, hormone sensitive lipase; 11β HSD-1, 11β hydroxysteroid dehydrogenase type-1; ATGL, adipose triglyceride lipase (= desnutrin or iPLA2ζ); LepR, leptin receptor; DGAT-1, diacylglycerol acyltransferase 1; IR, insulin receptor; ASP, acylation-stimulating protein; UCP1, uncoupling protein 1.

adipocytes and macrophages at the cellular and molecular level will help to understand the adverse metabolic outcomes of obesity better.

Concluding Remarks

The increasing prevalence of obesity worldwide constitutes a major incentive driving intensive research in the field of body weight and fat mass regulation. Strikingly, alteration in adipose tissue weight is observed in numerous transgenic mouse models, reflecting the pleiotropic control of normal or pathological adipose tissue growth (Valet *et al.*, 2002; Rankinen *et al.*, 2006). It has to be considered that increased fat mass in obesity results inevitably from excessive caloric intake over energy expenditure. As reviewed herein, increasing experimental evidence suggests that besides reflecting energy imbalance, adipose tissue itself influences energy homeostasis. This is exemplified in transgenic mice bearing adipose-specific gene alterations, which represent invaluable models to address this question (Fig. 13.1). The phenotypic analysis of these genetically modified animals allows an integrative approach to adipose tissue biology under normal circumstances or challenged with a high-fat diet. It is clear that the mechanisms underlying the observed shift in energy balance driven by targeted gene alteration in adipose tissue are still poorly understood. Nevertheless, these transgenic studies provide potential targets to promote resistance to diet-induced obesity. Detailed phenotypic analysis of these mice has further contributed to point out that, depending on the targeted pathway, reducing adipose tissue mass could produce undesired side effects, such as fatty acid spillover and alterations in adipose-derived factors. Finally, the development of mouse models with conditional adipose gene alterations is needed to determine whether or not adipose-specific strategies aimed at promoting leanness might reverse the pathological status of adipose tissue, when obesity is fully established.

References

Abel, E.D., Peroni, O., Kim, J.K., Kim, Y.B., Boss, O., Hadro, E., Minnemann, T., Shulman, G.I. and Kahn, B.B. (2001) Adipose-selective targeting of the GLUT4 gene impairs insulin action in muscle and liver. *Nature* 409, 729–733.

Ahima, R.S. (2006) Adipose tissue as an endocrine organ. *Obesity* 14 (Suppl. 5), 242S–249S.

Alessi, M.C., Bastelica, D., Morange, P., Berthet, B., Leduc, I., Verdier, M., Geel, O. and Juhan-Vague, I. (2000) Plasminogen activator inhibitor 1, transforming growth factor-beta1, and BMI are closely associated in human adipose tissue during morbid obesity. *Diabetes* 49, 1374–1380.

Blüher, M., Michael, M.D., Peroni, O.D., Ueki, K., Carter, N., Kahn, B.B. and Kahn, C.R. (2002) Adipose tissue selective insulin receptor knockout protects against obesity and obesity-related glucose intolerance. *Developmental Cell* 3, 25–38.

Blüher, M., Kahn, B.B. and Kahn, C.R. (2003) Extended longevity in mice lacking the insulin receptor in adipose tissue. *Science* 299, 572–574.

Bouloumie, A., Drexler, H.C., Lafontan, M. and Busse, R. (1998) Leptin, the product of Ob gene, promotes angiogenesis. *Circulation Research* 83, 1059–1066.

Bouloumie, A., Sengenes, C., Portolan, G., Galitzky, J. and Lafontan, M. (2001) Adipocyte produces matrix metalloproteinases 2 and 9: involvement in adipose differentiation. *Diabetes* 50, 2080–2086.

Broch, M., Vendrell, J., Ricart, W., Richart, C. and Fernández-Real, J.M. (2007) Circulating retinol-binding protein-4, insulin sensitivity, insulin secretion, and insulin disposition index in obese and non-obese subjects. *Diabetes Care* 30, 1802–1806.

Cancello, R., Henegar, C., Viguerie, N., Taleb, S., Poitou, C., Rouault, C., Coupaye, M., Pelloux, V., Hugol, D., Bouillot, J.L., Bouloumie, A., Barbatelli, G., Cinti, S., Svensson, P.A., Barsh, G.S., Zucker, J.D., Basdevant, A., Langin, D. and Clement, K. (2005) Reduction of macrophage infiltration and chemoattractant gene expression changes in white adipose tissue of morbidly obese subjects after surgery-induced weight loss. *Diabetes* 54, 2277–2286.

Cases, S., Stone, S.J., Zhou, P., Yen, E., Tow, B., Lardizabal, K.D., Voelker, T. and Farese, R.V. Jr (2001) Cloning of DGAT2, a second mammalian diacylglycerol acyltransferase, and related family members. *Journal of Biological Chemistry* 276, 38870–38876.

Cassis, L.A., Saye, J. and Peach, M.J. (1988) Location and regulation of rat angiotensinogen messenger RNA. *Hypertension* 11, 591–596.

Catalán, V., Gómez-Ambrosi, J., Rotellar, F., Silva, C., Rodríguez, A., Gil, M.J., Cienfuegos, J.A., Salvador, J. and Frühbeck, G. (2008) Expression of caveolin-1 in human adipose tissue is upregulated in obesity and obesity-associated type 2 diabetes mellitus and related to inflammation. *Clinical Endocrinology* 68, 213–219.

Charriere, G., Cousin, B., Arnaud, E., Andre, M., Bacou, F., Penicaud, L. and Castiella, L. (2003) Preadipocyte conversion to macrophage. Evidence of plasticity. *Journal of Biological Chemistry* 278, 9850–9855.

Chen, H.C. and Farese, R.V. Jr (2005) Inhibition of triglyceride synthesis as a treatment strategy for obesity: lessons from DGAT1-deficient mice. *Arteriosclerosis, Thrombosis and Vascular Biology* 25, 482–486.

Chen, H.C., Jensen, D.R., Myers, H.M., Eckel, R.H. and Farese, R.V. Jr (2003) Obesity resistance and enhanced glucose metabolism in mice transplanted with white adipose tissue lacking acyl CoA:diacylglycerol acyltransferase 1. *Journal of Clinical Investigation* 111, 1715–1722.

Chiu, C.H., Lin, W.D., Huang, S.Y. and Lee, Y.H. (2004) Effect of a C/EBP gene replacement on mitochondrial biogenesis in fat cells. *Genes and Development* 18, 1970–1975.

Cho, Y.M., Youn, B.S., Lee, H., Lee, N., Min, S.S., Kwak, S.H., Lee, H.K. and Park, K.S. (2006) Plasma retinol-binding protein-4 concentrations are elevated in human subjects with impaired glucose tolerance and type 2 diabetes. *Diabetes Care* 29, 2457–2461.

Cianflone, K., Xia, Z. and Chen, L.Y. (2003) Critical review of acylation-stimulating protein physiology in humans and rodents. *Biochemical Biophysical Acta* 1609, 127–143.

Clement, K., Viguerie, N., Poitou, C., Carette, C., Pelloux, V., Curat, C.A., Sicard, A., Rome, S., Benis, A., Zucker, J.D., Vidal, H., Laville, M., Barsh, G.S., Basdevant, A., Stich, V., Cancello, R. and Langin, D. (2004) Weight loss regulates inflammation-related genes in white adipose tissue of obese subjects. *FASEB Journal* 18, 1657–1669.

Cohen, A.W., Razani, B., Wang, X.B., Combs, T.P., Williams, T.M., Scherer, P.E. and Lisanti, M.P. (2003) Caveolin-1-deficient mice show insulin resistance and defective insulin receptor protein expression in adipose tissue. *American Journal of Physiology – Cell Physiology* 285, C222–C235.

Cohen, H.Y., Miller, C., Bitterman, K.J., Wall, N.R., Hekking, B., Kessler, B., Howitz, K.T., Gorospe, M., Cabo, R. de and Sinclair, D.A. (2004) Calorie restriction promotes mammalian cell survival by inducing the SIRT1 deacetylase. *Science* 305, 390–392.

Cohen, P., Zhao, C., Cai, X., Montez, J.M., Rohani, S.C., Feinstein, P., Mombaerts, P. and Friedman, J.M. (2001) Selective deletion of leptin receptor in neurons leads to obesity. *Journal of Clinical Investigation* 108, 1113–1121.

Coll, A.P., Farooqi, I.S. and O'Rahilly, S. (2007) The hormonal control of food intake. *Cell* 129, 251–262.

Cook, K.S., Min, H.Y., Johnson, D., Chaplinsky, R.J., Flier, J.S., Hunt, C.R. and Spiegelman, B.M. (1987) Adipsin: a circulating serine protease homolog secreted by adipose tissue and sciatic nerve. *Science* 237, 402–405.

Curat, C.A., Miranville, A., Sengenes, C., Diehl, M., Tonus, C., Busse, R. and Bouloumie, A. (2004) From blood monocytes to adipose tissue-resident macrophages: induction of diapedesis by human mature adipocytes. *Diabetes* 53, 1285–1292.

Drab, M., Verkade, P., Elger, M., Kasper, M., Lohn, M., Lauterbach, B., Menne, J., Lindschau, C., Mende, F., Luft, F.C., Schedl, A., Haller, H. and Kurzchalia, T.V. (2001) Loss of caveolae, vascular dysfunction, and pulmonary defects in caveolin-1 gene-disrupted mice. *Science* 293, 2449–2452.

Fasshauer, M., Klein, J., Neumann, S., Eszlinger, M. and Paschke, R. (2002) Hormonal regulation of adiponectin gene expression in 3T3-L1 adipocytes. *Biochemical Biophysical Research Communications* 290, 1084–1089.

Fasshauer, M., Kralisch, S., Klier, M., Lossner, U., Bluher, M., Klein, J. and Paschke, R. (2003) Adiponectin gene expression and secretion is inhibited by interleukin-6 in 3T3-L1 adipocytes. *Biochemical Biophysical Research Communications* 301, 1045–1050.

Frederich, R.C.J., Kahn, B.B., Peach, M.J. and Flier, J.S. (1992) Tissue-specific nutritional regulation of angiotensinogen in adipose tissue. *Hypertension* 19, 339–344.

Frühbeck, G. (2006) Hunting for new pieces to the complex puzzle of obesity. *Proceedings of the Nutrition Society* 65, 329–347.

Garg, A. (2004) Acquired and inherited lipodystrophies. *New England Journal of Medicine* 350, 1220–1234.

Gómez-Ambrosi, J., Rodríguez, A., Catalán, V., Ramírez, B., Silva, C., Rotellar, F., Gil, M.J., Salvador, J. and Frühbeck, G. (2008) Serum retinol-binding protein 4 is not increased in obesity or obesity-associated type 2 diabetes mellitus, but is reduced after relevant reductions in body fat following gastric bypass. *Clinical Endocrinology* 69, 208–219.

Graham, T.E., Yang, Q., Blüher, M., Hammarstedt, A., Ciaraldi, T.P., Henry, R.R., Wason, C.J., Oberbach, A., Jansson, P.A., Smith, U. and Kahn, B.B. (2006) Retinol-binding protein 4 and insulin resistance in lean, obese, and diabetic subjects. *New England Journal of Medicine* 354, 2552–2563.

Guarente, L. and Picard, F. (2005) Calorie restriction – the SIR2 connection. *Cell* 120, 473–482.

Guerre-Millo, M. (2004) Adipose tissue and adipokines: for better or worse. *Diabetes and Metabolism* 30, 13–19.

Guerre-Millo, M. (2008) Adiponectin: an update. *Diabetes and Metabolism* 34, 12–18.

Haemmerle, G., Lass, A., Zimmermann, R., Gorkiewicz, G., Meyer, C., Rozman, J., Heldmaier, G., Maier, R., Theussl, C., Eder, S., Kratky, D., Wagner, E.F., Klingenspor, M., Hoefler, G. and Zechner, R. (2006) Defective lipolysis and altered energy metabolism in mice lacking adipose triglyceride lipase. *Science* 312, 734–737.

Hotamisligil, G.S., Murray, D.L., Choy, L.N. and Spiegelman, B.M. (1994) Tumor necrosis factor alpha inhibits signaling from the insulin receptor. *Proceedings of the National Academy of Sciences of the United States of America* 91, 4854–4858.

Huan, J.N., Li, J., Han, Y., Chen, K., Wu, N. and Zhao, A.Z. (2003) Adipocyte-selective reduction of the leptin receptors induced by antisense RNA leads to increased adiposity, dyslipidemia, and insulin resistance. *Journal of Biological Chemistry* 278, 45638–45650.

Imai, S., Armstrong, C.M., Kaeberlein, M. and Guarente, L. (2000) Transcriptional silencing and longevity protein Sir2 is an NAD-dependent histone deacetylase. *Nature* 403, 795–800.

Janke, J., Engeli, S., Boschmann, M., Adams, F., Bohnke, J., Luft, F.C., Sharma, A.M. and Jordan, J. (2006) Retinol-binding protein 4 in human obesity. *Diabetes* 55, 2805–2810.

Jenkins, C.M., Mancuso, D.J., Yan, W., Sims, H.F., Gibson, B. and Gross, R.W. (2004) Identification, cloning, expression, and purification of three novel human calcium-independent phospholipase A2 family members possessing triacylglycerol lipase and acylglycerol transacylase activities. *Journal of Biological Chemistry* 279, 48968–48975.

Kalant, D., Cain, S.A., Maslowska, M., Sniderman, A.D., Cianflone, K. and Monk, P.N. (2003) The chemoattractant receptor-like protein C5L2 binds the C3a des-Arg77/acylation-stimulating protein. *Journal of Biological Chemistry* 278, 11123–11129.

Kershaw, E.E., Morton, N.M., Dhillon, H., Ramage, L., Seckl, J.R. and Flier, J.S. (2005) Adipocyte-specific glucocorticoid inactivation protects against diet-induced obesity. *Diabetes* 54, 1023–1031.

Kersten, S., Mandard, S., Tan, N.S., Escher, P., Metzger, D., Chambon, P., Gonzalez, F.J., Desvergne, B. and Wahli, W. (2000) Characterization of the fasting-induced adipose factor FIAF, a novel peroxisome proliferator-activated receptor target gene. *Journal of Biological Chemistry* 275, 28488–28493.

Kim, C.A., Delépine, M., Boutet, E., El Mourabit, H., Le Lay, S., Meier, M., Nemani, M., Bridel, E., Leite, C.C., Bertola, D.R., Semple, R.K., O'Rahilly, S., Dugail, I., Capeau, J., Lathrop, M. and Magré, J. (2008) Association of a homozygous nonsense caveolin-1 mutation with Berardinelli-Seip congenital lipodystrophy. *Journal of Clinical Endocrinology and Metabolism* 93, 1129–1134.

Kolonin, M.G., Saha, P.K., Chan, L., Pasqualini, R. and Arap, W. (2004) Reversal of obesity by targeted ablation of adipose tissue. *Nature Medicine* 10, 625–632.

Lagathu, C., Bastard, J.P., Auclair, M., Maachi, M., Capeau, J. and Caron, M. (2003) Chronic interleukin-6 (IL-6) treatment increased IL-6 secretion and induced insulin resistance in adipocyte: prevention by rosiglitazone. *Biochemical and Biophysical Research Communications* 311, 372–379.

Le Lay, S., Krief, S., Farnier, C., Lefrere, I., Le Lievpre, X., Bazin, R., Ferre, P. and Dugail, I. (2001) Cholesterol, a cell size-dependent signal that regulates glucose metabolism and gene expression in adipocytes. *Journal of Biological Chemistry* 276, 16904–16910.

Le Lay, S., Hajduch, E., Lindsay, M.R., Le Lievpre, X., Thiele, C., Ferre, P., Parton, R.G., Kurzchalia, T., Simons, K. and Dugail, I. (2006) Cholesterol-induced caveolin targeting to lipid droplets in adipocytes: a role for caveolar endocytosis. *Traffic* 7, 549–561.

Leonardsson, G., Steel, J.H., Christian, M., Pocock, V., Milligan, S., Bell, J., So, P.W., Medina-Gomez, G., Vidal-Puig, A., White, R. and Parker, M.G. (2004) Nuclear receptor corepressor RIP140 regulates fat accumulation. *Proceedings of the National Academy of Sciences of the United States of America* 101, 8437–8442.

Lijnen, H.R., Maquoi, E., Morange, P., Voros, G., Van Hoef, B., Kopp, F., Collen, D., Juhan-Vague, I. and Alessi, M.C. (2003) Nutritionally induced obesity is attenuated in transgenic mice overexpressing plasminogen activator inhibitor-1. *Arteriosclerosis, Thrombosis and Vascular Biology* 23, 78–84.

Londos, C., Sztalryd, C., Tansey, J.T. and Kimmel, A.R. (2005) Role of PAT proteins in lipid metabolism. *Biochimie* 87, 45–49.

Ma, L.J., Mao, S.L., Taylor, K.L., Kanjanabuch, T., Guan, Y., Zhang, Y., Brown, N.J., Swift, L.L., McGuinness, O.P., Wasserman, D.H., Vaughan, D.E. and Fogo, A.B. (2004) Prevention of obesity and insulin resistance in mice lacking plasminogen activator inhibitor 1. *Diabetes* 53, 336–346.

Martinez-Botas, J., Anderson, J.B., Tessier, D., Lapillonne, A., Chang, B.H., Quast, M.J., Gorenstein, D., Chen, K.H. and Chan, L. (2000) Absence of perilipin results in leanness and reverses obesity in Lepr(*db/db*) mice. *Nature Genetics* 26, 474–479.

Massiera, F., Bloch-Faure, M., Ceiler, D., Murakami, K., Fukamizu, A., Gasc, J.M., Quignard-Boulange, A., Negrel, R., Ailhaud, G., Seydoux, J., Meneton, P. and Teboul, M. (2001a) Adipose angiotensinogen is involved in adipose tissue growth and blood pressure regulation. *FASEB Journal* 15, 2727–2729.

Massiera, F., Seydoux, J., Geloen, A., Quignard-Boulange, A., Turban, S., Saint-Marc, P., Fukamizu, A., Negrel, R., Ailhaud, G. and Teboul, M. (2001b) Angiotensinogen-deficient mice exhibit impairment of diet-induced weight gain with alteration in adipose tissue development and increased locomotor activity. *Endocrinology* 142, 5220–5225.

Masuzaki, H., Paterson, J., Shinyama, H., Morton, N.M., Mullins, J.J., Seckl, J.R. and Flier, J.S. (2001) A transgenic model of visceral obesity and the metabolic syndrome. *Science* 294, 2166–2170.

Morange, P.E., Lijnen, H.R., Alessi, M.C., Kopp, F., Collen, D. and Juhan-Vague, I. (2000) Influence of PAI-1 on adipose tissue growth and metabolic parameters in a murine model of diet-induced obesity. *Arteriosclerosis, Thrombosis and Vascular Biology* 20, 1150–1154.

Murray, I., Havel, P.J., Sniderman, A.D. and Cianflone, K. (2000) Reduced body weight, adipose tissue, and leptin levels despite increased energy intake in female mice lacking acylation-stimulating protein. *Endocrinology* 141, 1041–1049.

Orci, L., Cook, W.S., Ravazzola, M., Wang, M.Y., Park, B.H., Montesano, R. and Unger, R.H. (2004) Rapid transformation of white adipocytes into fat-oxidizing machines. *Proceedings of the National Academy of Sciences of the United States of America* 101, 2058–2063.

Osuga, J., Ishibashi, S., Oka, T., Yagyu, H., Tozawa, R., Fujimoto, A., Shionoiri, F., Yahagi, N., Kraemer, F.B., Tsutsumi, O. and Yamada, N. (2000) Targeted disruption of hormone-sensitive lipase results in male sterility and adipocyte hypertrophy, but not in obesity. *Proceedings of the National Academy of Sciences of the United States of America* 97, 787–792.

Pfluger, P.T., Herranz, D., Velasco-Miguel, S., Serrano, M. and Tschöp, M.H. (2008) Sirt1 protects against high-fat diet-induced metabolic damage. *Proceedings of the National Academy of Sciences of the United States of America* 105, 9793–9798.

Picard, F., Kurtev, M., Chung, N., Topark-Ngarm, A., Senawong, T., Machado, D.O., Leid, M., McBurney, M.W. and Guarente, L. (2004) Sirt1 promotes fat mobilization in white adipocytes by repressing PPAR-gamma. *Nature* 429, 771–776.

Qi, Y., Takahashi, N., Hileman, S.M., Patel, H.R., Berg, A.H., Pajvani, U.B., Scherer, P.E. and Ahima, R.S. (2004) Adiponectin acts in the brain to decrease body weight. *Nature Medicine* 10, 524–529.

Rankinen, T., Zuberi, A., Chagnon, Y.C., Weisnagel, S.J., Argyropoulos, G., Walts, B., Pérusse, L. and Bouchard, C. (2006) The human obesity gene map: the 2005 update. *Obesity* 14, 529–644.

Razani, B., Engelman, J.A., Wang, X.B., Schubert, W., Zhang, X.L., Marks, C.B., Macaluso, F., Russell, R.G., Li, M., Pestell, R.G., Di Vizio, D., Hou, H. Jr, Kneitz, B., Lagaud, G.,

Christ, G.J., Edelmann, W. and Lisanti, M.P. (2001) Caveolin-1 null mice are viable but show evidence of hyperproliferative and vascular abnormalities. *Journal of Biological Chemistry* 276, 38121–38138.

Razani, B., Combs, T.P., Wang, X.B., Frank, P.G., Park, D.S., Russell, R.G., Li, M., Tang, B., Jelicks, L.A., Scherer, P.E. and Lisanti, M.P. (2002) Caveolin-1-deficient mice are lean, resistant to diet-induced obesity, and show hypertriglyceridemia with adipocyte abnormalities. *Journal of Biological Chemistry* 277, 8635–8647.

Ruan, H., Miles, P.D., Ladd, C.M., Ross, K., Golub, T.R., Olefsky, J.M. and Lodish, H.F. (2002) Profiling gene transcription *in vivo* reveals adipose tissue as an immediate target of tumor necrosis factor-alpha: implications for insulin resistance. *Diabetes* 51, 3176–3188.

Saint-Marc, P., Kozak, L.P., Ailhaud, G., Darimont, C. and Negrel, R. (2001) Angiotensin II as a trophic factor of white adipose tissue: stimulation of adipose cell formation. *Endocrinology* 142, 487–492.

Samad, F., Uysal, K.T., Wiesbrock, S.M., Pandey, M., Hotamisligil, G.S. and Loskutoff, D.J. (1999) Tumor necrosis factor alpha is a key component in the obesity-linked elevation of plasminogen activator inhibitor 1. *Proceedings of the National Academy of Sciences of the United States of America* 96, 6902–6907.

Schafer, K., Fujisawa, K., Konstantinides, S. and Loskutoff, D.J. (2001) Disruption of the plasminogen activator inhibitor 1 gene reduces the adiposity and improves the metabolic profile of genetically obese and diabetic *ob/ob* mice. *FASEB Journal* 15, 1840–1842.

Shepherd, P.R., Gnudi, L., Tozzo, E., Yang, H., Leach, F. and Kahn, B.B. (1993) Adipose cell hyperplasia and enhanced glucose disposal in transgenic mice overexpressing GLUT4 selectively in adipose tissue. *Journal of Biological Chemistry* 268, 22243–22246.

Skurk, T., Alberti-Huber, C., Herder, C. and Hauner, H. (2007) Relationship between adipocyte size and adipokine expression and secretion. *Journal of Clinical Endocrinology and Metabolism* 92, 1023–1033.

Smith, S.J., Cases, S., Jensen, D.R., Chen, H.C., Sande, E., Tow, B., Sanan, D.A., Raber, J., Eckel, R.H. and Farese, R.V. Jr (2000) Obesity resistance and multiple mechanisms of triglyceride synthesis in mice lacking Dgat. *Nature Genetics* 25, 87–90.

Soukas, A., Cohen, P., Socci, N.D. and Friedman, J.M. (2000) Leptin-specific patterns of gene expression in white adipose tissue. *Genes and Development* 14, 963–980.

Spalding, K.L., Arner, E., Westermark, P.O., Bernard, S., Buchholz, B.A., Bergmann, O., Blomqvist, L., Hoffstedt, J., Näslund, E., Britton, T., Concha, H., Hassan, M., Rydén, M., Frisén, J. and Arner, P. (2008) Dynamics of fat cell turnover in humans. *Nature* 453, 783–787.

Tansey, J.T., Sztalryd, C., Gruia-Gray, J., Roush, D.L., Zee, J.V., Gavrilova, O., Reitman, M.L., Deng, C.X., Li, C., Kimmel, A.R. and Londos, C. (2001) Perilipin ablation results in a lean mouse with aberrant adipocyte lipolysis, enhanced leptin production, and resistance to diet-induced obesity. *Proceedings of the National Academy of Sciences of the United States of America* 98, 6494–6499.

Tartare-Deckert, S., Chavey, C., Monthouel, M.N., Gautier, N. and Van Obberghen, E. (2001) The matricellular protein SPARC/osteonectin as a newly identified factor upregulated in obesity. *Journal of Biological Chemistry* 276, 22231–22237.

Tozzo, E., Shepherd, P.R., Gnudi, L. and Kahn, B.B. (1995) Transgenic GLUT-4 overexpression in fat enhances glucose metabolism: preferential effect on fatty acid synthesis. *American Journal of Physiology – Endocrinology and Metabolism* 268, E956–E964.

Valet, P., Tavernier, G., Castan-Laurell, I., Saulnier-Blache, J.S. and Langin, D. (2002) Understanding adipose tissue development from transgenic animal models. *Journal of Lipid Research* 43, 835–860.

Villena, J.A., Roy, S., Sarkadi-Nagy, E., Kim, K.H. and Sul, H.S. (2004) Desnutrin, an adipocyte gene encoding a novel patatin domain-containing protein, is induced by fasting and glucocorticoids: ectopic expression of desnutrin increases triglyceride hydrolysis. *Journal of Biological Chemistry* 279, 47066–47075.

Wang, S.P., Laurin, N., Himms-Hagen, J., Rudnicki, M.A., Levy, E., Robert, M.F., Pan, L., Oligny, L. and Mitchell, G.A. (2001) The adipose tissue phenotype of hormone-sensitive lipase deficiency in mice. *Obesity Research* 9, 119–128.

Wang, Y.X., Lee, C.H., Tiep, S., Yu, R.T., Ham, J., Kang, H. and Evans, R.M. (2003) Peroxisome-proliferator-activated receptor delta activates fat metabolism to prevent obesity. *Cell* 113, 159–170.

Weisberg, S.P., McCann, D., Desai, M., Rosenbaum, M., Leibel, R.L. and Ferrante, A.W. Jr (2003) Obesity is associated with macrophage accumulation in adipose tissue. *Journal of Clinical Investigation* 112, 1796–1808.

Wellen, K.E. and Hotamisligil, G.S. (2003) Obesity-induced inflammatory changes in adipose tissue. *Journal of Clinical Investigation* 112, 1785–1788.

Xia, Z., Sniderman, A.D. and Cianflone, K. (2002) Acylation-stimulating protein (ASP) deficiency induces obesity resistance and increased energy expenditure in *ob/ob* mice. *Journal of Biological Chemistry* 277, 45874–45879.

Xu, H., Barnes, G.T., Yang, Q., Tan, G., Yang, D., Chou, C.J., Sole, J., Nichols, A., Ross, J.S., Tartaglia, L.A. and Chen, H. (2003) Chronic inflammation in fat plays a crucial role in the development of obesity-related insulin resistance. *Journal of Clinical Investigation* 112, 1821–1830.

Yamada, T., Katagiri, H., Ishigaki, Y., Ogihara, T., Imai, J., Uno, K., Hasegawa, Y., Gao, J., Ishihara, H. and Niijima, A. (2006) Signals from intra-abdominal fat modulate insulin and leptin sensitivity through different mechanisms: neuronal involvement in food-intake regulation. *Cell Metabolism* 3, 223–229.

Yang, Q., Graham, T.E., Mody, N., Preitner, F., Peroni, O.D., Zabolotny, J.M., Kotani, K., Quadro, L. and Kahn, B.B. (2005) Serum retinol binding protein 4 contributes to insulin resistance in obesity and type 2 diabetes. *Nature* 436, 356–362.

Yoon, J.C., Chickering, T.W., Rosen, E.D., Dussault, B., Qin, Y., Soukas, A., Friedman, J.M., Holmes, W.E. and Spiegelman, B.M. (2000) Peroxisome proliferator-activated receptor gamma target gene encoding a novel angiopoietin-related protein associated with adipose differentiation. *Molecular and Cellular Biology* 20, 5343–5349.

Yvan-Charvet, L., Even, P., Bloch-Faure, M., Guerre-Millo, M., Moustaid-Moussa, N., Ferre, P. and Quignard-Boulange, A. (2005) Deletion of the angiotensin type 2 receptor (AT2R) reduces adipose cell size and protects from diet-induced obesity and insulin resistance. *Diabetes* 54, 991–999.

Zimmermann, R., Strauss, J.G., Haemmerle, G., Schoiswohl, G., Birner-Gruenberger, R., Riederer, M., Lass, A., Neuberger, G., Eisenhaber, F., Hermetter, A. and Zechner, R. (2004) Fat mobilization in adipose tissue is promoted by adipose triglyceride lipase. *Science* 306, 1383–1386.

14 Application of 'Omic' Strategies to Obesity Research

CORNELIU HENEGAR,[1] SORAYA TALEB,[1] DOMINIQUE LANGIN,[2] JEAN-DANIEL ZUCKER[1] AND KARINE CLÉMENT[1]

[1]INSERM, Nutriomique, Paris; University Paris, France; CHRU Pitié Salpétrière, Hôtel-Dieu Nutrition Department, Centre de Recherche en Nutrition Humaine, Paris, France; [2]INSERM, Laboratoire de Recherches sur les Obésités, Institut de Médecine Moléculaire de Rangueil, Toulouse, France

Introduction

Genetic advances have made remarkable progress in our understanding of body weight regulation. Much of our current knowledge has come from the cloning and characterization of the genes responsible for obesity syndromes and the identification of mutations causing rare forms of obesity in humans (Mutch and Clément, 2006; Yang *et al.*, 2007; Ichihara and Yamada, 2008; Lindgren and McCarthy, 2008). The genetic approach has been successful in identifying genes implicated in rare monogenic syndromes, but remains insufficient to identify master genes involved in polygenic human obesity, despite recent discoveries of candidate genes by genome-wide scan approaches in very large human populations. Indeed, the genetic determinants that underlie common forms of human obesity are largely polygenic, with a single gene producing only a discreet effect on body weight control, with common obesity resulting from interactions of multiple susceptibility alleles with environmental factors. Thus, elucidating the genetic determinants of common obesity remains a major challenge for researchers. In spite of the inherent technical difficulties, progress is being made by the use of 'omic' strategies to study the influence of nutrition and gene–environment interactions in human obesity (Brandacher *et al.*, 2008; Herbert, 2008; Ordovas and Tai, 2008; Rasche *et al.*, 2008; Stylianou *et al.*, 2008; Twigger *et al.*, 2008). The purpose of this chapter is to provide some examples to explain how the 'omic' approaches, and especially transcriptomics, contribute to advancing our knowledge in the field of obesity at the molecular, cellular, tissue and whole-body physiology level. Particular emphasis is placed on how adipose tissue microarrays are now used to identify novel tissue candidates involved in energy homeostasis and obesity-related complications and, more generally, to better understand the complex pathophysiological mechanisms that could play a role in the natural history of human obesity (Corthésy-Theulaz *et al.*, 2005; Viguerie *et al.*, 2005a; Ferguson, 2006; Jiao *et al.*, 2008). In the near future, information derived from diverse fields including nutrigenomics, transcriptomics, proteomics, metabolomics and lipidomics will be combined with data obtained from genetic approaches to improve our knowledge and management of a highly complex disease such as obesity.

From Genetic to 'Omic' Approaches

Genetic approach

The most evident success of the molecular approach has been to unravel the mechanisms of some monogenic forms of obesity. The crucial roles of the leptin and melanocortin pathways in controlling food intake have been individualized within a redundant system of feeding control (Coll *et al.*, 2007). It has been shown that 2–4% of morbidly obese subjects have a mutation in the MC4R gene (Vaisse *et al.*, 1998; Farooqi and O'Rahilly, 2007; Farooqi, 2008). However, in less frequent forms of obesity, the situation is much more complex, as is the case for other multifactorial diseases. The recognized aim of the molecular approach is to define the contribution of specific genes to the risk of developing the disease, taking into account the inter-individual variance of clinical traits. European and North American research groups have established cohorts of obese families (Mutch and Clément, 2006; Rankinen *et al.*, 2006; Farooqi and O'Rahilly, 2007). Some are directed towards ethnic 'isolates' supposed to be less heterogeneous, or towards representative samples of the general population. In these different populations, the role of numerous genetic polymorphisms in the multifactorial components of obesity has been proposed, showing sometimes, but not always, overlapping results (Dahlman and Arner, 2007; Dolley *et al.*, 2008). These genes code for a wide number of proteins associated with, for example, central and peripheral control of feeding, energy expenditure, the biology of adipose tissue, skeletal muscle and the liver. The genome-wide scan technique, starting with no 'a priori' hypothesis, has offered a new way of identifying candidates that subsequently can be examined using the candidate gene approach. Genetic regions linked to obesity have been identified both in obese adults and children using high throughput genome study techniques. Polymorphisms of candidate genes located in the regions of linkage to obesity have been identified by this approach, such as GAD2 (Boutin *et al.*, 2003), adiponectin (Vasseur *et al.*, 2003), an amino acid transporter protein, SLC4 (Suviolahti *et al.*, 2003; Durand *et al.*, 2004) and, more recently, in the vicinity of the FTO, Insig2 and MC4R genes. The recently identified genes related to obesity and diabetes have been summarized clearly in extensive reviews (English and Butte, 2007; Sun, 2007; Lindgren and McCarthy, 2008).

Some individual single nucleotide polymorphisms (SNPs) or candidate haplotypes are associated with either an increased or a decreased relative risk of obesity or diabetes development in humans. Genetic maps periodically record the genes and polymorphisms implicated in obesity of various European and

American populations (Mutch and Clément, 2006; Rankinen *et al.*, 2006). In spite of the power of current analysis, it has been difficult for some studies to draw definitive conclusions concerning the role of these tested candidates to explain fully the links observed in the genome, which generally include thousands of bases. Among the difficulties in identifying molecular targets, the statistical power of genetic analysis and the necessity to combine several methods in heterogeneous populations have to be pointed out (Stylianou *et al.*, 2008). While the associations between body mass index (BMI) and SNPs located in the vicinity of newly identified candidates appear to be replicated more in independent and very large populations than the first ones (Dina *et al.*, 2007; Frayling *et al.*, 2007; Hinney *et al.*, 2007; Loos *et al.*, 2008), the precise biological role of the candidate genes involved in the complex pathophysiology of obesity is mostly unknown.

In addition, strategies for addressing the possible interaction between genetic background and diet composition, as well as the level of physical activity in the development of obesity, are currently starting to be used, most notably in large European cohorts. The number of these loci increases regularly and it is becoming a Sysiphean task to get an updated view of all susceptibility genetic regions or genes. Despite this increasing mass of genetic information, the underlying molecular mechanisms explaining the development and progression of obesity in the majority of individuals are mostly unknown.

Obesity complexity

Obesity is characterized by a high phenotype heterogeneity linked to differences in stages of evolution (Fig. 14.1). The development of obesity is a process, ongoing for years and considered classically to result from an inappropriate adaptation of the systems involved in energy balance control to either a primarily increased energy intake or a reduced energy expenditure, leading to a passive accumulation of surplus energy as fat in adipose tissue. However, the concept of obesity due to a passive storage of the surplus of energy is not sufficient. Adipose tissue is no longer considered a passive bystander in energy balance regulation. On the contrary, it is viewed as an active player in energy homeostasis, due to its intense metabolic activity and its capacity to produce a large number of biomolecules. Advances in the cellular and molecular biology of adipose tissue, the neurobiology of energy balance and the biology of tissue plasticity led to the concept of a specific 'organ disease' (Ailhaud, 1999; Holst and Grimaldi, 2002; Ricquier, 2006; Fricker, 2007; Lago *et al.*, 2007; Guilherme *et al.*, 2008). At some stage of the disease (i.e. the maintenance phase), obesity can be considered as a pathology of adipose tissue and of its relations with other organs implicated in energy balance, such as the liver, skeletal muscle and the hypothalamus (Zigman and Elmquist, 2003). The mechanisms of weight regulation involve an 'inter-organ' dialogue mediated by some known actors (e.g. leptin and adiponectin) and also by other unknown molecules. This stage of obesity maintenance can be illustrated by the fact that obesity after several years of evolution is characterized by a high resistance to weight loss and facility to weight regain. This phenomenon might be related to a profound modification of the biology

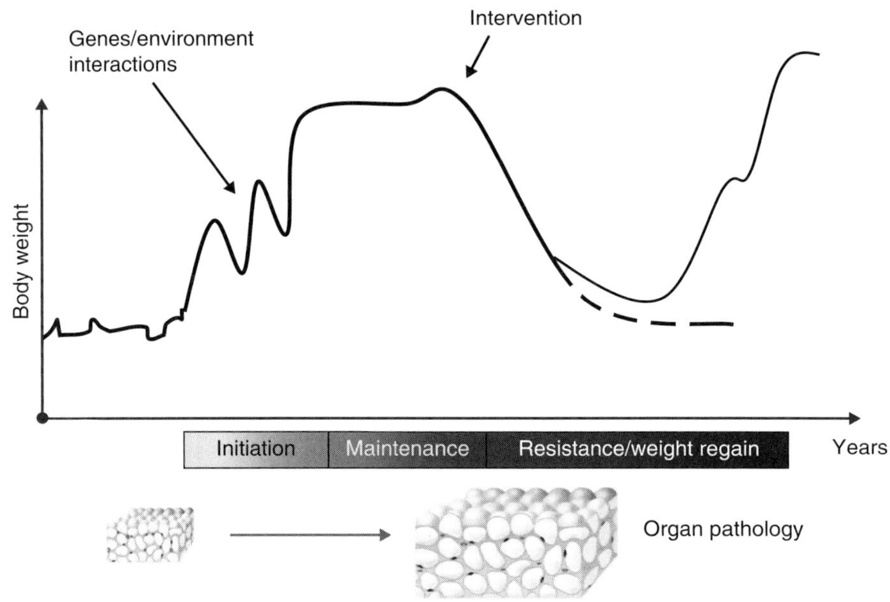

Fig. 14.1. Schematic representation of the different stages of the development of obesity. The body weight increase curve along the years shows that obesity initiation classically results from interactions of susceptibility genes with environmental factors (i.e. overeating and reduction in physical activity). The second phase consists of weight stabilization characterized by modifications of both biology and architecture of adipose tissue. Successful long-term weight loss is difficult to achieve in obese subjects and most of them develop resistance to weight loss finally to regain weight, and even more than before.

and architecture of adipose tissue (Lacasa *et al.*, 2007). Active accumulation of triglycerides in adipocytes, due to a dysregulation between release and accumulation, followed by a subsequent regulatory energy balance adjustment, should be considered (Sorensen, 2003). Each stage in the development of obesity, weight gain and maintenance, as well as in the variable response to treatment, probably can be associated with different molecular mechanisms. No clear biological or molecular predictors (biomarkers) of transition from one stage to the other have been identified (Kussmann *et al.*, 2006). The need to tackle multifactorial diseases including obesity through integration of all available sources of information is now widely recognized (Herbert, 2008; Jiao *et al.*, 2008).

Rationale for using 'omic' approaches

In this complex picture of the pathophysiology of obesity at its different stages of evolution, studies of tissular genes, proteins and metabolites may contribute to revealing the role of certain signals and thus may help to provide a better understanding of the mechanisms of energy homeostasis. Some pathways of fat expansion can be identified in the groups of genes/proteins/metabolites, which are

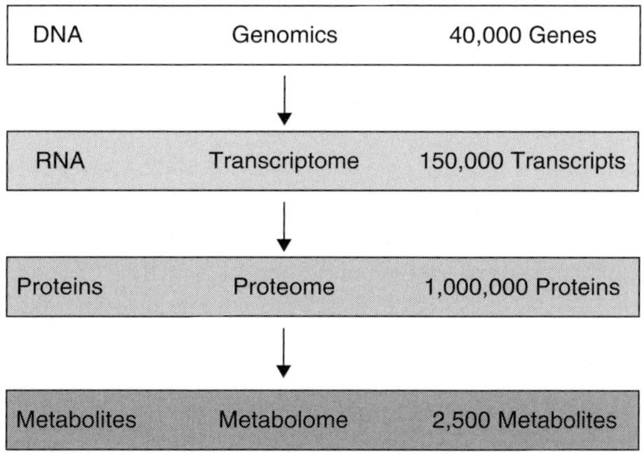

Fig. 14.2. Information stored in 'omic'-derived components. In agreement with current molecular biology dogma, DNA is transcribed into RNA, which is translated into proteins, while the metabolome appears to be smaller.

modified in response to different conditions and environmental changes (Arner, 2000). For gene and protein expression studies, key peripheral tissues (adipose tissues, skeletal muscle and, to a lesser extent, the liver and heart) are accessible by biopsies in humans during clinical investigation protocols. The 'omic' technologies encompassing genomics, transcriptomics, proteomics and metabolomics allow assessment of genes, gene transcripts, proteins and metabolites, respectively, making it possible to analyse a large number of pathways simultaneously (Fig. 14.2).

The field of proteomics and metabolomics is an evolving area, which will help to shed light on the protein and metabolite variations associated with complex diseases (Wang *et al.*, 2008). Although proteomic technology has advanced tremendously in the past years, mainly in the field of cancer research, there are significant technical challenges that pose limitations to the routine application of current proteomic methods. For example, an important challenge when producing protein microarrays is to maintain protein characteristics, such as post-translational modifications and phosphorylations, that are now difficult to achieve with the same protein-array surface chemistry. Another relevant step in understanding cell function is the analysis of the metabolites that are produced as the end products of cellular function (metabolome). The metabolome encompasses a diverse array of chemotypes, including peptides, carbohydrates, lipids, nucleosides and catabolic products of exogenous compounds. These metabolites, or improper degradation of cellular proteins, may contribute to disease development. However, in the field of proteomics and metabolomics, a substantial improvement in the breadth, sensitivity and throughput of global-profiling technologies is still needed. Due to technological advances and the intense development of informatic tools, DNA microarrays have been applied rapidly in large-scale studies and the expression profiles of thousands of genes that change in different cellular or tissular types and environmental conditions have been obtained.

Use of Transcriptomics in the Obesity Field

Transcriptomics: toolbox and goals

Transcriptomics is now one of the most widely used techniques in the field of obesity research due to its profitability and capacity of quantifying thousands of genes simultaneously in a single experiment (Viguerie *et al.*, 2005a). Microarray technology is based on the reverse concept of dot blot and Northern blot analysis. The DNA is attached to the solid phase, whereas labelled cDNA or RNA are in solution. Large numbers of cDNA sequences or synthetic DNA oligomers can be fixed on to a glass slide (or other substrates like filters) in known locations on a grid. Arrays with thousands of cDNA probes or oligomers have been developed by academic consortia or companies. The measured amount of labelled cDNA or RNA bound to each spot reflects the level of expression of the gene. While used initially for simple organisms (e.g. yeast), this approach now analyses thousands of known and newly identified genes in various large groups defined by expression similarities in terms of physiological pathways, such as, for example, respiration and cell division, and in response to chemical or thermal stress.

Microarray DNA screening is now applied to the understanding of complex pathophysiological processes including ageing, cancer and, more recently, metabolic diseases such as diabetes and obesity. The key objective is to dissect and characterize the regulatory pathways and networks involved in energy balance and to define the resulting signalling patterns in gene expression. Goals of future projects represent the identification of clusters of genes that are recruited or modified by given nutritional conditions, together with their links in families of biological processes and their co-regulation in different tissues, as well as gene markers specific for some nutrients, differences/similarities in different models of obesity and, eventually, the patterns of tissue expression in individuals with different genetic polymorphisms located in these genes. Adipose tissue gene expression profile studies may help to understand better what happens during weight changes and the key factors underlying the improvement in obesity complications.

Examples of transcriptomic applications in obesity

Studies of adipocyte differentiation

As already described, adipose mass increases in part through the recruitment and differentiation of existing pools of preadipocytes into adipocytes (Rosen and MacDougald, 2006). DNA microarrays enable the examination of gene expression profiles of cells across differentiation and should allow the discovery of novel adipogenic mediators and biomarkers of adipogenesis. Microarray studies have been performed using 3T3-L1 murine cells, the most widely used model for adipocyte differentiation studies *in vitro* (Jessen and Stevens, 2002; Burton *et al.*, 2004; Sun, 2007). The first list of cellular biomarkers in humans was provided by comparing *in vivo* the gene expression profiles of mature adipocytes and preadipocytes (Urs *et al.*, 2004). Two distinct gene expression profiles were evident, with several genes involved in lipid metabolism being overexpressed in adipocytes while genes encoding extracellular matrix components, such as fibronectin, matrix metalloproteins and other proteins such as lysyl oxidase, were expressed predominantly in preadipocytes. Microarray studies comparing the gene expression of *in vitro* preadipocytes and adipocytes to their *in vivo* counterparts revealed differences in the expression of certain genes in the two systems. It is noteworthy that preadipocytes and adipocytes *in vivo* display a particular phenotype that is not mimicked entirely by preadipocytes and adipocytes *in vitro* (Soukas *et al.*, 2001), confirming the importance of *in vivo* studies.

Studies of animal models

The use of animal models of obesity such as diet-induced obesity (DIO) in rodents represents a useful tool for studying human obesity since a high-fat intake is well known to promote fat mass development. In adipose tissue of DIO mice, the majority of genes, including genes encoding enzymes of the lipid metabolism or markers of adipocyte differentiation, were downregulated in comparison with wild-type mice (Moraes *et al.*, 2003). Similar observations were made in leptin-deficient *ob/ob* mice, suggesting that adipocytes from enlarged adipose tissue mice exhibit reduced lipogenic skills (Nadler *et al.*, 2000; Soukas *et al.*, 2000). In DIO and *ob/ob* mice, other genes, such as those encoding inflammatory markers, displayed an increased expression. In this sense, studies in rodents using microarrays have shown that genes related to inflammatory pathways are overexpressed in adipose tissue of obese mice and that these factors are produced by infiltrating macrophages (Weisberg *et al.*, 2003; Xu *et al.*, 2003). These changes in gene expression probably reflect adaptative mechanisms in adipose tissue of severely obese mice.

Studies of gene expression in subcutaneous and visceral adipose tissues

The anatomical distribution of adipose tissue is a key indicator of metabolic alterations and cardiovascular diseases. Excess fat mass in the upper parts of the body constitutes a classic risk factor to develop diabetes and cardiovascular diseases as compared to the accumulation of adiposity in lower body parts (Donahue and Abbott, 1987; Ducimetiere and Richard, 1989). There are marked differences between subcutaneous and visceral adipose tissues in the expression and secretion of key adipose genes such as leptin (Van Harmelen *et al.*, 1998) and adiponectin (Motoshima *et al.*, 2002), as well as other proinflammatory factors (Klimcakova *et al.*, 2007; Lacasa *et al.*, 2007). Plasminogen activator inhibitor-1, for example, has been related to the pathogenic effects of visceral fat (Bastelica *et al.*, 2002). The large-scale screening of genes differentially expressed in human or animal subcutaneous and visceral adipose depots has allowed the identification of biomarkers of visceral obesity that may represent the mediators of metabolic alterations (Gabrielsson *et al.*, 2000; Yang *et al.*, 2003; Fukuhara *et al.*, 2005; Lacasa *et al.*, 2007). A pangenomic approach based on the combination of two methods, representational difference analysis and microarrays, was performed on cDNA from subcutaneous and omental fat tissues in men with severe abdominal obesity. Forty-four putatively differentially expressed genes were identified and differential expression of genes such as calcyclin and adipsin was found (Linder *et al.*, 2004). Calcyclin belongs to the family of calcium-binding proteins that previously have been implicated in several inflammatory diseases. The adipsins were assimilated to immune complement factors. They have been implicated in the regulation of lipid turnover in human fat cells. In another study, carboxypeptidase-E and thrombospondin-1 were also found to be overexpressed in visceral adipose tissue (Ramis *et al.*, 2002). Moreover, these studies have led to the identification of a large number of genes expressed in adipose tissue whose function still needs to be unravelled (Mazzucotelli *et al.*, 2007).

Biomarkers of increased fat mass

Gene profiling comparison of adipose tissue of obese and non-obese subjects may help to identify the key molecules implicated in the development or consequences of obesity development. Indeed, the molecular links between expanded adipose tissue and obesity complications are still being disentangled (Henegar *et al.*, 2008). Recent large-scale analysis highlighted a significant upregulation of genes and biological functions related to extracellular matrix (ECM) constituents, including members of the integrins family, and suggested that these elements could play a major mediating role in a chain of interactions which connects local inflammatory phenomena to the alteration of WAT metabolic functions in obese subjects (Henegar *et al.*, 2008). The comparison of the gene expression in omental adipose tissue of obese and non-obese subjects has revealed the modified expression of numerous genes, such as those implicated in cellular signalling and immunity processes (Gómez-Ambrosi *et al.*, 2004). Interestingly, receptors of the Fc fragment of IgG are reportedly upregulated in adipose tissue of obese subjects. Fc receptors are known to mediate antibody-dependent inflammatory responses. These results suggest a relevant link between adipose tissue and immunity, as suggested previously in another gene expression study (Gabrielsson *et al.*, 2003).

Adipose tissue gene expression profiling during weight loss

Energy restriction is still one of the most effective methods to study the effect of energy imbalance, as it induces considerable fat mass loss and modifies clinical and metabolic variables (Hainer *et al.*, 1992). Highlighting transcriptional modifications of genes in adipose tissue during modulations of energy balance helps to understand the role of these genes in the adaptation to calorie restriction and weight change (Viguerie *et al.*, 2005a,b). Little is known about the effect of energy restriction and macronutrients on the regulation of adipose tissue gene expression. A multicentric study of the effect of low-fat and moderate-fat low-calorie diets shows that, as observed for anthropometric variables, there is no difference in adipose tissue gene expression between the two diets. However, the energy restriction induced an increase in the expression of the transcriptional coactivator, PGC-1α, which is linked to energy metabolism, as well as a decrease in genes involved in lipid metabolism. The study provides evidence that, during a 10-week low-calorie diet, energy restriction rather than the fat/carbohydrate ratio is of importance to modify the transcriptional programme in human adipose tissue

(Viguerie *et al.*, 2005b). Thus, the beneficial effect of weight loss on obesity-related complications may be associated with the modification of the inflammatory profile in adipose tissue (Clément and Langin, 2007). Our group has used DNA microarrays to investigate the gene expression profile change after 4 weeks of a very low-calorie diet (VLCD) (Clément *et al.*, 2004). The comparison of the gene expression profiles of subcutaneous white adipose tissue from obese subjects before and after the VLCD, as well as with that of non-obese subjects, revealed the importance of inflammation. The mRNA of inflammatory genes was increased in adipose tissue of obese subjects compared to that obtained from non-obese patients. The gene expression of these factors decreased in adipose tissue of obese individuals following a VLCD to a level similar to that of non-obese subjects (Fig. 14.3). We highlighted the fact that subcutaneous adipose tissue of obese subjects also expressed a wide range of factors related to inflammation and, thereby, might participate in metabolic alterations observed in obesity. Furthermore, we and others observed the decrease of the expression of genes encoding acute phase reactants such as serum amyloid A (SAA) after weight loss (Sjöholm *et al.*, 2004; Poitou *et al.*, 2005, 2006). SAA are apolipoprotein A molecules involved in cholesterol transport and in early response to injury. There are also the precursors of amyloid A, a constituent of amyloid fibrils involved in AA amyloidosis (Urieli-Shoval *et al.*, 2000; Merlini and Bellotti, 2003). AA amyloidosis is a complication of many inflammatory conditions (including rheumatoid arthritis and neoplasia). We have shown that the gene expression of SAA is increased in adipose tissue of obese subjects. This finding was confirmed by immunohistochemical analysis of

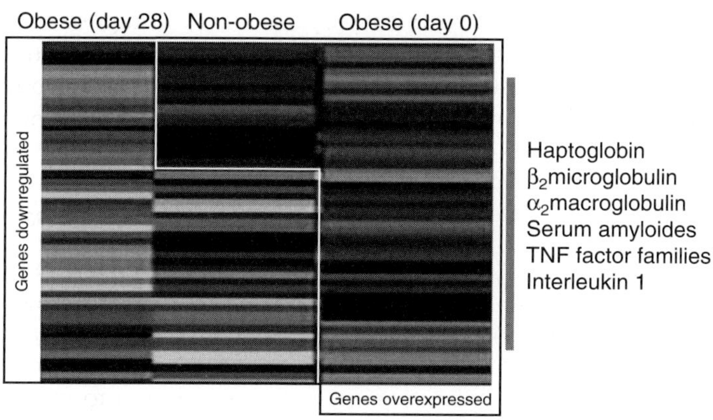

Fig. 14.3. Cluster of inflammation-related gene expression in adipose tissue of non-obese subjects and obese patients before and after following a very low-calorie diet (VLCD) for 4 weeks. Each row represents one gene and each column represents one clinical situation or nutritional condition. The increased darkness of each gene represents the mean in $\log_2$ of gene expression ratios of seven non-obese subjects and ten obese subjects before and after VLCD. The mainly downregulated and overexpressed genes are clearly indicated. Interspersed (or scattered) among the either mainly downregulated or mainly overexpressed genes, the black lines correspond to genes with equal to median (or unchanged) gene expression. An example of inflammation-related genes is given alongside the cluster.

adipose tissue. The SAA gene expression was also shown to be increased in adipose tissue of two leptin receptor-deficient patients characterized by extreme adiposity of very early onset. Surprisingly, in one patient, we found systemic AA amyloidosis associated with kidney dysfunction of unknown origin. We suggested that in obesity, the increase of SAA in adipose tissue might, in some circumstances, be associated with systemic amyloidosis and play a key role in the development of renal disease. Our studies showed that the beneficial effect of weight loss on obesity-related complications might be associated with the modification of the inflammatory profile in subcutaneous adipose tissue. The way is paved for future clinical and cellular studies aimed at determining the impact of these molecular adaptations on the development of obesity and its comorbidities.

Limitations of Microarray Use

Although information and resources are growing, it should be kept in mind that there are still many difficulties and pitfalls in the interpretation of microarray data. A relevant issue faced constantly in research using human tissues is the small number of samples studied compared to the potentially huge number of genes that can be evaluated at the same time. The cost of these techniques is considerable and there is also limited availability of human tissues that are less accessible to biopsy, such as the heart, liver or skeletal muscle (and obviously the brain). However, this later hurdle is being overcome as improvements in mRNA amplification and microarray signal sensitivity allow the use of minute quantities of tissues. A number of studies have been published on the hormonal control of skeletal muscle gene expression (Clément *et al.*, 2002; Viguerie *et al.*, 2004; Larrouy *et al.*, 2008). Moreover, distinctive patterns of expression induced by obesity and hypertension have been described in the heart (Philip-Couderc *et al.*, 2004). It is critical to obtain at some stage an integrated view of gene signatures in different tissues under different conditions (Schadt *et al.*, 2008). The development of large human tissue banks appears to be necessary. Furthermore, in order to identify gene predictors of clinical changes after environmental modifications, large experiments are mandatory to get enough power in the informatic analysis; this is the objective of several ongoing studies (Mutch *et al.*, 2007) funded by the European Union (DioGenes: www.diogenes-eu.org; Nugenob: www.nugenob.org). The tremendous source of variability at different levels coupled to using these techniques has to be mentioned. Measurements of mRNA are subjected inherently to a high biological variability, which also depends on the level of expression of the gene (low versus highly expressed genes). The methods themselves are accompanied by variability: mRNA extraction, hybridization (variability due to temperature, time, mixing), probe labelling (differences in the chemistry of the fluorescent label), image analysis and scanning (laser and detector limitations). In addition, analysis of thousands of results finds large differences that merely are attributable to the random normal distribution of the data. Thus, adequate procedures for multiple testing have to be developed specifically and applied.

Another relevant aspect is related to the standardization procedures not only due to experimental variations but also as regards the reported gene information

(Tables 14.1 and 14.2). An assessment of a selection of DNA microarray literature shows scarce standardization in the field with regard to the methods, analysis and controls used, as well as with data validation (Table 14.1). For example, most functional gene annotations in many publications have been made manually, enabling bias in interpretation and limiting the possibility of comparing information among different independent studies (Table 14.2). The examples of transcriptomic approaches described above already emphasize the need for standardization. Working groups have proposed different standardization procedures recommended for the representation of microarray information. The use of these procedures facilitates the exchange of information between different data systems and research groups. The MicroArray Gene Expression (MAGE) consortium represents a good example. Several types of automatic annotations of genes can also be used. Furthermore, it has to be kept in mind that the resulting gene expression does not necessarily reflect the proteins that serve as the functional effectors of cellular processes. The use of other emerging technologies such as proteomics and metabolomics offers additional and complementary opportunities.

Future Directions

Since the pathophysiology of obesity is complex, it becomes evident that a multidisciplinary research effort involving the combination of various fields (e.g. clinical, biochemical, genetic, transcriptomic, proteomic and metabolomic) is necessary in order to increase our knowledge of the complexity of the biological

Table 14.1. Examples of limitations in microarray data analysis performed in animal adipose tissue studies.

Reference	Type/ condition	Number of genes on microarrays	% Mobilized genes	Validation	Test of multiplicity	Annotation
Nadler *et al.*, 2000	Obesity	11,000	10%	No	No	Manual
Soukas *et al.*, 2000	Obesity	6,500	25%	Yes ($n = 20$)	No	Manual
López *et al.*, 2003	High-fat diet	12,500	15%	Yes ($n = 3$)	No	Manual
Moraes *et al.*, 2003	High-fat diet	12,488	6%	Yes ($n = 6$)	No	Manual
Takahashi *et al.*, 2003	Obesity	12,000	0.1%	–	No	–

Note: The number of genes on the microarray represents the number of cDNA or genes spotted on the array. The % mobilized genes means the % of genes that were significantly selected under the tested condition. Validation of microarrays was performed by quantitative RT-PCR or Northern blot. The test of multiplicity refers to the statistical test adapted to multiple data. Annotation means to assign a gene into a functional class.

Table 14.2. Examples of limitations in microarray data analysis performed in studies with human adipose tissue or adipocytes.

Reference	Type/ condition	Number of genes on microarray	% Mobilized genes	Validation	Test of multiplicity	Annotation
Gabrielsson *et al.*, 2003	Subcutaneous/ visceral fat	–	–	No	No	Manual
Clément *et al.*, 2004	Caloric restriction	40,000	~5%	Yes ($n = 10$)	Yes	Manual
Linder *et al.*, 2004	Subcutaneous/ visceral fat	44	9%	No	No	Celera database
Gómez-Ambrosi *et al.*, 2004	Visceral obesity	1,152	13%	Yes ($n = 6$)	No	Manual
Urs *et al.*, 2004	Preadipocyte/ adipocyte	~9,000	~1%	Yes ($n = 5$)	Yes	Gene ontology tree machine
Poitou *et al.*, 2005	Obesity	40,000	~2%	–	Yes	Manual

Note: The number of genes on the microarray represents the number of cDNA or genes spotted on the array. The % mobilized genes means the % of genes that were significantly selected under the tested condition. Validation of microarrays was performed by quantitative RT-PCR or Northern blot. The test of multiplicity refers to the statistical test adapted to multiple data. Annotation means to assign a gene into a functional class.

traits and processes of the disease. The development of data mining tools is essential to exploit fully the enormous amount of information yielded by these techniques. One of the future challenges is to process the mass of information generated from diverse phenotypic and genotypic analyses, as well as different nutritional conditions.

Advances in technology now allow for combining the search for gene variation/mutation and gene expression profiling on a genome-wide basis. Thus, the overlap between gene profiling studies, whole genome scans and the candidate gene maps available in humans and rodents will constitute important steps in combining information. A proof of concept of this approach has been provided by a study in which gene expression data and genome-wide scans were combined in standard inbred mice strains (Schadt *et al.*, 2003). The strains were crossed together and the F2 generation yielded was fed with a high-fat diet for 4 months. The animals were phenotyped with regard to obesity-related traits and metabolic variables. The comparison of the differential hepatic gene expression levels of obese and lean animals showed that 30% of the genes affected differentially in the rodents could represent molecular signatures of the lean and the obese status. Moreover, two different hepatic gene expression profiles were observed in the obese mice. A second step aimed at identifying genes or chromosomal regions linked to obesity phenotype has been achieved by using genome-wide scans. The variation of liver gene expression in the animal strains was used as a quantitative trait (eQTL). Levels of mRNA proved to be a highly valuable trait that gave strong association with chromosomal loci. The linkage study not only identified chromosomal regions involved in the control of adiposity, but also in the regulation of hepatic gene expression. In fact, specific chromosomal regions discriminated the two types of obese mice with different patterns of liver expression. Some genes or chromosomal regions that may be involved both in the regulation of genes expressed in the liver (with a different pattern of expression in lean and obese rodents) and in adiposity probably will be better identified. Another study using the same approach combining gene expression profiling in fat and kidney with linkage analysis was used to identify the genes involved in the metabolic syndrome (Hubner *et al.*, 2005). This technology combining large-scale analysis of the genome and of pangenomic expression has been applied recently to human blood and adipose tissue and points out the important contribution of inflammatory genes in obese individuals (Emilsson *et al.*, 2008).

The future will tell us whether the identified genes and pathways are good targets for intervention. Among the limitations to this integrated approach, the difficulty of having large enough samples, as well as the stage of development of biocomputing tools that are still in their infancy, for addressing the question of multiple interactions with no 'a priori hypotheses' have to be mentioned. None the less, it can be expected that this rapidly changing and expanding technology will open up new avenues in this exciting and complex research field.

Acknowledgements

The works cited as having been carried out by the authors' teams were promoted by Direction de la Recherche Clinique (DRC)/Assistance Publique Hôpitaux de Paris (PHRC Programme 02076) and grants were obtained by ALFEDIAM, INSERM (PRNH No 4NU10G), Agence Nationale de la Recherche (RIOMA Programme No ANR-05-PCOD-030-02) and the European programmes, 'NUGENOB' and 'Diogenes'.

References

Ailhaud, G. (1999) Cell surface receptors, nuclear receptors and ligands that regulate adipose tissue development. *Clinica Chimica Acta* 286, 181–190.

Arner, P. (2000) Obesity – a genetic disease of adipose tissue? *British Journal of Nutrition* 83 (Suppl. 1), S9–16.

Bastelica, D., Morange, P., Berthet, B., Borghi, H., Lacroix, O., Grino, M., Juhan-Vague, I. and Alessi, M.C. (2002) Stromal cells are the main plasminogen activator inhibitor-1-producing cells in human fat: evidence of differences between visceral and subcutaneous deposits. *Arteriosclerosis, Thrombosis, and Vascular Biology* 22, 173–178.

Boutin, P., Dina, C., Vasseur, F., Dubois, S., Corset, L., Séron, K., Bekris, L., Cabellon, J., Neve, B., Vasseur-Delannoy, V., Chikri, M., Charles, M.A., Clement, K., Lernmark, A. and Froguel, P. (2003) GAD2 on chromosome 10p12 is a candidate gene for human obesity. *PLoS Biology* 1, E68.

Brandacher, G., Golderer, G., Kienzl, K., Werner, E.R., Margreiter, R. and Weiss, H.G. (2008) Potential applications of global protein expression analysis (proteomics) in morbid obesity and bariatric surgery. *Obesity Surgery* 18, 905–910.

Burton, G.R., Nagarajan, R., Peterson, C.A. and McGehee, R.E. Jr (2004) Microarray analysis of differentiation-specific gene expression during 3T3-L1 adipogenesis. *Gene* 329, 167–185.

Clément, K. and Langin, D. (2007) Regulation of inflammation-related genes in human adipose tissue. *Journal of Internal Medicine* 262, 422–430.

Clément, K., Viguerie, N., Diehn, M., Alizadeh, A., Barbe, P., Thalamas, C., Storey, J.D., Brown, P.O., Barsh, G.S. and Langin, D. (2002) *In vivo* regulation of human skeletal muscle gene expression by thyroid hormone. *Genome Research* 12, 281–291.

Clément, K., Viguerie, N., Poitou, C., Carette, C., Pelloux, V., Curat, C.A., Sicard, A., Rome, S., Benis, A., Zucker, J.D., Vidal, H., Laville, M., Barsh, G.S., Basdevant, A., Stich, V., Cancello, R. and Langin, D. (2004) Weight loss regulates inflammation-related genes in white adipose tissue of obese subjects. *FASEB Journal* 18, 1657–1669.

Coll, A.P., Farooqi, I.S. and O'Rahilly, S. (2007) The hormonal control of food intake. *Cell* 129, 251–262.

Corthésy-Theulaz, I., Dunnen, J.T. den, Ferré, P., Geurts, J.M., Müller, M., Belzen, N. van and Ommen, B. van (2005) Nutrigenomics: the impact of biomics technology on nutrition research. *Annals of Nutrition and Metabolism* 49, 355–365.

Dahlman, I. and Arner, P. (2007) Obesity and polymorphisms in genes regulating human adipose tissue. *International Journal of Obesity* 31, 1629–1641.

Dina, C., Meyre, D., Gallina, S., Durand, E., Körner, A., Jacobson, P., Carlsson, L.M., Kiess, W., Vatin, V., Lecoeur, C., Delplanque, J., Vaillant, E., Pattou, F., Ruiz, J., Weill, J., Levy-Marchal, C., Horber, F., Potoczna, N., Hercberg, S., Le Stunff, C., Bougnères, P., Kovacs, P., Marre, M., Balkau, B., Cauchi, S., Chèvre, J.C. and Froguel, P. (2007) Variation in FTO contributes to childhood obesity and severe adult obesity. *Nature Genetics* 39, 724–726.

Dolley, G., Bertrais, S., Frochot, V., Bebel, J.F., Guerre-Millo, M., Tores, F., Rousseau, F., Hager, J., Basdevant, A., Hercberg, S., Galan, P., Oppert, J.M., Lacorte, J.M. and Clément, K. (2008) Promoter adiponectin polymorphisms and waist/hip ratio variation in a prospective French adults study. *International Journal of Obesity* 32, 669–675.

Donahue, R.P. and Abbott, R.D. (1987) Central obesity and coronary heart disease in men. *Lancet* 2, 1215.

Ducimetiere, P. and Richard, J.L. (1989) The relationship between subsets of anthropometric upper versus lower body measurements and coronary heart disease risk in middle-aged men. The Paris Prospective Study. I. *International Journal of Obesity* 13, 111–121.

Durand, E., Boutin, P., Meyre, D., Charles, M.A., Clement, K., Dina, C. and Froguel, P. (2004) Polymorphisms in the amino acid transporter solute carrier family 6 (neurotransmitter transporter) member 14 gene contribute to polygenic obesity in French Caucasians. *Diabetes* 53, 2483–2486.

Emilsson, V., Thorleifsson, G., Zhang, B., Leonardson, A.S., Zink, F., Zhu, J., Carlson, S., Helgason, A., Walters, G.B., Gunnarsdottir, S., Mouy, M., Steinthorsdottir, V., Eiriksdottir, G.H., Bjornsdottir, G., Reynisdottir, I., Gudbjartsson, D., Helgadottir, A., Jonasdottir, A., Jonasdottir, A., Styrkarsdottir, U., Gretarsdottir, S., Magnusson, K.P., Stefansson, H., Fossdal, R., Kristjansson, K., Gislason, H.G., Stefansson, T., Leifsson, B.G., Thorsteinsdottir, U., Lamb, J.R., Gulcher, J.R., Reitman, M.L., Kong, A., Schadt, E.E. and Stefansson, K. (2008) Genetics of gene expression and its effect on disease. *Nature* 452, 423–428.

English, S.B. and Butte, A.J. (2007) Evaluation and integration of 49 genome-wide experiments and the prediction of previously unknown obesity-related genes. *Bioinformatics* 23, 2910–2917.

Farooqi, I.S. (2008) Monogenic human obesity. *Frontiers of Hormone Research* 36, 1–11.

Farooqi, I.S. and O'Rahilly, S. (2007) Genetic factors in human obesity. *Obesity Reviews* 8 (Suppl. 1), 37–40.

Ferguson, L.R. (2006) Nutrigenomics: integrating genomic approaches into nutrition research. *Molecular Diagnosis and Therapeutics* 10, 101–108.

Frayling, T.M., Timpson, N.J., Weedon, M.N., Zeggini, E., Freathy, R.M., Lindgren, C.M., Perry, J.R., Elliott, K.S., Lango, H., Rayner, N.W., Shields, B., Harries, L.W., Barrett, J.C., Ellard, S., Groves, C.J., Knight, B., Patch, A.M., Ness, A.R., Ebrahim, S., Lawlor, D.A., Ring, S.M., Ben-Shlomo, Y., Jarvelin, M.R., Sovio, U., Bennett, A.J., Melzer, D., Ferrucci, L., Loos, R.J., Barroso, I., Wareham, N.J., Karpe, F., Owen, K.R., Cardon, L.R., Walker, M., Hitman, G.A., Palmer, C.N., Doney, A.S., Morris, A.D., Smith, G.D., Hattersley, A.T. and McCarthy, M.I. (2007) A common variant in the FTO gene is associated with body mass index and predisposes to childhood and adult obesity. *Science* 316, 889–894.

Fricker, L.D. (2007) Neuropeptidomics to study peptide processing in animal models of obesity. *Endocrinology* 148, 4185–4190.

Fukuhara, A., Matsuda, M., Nishizawa, M., Segawa, K., Tanaka, M., Kishimoto, K., Matsuki, Y., Murakami, M., Ichisaka, T., Murakami, H., Watanabe, E., Takagi, T., Akiyoshi, M., Ohtsubo, T., Kihara, S., Yamashita, S., Makishima, M., Funahashi, T., Yamanaka, S., Hiramatsu, R., Matsuzawa, Y. and Shimomura, I. (2005) Visfatin: a protein secreted by visceral fat that mimics the effects of insulin. *Science* 307, 426–430.

Gabrielsson, B.G., Johansson, J.M., Lönn, M., Jernås, M., Olbers, T., Peltonen, M., Larsson, I., Lönn, L., Sjöström, L., Carlsson, B. and Carlsson, L.M. (2003) High expression of complement components in omental adipose tissue in obese men. *Obesity Research* 11, 699–708.

Gabrielsson, B.L., Carlsson, B. and Carlsson, L.M. (2000) Partial genome-scale analysis of gene expression in human adipose tissue using DNA array. *Obesity Research* 8, 374–384.

Gómez-Ambrosi, J., Catalán, V., Diez-Caballero, A., Martínez-Cruz, A., Gil, M.J., García-Foncillas, J., Cienfuegos, J.A., Salvador, J., Mato, J.M. and Frühbeck, G. (2004) Gene expression profile of omental adipose tissue in human obesity. *FASEB Journal* 18, 215–217.

Guilherme, A., Virbasius, J.V., Puri, V. and Czech, M.P. (2008) Adipocyte dysfunctions linking obesity to insulin resistance and type 2 diabetes. *Nature Reviews Molecular and Cellular Biology* 9, 367–377.

Hainer, V., Stich, V., Kunesová, M., Parízková, J., Zák, A., Wernischová, V. and Hrabák, P. (1992) Effect of 4-wk treatment of obesity by very-low-calorie diet on anthropometric, metabolic, and hormonal indexes. *American Journal of Clinical Nutrition* 56 (Suppl. 1), 281S–282S.

Henegar, C., Tordjman, J., Achard, V., Lacasa, D., Cremer, I., Guerre-Millo, M., Poitou, C., Basdevant, A., Stich, V., Viguerie, N., Langin, D., Bedossa, P., Zucker, J.D. and Clement, K. (2008) Adipose tissue transcriptomic signature highlights the pathological relevance of extracellular matrix in human obesity. *Genome Biology* 9, R14.

Herbert, A. (2008) The fat tail of obesity as told by the genome. *Current Opinion in Clinical and Nutritional Metabolic Care* 11, 366–370.

Hinney, A., Nguyen, T.T., Scherag, A., Friedel, S., Brönner, G., Müller, T.D., Grallert, H., Illig, T., Wichmann, H.E., Rief, W., Schäfer, H. and Hebebrand, J. (2007) Genome wide association (GWA) study for early-onset extreme obesity supports the role of fat mass and obesity associated gene (FTO) variants. *PLoS ONE* 2, e1361.

Holst, D. and Grimaldi, P.A. (2002) New factors in the regulation of adipose differentiation and metabolism. *Current Opinion in Lipidology* 13, 241–245.

Hubner, N., Wallace, C.A., Zimdahl, H., Petretto, E., Schulz, H., Maciver, F., Mueller, M., Hummel, O., Monti, J., Zidek, V., Musilova, A., Kren, V., Causton, H., Game, L., Born, G., Schmidt, S., Müller, A., Cook, S.A., Kurtz, T.W., Whittaker, J., Pravenec, M. and Aitman, T.J. (2005) Integrated transcriptional profiling and linkage analysis for identification of genes underlying disease. *Nature Genetics* 37, 243–253.

Ichihara, S. and Yamada, Y. (2008) Genetic factors for human obesity. *Cellular and Molecular Life Sciences* 65, 1086–1098.

Jessen, B.A. and Stevens, G.J. (2002) Expression profiling during adipocyte differentiation of 3T3-L1 fibroblasts. *Gene* 299, 95–100.

Jiao, H., Kaaman, M., Dungner, E., Kere, J., Arner, P. and Dahlman, I. (2008) Association analysis of positional obesity candidate genes based on integrated data from transcriptomics and linkage analysis. *International Journal of Obesity* 32, 816–825.

Klimcakova, E., Moro, C., Mazzucotelli, A., Lolmède, K., Viguerie, N., Galitzky, J., Stich, V. and Langin, D. (2007) Profiling of adipokines secreted from human subcutaneous adipose tissue in response to PPAR agonists. *Biochemical and Biophysical Research Communications* 358, 897–902.

Kussmann, M., Raymond, F. and Affolter, M. (2006) OMICS-driven biomarker discovery in nutrition and health. *Journal of Biotechnology* 124, 758–787.

Lacasa, D., Taleb, S., Keophiphath, M., Miranville, A. and Clément, K. (2007) Macrophage-secreted factors impair human adipogenesis: involvement of proinflammatory state in preadipocytes. *Endocrinology* 148, 868–877.

Lago, F., Diéguez, C., Gómez-Reino, J. and Gualillo, O. (2007) The emerging role of adipokines as mediators of inflammation and immune responses. *Cytokine and Growth Factor Reviews* 18, 313–325.

Larrouy, D., Barbe, P., Valle, C., Déjean, S., Pelloux, V., Thalamas, C., Bastard, J.P., Le Bouil, A., Diquet, B., Clément, K., Langin, D. and Viguerie, N. (2008) Gene expression profiling of human skeletal muscle in response to stabilized weight loss. *American Journal of Clinical Nutrition* 88, 125–132.

Linder, K., Arner, P., Flores-Morales, A., Tollet-Egnell, P. and Norstedt, G. (2004) Differentially expressed genes in visceral or subcutaneous adipose tissue of obese men and women. *Journal of Lipid Research* 45, 148–154.

Lindgren, C.M. and McCarthy, M.I. (2008) Mechanisms of disease: genetic insights into the etiology of type 2 diabetes and obesity. *Nature Clinical Practice Endocrinology and Metabolism* 4, 156–163.

Loos, R.J., Lindgren, C.M., Li, S., Wheeler, E., Zhao, J.H., Prokopenko, I., Inouye, M., Freathy, R.M., Attwood, A.P., Beckmann, J.S., Berndt, S.I., Prostate, Lung, Colorectal, and Ovarian (PLCO) Cancer Screening Trial, Jacobs, K.B., Chanock, S.J., Hayes, R.B., Bergmann, S., Bennett, A.J., Bingham, S.A., Bochud, M., Brown, M., Cauchi, S., Connell, J.M., Cooper, C., Smith, G.D., Day, I., Dina, C., De, S., Dermitzakis, E.T., Doney, A.S., Elliott, K.S., Elliott, P., Evans, D.M., Sadaf Farooqi, I., Froguel, P., Ghori, J., Groves, C.J., Gwilliam, R., Hadley, D., Hall, A.S., Hattersley, A.T., Hebebrand, J., Heid, I.M., KORA, Lamina, C., Gieger, C., Illig, T., Meitinger, T., Wichmann, H.E., Herrera, B., Hinney, A., Hunt, S.E., Jarvelin, M.R., Johnson, T., Jolley, J.D., Karpe, F., Keniry, A., Khaw, K.T., Luben, R.N., Mangino, M., Marchini, J., McArdle, W.L., McGinnis, R., Meyre, D., Munroe, P.B., Morris, A.D., Ness, A.R., Neville, M.J., Nica, A.C., Ong, K.,K., O'Rahilly, S., Owen, K.R., Palmer, C.N., Papadakis, K., Potter, S., Pouta, A., Qi, L., Nurses' Health Study, Randall, J.C., Rayner, N.W., Ring, S.M., Sandhu, M.S., Scherag, A., Sims, M.A., Song, K., Soranzo, N., Speliotes, E.K., Diabetes Genetics Initiative, Syddall, H.E., Teichmann, S.A., Timpson, N.J., Tobias,

J.H., Uda, M., SardiNIA Study, Vogel, C.I., Wallace, C., Waterworth, D.M., Weedon, M.N., Wellcome Trust Case Control Consortium, Willer, C.J., FUSION, Wraight, V.L., Yuan, X., Zeggini, E., Hirschhorn, J.N., Strachan, D.P., Ouwehand, W.H., Caulfield, M.J., Samani, N.J., Frayling, T.M., Vollenweider, P., Waeber, G., Mooser, V., Deloukas, P., McCarthy, M.I., Wareham, N.J., Barroso, I., Jacobs, K.B., Chanock, S.J., Hayes, R.B., Lamina, C., Gieger, C., Illig, T., Meitinger, T., Wichmann, H.E., Kraft, P., Hankinson, S.E., Hunter, D.J., Hu, F.B., Lyon, H.N., Voight, B.F., Ridderstrale, M., Groop, L., Scheet, P., Sanna, S., Abecasis, G.R., Albai, G., Nagaraja, R., Schlessinger, D., Jackson, A.U., Tuomilehto, J., Collins, F.S., Boehnke, M. and Mohlke, K.L. (2008) Common variants near MC4R are associated with fat mass, weight and risk of obesity. *Nature Genetics* 40, 768–775.

López, I.P., Marti, A., Milagro, F.I., Zulet, M., Moreno-Aliaga, M.J., Martinez, J.A. and De Miguel, C. (2003) DNA microarray analysis of genes differentially expressed in diet-induced (cafeteria) obese rats. *Obesity Research* 11, 188–194.

Mazzucotelli, A., Viguerie, N., Tiraby, C., Annicotte, J.S., Mairal, A., Klimcakova, E., Lepin, E., Delmar, P., Dejean, S., Tavernier, G., Lefort, C., Hidalgo, J., Pineau, T., Fajas, L., Clément, K. and Langin, D. (2007) The transcriptional coactivator peroxisome proliferator activated receptor (PPAR)gamma coactivator-1 alpha and the nuclear receptor PPAR alpha control the expression of glycerol kinase and metabolism genes independently of PPAR gamma activation in human white adipocytes. *Diabetes* 56, 2467–2475.

Merlini, G. and Bellotti, V. (2003) Molecular mechanisms of amyloidosis. *New England Journal of Medicine* 349, 583–596.

Moraes, R.C., Blondet, A., Birkenkamp-Demtroeder, K., Tirard, J., Orntoft, T.F., Gertler, A., Durand, P., Naville, D. and Bégeot, M. (2003) Study of the alteration of gene expression in adipose tissue of diet-induced obese mice by microarray and reverse transcription-polymerase chain reaction analyses. *Endocrinology* 144, 4773–4782.

Motoshima, H., Wu, X., Sinha, M.K., Hardy, V.E., Rosato, E.L., Barbot, D.J., Rosato, F.E. and Goldstein, B.J. (2002) Differential regulation of adiponectin secretion from cultured human omental and subcutaneous adipocytes: effects of insulin and rosiglitazone. *Journal of Clinical Endocrinology and Metabolism* 87, 5662–5667.

Mutch, D.M. and Clément, K. (2006) Unraveling the genetics of human obesity. *PLoS Genetics* 2, e188.

Mutch, D.M., Temanni, M.R., Henegar, C., Combes, F., Pelloux, V., Holst, C., Sørensen, T.I., Astrup, A., Martinez, J.A., Saris, W.H., Viguerie, N., Langin, D., Zucker, J.D. and Clément, K. (2007) Adipose gene expression prior to weight loss can differentiate and weakly predict dietary responders. *PLoS ONE* 2, e1344.

Nadler, S.T., Stoehr, J.P., Schueler, K.L., Tanimoto, G., Yandell, B.S. and Attie, A.D. (2000) The expression of adipogenic genes is decreased in obesity and diabetes mellitus. *Proceedings of the National Academy of Sciences of the United States of America* 97, 11371–11376.

Ordovas, J.M. and Tai, E.S. (2008) Why study gene–environment interactions? *Current Opinion in Lipidology* 19, 158–167.

Philip-Couderc, P., Pathak, A., Smih, F., Dambrin, C., Harmancey, R., Buys, S., Galinier, M., Massabuau, P., Roncalli, J., Senard, J.M. and Rouet, P. (2004) Uncomplicated human obesity is associated with a specific cardiac transcriptome: involvement of the Wnt pathway. *FASEB Journal* 18, 1539–1540.

Poitou, C., Viguerie, N., Cancello, R., De Matteis, R., Cinti, S., Stich, V., Coussieu, C., Gauthier, E., Courtine, M., Zucker, J.D., Barsh, G.S., Saris, W., Bruneval, P., Basdevant, A., Langin, D. and Clément, K. (2005) Serum amyloid A: production by human white adipocyte and regulation by obesity and nutrition. *Diabetologia* 48, 519–528.

Poitou, C., Coupaye, M., Laaban, J.P., Coussieu, C., Bedel, J.F., Bouillot, J.L., Basdevant, A., Clément, K. and Oppert, J.M. (2006) Serum amyloid A and obstructive sleep apnea syndrome before and after surgically-induced weight loss in morbidly obese subjects. *Obesity Surgery* 16, 1475–1481.

Ramis, J.M., Franssen-van Hal, N.L., Kramer, E., Llado, I., Bouillaud, F., Palou, A. and Keijer, J. (2002) Carboxypeptidase E and thrombospondin-1 are differently expressed in subcutaneous and visceral fat of obese subjects. *Cellular and Molecular Life Sciences* 59, 1960–1971.

Rankinen, T., Zuberi, A., Chagnon, Y.C., Weisnagel, S.J., Argyropoulos, G., Walts, B., Pérusse, L. and Bouchard, C. (2006) The human obesity gene map: the 2005 update. *Obesity* 14, 529–644.

Rasche, A., Al-Hasani, H. and Herwig, R. (2008) Meta-analysis approach identifies candidate genes and associated molecular networks for type-2 diabetes mellitus. *BMC Genomics* 9, 310.

Ricquier, D. (2006) Fundamental mechanisms of thermogenesis. *Comptes Rendus Biologie* 329, 578–586.

Rosen, E.D. and MacDougald, O.A. (2006) Adipocyte differentiation from the inside out. *Nature Reviews Molecular and Cellular Biology* 7, 885–896.

Schadt, E.E., Monks, S.A., Drake, T.A., Lusis, A.J., Che, N., Colinayo, V., Ruff, T.G., Milligan, S.B., Lamb, J.R., Cavet, G., Linsley, P.S., Mao, M., Stoughton, R.B. and Friend, S.H. (2003) Genetics of gene expression surveyed in maize, mouse and man. *Nature* 422, 297–302.

Schadt, E.E., Molony, C., Chudin, E., Hao, K., Yang, X., Lum, P.Y., Kasarskis, A., Zhang, B., Wang, S., Suver, C., Zhu, J., Millstein, J., Sieberts, S., Lamb, J., GuhaThakurta, D., Derry, J., Storey, J.D., Avila-Campillo, I., Kruger, M.J., Johnson, J.M., Rohl, C.A., Nas, A. van, Mehrabian, M., Drake, T.A., Lusis, A.J., Smith, R.C., Guengerich, F.P., Strom, S.C., Schuetz, E., Rushmore, T.H. and Ulrich R. (2008) Mapping the genetic architecture of gene expression in human liver. *PLoS Biology* 6, e107.

Sjöholm, K., Palming, J., Olofsson, L.E., Gummesson, A., Svensson, P.A., Lystig, T.C., Jennische, E., Brandberg, J., Torgerson, J.S., Carlsson, B. and Carlsson, L.M. (2004) A microarray search for genes predominantly expressed in human omental adipocytes: adipose tissue as a major production site of serum amyloid A. *Journal of Clinical Endocrinology and Metabolism* 90, 2233–2239.

Sorensen, T.I. (2003) Weight loss causes increased mortality: pros. *Obesity Reviews* 4, 3–7.

Soukas, A., Cohen, P., Socci, N.D. and Friedman, J.M. (2000) Leptin-specific patterns of gene expression in white adipose tissue. *Genes and Development* 14, 963–980.

Soukas, A., Socci, N.D., Saatkamp, B.D., Novelli, S. and Friedman, J.M. (2001) Distinct transcriptional profiles of adipogenesis *in vivo* and *in vitro*. *Journal of Biological Chemistry* 276, 34167–34174.

Stylianou, I.M., Affourtit, J.P., Shockley, K.R., Wilpan, R.Y., Abdi, F.A., Bhardwaj, S., Rollins, J., Churchill, G.A. and Paigen, B. (2008) Applying gene expression, proteomics and single-nucleotide polymorphism analysis for complex trait gene identification. *Genetics* 178, 1795–1805.

Sun, G. (2007) Application of DNA microarrays in the study of human obesity and type 2 diabetes. *OMICS* 11, 25–40.

Suviolahti, E., Oksanen, L.J., Ohman, M., Cantor, R.M., Ridderstrale, M., Tuomi, T., Kaprio, J., Rissanen, A., Mustajoki, P., Jousilahti, P., Vartiainen, E., Silander, K., Kilpikari, R., Salomaa, V., Groop, L., Kontula, K., Peltonen, L. and Pajukanta, P. (2003) The SLC6A14 gene shows evidence of association with obesity. *Journal of Clinical Investigation* 112, 1762–1772.

Takahashi, K., Mizuarai, S., Araki, H., Mashiko, S., Ishihara, A., Kanatani, A., Itadani, H. and Kotani, H. (2003) Adiposity elevates plasma MCP-1 levels leading to the increased CD11b-positive monocytes in mice. *Journal of Biological Chemistry* 278, 46654–46660.

Twigger, S.N., Pruitt, K.D., Fernández-Suárez, X.M., Karolchik, D., Worley, K.C., Maglott, D.R., Brown, G., Weinstock, G., Gibbs, R.A., Kent, J., Birney, E. and Jacob, H.J. (2008) What everybody should know about the rat genome and its online resources. *Nature Genetics* 40, 523–527.

Urieli-Shoval, S., Linke, R.P. and Matzner, Y. (2000) Expression and function of serum amyloid A, a major acute-phase protein, in normal and disease states. *Current Opinion in Hematology* 7, 64–69.

Urs, S., Smith, C., Campbell, B., Saxton, A.M., Taylor, J., Zhang, B., Snoddy, J., Jones Voy, B. and Moustaid-Moussa, N. (2004) Gene expression profiling in human preadipocytes and adipocytes by microarray analysis. *Journal of Nutrition* 134, 762–770.

Vaisse, C., Clement, K., Guy-Grand, B. and Froguel, P. (1998) A frameshift mutation in human MC4R is associated with a dominant form of obesity. *Nature Genetics* 20, 113–114.

Van Harmelen, V., Reynisdottir, S., Eriksson, P., Thörne, A., Hoffstedt, J., Lönnqvist, F. and Arner, P. (1998) Leptin secretion from subcutaneous and visceral adipose tissue in women. *Diabetes* 47, 913–917.

Vasseur, F., Leprêtre, F., Lacquemant, C. and Froguel, P. (2003). The genetics of adiponectin. *Current Diabetes Reports* 3, 151–158.

Viguerie, N., Clement, K., Barbe, P., Courtine, M., Benis, A., Larrouy, D., Hanczar, B., Pelloux, V., Poitou, C., Khalfallah, Y., Barsh, G.S., Thalamas, C., Zucker, J.D. and Langin, D. (2004) *In vivo* epinephrine-mediated regulation of gene expression in human skeletal muscle. *Journal of Clinical Endocrinology and Metabolism* 89, 2000–2014.

Viguerie, N., Poitou, C., Cancello, R., Stich, V., Clément, K. and Langin, D. (2005a) Transcriptomics applied to obesity and caloric restriction. *Biochimie* 87, 117–123.

Viguerie, N., Vidal, H. Arner, P., Holst, C., Verdich, C., Avizou, S., Astrup, A., Saris, W.H., Macdonald, I.A., Klimcakova, E., Clément, K., Martinez, A., Hoffstedt, J., Sørensen, T.I., Langin, D. and Nutrient–Gene Interactions in Human Obesity – Implications for Dietary Guideline (NUGENOB) Project (2005b) Adipose tissue gene expression in obese subjects during low-fat and high-fat hypocaloric diets. *Diabetologia* 48, 123–131.

Wang, P., Mariman, E., Renes, J. and Keijer, J. (2008) The secretory function of adipocytes in the physiology of white adipose tissue. *Journal of Cell Physiology* 216, 3–13.

Weisberg, S.P., McCann, D., Desai, M., Rosenbaum, M., Leibel, R.L. and Ferrante, A.W. Jr (2003) Obesity is associated with macrophage accumulation in adipose tissue. *Journal of Clinical Investigation* 112, 1796–1808.

Xu, H., Barnes, G.T., Yang, Q., Tan, G., Yang, D., Chou, C.J., Sole, J., Nichols, A., Ross, J.S., Tartaglia, L.A. and Chen, H. (2003) Chronic inflammation in fat plays a crucial role in the development of obesity-related insulin resistance. *Journal of Clinical Investigation* 112, 1821–1830.

Yang, W., Kelly, T. and He, J. (2007) Genetic epidemiology of obesity. *Epidemiological Reviews* 29, 49–61.

Yang, Y.S., Song, H.D., Li, R.Y., Zhou, L.B., Zhu, Z.D., Hu, R.M., Han, Z.G. and Chen, J.L. (2003) The gene expression profiling of human visceral adipose tissue and its secretory functions. *Biochemical and Biophysical Research Communications* 300, 839–846.

Zigman, J.M. and Elmquist, J.K. (2003) Minireview: from anorexia to obesity – the yin and yang of body weight control. *Endocrinology* 144, 3749–3756.

15 Implications for the Future of Obesity Management

George N. Chaldakov,[1] Anton B. Tonchev,[1,2] Marco Fiore,[3] Mariyana G. Hristova,[1] Rouzha Pancheva,[2] Gorana Rancic[4] and Luigi Aloe[3]

[1]Division of Cell Biology and [2]Nutrigenomics Centre, Medical University, Varna, Bulgaria; [3]Institute of Neurobiology and Molecular Medicine, National Research Council-European Brain Research Institute, Rome, Italy; [4]Department of Histology and Embryology, Medical Faculty, Niŝ, Serbia

Introduction

Life at both the local and systemic level requires metabotrophic, neurotrophic and nutritional supports. This chapter presents the current knowledge regarding the pathogenesis and therapy of what can be called *Homo obesus* (Chaldakov *et al.*, 2007). Arguably, any interventions curbing morbid glucose, lipid and energy metabolism and also adipose inflammation will be beneficial for *Homo obesus*, characterized by a deficiency in metabotrophic factors (MTF) or metabotrophins, these latter referring to endogenous proteins involved in the maintenance of vasculometabolic homeostasis. Special attention is paid to adiponectin, nerve growth factor (NGF), brain-derived neurotrophic factor (BDNF) and sirtuins. Pharmaceuticals, nutraceuticals, xenohormetics and caloric restriction mimetics targeting transcriptional, secretory and/or signalling pathways of metabotrophic factors might provide novel potential tools for phenotypic modulation of *Homo obesus* into *Homo sanus*.

The terms *Homo obesus* and MTF are relatively new and both have grown out of recent developments in adipobiology and related topics including metabolism, feeding behaviour and nutrition (Shimomura *et al.*, 1996, 2006; Funahashi *et al.*, 1999; Chaldakov *et al.*, 2000, 2003b, 2006a, 2007; Pond, 2003, 2005; Trayhurn and Wood, 2004; Cinti *et al.*, 2005; Fantuzzi, 2005, 2007; Frühbeck, 2005, 2006; Xu *et al.*, 2005a,b; Beltowski, 2006; Fain, 2006; Miggiano and De Sanctis, 2006; Pradova and Fickova, 2006; Rodriguez *et al.*, 2006; Alvarez-Llamas *et al.*, 2007; Töre *et al.*, 2007; Klein *et al.*, 2007; Zvonic *et al.*, 2007).

In its core, obesity is a disease of both accumulation and inflammation of adipose tissue (Zigman and Elmquest, 2003; Fantuzzi, 2005, 2007; Viguerie *et al.*, 2005; Cancello and Clement, 2006; Fain, 2006; Neels and Olefsky, 2006;

Permana, 2006; Goldstein and Scalia, 2007; Moschen *et al.*, 2007; Waki and Tontonoz, 2007), a process being disseminated to many organs of the body, and thus resulting in cardiometabolic diseases (atherosclerosis, hypertension, type 2 diabetes and metabolic syndrome) (Shimomura *et al.*, 1996, 2006; Funahashi *et al.*, 1999; Yamori *et al.*, 2001, 2004; Chaldakov *et al.*, 2001a, 2004, 2005, 2007; Maeda *et al.*, 2002; Yamori, 2004; Berg and Scherer, 2005; Iacobellis *et al.*, 2005; Manni *et al.*, 2005; Yudkin *et al.*, 2005; Baker *et al.*, 2006; Funahashi and Matsuzawa, 2006; Matsuzawa, 2006; Okamoto *et al.*, 2006; Kim *et al.*, 2007; Kralisch *et al.*, 2007; Tilg and Moschen, 2008), non-alcoholic steatohepatitis (Schaffler *et al.*, 2005), polycystic ovary syndrome, obstructive sleep apnea syndrome (see Cheng *et al.*, 2006), inflammatory bowel disease (Moschen *et al.*, 2007), endometriosis and thyroid-associated ophthalmopathy (Gianoukakis and Smith, 2008), several forms of cancer (Celis *et al.*, 2005; Shimomura *et al.*, 2006), AIDS (Pond, 2003), periodontal disease (Nishimura *et al.*, 2003) and Alzheimer's disease (Franceschi *et al.*, 2001; Sjogren and Blennow, 2005; Sun and Alkon, 2006). Each of these diseases increases the risk of having an unsuccessful ageing and a shortened life expectancy (Kalra and Kalra, 2005; Candore *et al.*, 2006; Dimitrov *et al.*, 2006; Grimaldi *et al.*, 2006).

The cost associated with obesity and related diseases is enormous (e.g. about US$100 billion per year in the USA). An effective way to reduce costs and increase the quality of life (QOL) would be to treat one of the major underlying causes, obesity. Currently, available therapies employed to combat obesity have ranged from modifications of lifestyle factors such as nutrition and physical activity to pharmacotherapy and bariatric surgery through to gene-transfer technology (Bray and Bouchard, 2004; Dimitrov *et al.*, 2005; Kalra and Kalra, 2005; Boss and Bergenhem, 2006; Chaldakov *et al.*, 2006a; Foster-Schubert and Cummings, 2006; Xavier Pi-Sunyer, 2006; Hofbauer *et al.*, 2007). However, these treatments have disadvantages, such as poor compliance for lifestyle modifications, transient effectiveness and undesirable side effects of some pharmacological products, together with insufficient experience and controversial findings for certain recently applied biologicals (Del Porto *et al.*, 2006) and gene/cell therapy (Kalra and Kalra, 2005).

This chapter presents a set of conceptual principles that might help to prioritize specific approaches in the development of obesity therapeutics, based on the current understanding of adipose tissues as a potent secretory organ, especially when hypertrophied and inflamed. Focus is laid primarily on *Homo obesus* as a metabotrophin-deficient species and, respectively, on metabotrophin-targeted pharmacology and (adipo)nutrigenomics in obesity-linked cardiometabolic diseases, including unsuccessful ageing.

Homo obesus: A Memory of Thrifty Genes or an Adipochronobiological Disorder?

At the evolutionary level, the survival of biological species is mediated by growth, fertility and longevity phenotypes. However, the human race has evolved in an environment of extremely difficult periods when food was scarce. Hence, hunting or gathering food was laborious and required higher energy expenditure than that required to obtain food nowadays. Such circumstances promoted the ability to eat (and possibly drink) as much as was available. The thrifty genes thus evolved to promote human survival in a life characterized by famine–feast cycles (Neel, 1962; Boss and Bergenhem, 2006). This scenario, especially in the economically advanced countries, has changed radically in the past decades, with the abundance of fast-food meals combined with a marked decrease in daily physical activity. This lifestyle collides with our genome, which was most likely selected in the late Palaeolithic era (50,000–10,000 BC) by criteria oriented to benefit survival in environments with marked fluctuations of famine and feast (FFF) (Halberg *et al.*, 2005). Briefly, these thrifty genes become obesogenic in today's society with a surplus amount of food in parallel with more sedentary lifestyles. It is noteworthy that Halberg *et al.* (2005) mimicked the FFF scenario in healthy young males by subjecting them to intermittent fasting every second day for 20 h over 15 days. After these fasting periods, both insulin sensitivity and plasma adiponectin levels increased compared to the basal level before and after the FFF scenario. Interestingly, FFF induced an upregulation of NGF and BDNF in animal models (see below).

Neel's thrifty gene hypothesis was revisited recently, implicating a hibernation-like motif in the development of insulin resistance and related disorders; it is noteworthy that adipocytes reportedly exhibit melatonin receptors (Scott and Grant, 2006).

The Concept of Adipobiology

The seminal findings of leptin and adiponectin (see Chapter 5) triggered a period of intense interest in the elucidation of the endocrine and paracrine activity of adipose tissue and its potential involvement in the molecular mechanisms of obesity and related diseases (Tables 15.1 and 15.2) leading to *adipoendocrinology* (Chaldakov *et al.*, 2001b) and *adipobiology* (Chaldakov *et al.*, 2003b).

Different types of adipose tissue can be distinguished, namely white and brown adipose, transdifferentiating white–brown adipocytes, and adipose tissue related to various organs such as perivascular, epicardial, perinodal (lymph node-associated), orbital (eye) and striated muscle-, breast- and bone marrow-associated adipose tissue. The predominant adipose cell type in adult humans is the white adipocyte.

The presence of an adequate amount at the same time as functionally active adipose tissue is important for good health, including the control over lipid and energy homeostasis. The importance of adipose tissue for health has been emphasized by observations made in lipodystrophic situations such as sterol element-binding protein-1c-deficient mice (Shimomura *et al.*, 1998) and highly active antiretroviral therapy in AIDS patients (Pond, 2003). Thus, both too much and too little fat is detrimental. Interestingly, adipose-derived adult stem cells may differentiate into chondrocytes, myocytes, osteoblasts and neurones, representing a huge potential source for intervention and regenerative medicine (Kokai *et al.*, 2005).

Table 15.1. Selected list of adipokines.

Cytokines
Leptin, Interleukin-1 (IL-1), IL-6, IL-10, IL-1 receptor antagonist, IL-18
Tumour necrosis factor-α (TNF-α), TNF-like weak inducer of apoptosis (TWEAK)

Chemokines
MCP-1 (CCL2), IL-8 (CXCL8), Eotaxin (CCL11)
RANTES (CCL5), IP-10, SDF-1 (CXCL12)

Growth factors
FGF, TGF-β, NGF, CNTF, GDNF, MCSF, BDNF,
HB-EGF, IGF, HGF, BMP-2, LIF

Angiogenic factors
VEGF, Angiogenin, Angiopoietin-2, Angiopoietin-like protein-4

Renin–angiotensin system
Renin, Angiotensinogen, Angiotensin I, II, Aldosterone, Chymase, Cathepsins

Acute phase reactants
Serum amyloid A, Lipocalin, Ceruloplasmin, Haptoglobin

Haemostatic factors
Plasminogen activator inhibitor type 1, Tissue factor

Serpins (serine protease inhibitors)
Plasminogen activator inhibitor-1, Pigment epithelium-derived factor, Vaspin
Placental thrombin inhibitor, Pregnancy zone protein, Protease C1 inhibitor

Enzymes
Lipoprotein lipase, Adipsin, Matrix metalloproteinases, Tryptase

Others
Adiponectin, Acylation-stimulating protein, FIZZ-1, Resistin (FIZZ-3), SPARC (Osteonectin), Omentin, Apelin, Visfatin, Prolactin, Adrenomedullin, Calcitonin, Somatostatin, *agouti* protein, Prohibitin, Tissue inhibitors of matrix metalloproteinases, Calcitonin gene-related protein, Urocortin, Metallothioneins, Retinol-binding protein-4, Hypoxia-inducible factor-1α, Autotaxin, Cholesterol ester transfer protein, Zinc-alpha2 glycoprotein, Complement 3, Cystatin C, Fibrinogen, Hevin.

Note: MCP-1 or CCL2, monocyte chemoattractant protein-1 or cysteine–cysteine motif chemokine ligand 2; RANTES, regulated on activated normal T-cell expressed and secreted; NGF, nerve growth factor; GDNF, glial cell line-derived neurotrophic factor; IP-10, interferon-γ-inducible protein-10; SDF-1, stromal cell-derived factor-1; FGF, fibroblast growth factor; TGF-β, transforming growth factor-β; CNTF, ciliary neurotrophic factor; MCSF, macrophage colony-stimulating factor; BDNF, brain-derived neurotrophic factor; HB-EGF, heparin-binding EGF-like growth factor; IGF, insulin-like growth factor; HGF, hepatocyte growth factor; BMP-2, bone morphogenetic protein-2; LIF, leukaemia inhibitory factor; VEGF, vascular endothelial growth factor; FIZZ, found in inflammatory zone; SPARC, secretory protein, acidic and rich in cysteine.

Table 15.2. Adipokines as 'yin-and-yang' modulators of inflammation.

Anti-inflammatory factors	Proinflammatory molecules
Adiponectin	TNF-α, TWEAK
Interleukin-10	Interleukin-1, IL-6, IL-18
Interleukin-1 receptor antagonist	Leptin
Nerve growth factor	FIZZ-1, Resistin (FIZZ-3)
Metallothionein-I and II	Visfatin
Tissue inhibitor of matrix metalloproteinases	Matrix metalloproteinases
Prohibitin	MCP-1 (CCL2)
Adrenomedullin	Interleukin-8 (CXCL8)
Urocortin	Eotaxin (CCL11)
Calcitonin gene-related peptide	RANTES (CCL5)

Adipose Protein Secretion

In the middle of the 1990s, several research groups investigated the gene expression profiles in different human adipose tissue depots and compared the data with those of other tissues and organs (Shimomura *et al.*, 1996, 2006; Funahashi *et al.*, 1999). The results from this body mapping revealed that adipose tissue, especially visceral fat, abundantly expressed a variety of genes coding for secretory proteins: around 30% of the genes expressed in human visceral fat and 20% of those of the subcutaneous fat. Recent analysis of the human adipose tissue secretome has revealed a total of 259 proteins, 108 of them being secretory proteins (Alvarez-Llamas *et al.*, 2007), including serpins (serine protease inhibitors) (Zvonic *et al.*, 2007; Wang *et al.*, 2008).

The protein secretory pathway encompasses several intracellular steps, including synthesis, translocation, targeting, sorting, storage (in case of regulated versus constitutive secretion) and, finally, exocytosis. Generally, the secretory proteins are of four major types: lysosomal, plasmalemmal, recycled and exported. The vast majority of extracellular proteins are exported by the classical rough endoplasmic reticulum-Golgi complex-dependent secretory route (Chaldakov and Vankov, 1986). However, other proteins such as angiogenic growth factors, inflammatory cytokines and extracellular matrix proteins use unconventional protein secretion, also known as non-classical protein export or rough endoplasmic reticulum/Golgi-independent protein secretion (Töre *et al.*, 2007). This secretion, for example, results in the release of exosomes, 50–60 nm vesicles derived from multivesicular bodies and carrying important bioactive molecules to communicate, via endo- and paracrine ways, with other cells. Although the presence of multivesicular structures in adipocytes has been described earlier, this type of (nano)secretion has not yet been evaluated fully in this cell type (Töre *et al.*, 2007).

Adipokines, adipocytokines, adipokinome and secretome

The multiple proteins synthesized, stored and released by adipose tissue cells (adipocytes, stromovascular and matrix cells and associated macrophages and mast cells) collectively have been designated 'adipocytokines' (Shimomura *et al.*, 1996; Funahashi *et al.*, 1999) or 'adipokines' (Chaldakov *et al.*, 2000; 2003b; Trayhurn and Wood, 2004).

While at the adipobiological level both names articulate clearly the secretory nature of adipose tissue cells, the term 'adipokines' is more accurate than the name 'adipocytokines' ('adipocyto-kines' or 'adipo-cytokines'), as it includes the proteins secreted by both adipocytes and non-adipocyte cell types of adipose tissue, as well as both the cytokine and non-cytokine proteins (Table 15.1). At the functional level, adipokines control multiple biological processes beyond lipid and carbohydrate metabolism (Table 15.2).

Trayhurn and Wood (2004) conceptualized the secretory proteome of adipose tissue as 'adipokinome', whereas the whole spectrum of adipose secretory products was designated 'secretome', the latter embodying both proteins (adipokines) and non-proteins (Kratchmarova *et al.*, 2002; Celis *et al.*, 2005; Viguerie *et al.*, 2005; Alvarez-Llamas *et al.*, 2007; Zvonic *et al.*, 2007). It is noteworthy that adipocytes are not the sole secretory cell type of adipose tissue. Notably, non-fat cells including those of the stromovascular fraction (Fain, 2006) and associate macrophages (Cinti *et al.*, 2005; Cancello and Clement, 2006; Permana, 2006) and mast cells (Hristova *et al.*, 2001; Chaldakov *et al.*, 2004, 2006b) secrete (especially in an inflamed adipose environment) a major part of the known adipokines.

Concept of Adipopharmacology

'Adipopharmacology' connotes the adipotargeting studies aimed at drug discovery. The following major adipopharmacological targets could be defined: (i) nuclear transcription factors, especially peroxisome proliferator-activated receptors and sterol regulatory element-binding protein-1; (ii) adipokines as products of intracellular secretory pathways, including endoplasmic reticulum stress and the unfolded protein response, Golgi complex, microtubules and various exocytosis-mediated molecular complexes; (iii) adipokines as signalling molecules; (iv) insulin downstream components controlling trafficking of glucose transporter type 4-containing vesicles; (v) uncoupling proteins, including mitochondrial biogenesis; (vi) steroidogenesis mediated by adipofibroblasts; (vii) metabolic pathways of triacyglycerol and fatty acids, including lipases (adiponutrin, desnutrin) and lipid droplet-associated proteins (perilipin, adipophilin, caveolin-1); and (viii) adipose-derived stem cells (Chaldakov *et al.*, 2006a; Töre *et al.*, 2007).

Concept of Metabotrophic Factors

In analogy to Levi-Montalcini's terminology for neurotrophic factors and neurotrophins (see Aloe and Calza, 2004; Allen and Dawbarn, 2006; Chaldakov *et al.*, 2007), the terms 'metabotrophic factors' and 'metabotrophins' (from the Greek *metabole* and *trophe*, meaning 'nutritious for metabolism') were introduced (Chaldakov *et al.*, 2003a, 2004, 2006a). As indicated, *endogenous* MTF

Table 15.3. Selected list of endogenous metabotrophic factors.

Secretory proteins
Adiponectin, NGF, BDNF
IL-10, IL-1 receptor antagonist
Ciliary neurotrophic factor
Glial cell line-derived neurotrophic factor
Transforming growth factor-β
Insulin-like growth factor-1
Bone morphogenetic protein-2 and -9
Leukaemia inhibitory factor
Metallothioneins
Angiopoietin-like protein 4
Incretins (glucagon-like protein-1, glucose-dependent insulinotropic protein)
Intracellular or plasma membrane proteins
Peroxisome proliferator-activated receptor-γ
Glucose transporters
Aquaporin-7 (AQP7)[a], AQP9[a]
Sirtuins
Pyrin[b]
Prohibitin

Note: [a]Discovered in 1986 by Gheorghe Benga (Benga, 2006) as water channel integral membrane protein of erythrocytes, this family of proteins today includes 13 members, AQP7 and AQP9 being aquaglyceroporins in adipocytes and hepatocytes, respectively. AQP7 gene knockout mice develop insulin resistance and obesity. Both AQP7 and AQP9 may be new targets for pharmacological studies in obesity and related diseases (Frühbeck, 2005; Frühbeck *et al.*, 2006; Rodríguez *et al.*, 2006; Wintour and Henry, 2006). [b]An anti-inflammatory protein in leukocytes, which is encoded in the gene responsible for familial Mediterranean fever; wild-type pyrin genotype predisposes to a longer lifespan (Candore *et al.*, 2006; Grimaldi *et al.*, 2006).

comprise secretory and intracellular proteins (Table 15.3) derived from adipose and non-adipose cellular sources. They are pleiotropic proteins involved essentially in the maintenance of glucose, lipid, energy and vascular homeostasis, as well as inflammation and wound healing. Adiponectin, NGF and BDNF are discussed below in more detail as characteristic examples of MTF.

Adiponectin

In 1995, Matsuzawa's research group in Osaka discovered a novel, adipose most abundant gene transcript 1 (apM1) encoding a secretory protein named adiponectin (Funahashi *et al.*, 1999; Kadowaki and Yamauchi, 2005; Cheng *et al.*, 2006; Funahashi and Matsuzawa, 2006; Matsuzawa, 2006; Shimomura *et al.*, 2006; Tilg and Moschen, 2008). Adiponectin consists of 244 amino acids sharing significant similarity with collagens type VIII and X, as well as complement protein C1q. Today, adiponectin is one of the best-characterized metabotrophic adipokines with a great potential for developing novel therapeutic approaches for various adiponectin-deficient disorders, including cardiometabolic diseases. Initially, Matsuzawa and colleagues were surprised by two facts: (i) plasma adiponectin levels were extremely high (up to 5–20 μg/ml in healthy humans); and (ii) plasma adiponectin levels decreased with the accumulation of body fat, especially in visceral adipose tissue, while the blood concentrations of other adipokines known to date, like leptin, plasminogen activator inhibitor type-1 (PAI-1) and tumour necrosis factor-alpha (TNF-α), increased in parallel with fat accumulation. Numerous studies by diverse research groups revealed that hypoadiponectinaemia associated with visceral adipose tissue hypertrophy was an essential pathological condition of many lifestyle-related diseases.

Pharmacology of adiponectin: targeting an 'anti-kine'

Along with being the major endogenous insulin-sensitizing factor, adiponectin exerts a multitude of anti-inflammatory, antiatherogenic, antidiabetic, antiobesity, antifibrotic, antiangiogenic and anticancer effects. Further studies may validate whether boosting adiponectin's beneficial properties might indeed contribute to the adipopharmacology of disease. The cloning of three adiponectin receptors, AdipoR1, AdipoR2 and T-cadherin, further reinforces the potential usefulness of adiponectin/AdipoR-targeting. For instance, upregulation of AdipoRs and development of AdipoR agonists represents a promising approach for developing new drugs for *Homo obesus* and his associated diseases (Kadowaki and Yamauchi, 2005; Cheng *et al.*, 2006; Tilg and Moschen, 2008).

In perspective, genetically engineered cell lines for production of adiponectin may be used to screen for secretagogues of adiponectin. For instance, treatment with antiobesity drugs such as sibutramine (a selective serotonin/norepinephrin reuptake inhibitor), orlistat (a lipase inhibitor), rimonabant (a selective cannabinoid 1 receptor antagonist) (Xavier Pi-Sunyer, 2006), or with thiazolidinediones (insulin-sensitizing drugs) (Okamoto *et al.*, 2006) results in an enhanced adiponectin secretion. At a clinical level, a recombinant adiponectin fragment (Famoxin) has been tested in the past few years (Boss and Bergenhem, 2006).

The intracellular secretory pathway of adiponectin in cardiomyocytes and skeletal muscle (Tilg and Moschen, 2008) may also represent a potential pharmacological target; in particular, a combinatorial boosting of both adiponectin secretion and signalling. In this context, osmotin, which is a plant relative of adiponectin, is a ligand for the yeast homologue of AdipoR. Further research examining similarities shared by adiponectin and osmotin may unravel potential AdipoR agonists (Narasimhan *et al.*, 2005).

The anti-inflammatory activity of adiponectin is mediated by inhibition of a proinflammatory cascade involving TNF-α and IL-6, as well as by induction of anti-inflammatory cytokines such as IL-10 and IL-1 receptor antagonist (Shimomura *et al.*, 2006). Altogether, these effects may contribute to adiponectin's antiatherogenic and antidiabetic effects. Intriguingly, low amounts of adiponectin are found in epicardial adipose tissue of patients with coronary atherosclerosis (Iacobellis *et al.*, 2005), and hypoadiponectinaemia is observed in cardiometabolic diseases (Funahashi *et al.*, 1999; Matsuzawa, 2006; Okamoto *et al.*, 2006; Shimomura *et al.*, 2006). Since patients with the metabolic syndrome (Chaldakov *et al.*, 2001a, 2004) and acute coronary syndromes (Manni *et al.*, 2005) also express low plasma levels of NGF and BDNF, one may ask whether such a triple (adiponectin, NGF and BDNF) deficit might be involved interactively in the pathogenesis of these diseases (see below). Furthermore, since adiponectin inhibits TNF-α secretion (Shimomura *et al.*, 2006) and treatment with colchicine, a microtubule-disassembling (antitubulin) agent (Chaldakov and Vankov, 1986), inhibits TNF-α secretion (see Chaldakov *et al.*, 2003b, 2007), it may be speculated that adiponectin also has an antitubulin activity of its own.

Given the apparent complexity of adiponectin biology and the cost and inconvenience associated with recombinant adiponectin treatment, the development of low molecular weight (small) molecules of adiponectin with secretion stimulatory effects would seem preferable. Importantly, recent evidence indicates that IL-15, a cytokine (myokine) highly expressed in skeletal muscle, inhibits white adipose tissue (WAT) deposition and stimulates adiponectin secretion (Quinn *et al.*, 2005). Contrarily, testosterone inhibits adiponectin secretion (Xu *et al.*, 2005a). Although it is not clear which particular step(s) of the intracellular secretory pathway is promoted by IL-15, or inhibited by testosterone, searching for new secretagogues targeting adiponectin secretion might represent a novel therapeutic approach for adiponectin-deficient states. Further studies are also required to evaluate the potential relevance of muscle–adipose interactions (Petersen and Pedersen, 2005).

NGF and BDNF

The NGF, a prototypic member of the protein family of neurotrophins, was discovered by Rita Levi-Montalcini in 1951 (reviewed in her Nobel Prize lecture published in *Science*, 1987). Conventional wisdom held that neurotrophic factors such as NGF, BDNF, ciliary neurotrophic factor (CNTF) and glial cell line-derived neurotrophic factor (GNTF) were only for the promotion of neuronal differentiation and survival. However, studies in the past three decades (Aloe and Calza, 2004; Chaldakov *et al.*, 2007) have revealed that 'neurotrophins', particularly NGF and BDNF, are not only stimulating for nerve growth and survival, but also exert trophic effects over: (i) immune cells, being named 'immunotrophins' (Fainzilber and Carter, 2002); (ii) keratinocytes, endothelial cells and enterocytes, named 'epitheliotrophins' (Botchkarev *et al.*, 2004); and (iii) glucose, lipid, energy and vascular homeostasis, as well as inflammation and wound healing, and thus being designated 'metabotrophins' (Table 15.3). The metabotrophic effects of both NGF and BDNF in health and disease are summarized in Table 15.4 (Chaldakov *et al.*, 2000, 2001a, 2004; Nakagawa *et al.*, 2003; Seki *et al.*, 2004; Trayhurn and Wood, 2004; Manni *et al.*, 2005; Geroldi *et al.*, 2006; Krabbe *et al.*, 2006).

Further evaluation of NGF and BDNF may lead to the development of novel antiobesity, antidiabetic and antiatherosclerotic drugs (Allen and Dawbarn, 2006; Hefti *et al.*, 2006). In this context, it has to be mentioned that a synthetic analogue of the CNTF is being tested in clinical trials as an antiobesity drug under the name Axokine (see Chapter 3). Interestingly, the effects of CNTF are

Table 15.4. Metabotrophic characteristics of NGF and BDNF.

NGF and BDNF are synthesized and released from pancreatic beta cells
NGF and BDNF exert insulinotropic effects
NGF improves transplantation of Langerhans' islets
BDNF ameliorates glucose and lipid profile in experimental diabesity
NGF upregulates the expression of LDL receptor-related proteins
NGF upregulates the expression of PPAR-γ
NGF exerts antioxidant effects
NGF and BDNF suppress food intake
Mutation of TrkB, the main receptor for BDNF, results in hyperphagia and obesity
BDNF-deficient mice develop metabolic abnormalities similar to the metabolic syndrome
An atherogenic diet decreases brain BDNF levels
Treatment with NGF improves experimentally induced cardiac ischaemia
Caloric restriction increases brain BDNF levels and improves the metabolic profile in experimetal metabolic syndrome
NGF accelerates skin and corneal wound healing
Involvement in the following diseases:
Coronary atherosclerosis (reduced vascular tissue NGF levels)
Heart failure (reduced myocardial NGF[a] levels)
Metabolic syndrome (reduced circulating levels of NGF and BDNF)
Acute coronary syndromes (reduced circulating levels of NGF and BDNF[b])
Type 2 diabetes mellitus (reduced circulating levels of BDNF[c])
Diabetic neuropathy (decreased tissue and circulating levels of NGF)
Migraine and cluster headache (decreased platelet NGF and/or BDNF levels)
Sudden cardiac death[d]
Transient cerebral ischaemia
Skin wound healing

Note: NGF, nerve growth factor; BDNF, brain-derived neurotrophic factor. [a]For increased BDNF levels (Cai *et al.*, 2006). [b]For increased BDNF levels (Ejiri *et al.*, 2005). [c]For increased BDNF levels (Suwa *et al.*, 2006). [d]See Chen *et al.* (2001).

not mediated at the hypothalamic level only, but also directly at the level of adipose tissue as adipocytes express CNTF receptors, and CNTF affects adipocyte leptin secretion directly (Chaldakov *et al.*, 2006a).

Obesity and Related Diseases: A Dysfunction of Secretion and Signalling in Metabotrophic Factors

Contrary to the increased bioavailability of most known adipokines, endocrine release of adiponectin (Matsuzawa, 2006; Shimomura *et al.*, 2006) and NGF/BDNF (Chaldakov *et al.*, 2004; Geroldi *et al.*, 2006; Krabbe *et al.*, 2006), and also paracrine presence of adiponectin (Iacobellis *et al.*, 2005), are decreased in obesity and related cardiometabolic diseases. Mechanistically, it was suggested that a reduced secretion of these MTF (together with a dysregulation of GLUT4 traffic) might lie at the heart of a complex network of factors involved in the pathogenesis of obesity. If so, simultaneous administration of NGF–BDNF–adiponectin may indeed be a novel pharmacological approach, with polypills representing an attractive approach. While diabetes is described as a protein misfolding disease, less is known about the adipobiology of the protein folding capacity of endoplasmic reticulum (ER) and ER stress and related unfolded protein response (UPR). Obesity indeed causes ER stress (Töre *et al.*, 2007), which in turn leads to suppression of insulin receptor signalling. Mice deficient in the X-box-binding protein-1, a transcription factor that modulates UPR, as well as mice with mutations affecting the ER stress-activated pancreatic ER kinase, develop insulin resistance and type 2 diabetes. These findings suggest that drug targeting aimed at the protein folding–ER stress–UPR complex might offer novel therapeutic opportunities for insulin resistance and type 2 diabetes, with both NGF and BDNF potentially exerting an ER stress-preventing effect (see Töre *et al.*, 2007).

Homo Obesus as a Metabotrophin-deficient Species

Since an endogenous metabotrophic deficit in the development of obesity and related diseases is observed, the search for exogenous factors targeting MTF may be useful. For example: (i) agents boosting secretory and/or signalling pathways of metabotrophins; (ii) selective inhibitors of the incretin degradating enzyme dipeptidyl peptidase-4 (Barnett, 2006); (iii) incretin mimetics (Gallwitz, 2006); (iv) agonists of peroxisomal proliferator-activated receptor-gamma (PPAR-γ), such as thiazolidinediones, which stimulate adiponectin secretion (Xavier Pi-Sunyer, 2006); (v) treatment with pitavastatin (an HMG-CoA reductase inhibitor), which upregulates both NGF and BDNF in the ischaemic hippocampus (Himeda *et al.*, 2007); and (vi) phosphodiesterase inhibitors such as ibudilast, which exerts anti-inflammatory and neuroprotective effects via downregulation of the production of reactive oxygen species, IL-1β, IL-6 and TNF-α, as well as upregulation of IL-10, NGF and GDNF in activated microglia (Mizuno *et al.*, 2004).

Another tempting approach may be measuring local and systemic levels of MTF in centenarians (Franceschi *et al.*, 2001 for other biomarkers). Interestingly: (i) the postnatal brain deletion of *Bdnf* (the gene encoding for BDNF) in mice leads to a dramatic (80–150%) increase in body weight accompanied by increased plasma levels of leptin, insulin, glucose and cholesterol (Rios *et al.*, 2001); and (ii) the mutation of *Ntrk2* (the gene encoding for the high-affinity BDNF receptor TrkB) is associated with hyperphagia and severe obesity (Yeo *et al.*, 2004; Gray *et al.*, 2007).

(Adipo)nutrigenomics of Lifespan

The Human Genome Project has estimated over 30,000 genes encoding more than 100,000 functionally distinct proteins. Understanding the interactions among genes, proteins and nutrients is fundamental to outlining a personalized 'nutritional phenotype'. Altogether, such an intellectual process opens new areas for the 'omics' sciences such as nutrigenomics, proteomics and metabolomics, among others (Dimitrov *et al.*, 2006; Miggiano and De Sanctis, 2006; Ghosh *et al.*, 2007) (see Chapter 14). Nutrigenomics addresses the nutrient-induced gene and protein expression relative to either prevention/therapy or pathogenesis of various diseases, including obesity and its relatives. These are viewed as lifestyle- or QOL-related disorders, which increase the risk of having a decreased life expectancy (Yamori *et al.*, 2004). While susceptibility to becoming *Homo obesus* is determined largely by genetic factors, the current obesity epidemic is influenced significantly by adverse lifestyle factors, including increased consumption of energy-rich diets and a sedentary life.

Exciting findings obtained in very different animal models are converging rapidly and suggest that animal lifespan may not only be subject to genetic background but also related strongly to lifestyle, particularly physical activity and healthy nutrition (Tsuji-Hayashi *et al.*, 2005; Kim *et al.*, 2006).

The ageing process and its associated diseases involve an altered energy metabolism, overproduction of reactive oxygen species (ROS), preservation of the activities of antioxidant enzymes and an impaired ability of the organism and its cells to cope with dysregulation. Calorie restriction (CR) is the most robust, non-genetic intervention that increases lifespan and reduces the rate of ageing in a variety of species (Roth *et al.*, 2005; Kim *et al.*, 2006). Mechanisms responsible for the anti-ageing effects of CR remain uncertain, but reduction of ROS within mitochondria remains a major focus of research. Specifically, both *in vivo* and *in vitro* analyses have demonstrated that CR attenuates ROS-mediated damages (Armeni *et al.*, 2003) at the same time that it stimulates mitochondrial biogenesis through induction of endothelial nitric oxide synthase (eNOS) expression and 3',5'-cyclic guanosine monophosphate formation in various tissues of male mice (Nisoli *et al.*, 2005, also via sirtuin-1 upregulation) and through a PPAR coactivator 1 alpha signalling pathway (Lopez-Lluch *et al.*, 2006). Likewise, intermittent fasting, a dietary regimen in which food is available only every other day, exerts cardioprotective effects involving antiapoptotic and anti-inflammatory pathways (Ahmet *et al.*, 2005).

Polyphenolic flavonoids and phytoestrogens have captured interest recently by virtue of their reduction of the risk of cardiometabolic diseases and the extension of lifespan. The polyphenol resveratrol (3,5,4´-tri-hydroxy stilbene) requires special nutritional and nutrigenomic attention. Resveratrol and its relatives are synthesized in a response-to-injury manner in grapes (also nuts) as a defence reaction to environmental risks. This phenomenon was dubbed xenohormesis, with xenohormetic molecules becoming a current target for ageing and lifespan studies. The resveratrol-mediated benefits include stimulation of nicotinamide adenine dinucleaotide (NAD+) histon (and α-tubulin) deacetylases, termed collectively sirtuins. Resveratrol makes DNA more resistant to various disease-causing factors and thus extends life expectancy of treated animals.

Recent data link adipose tissue, nutrition and xenohormesis as involved pivotally in the processes of ageing and longevity (Armeni *et al.*, 2003; Lamming *et al.*, 2004; Guarente and Picard, 2005; Kim *et al.*, 2006; Trapp and Jung, 2006; Yun *et al.*, 2006; Chen and Guarente, 2007). Encouraging results obtained so far include:

1. Yeast, worms, flies and rodents kept on a CR diet enjoy 60–70% longer lifespan than those on an *ad libitum* diet.
2. Treatment with resveratrol inhibits apoptosis and stimulates mitochondrial biogenesis and nitric oxide synthesis (Nisoli *et al.*, 2005).
3. CR increases brain BDNF levels and extends lifespan (Chaldakov *et al.*, 2007); it is noteworthy that resveratrol mimics the effects of CR (Chen and Guarente, 2007) since for *Homo obesus* it is extremely difficult to maintain a long-term CR, resveratrol and other products mimicking CR might be potential nutraceuticals to combat the deleterious effects of obesity and related diseases.
4. *n*-3 Polyunsaturated fatty acids increase circulating levels of adiponectin (Flachs *et al.*, 2006).
5. 2-Deoxyglucose, an analogue of native sugar, acts as a glycolytic inhibitor, and reduces overall energy flow similarly to CR (Roth *et al.*, 2005).

According to present paradigms, a proinflammatory phenotype seems to contribute to the risk of developing atherosclerosis (Ross, 1999) and other cardiometabolic diseases (Funahashi *et al.*, 1999; Fain, 2006; Matsuzawa, 2006; Okamoto *et al.*, 2006; Shimomura *et al.*, 2006). In the same way, ageing involves a complex rearrangement of the cytokine pattern towards a proinflammatory status, a phenomenon that can be called inflamm-ageing (Franceschi *et al.*, 2001; Candore *et al.*, 2006; Grimaldi *et al.*, 2006). Because adipose tissue is a potent source of pro- and anti-inflammatory adipokines (Table 15.2), it is logical to consider the potential contribution of adipokines to ageing and longevity.

Conclusion and Future Directions

This chapter provides a conceptualized update about the pathogenesis and therapy of obesity and its comorbidities. Many routes may lead to a transition from a healthy to an obese phenotype. Which and how many obesogenic routes may be considered most 'vulnerable' to antiobesity therapy? Therapy might focus on different approaches such as polypills, pharmaceuticals, nutraceuticals, xenohormetics, CR mimetics and gene-transfer technology.

The present chapter underlines that a metabotrophin deficit may represent a major obesogenic route. Mechanistically, targeting the transcriptional, secretory and/or signalling pathways of MTF may be the core of (adipo)pharmacology. In addition, products that mimic the beneficial effects of CR appear to be promising preventive and therapeutic approaches. Furthermore, various nutraceuticals may be examined for their potential impact on MTF, in particular on adiponectin, NGF, BDNF, IL-10, IL-1Ra and aquaporin-7.

The chapter also highlights that dysfunctional secretory and signalling pathways of MTF may be linked to obesity and its related diseases. It is noteworthy that each step of the intracellular secretory pathway of MTF at the adipose and non-adipose cellular level might be a potential target for drug development. A detailed molecular understanding of adiposecretion may open new avenues for discovering effective antiobesity drugs. As we move from disease treatment to prevention, the 'omics' technologies will be incorporated routinely into practice to understand health and disease better, including obesity. One of the present challenges is therefore to cultivate a metabotrophic, adipocentric and nutrigenomic approach to obesity management. Teleologically, a link to systems biology is foreseeable (Bocsi *et al.*, 2006; Keusch, 2006).

Acknowledgements

Supported by grants from the Bulgarian Ministry of Education and Science and from the CNR-European Brain Research Institute of Rome, Italy. The collaboration of Peter Ghenev, Kamen Valchanov, Ivan Stankulov, Paola Tirassa, Stoyan Stoev, Antoniya Kisheva, Dimiter Kostov, Pepa Atanassova and Stansilav Yanev is greatly appreciated.

References

Ahmet, I., Wan, R., Mattson, M.P., Lakatta, E.G. and Talan, M. (2005) Cardioprotection by intermittent fasting in rats. *Circulation* 112, 3115–3121.

Allen, S.J. and Dawbarn, D. (2006) Clinical relevance of the neurotrophins and their receptors. *Clinical Science (London)* 11, 175–191.

Aloe, L. and Calza, L. (eds) (2004) *NGF and Related Molecules in Health and Disease. Progress in Brain Research 146*. Elsevier Science B.V., Amsterdam, pp. 3–527.

Alvarez-Llamas, G., Szalowska, E., Vries, M.P. de, Weening, D., Landman, K., Hoek, A., Wolffenbuttel, B.H., Roelofsen, H. and Vonk, R.J. (2007) Characterization of the human adipose tissue secretome. *Molecular and Cellular Proteomics* 6, 589–600.

Armeni, T., Pricipato, G., Quiles, J.L., Pieri, C., Bompadre, S. and Battino, M. (2003) Mitochondrial dysfunctions during aging: vitamin E deficiency or caloric restriction – two different ways of modulating stress. *Journal of Bioenergy and Biomembranes* 35, 181–191.

Baker, A.R., Silva, N.F., Quinn, D.W., Harte, A.L., Pagano, D., Bonser, R.S., Kumar, S. and McTernan, P.G. (2006) Human epicardial adipose tissue expresses a pathogenic profile of adipocytokines in patients with cardiovascular disease. *Cardiovascular Diabetology* 5, 1.

Barnett, A. (2006) DPP-4 inhibitors and their potential role in the management of type 2 diabetes. *International Journal of Clinical Practice* 60, 1454–1470.

Beltowski, J. (2006) Apelin and visfatin: Unique 'beneficial' adipokines upregulated in obesity? *Medical Science Monitor* 12, RA112–119.

Benga, G. (2006) Aquaporin7 and adipose tissue. *Biomedical Reviews* 17, 102–108.

Berg, A.H. and Scherer, P.E. (2005) Adipose tissue, inflammation, and cardiovascular diseases. *Circulation Research* 96, 939–949.

Bocsi, J., Mittag, A., Sack, U., Gerstner, A.O., Barten, M.J. and Tarnok, A. (2006) Novel aspects of systems biology and clinical cytomics. *Cytometry* 69, 105–108.

Boss, O. and Bergenhem, N. (2006) Adipose targets for obesity drug development. *Expert Opinion and Therapeutic Targets* 10, 119–134.

Botchkarev, V.A., Botchkareva, N.V., Peters, E.M.J. and Paus, R. (2004) Epithelial growth control by neurotrophins: leads and lessons from the hair follicle. *Progress in Brain Research* 146, 493–514.

Bray, G.A. and Bouchard, C. (eds) (2004) *Handbook of Obesity*, 2nd edition. Marcel Dekker Inc., New York.

Cai, D., Holm, J.M., Duignan, I.J., Zheng, J., Xaymardan, M., Chin, A., Ballard, V.L.T., Bella, J.N. and Edelberg, J.M. (2006) BDNF-mediated enhancement of inflammation and injury in the aging heart. *Physiological Genomics* 24, 191–197.

Cancello, R. and Clement, K. (2006) Is obesity an inflammatory illness? Role of low-grade inflammation associated with macrophage infiltration in human white adipose tissue. *British Journal of Obstetrics and Gynaecology* 113, 1141–1147.

Candore, G., Balisteri, C.R., Grimaldi, M.P., Listi, F., Vasto, S., Caruso, M., Caimi, G., Hoffmann, E., Colonna-Romano, G., Lio, D., Paolisso, G., Franceschi, C. and Caruso, C. (2006) Opposite role of pro-inflammatory alleles in acute myocardial infarction and longevity: results of studies performed in a Sicilian population. *Annals of the New York Academy of Sciences* 1067, 270–275.

Celis, J.E., Moreira, J.M., Cabezon, T., Gromov, P., Friis, E., Rank, F. and Gromova, I. (2005) Identification of extracellular and intracellular components of the mammary adipose tissue and its interstitial fluid in high risk breast cancer patients: towards dissecting the molecular circuitry of epithelial–adipocyte stromal cell interactions. *Molecular and Cellular Proteomics* 4, 492–522.

Chaldakov, G.N. and Vankov, V.N. (1986) Morphological aspects of the secretion in the arterial smooth muscle cell, with special reference to the Golgi complex and microtubular cytoskeleton. *Atherosclerosis* 61, 175–192.

Chaldakov, G.N., Fiore, M., Ghenev, P.I., Stankulov, I.S. and Aloe, L. (2000) Atherosclerotic lesions: possible interactive involvement of intima, adventition and associated adipose tissue. *International Medical Journal* 7, 43–49.

Chaldakov, G.N., Fiore, M., Stankulov, I.S., Hristova, M., Antonelli, A., Manni, L., Ghenev, P.I., Angelucci, F. and Aloe, L. (2001a) NGF, BDNF, leptin, and mast cells in human coronary atherosclerosis and metabolic syndrome. *Archives of Physiology and Biochemistry* 109, 357–360.

Chaldakov, G.N., Stankulov, I.S., Fiore, M., Hristova, M.G., Rancic, G., Ghenev, P.I. and Pavlov, P.S. (2001b) Adipoendocrinology and adipoparacrinology: emerging fields of study on the adipose tissue. *Biomedical Reviews* 12, 31–39.

Chaldakov, G.N., Fiore, M., Hristova, M. and Aloe, L. (2003a) Metabotrophic potential of neurotrophins: implication in obesity and related diseases. *Medical Science Monitor* 21, H19–21.

Chaldakov, G.N., Stankulov, I.S., Hristova, M. and Ghenev, P.I. (2003b) Adipobiology of disease: adipokines and adipokine-targeted pharmacology. *Current Pharmaceutical Design* 9, 1023–1031.

Chaldakov, G.N., Fiore, M., Stankulov, I.S., Manni, L., Hristova, M.G., Antonelli, A., Ghenev, P.I. and Aloe, L. (2004) Neurotrophin presence in human coronary atherosclerosis and metabolic syndrome: a role for NGF and BDNF in cardiovascular disease? *Progress of Brain Research* 146, 279–289.

Chaldakov, G.N., Tonchev, A.B., Georgieva, Z., Ghenev, P.I. and Stankulov, I.S. (2005) Adipobiology of inflammation. *Biomedical Reviews* 16, 83–88.

Chaldakov, G.N., Fore, M., Tonchev, A.B. and Aloe, L. (2006a) Adipopharmacology, a novel drug discovery approach. A metabotrophic perspective. *Letters of Drug Design and Discovery* 3, 503–505.

Chaldakov, G.N., Tonchev, A.B., Tuncel, N., Atanassova, P. and Aloe, L. (2006b) Adipose tissue and mast cells. Adipokines as yin-yang modulators of inflammation. In: Fantuzzi, G. and Mazzone, T. (eds) *Nutrition and Health: Adipose Tissue and Adipokines in Health and Disease*. Humana Press Inc., Totowa, New Jersey, pp. 147–154.

Chaldakov, G.N., Fiore, M., Tonchev, A.B., Dimitrov, D., Pancheva, R., Rancic, G. and Aloe, L. (2007) *Homo obesus*: a metabotrophin-deficient species. Pharmacology and nutrition insight. *Current Pharmaceutical Design* 13, 2176–2179.

Chen, P.S., Chen, L.S., Cao, J.M., Sharifi, B., Karagueuzian, H.S. and Fishbein, M.C. (2001) Sympathetic nerve sprouting, electrical remodeling and the mechanisms of sudden cardiac death. *Cardiovascular Research* 50, 409–416.

Chen, D. and Guarente, L. (2007) SIR2: a potential target for calorie restriction mimetics. *Trends in Molecular Medicine* 13, 64–71.

Cheng, K.K.Y., Lam, K.S.L., Wang, Y. and Xu, A. (2006) Adiponectin as a key player in inflammation. *Biomedical Reviews* 17, 3–12.

Cinti, S., Mitchell, G., Barbatelli, G., Murano, I., Ceresi, E., Faloia, E., Wang, S., Fortier, M., Greenberg, A.S. and Obin, M.S. (2005) Adipocyte death defines macrophage location and function in adipose tissue of obese mice and humans. *Journal of Lipid Research* 46, 2347–2355.

Del Porto, F., Aloe, L., Lagana, B., Triaca, V., Nofroni, I. and D'Amelio, R. (2006) NGF and BDNF levels in patients with rheumatoid arthritis treated with TNF-α blockers. *Annals of the New York Academy of Sciences* 1069, 438–443

Dimitrov, D., Bohchelian, H. and Koeva, L. (2005) Effect of orlistat on plasma leptin levels and risk factors for the metabolic syndrome. *Metabolic Syndrome and Related Disorders* 3, 149–156.

Dimitrov, D., Pancheva, R., Tonchev, A.B., Kosseva, K., Kostov, D., Georgieva, Z. and Chaldakov, G.N. (2006) Nutrigenomics: DNA-based individualized nutrition. *Biomedical Reviews* 17, 109–114.

Ejiri, J., Inoue, N., Kobayashi, S., Shiraki, R., Otsui, K., Honjo, T., Takahashi, M., Ohashi, Y., Ichikawa, S., Terashima, M., Mori, T., Awano, K., Shinke, T., Shite, J., Hirata, K., Yokozaki, H., Kawashima, S. and Yokoyama, M. (2005) Possible role of brain-derived neurotrophic factor in the pathogenesis of coronary artery disease. *Circulation* 112, 2114–2120.

Fain, J.N. (2006) Release of interleukins and other inflammatory cytokines by human adipose tissue is enhanced in obesity and primarily due to the non-fat cells. *Vitamins and Hormones* 74, 443–477.

Fainzilber, M. and Carter, B.D. (2002) From neurotrophins to immunotrophins. *EMBO Reports* 3, 1029–1034.

Fantuzzi, G. (2005) Adipose tissue, adipokines, and inflammation. *Journal of Allergy and Clinical Immunology* 115, 911–919.

Fantuzzi, G. (2007) Adipose tissue and the regulation of inflammation. *Immunology Endocrine and Metabolic Agents in Medicinal Chemistry* 7, 129–136.

Flachs, P., Mohamed-Ali, V., Horakova, D., Rossmeisl, M., Hosseinzadeh-Attar, M.J., Hensler, M., Ruzickova, J. and Kopecky, J. (2006) Polyunsaturated fatty acids of marine origin induce adiponectin in mice fed a high-fat diet. *Diabetologia* 49, 394–397.

Foster-Schubert, K.E. and Cummings, D.E. (2006) Emerging therapeutic strategies for obesity. *Endocrine Reviews* 27, 779–793.

Franceschi, C., Valensin, S., Lescai, F., Olivieri, F., Licastro, F., Grimaldi, L.M., Monti, D., De Benedictis, G. and Bonafè, M. (2001) Neuroinflammation and the genetics of Alzheimer's disease: the search for a pro-inflammatory phenotype. *Aging* (Milano) 13, 163–170.

Frühbeck, G. (2005) Aquaporin enters the picture. *Nature* 438, 436–437.

Frühbeck, G. (2006) Vasoactive factors and inflammatory mediators produced in adipose tissue. In: Fantuzzi, G. and Mazzone, T. (eds) *Health and Nutrition. Adipose Tissue and Adipokines in Health and Disease.* Humana Press, Inc., Totowa, New Jersey, pp. 61–75.

Frühbeck, G., Catalan, V., Gomez-Ambrosi, J. and Rodriguez, A. (2006) Aquaporin-7 and glycerol permeability as novel obesity drug-target pathways. *Trends of Pharmacological Sciences* 27, 345–347.

Funahashi, T. and Matsuzawa, Y. (2006) Hypoadiponectinemia: a common basis for diseases associated with overnutrition. *Current Atherosclerosis Reports* 8, 433–438.

Funahashi, T., Nakamura, T., Shimomura, I., Maeda, K., Kuriyama, H., Takahashi, M., Arita, Y., Kihara, S. and Matsuzawa, Y. (1999) Role of adipocytokines on the pathogenesis of atherosclerosis in vascular obesity. *Internal Medicine* 38, 202–206.

Gallwitz, B. (2006) Therapies for the treatment of type 2 diabetes mellitus based on incretin action. *Minerva Endocrinology* 31, 133–147.

Geroldi, D., Minoretti, P. and Emanuele, E. (2006) Brain-derived neurotrophic factor and metabolic syndrome: more than just a hypothesis. *Medical Hypotheses* 67, 195–196.

Ghosh, D., Skinner, M.A. and Laing, W.A. (2007) Pharmacogenomics and nutrigenomics: synergies and differences. *European Journal of Clinical Nutrition* 61, 567–574.

Gianoukakis, A.G. and Smith, T.J. (2008) Recent insights into the pathogenesis and management of thyroid-associated ophthalmopathy. *Current Opinion in Endocrinology, Diabetes, and Obesity* 15, 446–452.

Goldstein, B.J. and Scalia, R. (2007) Adipocytokines and vascular disease in diabetes. *Current Diabetes Reports* 7, 25–32.

Gray, J., Yeo, G., Hung, C., Keogh, J., Clayton, P., Banerjee, K., McAulay, A., O'Rahilly, S. and Farooqi, S. (2007) Functional characterization of human *NTRK2* mutations identified in patients with severe early-onset obesity. *International Journal of Obesity* 31, 359–364.

Grimaldi, M.P., Candore, G., Vasto, S., Caruso, M., Gaimi, G., Hoffmann, E., Colonna-Romano, G., Lio, D., Shinar, Y., Franceschi, C. and Caruso, C. (2006) Role of the pyrin M694V (A2080G) allele in acute myocardial infarction and longevity: a study in the Sicilian population. *Journal of Leukocyte Biology* 79, 611–615.

Guarente, L. and Picard, F. (2005) Calorie restriction – the SIR2 connection. *Cell* 120, 473–482.

Halberg, N., Henrikson, M., Soderhamn, N., Stallkecht, B., Plough, T., Schjerling, P. and Dela, F. (2005) The effect of intermittent fasting and re-feeding on insulin action in healthy men. *Journal of Applied Physiology* 99, 2128–2136.

Hefti, F.F., Rosenthal, A., Walicke, P.A., Wyatt, S., Vergata, G., Shelton, D.L. and Davies, A.M. (2006) Novel class of pain drugs based on antagonism of nerve growth factor. *Trends in Pharmacological Sciences* 27, 85–91.

Himeda, T., Tounai, H., Hayakawa, N. and Araki, T. (2007) Post-ischemic alterations of BDNF, NGF, HSP 70 and ubiquitin immunoreactivity in the gerbil hypothalamus: pharmacological approach. *Cellular and Molecular Neurobiology* 27, 229–250.

Hofbauer, K.G, Nicholson, J.K. and Boss, O. (2007) The obesity epidemics: current and future pharmacological treatments. *Annual Review of Pharmacology and Toxicology* 47, 565–592.

Hristova, M., Aloe, L., Ghenev, P.I., Fiore, M. and Chaldakov, G.N. (2001) Leptin and mast cells: a novel adipoimmune link. *Turkish Journal of Medical Sciences* 31, 581–583.

Iacobellis, G., Corradi, D. and Sharma, A.M. (2005) Epicardial adipose tissue: anatomic, biomolecular and clinical relationships with the heart. *Nature Clinical Practice and Cardiovascular Medicine* 2, 536–543.

Kadowaki, T. and Yamauchi, T. (2005) Adiponectin and adiponectin receptors. *Endocrine Reviews* 26, 439–451.

Kalra, S.P. and Kalra, P.S. (2005) Gene-transfer technology: a preventive neurotherapy to curb obesity, ameliorate metabolic syndrome and extend life expectancy. *Trends in Pharmacological Sciences* 26, 488–495.

Keusch, G.T. (2006) What do -omics mean for the science and policy of the nutritional sciences? *American Journal of Clinical Nutrition* 83, 520S–522S.

Kim, Y.J., Kim, H.J., No, J.K., Chung, H.Y., Fernandes, G. (2006) Anti-inflammatory action of dietary fish oil and calorie restriction. *Life Sciences* 78, 2523–2532.

Kim, M.J., Lee, E.Y., Lee, M.Y. and Chung, C.H. (2007) Adipobiology of diabetes mellitus. *Immunology Endocrine and Metabolic Agents in Medicinal Chemistry* 7, 123–127.

Klein, J., Permana, P.A., Owecki, M., Chaldakov, G.N., Böhm, M., Hausman, G., Lapière, C.M., Atanassova, P., Sowiński, J., Fasshauer, M., Hausman, D.B., Maquoi, E., Tonchev, A.B., Peneva, V.N., Vlachanov, K.P., Fiore, M., Aloe, L., Slominski, A., Reardon, C.L., Ryan, T.J., Pond, C.M. and Ryan, T.J. (2007) What are subcutaneous adipocytes really good for? *Experimental Dermatology* 16, 45–70.

Kokai, L.E., Rubin, J.P. and Marra, K.G. (2005) The potential of adipose-derived adults stem cells as a source of neuronal progenitor cells. *Plastic and Reconstructive Surgery* 116, 1453–1460.

Krabbe, K.S., Nielsen, A.R., Krogh-Madsen, R., Plomgaard, P., Rasmussen, P., Erikstrup, C., Fischer, C.P., Lindegaard, B., Peteresen, A.M.W., Taudorf, S., Secher, N.H., Pilegaard, H., Bruunsgaard, H. and Pedersen, B.K. (2006) Brain-derived neurotrophic factor (BDNF) and type 2 diabetes. *Diabetologia.* 50, 431–438.

Kralisch, S., Blüher, M., Paschke, R., Stumvoll, M. and Fasshauer, M. (2007) Adipokines and adipocyte targets in the future management of obesity and the metabolic syndrome. *Mini Reviews Medicinal Chemistry* 7, 39–45.

Kratchmarova, I., Kalume, D.E., Blagoev, B., Scherer, P.E., Podtelejnikov, A.V., Molina, H., Bickel, P.E., Andersen, J.S., Ferandez, M.M., Bunkenborg, J., Roepstorff, P., Kristiansen, K., Lodish, H.F., Mann, M. and Pandey, A. (2002) A proteomic approach for identification of secreted proteins during the differentiation of 3T3-L1 preadipocytes to adipocytes. *Molecular and Cellular Proteomics* 1, 213–222.

Lamming, D.W., Wood, J.G. and Sinclair, D.A. (2004) Small molecules that regulate lifespan: evidence for xenohormesis. *Molecular Microbiology* 53, 1003–1009.

Lopez-Lluch, G., Hunt, N., Jones, B., Zhu, M., Jamieson, H., Hilmer, S., Cascajo, M.V., Allard, J., Ingram, D.K., Navas, P. and Cabo, R. de (2006) Calorie restriction induces mitochondrial biogenesis and bioenergetic efficiency. *Proceedings of National Academy of Sciences of the United States of America* 103, 1768–1773.

Maeda, N., Shimomura, I., Kishida, K., Nishizawa, H., Matsuda, M., Nagaretani, H., *et al.* (2002) Diet-induced insulin resistance in mice lacking adiponectin/ACRP30. *Nature Medicine* 8, 731–737.

Manni, L., Nikolova, V., Vyagova, D., Chaldakov, G.N. and Aloe, L. (2005) Reduced plasma levels of NGF and BDNF in patients with acute coronary syndromes. *International Journal of Cardiology* 102, 169–171.

Matsuzawa, Y. (2006) Therapy insight: adipocytokines in metabolic syndrome and related cardiovascular disease. *Nature Clinical Practice and Cardiovascular Medicine* 3, 35–42.

Miggiano, G.A. and De Sanctis, R. (2006) Nutritional genomics: toward a personalized diet. *La Clinica Terapeutica* 157, 355–361.

Mizuno, T., Kurotani, T., Komatsu, Y., Kawanokuchi, J., Kato, H., Mitsuma, N. and Suzumura, A. (2004) Neuroprotective role of phosphodiesterase inhibitor ibudilast on neuronal cell death induced by activated microglia. *Neuropharmacology* 46, 404–411.

Moschen, A.R., Kaser, A., Enrich, B., Mosheimer, B., Theurl, M., Niebergger, H. and Tilg, H. (2007) Visfatin, an adipocytokine with proinflammatory and immunomodulatory properties. *Journal of Immunology* 178, 1748–1758.

Nakagawa, T., Ogawa, Y., Ebihara, K., Yamanaka, M., Tsuchida, A., Taiji, M., Noguchi, H. and Nakao, K. (2003) Anti-obesity and anti-diabetic effects of brain-derived neurotrophic factor in rodent models of leptin resistance. *International Journal of Obesity and Related Metabolic Disorders* 27, 557–565.

Narasimhan, I.M., Coca, M.A., Jin, J., Yamauchi, T., Ito, Y., Kadowaki, T., Kim, K.K., Pardo, J.M., Damsz, B., Hasegawa, P.M., Yun, D.J. and Bressan, R.A. (2005) Osmotin is a homolog of mammalian adiponectin and controls apoptosis in yeast through a homolog of mammalian adiponectin receptor. *Molecular Cell* 17, 171–180.

Neel, J.V. (1962) Diabetes mellitus: a 'thrifty' genotype rendered detrimental by 'progress'? *American Journal of Human Genetics* 14, 353–362.

Neels, J.G. and Olefsky, J.M. (2006) Inflamed fat: what starts the fair? *Journal of Clinical Investigation* 116, 33–35.

Nishimura, F., Iwamoto, Y., Mineshiba, J., Shimizu, A., Soga, Y. and Murayama, Y. (2003) Periodontal disease and diabetes mellitus: the role of tumor necrosis factor-alpha in a 2-way relationship. *Journal of Periodontology* 74, 97–102.

Nisoli, E., Tonello, C., Cardile, A., Cozzi, V., Bracale, R., Tedesco, L., Falcone, S., Valerio, A., Cantoni, O., Clementi, E., Moncada, S. and Carruba, M.O. (2005) Calorie restriction promotes mitochondrial biogenesis by inducing the expression of eNOS. *Science* 310, 314–317.

Okamoto, Y., Kihara, S., Funahashi, T., Matsuzawa, Y. and Libby, P. (2006) Adiponectin: a key adipocytokine in metabolic syndrome. *Clinical Science (London)* 110, 267–278.

Permana, P.A. (2006) Adipokine expression and secretion: a target for pharmacologic treatment. *Biomedical Reviews* 17, 72–81.

Petersen, A.M. and Pedersen, B.K. (2005) The anti-inflammatory effect of exercise. *Journal of Applied Physiology* 98, 1154–1162.

Pond, C. (2003) Paracrine relationships between adipose and lymphoid tissue: implications for the mechanisms of HIV-associated adipose redistribution syndrome. *Trends of Immunology* 24, 13–18.

Pond, C.M. (2005) Adipose tissue and the immune system. *Prostaglandins Leukocytes and Essential Fatty Acids* 73, 17–30.

Pradova, E. and Fickova, M. (2006) Alcohol intake modulates hormonal activity of adipose tissue. *Endocrine Regulation* 40, 91–104.

Quinn, L.S., Strait-Bodey, L., Anderson, B.G., Argiles, J.M. and Havel, P.J. (2005) Interleukin-15 stimulates adiponectin secretion by 3T3-L1 adipocytes: evidence for a skeletal muscle-to-fat signaling pathway. *Cell Biology International* 29, 449–457.

Rios, M., Fan, G., Fekete, C., Kelly, J., Bates, B., Kuehn, R., Lechan, R.M. and Jaenisch, R. (2001) Conditional deletion of brain-derived neurotrophic factor in post-natal brain leads to obesity and hyperactivity. *Molecular Endocrinology* 15, 1748–1757.

Rodriguez, A., Catalan, V., Gomez-Ambrosi, J. and Frühbeck, G. (2006) Role of AQP7 in the pathophysiological control of fat accumulation in mice. *FEBS Letters* 580, 4771–4776.

Ross, R. (1999) Mechanisms of disease: atherosclerosis – an inflammatory disease. *New England Journal of Medicine* 340, 115–126.

Roth, G.S., Lane, M.A. and Ingram, D.K. (2005) Caloric restriction mimetics: the next phase. *Annals of the New York Academy of Sciences* 57, 365–371.

Schaffler, A., Scholmerich, J. and Buchler, C. (2005) Mechanisms of disease: adipocytokines and visceral adipose tissue – emerging role in non-alcoholic liver disease. *Nature Clinical Practice Gastroenterology and Hepatology* 2, 273–280.

Scott, E.M. and Grant, P.J. (2006) Neel revisited: the adipocyte, seasonality and type 2 diabetes. *Diabetologia* 49, 1462–1466.

Seki, M., Tanaka, T., Nawa, H., Usui, T., Fukuchi, T., Ikeda, K., Abe, H. and Takei, N. (2004) Involvement of brain-derived neurotrophic factor in early retinal neuropathy of streptozotocin-induced diabetes in rats: therapeutic potential of brain-derived neurotrophic factor for dopaminergic amacrine cells. *Diabetes* 53, 2412–2419.

Shimomura, I., Funahashi, T., Takahashi, M., Maeda, K., Kotani, K., Nakamura, T., Yamashita, S., Miura, M., Fukuda, Y., Takemura, K., Tokunaga, K. and Matsuzawa, Y. (1996) Enhanced expression of PAI-1 in visceral fat: possible contributor to vascular disease in obesity. *Nature Medicine* 2, 800–803.

Shimomura, I., Hammer, R.E., Richardson, J.A., Ikemo, S., Bashmakov, Y., Goldstein, J.L. and Brown, M.S. (1998) Insulin resistance and diabetes mellitus in transgenic mice expressing nuclear SREBP-1c in adipose tissue: model for congenital generalized lipodystrophy. *Genes and Development* 12, 3182–3194.

Shimomura, I., Funahashi, T. and Matsuzawa, Y. (2006) Metabolic syndrome, adiponectin and fat ROS. *Biomedical Reviews* 17, 1–10.

Sjogren, M. and Blennow, K. (2005) The link between cholesterol and Alzheimer's disease. *World Journal of Biological Psychiatry* 6, 85–97.

Sun, M.K. and Alkon, D.L. (2006) Links between Alzheimer's disease and diabetes. *Drugs Today (Barc)* 42, 481–489.

Suwa, M., Kishimoto, H., Nofuji, Y., Nakano, H., Sasaki, H., Radak, Z. and Kumagai, S. (2006) Serum brain-derived neurotrophic factor level is increased and associated with obesity in newly diagnosed female patients with type 2 diabetes mellitus. *Metabolism Clinical and Experimental* 55, 852–857.

Tilg, H. and Moschen, A.R. (2008) Role of adiponectin and PBEF/visfatin as regulators of inflammation: involvement in obesity-associated diseases. *Clinical Science (London)* 114, 275–288.

Töre, F., Tonchev, A.B, Fiore, M., Atanassova, P., Tuncel, N., Aloe, L. and Chaldakov, G.N. (2007) From adipose tissue protein secretion to adipopharmacology of disease. *Immunology Endocrine and Metabolic Agents in Medicinal Chemistry* 7, 149–155.

Trapp, J. and Jung, M. (2006) The role of NAD+ dependent histone deacetylases (sirtuins) in ageing. *Current Drug and Targets* 7, 1553–1560.

Trayhurn, P. and Wood, I.S. (2004) Adipokines: inflammation and the pleiotropic role of white adipose tissue. *British Journal of Nutrition* 92, 347–355.

Tsuji-Hayashi, Y., Young, B.A., Green, J., Tsuji, A., Hosoya, T., Fukuhara, S. and Blagg, C.R. (2005) Health-related quality of life among community-dwelling elderly people in the general populations of the US and Japan. *Japan Medical Association Journal* 48, 489–496.

Viguerie, N., Pottou, C., Cancello, R., Stich, V., Clement, K. and Langin, D. (2005) Transcriptomics applied to obesity and caloric restriction. *Biochemie* 87, 117–123.

Waki, H. and Tontonoz, P. (2007) Endocrine function of adipose tissue. *Annual Review of Pathology: Mechanisms of Disease* 2, 31–56.

Wang, P., Mariman, E., Renes, J. and Keijer, J. (2008) The secretary function of adipocytes in the physiology of white adipose tissue. *Journal of Cellular Physiology* 216, 3–13.

Wintour, E.M. and Henry, B.A. (2006) Glycerol transport: an additional target for obesity therapy? *Trends of Endocrinology and Metabolism* 17, 77–78.

Xavier Pi-Sunyer, F. (2006) The relation of adipose tissue to cardiometabolic risk. *Clinical Cornerstone* 8 (Suppl. 4), S14–23.

Xu, A., Chan, K.W., Hoo, R.L.C., Wang, Y., Tan, K.C.B., Zhang, J., Chen, B., Lam, M.C., Tse, C., Cooper, G.J.S. and Lam, K.S.L. (2005a) Testosterone selectively reduces the high molecular weight form of adiponectin by inhibiting its secretion from adipocytes. *Journal of Biological Chemistry* 280, 18073–18080.

Xu, A., Laam, M.C., Chan, K.W., Wang, Y., Zhang, J., Hoo, R.L.C., *et al.* (2005b) Angiopoietin-like protein 4 decreases blood glucose and improves glucose tolerance but induces hyperlipidemia and hepatic steatosis in mice. *Proceedings of National Academy of Sciences of the United States of America* 102, 6086–6092.

Yamori, Y. (2004) Worldwide epidemic of obesity: hope for Japanese diets. *Clinical Experimental Pharmacology and Physiology* 31 (Suppl. 2), S2–4.

Yamori, Y., Murakami, S., Ikeda, K. and Nara, Y. (2004) Fish and lifestyle-related disease prevention: experimental and epidemiological evidence for anti-atherogenic potential of taurine. *Clinical Experimental Pharmacology and Physiology* 31 (Suppl. 2), S20–23.

Yeo, G.S. Connie Hung, C.C., Rochford, J., Keogh, J., Gray, J., Sivaramakrishnan, S., O'Rahilly, S. and Farooqi, I.S. (2004) A *de novo* mutation affecting human TrkB associated with severe obesity and developmental delay. *Nature Neuroscience* 7, 1187–1189.

Yudkin, J.S., Eringa, E. and Stehouwer, C.D.A. (2005) 'Vasocrine' signalling from perivascular fat: a mechansim linking insulin resistance to vascular diseases. *Lancet* 365, 1817–1820.

Yun, A.J., Lee, P.Y. and Doux, J.D. (2006) Are we eating more than we think? Illegitimate signaling and xenohormesis as participants in the pathogenesis of obesity. *Medical Hypotheses* 67, 36–40.

Zigman, J.M. and Elmquist, J.K. (2003) Minireview: from anorexia to obesity – the yin and yang of body weight control. *Endocrinology* 144, 3749–3756.

Zvonic, S., Lefevre, M., Kilroy, G., Floyd, Z.E., Delany, J.P., Kheterpal, I., Gravos, A., Dow, R., White, A., Wu, X. and Gimble, J.M. (2007) Secretome of primary culture of human adipose stem cells: modulation of serpins by adipogenesis. *Molecular and Cellular Proteomics* 6, 18–28.

Index